Computational and Mathematical Methods for Neuroscience

Computational and Mathematical Methods for Neuroscience

Guest Editor

Alexander N. Pisarchik

MDPI

Basel • Beijing • Wuhan • Barcelona • Belgrade • Novi Sad • Cluj • Manchester

Guest Editor
Alexander N. Pisarchik
Center for Biomedical
Technology
Universidad Politécnica de
Madrid
Pozuelo de Alarcón, Madrid
Spain

Editorial Office
MDPI AG
Grosspeteranlage 5
4052 Basel, Switzerland

This is a reprint of the Special Issue, published open access by the journal *Applied Sciences* (ISSN 2076-3417), freely accessible at: https://www.mdpi.com/journal/applsci/special_issues/40RU0Z7J3K.

For citation purposes, cite each article independently as indicated on the article page online and as indicated below:

Lastname, A.A.; Lastname, B.B. Article Title. *Journal Name* **Year**, *Volume Number*, Page Range.

ISBN 978-3-7258-3493-8 (Hbk)
ISBN 978-3-7258-3494-5 (PDF)
https://doi.org/10.3390/books978-3-7258-3494-5

Contents

About the Editor

Alexander N. Pisarchik

Alexander N. Pisarchik is a Distinguished Researcher at the Center for Biomedical Technology, Universidad Politécnica de Madrid. He earned his PhD in Physics and Mathematics in 1990 from the Institute of Physics of the Belarus Academy of Science. In 1997, he completed advanced courses on "Nonlinear Dynamics in Physiology and Medicine" at McGill University, Canada, and "Time Evolution of Complex Systems" in Lisbon, Portugal. Since 1992, Dr. Pisarchik has secured research grants from various international governments and conducted research in Belgium, Spain, Iceland, and Mexico. In 1999, he was awarded the Chair of Excellence by the National Council of Science and Technology (CONACYT) and was employed at the Center for Research in Optics in Leon, Mexico. Since 2001, he has been admitted to the National System of Researchers (SNI), and in 2006, he received level III. In 2010, he was elected as a member of the SNI Evaluation Commission of CONACYT by the Government of Mexico. In 2013, he was appointed as the Isaac-Peral Chair in Computational Systems Biology at the Center for Biomedical Technology of the Universidad Politécnica de Madrid within the framework of the BBVA-UPM BioTech Chairs Program. He is the author of several monographs, numerous book chapters and patents, and has published over 300 papers in peer-reviewed scientific journals. Additionally, he is a member of the Board of Directors of the International Physics and Control Society (IPACS) and a member of the Biomedical Research Networking Centers (CIBER).

Preface

As our understanding of the brain deepens, so does the need for advanced computational tools to model, simulate, and interpret complex neural data. This interdisciplinary volume integrates biology, mathematics, AI, and physics to explore neural dynamics, cognition, and disease. By bridging theoretical frameworks with empirical research, it highlights cutting-edge methodologies that drive innovation in neuroscience. We extend our gratitude to the authors, reviewers, and editors whose efforts have made this reprint possible, inspiring further exploration of brain function. that focuses on one of the most studied and relevant food-associated mycotoxins.

Alexander N. Pisarchik
Guest Editor

applied sciences

MDPI

Editorial

Computational and Mathematical Methods for Neuroscience

Alexander N. Pisarchik

Center for Biomedical Technology, Universidad Politécnica de Madrid, Campus de Montegancedo, Pozuelo de Alarcón, 28223 Madrid, Spain; alexander.pisarchik@upm.es

1. Introduction

As our understanding of the brain continues to advance, so too does the demand for sophisticated tools that can model, simulate, and interpret the intricate data generated by contemporary neuroimaging and electrophysiological techniques. The interdisciplinary field of theoretical and computational neuroscience, drawing on biology, mathematics, computer science, and physics, seeks to capture the complexities of the nervous system through rigorous quantitative models and simulations. In recent years, this field has grown rapidly, with computational and mathematical methodologies becoming essential for probing the nuances of neural circuitry and cognitive function.

Computational approaches in neuroscience encompass diverse techniques, from advanced statistical methods to machine learning algorithms, each designed to identify meaningful patterns in high-dimensional data. Complementing these are mathematical models that provide a robust framework for understanding neural dynamics, connectivity, and information processing across various scales, from single neurons to vast networks. Together, these computational and mathematical strategies empower researchers to generate precise hypotheses, make quantitative predictions, and gain deeper insights into the fundamental principles that drive brain function, neural plasticity, and the mechanisms behind neurological disorders.

This Special Issue brings together the latest advancements in computational and mathematical methods in neuroscience, showcasing articles that address foundational concepts, established models, and emerging technologies at the forefront of the field. By bridging theoretical frameworks with empirical data, these approaches not only expand our knowledge of neural systems but also open new pathways for therapeutic innovation and applications in clinical neuroscience.

Citation: Pisarchik, A.N. Computational and Mathematical Methods for Neuroscience. *Appl. Sci.* **2024**, *14*, 11296. https://doi.org/10.3390/app142311296

Received: 9 November 2024
Revised: 21 November 2024
Accepted: 30 November 2024
Published: 4 December 2024

2. Fields of Neuroscience

Neuroscience is a vast and inherently interdisciplinary field dedicated to understanding the complexities of the nervous system. It encompasses a diverse array of subfields, each focusing on different levels of neural organization and function, as illustrated in Figure 1. At its core, neuroscience integrates both theoretical and experimental approaches, each bringing distinct methodologies and perspectives that, together, drive a more comprehensive understanding of brain mechanisms and behavior. The synergy between these approaches allows researchers to bridge molecular-, cellular-, and systems-level insights, advancing our knowledge of how neural processes underpin cognition, perception, and action.

2.1. Theoretical Neuroscience

Theoretical neuroscience focuses on developing mathematical, computational, and statistical models to represent neural processes across multiple scales, from an individual neuron to the brain.

Mathematical neuroscience applies mathematical theories, models, and equations to describe and analyze the mechanisms of the nervous system at various levels, from single neurons to whole-brain dynamics. This branch of neuroscience aims to build theoretical frameworks for neural activity, capturing phenomena such as neuronal electrical properties, network dynamics, and brain connectivity patterns. Some of the most widely used neural models include the Hodgkin–Huxley (HH) [1], FitzHugh–Nagumo (FHN) [2,3], Hindmarsh-Rose (HR) [4], Wilson–Cowan (WC) [5], and Izhikevich [6] models.

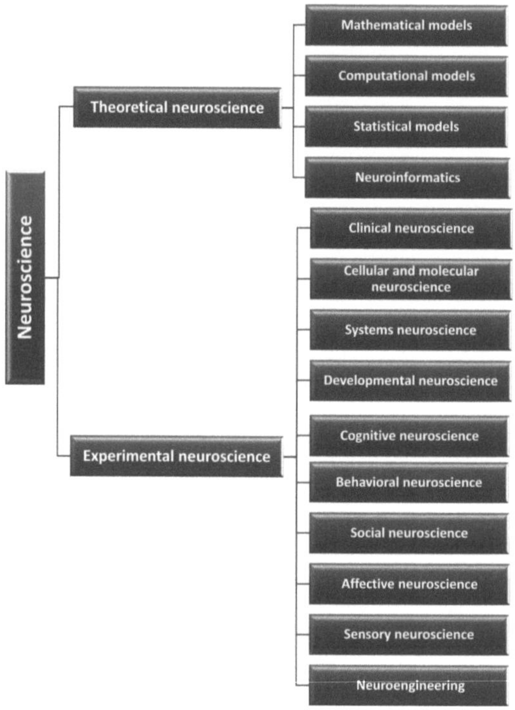

Figure 1. Fields and subfields of neuroscience.

The HH model provides a detailed description of neuronal electrical behavior based on ion channel dynamics, offering a foundation for understanding neuron excitability. The FHN and HR models, simplified versions of the HH model, are commonly used to simulate excitable systems due to their computational efficiency. The WC model, on the other hand, captures the collective dynamics of populations of excitatory and inhibitory neurons, making it useful for studying large-scale neural networks. The Izhikevich model combines biological realism with computational efficiency, enabling the simulation of a wide range of spiking and bursting patterns observed in neurons.

In addition to these continuous-time models, various discrete-time models are employed in theoretical neuroscience, such as the Leaky Integrate-and-Fire (LIF) model [7] and the Rulkov map [8]. The LIF model approximates biological neurons by simulating membrane potential decay in the absence of input spikes, while the Rulkov map generates spike patterns through the interplay of membrane potential dynamics and recovery variables. The Rulkov model also incorporates a reset mechanism, allowing neurons to recover after firing.

These mathematical models play a crucial role in predicting and explaining phenomena such as neural oscillations, wave propagation in the brain, and the synchronization of neuronal activity, processes essential for neural communication and understanding neurological conditions like epilepsy [9]. These models enable theoretical neuroscience to uncover fundamental principles of neural behavior, enhancing our capacity to analyze and interpret the intricate dynamics of neural systems.

Computational neuroscience aims to develop quantitative tools to analyze neural data and predict neural system dynamics, helping to uncover the principles that govern brain function. Computational neuroscience involves developing and using computer simulations, algorithms, and artificial neural network (ANN) models to investigate the functioning of the nervous system. It bridges theoretical models with experimental data, often serving as a testing ground for hypotheses. Computational neuroscience focuses on simulating neural circuits, analyzing large datasets from neural recordings, and predicting brain activity and behavior. Popular examples of computational techniques are neural network simulations, machine learning (ML), and data-driven models. Former methods simulate networks of neurons to study how they encode, process, and retrieve information. ML models are used to classify patterns in brain data, such as electroencephalography (EEG), magnetoencephalography (MEG), Magnetic Resonance Imaging (MRI), functional Magnetic Resonance Imaging (fMRI), and Positron Emission Tomography (PET), to model learning and adaptation in neural systems. Finally, data-driven models use real experimental data to create models of complex phenomena like sensory processing, decision-making, or motor control. Computational neuroscience helps us understand brain function (e.g., sensory processing, memory, and emotions), design brain–machine interfaces (BMIs), and develop treatments for neurological diseases through predictive modeling.

Statistical neuroscience applies advanced statistical techniques to analyze and interpret the complex data generated by neuroscience experiments, addressing challenges such as high dimensionality, noise, and the temporal structure of neural activity. By providing robust tools for managing the variability inherent in neural data, statistical neuroscience helps researchers identify patterns, relationships, and statistical dependencies, which are essential for testing hypotheses and making inferences about neural function.

Key methods in statistical neuroscience include Spike Train Analysis (STA), Dimensionality Reduction (DR), Bayesian inference, and information theory. STA encompasses statistical techniques for analyzing the timing and patterns of neuronal spikes, which carry critical information about neural signaling. DR techniques, such as Principal Component Analysis (PCA) and t-distributed Stochastic Neighbor Embedding (t-SNE), simplify high-dimensional neural data, making them more accessible for interpretation. Bayesian inference introduces a probabilistic approach to understanding neural data, commonly applied to decode sensory information and predict neural responses. Information theory, meanwhile, quantifies the amount of information transmitted within neural circuits, providing insights into the efficiency and mechanisms of neural coding.

These statistical approaches are essential for analyzing data from electrophysiological recordings, neuroimaging, and behavioral experiments. They allow researchers to draw reliable conclusions about brain activity, predict behaviors, and even forecast events like epileptic seizures [10]. Through statistical neuroscience, scientists gain critical insights into the principles of neural organization and function, advancing our understanding of brain dynamics.

Neuroinformatics plays a crucial role in neuroscience by developing algorithms, data management tools, and computational techniques to organize, integrate, and share large datasets, thereby facilitating discovery and enhancing collaboration across studies. This field focuses on establishing standardized formats and databases that enable diverse types of neuroscience data to be combined, compared, and interpreted across studies and institutions. By streamlining the organization and accessibility of complex datasets, neuroinformatics supports efficient data sharing, reproducibility, and broader analyses.

Popular neuroinformatics tools include Brainstorm, Brainsuite, Statistical Parametric Mapping (SPM), and FMRIB Software Library (FSL), containing image analysis and statistical tools for functional, structural, and diffusion MRI. Additionally, platforms such as Neuron and Brain Imaging Data Structure (BIDS) provide standardized data formatting and processing pipelines that enhance consistency in data analysis. Neuroinformatics also involves creating computational models that simulate brain processes, providing valuable insights into brain dynamics, neural circuits, and cognitive functions.

A key goal of neuroinformatics is to promote open science by facilitating data sharing across labs, institutions, and even international boundaries, accelerating discovery, enhancing reproducibility, and allowing for larger-scale and more diverse analyses. Platforms like the NeuroInformatics Framework (NIF) exemplify this approach, providing standardized access to a wide range of neuroscience datasets and tools.

One of the primary applications of neuroinformatics is brain mapping, which involves creating detailed maps of the brain's structural and functional connectivity to better understand regional interactions. By comparing large datasets from both healthy and diseased brains, neuroinformatics enables the identification of biomarkers and genetic markers associated with neurological and psychiatric disorders. Moreover, neuroinformatics contributes to the development of algorithms for processing data from Brain–Computer Interfaces (BCIs), facilitating direct communication between the brain and external devices [11,12]. Additionally, virtual brain models and curated databases serve as valuable educational tools, providing students, clinicians, and researchers with training in neuroanatomy, neurophysiology, and neural dynamics.

2.2. Experimental Neuroscience

Experimental neuroscience, distinct from theoretical approaches, is centered on empirical studies that directly observe, manipulate, and measure neural function and behavior. This hands-on branch of neuroscience encompasses several key subfields, each focused on specific aspects of the brain and nervous system.

Clinical neuroscience targets neurological, psychiatric, and neurodevelopmental disorders, facilitating collaborations among neurologists, psychiatrists, and neuroscientists to advance diagnostic methods and therapeutic strategies.

Cellular and molecular neuroscience delves into the structure and function of individual neurons, exploring processes such as synaptic transmission, plasticity, and the role of molecular components like neurotransmitters, ion channels, and genetic factors. This foundational research provides insights into the basic units of neural activity.

Systems neuroscience investigates how neural circuits and larger brain systems organize and function, analyzing interactions across brain regions and networks that enable complex capabilities like sensory processing, motor coordination, and emotional regulation.

Developmental neuroscience examines the processes governing nervous system development from embryonic stages through adulthood, including neurogenesis, cell differentiation, and synaptic formation, as well as the effects of genetic and environmental factors on brain maturation.

Cognitive neuroscience studies the neural basis of higher-order cognitive functions, including perception, memory, language, and decision-making. This field often employs neuroimaging techniques such as EEG, MEG, MRI, PET, and fMRI to link brain activity with cognitive processes.

Behavioral neuroscience explores the relationship between neural mechanisms and behavior, investigating how alterations in the brain—whether from injury, disease, or experimental manipulation—impact behavior and psychological processes.

Sensory neuroscience focuses on the neural interpretation of sensory information from the external environment, examining how sensory systems like vision, hearing, and touch process and respond to stimuli.

Social neuroscience examines the neural underpinnings of social behaviors, such as empathy, cooperation, and social decision-making, integrating methodologies from psychology, biology, and neuroscience to understand interpersonal and group dynamics.

Affective neuroscience investigates how the brain processes emotions such as admiration, adoration, aesthetic appreciation, amusement, anger, anxiety, awe, awkwardness, boredom, calm, caring, confusion, craving, disgust, empathic pain, fascination, excitement, fear, horror, interest, joy, lust, nostalgia, play, relief, romance, sadness, satisfaction, seeking, sexual desire, surprise, etc., which are common to all mammals and evolutionarily defined as tools for survival and, in general, fitness.

Social neuroscience and affective neuroscience apply traditional neuroimaging techniques used in experimental neuroscience (e.g., EEG, MEG, and fMRI) to better understand the neural and psychological mechanisms underlying human behavior.

Neuroengineering combines engineering principles with neuroscience to create innovative tools and technologies for studying and manipulating the nervous system, including brain–machine interfaces, neuroprosthetics, and neurostimulation devices.

Each of these branches can also benefit from theoretical neuroscience through the application of mathematical models, computational algorithms, and statistical analyses. Integrating theoretical and experimental approaches allows researchers to investigate the nervous system at all levels, from molecules to behavior, building a comprehensive picture of how the brain enables perception, thought, emotion, and action. Collectively, these subfields deepen our understanding of the brain and hold transformative potential for both medicine and technology, advancing our ability to address neurological disorders and improve human health.

3. Highlights and Key Contributions of Published Articles

This Special Issue comprises 16 papers, which can be broadly categorized into four main areas: medical applications (8 papers), cognitive neuroscience (3 papers), statistical methods (2 papers), and machine learning (5 papers, including 2 focused on medical applications). Below is a brief overview of each paper and its key contributions.

3.1. Medical Applications

Half of the papers in this Special Issue (8 out of 16) focus on medical applications, with 4 specifically addressing Alzheimer's disease (AD) (contributions 2, 9, 11, and 14). This focus is aligned with the critical importance of early and accurate AD diagnosis, which enables timely therapeutic intervention and management. Brain imaging technologies like MRI and PET scans facilitate the early detection of AD-related structural and functional changes in the brain, often identifying the disease before severe clinical symptoms appear. Early detection provides valuable opportunities for intervention that may slow disease progression. Additionally, brain imaging distinguishes AD from other dementias, such as Lewy body or vascular dementia, by detecting unique patterns like amyloid plaques or hippocampal atrophy. Moreover, imaging allows clinicians to track disease progression over time, informing treatment adjustments and helping to assess therapeutic efficacy.

Among available neuroimaging techniques, MRI is particularly popular in AD research, as it visualizes brain structures, allowing clinicians to detect hippocampal atrophy, a key marker of AD. In this issue, Altwijri et al. (**contribution 2**) introduce an innovative deep learning approach to automatically diagnose AD using MRI datasets. Leveraging the strengths of deep learning, which often outperforms human detection in assessing AD stages, the authors employ pre-trained convolutional neural networks (CNNs) to classify AD severity with high accuracy, even when dataset quality and quantity are limited. Their method, which preprocesses AD data through an advanced image processing module before training, achieves a 99.3% accuracy rate—an improvement over existing models.

Kozminski and Gniazdowska (**contribution 9**) review studies on tacrine and its derivatives labeled with radionuclides, exploring their potential as diagnostic radiotracers for AD. While AD is not curable, its progression and symptoms can be managed using treatments like acetylcholinesterase (AChE) inhibitors (e.g., tacrine, rivastigmine, galantamine, and donepezil). The authors analyze radiolabeled tacrine derivatives in early AD diagnosis, with a particular emphasis on computational molecular modeling to visualize tacrine's interaction with cholinesterase. Their review highlights the limitations of current radiopharmaceuticals based on tacrine derivatives and suggests a shift toward other biomolecules relevant to early AD stages.

Sait (**contribution 11**) presents a novel integrated model combining LeViT, EfficientNet B7, and Dartbooster XGBoost (DXB) for AD detection using MRI. LeViT is a vision transformer-based hybrid neural network, EfficientNet B7 is a high-performance CNN, and DXB is a robust model blending DART and XGBoost algorithms for predictive accuracy. Using MRI datasets totaling 86,390 images, Sait's approach achieved 99.8% average generalization accuracy, underscoring the potential of multi-model fusion for high-precision AD detection.

Mattle et al. (**contribution 14**) examine brain-wide structural connectomics in early AD stages. Analyzing a longitudinal diffusion-weighted imaging dataset of 264 subjects, they apply a tailored machine learning approach that combines exhaustive tractography with neuropsychological data to achieve high classification accuracy. Their model identifies early biomarkers of AD based on hemispheric lateralization of mean tract volume for specific tracts in the supramarginal and paracentral regions, demonstrating the predictive value of diffusion MRI and the importance of multi-modal data integration in neurodegenerative disease research.

Other medical applications in this issue cover cerebral palsy (CP), head tremor, sports medicine, and epilepsy (contributions 5, 6, 13, and 16, respectively).

Roy, Ehrlich, and Lampe (**contribution 5**) conducted an in-depth EEG study comparing the neural responses of seven patients with cerebral palsy (CP) to a control group of four healthy participants. CP, a movement disorder stemming from early, nonprogressive brain damage, often leads to additional cognitive, communicative, and behavioral symptoms. The study employed two types of tactile stimulation—'frequent' and 'infrequent'—applied to the ring finger and thumb of participants' left hands, respectively, to elicit event-related potentials (ERPs) recorded at frontal, central, and parietal scalp locations. In the control group, typical mismatch-related ERP responses were observed, while in CP patients, statistically significant differences were detected between the responses to the two stimuli on frontocentral and parietal channels within the 150–250 ms post-stimulus window. Additionally, a distinct late discriminative response appeared on frontal and parietal channels. These findings reveal the presence and potential observability of mismatch-related neural components in CP patients, providing insight into how CP impacts sensory processing. The authors acknowledged certain limitations, including the small sample size, suggesting future studies to build on this work with larger cohorts.

Rossi et al. (**contribution 6**) investigated head micromovements and body posture to assess vigilance and monitor changes in mental states—an area increasingly relevant due to global population aging trends. With the proportion of individuals over 60 expected to nearly double by 2050 [13], and head tremors being a prevalent symptom in age-related conditions such as Parkinson's disease, precise monitoring of head movements is increasingly important. Head tremors are commonly experienced by older adults, often as a result of Parkinson's disease. According to the American Parkinson Disease Association, tremors affect approximately 80% of individuals with Parkinson's, making them a defining feature of the condition [14]. The miniaturization and widespread use of inertial measurement units (IMUs) in devices like smart glasses have simplified tracking, but self-reports and simple performance measures alone do not provide reliable real-time indicators of vigilance. To address this, the authors examined the relationship between head micromovements, body posture changes, and vigilance reduction during a psychomotor vigilance task. Their results demonstrate that head micromovements are valuable markers for tracking prolonged vigilance decrement and can effectively distinguish between high and low vigilance states, highlighting the potential of IMUs in monitoring cognitive states in aging populations.

Billat et al. (**contribution 13**) explored the brain's role in limiting exercise capacity by analyzing EEG recordings taken during incremental exercise tests (IETs) with 42 participants. IETs assess maximal aerobic power and oxygen consumption ($\dot{V}O_2max$), key indicators in sports medicine. The study aimed to test whether the inability to reach a $\dot{V}O_2$ plateau ($\dot{V}O_2pl$) is primarily influenced by central (brain-based) rather than peripheral (muscle-based) factors. The authors observed a general EEG power decline across all frequency bands, irrespective of $\dot{V}O_2$ plateau occurrence, suggesting depletion of overall "EEG reserve", while alpha activity in the motor cortex remained relatively preserved. They hypothesize that fatigue-associated EEG changes may reflect the brain's attempts to conserve neural resources for motor function and that these changes might vary depending on individuals' sport experience levels. This study opens up the possibility of using EEG as a predictive indicator of exercise exhaustion, which could have applications in optimizing training and managing fatigue.

Ferri et al. (**contribution 16**) made a significant contribution to epilepsy research by using EEG to study cortical connectivity responses to hyperventilation (HV) in patients with focal epilepsy, a type of epilepsy where seizures originate in specific brain lobes. HV is routinely performed during EEG recording as an activation technique recommended by neurophysiology guidelines. The authors applied phase transfer entropy, an advanced connectivity analysis, to assess how HV affects cortical connectivity. They found that HV-induced connectivity significantly increases, similar to patterns observed during non-REM sleep, which is known to promote epileptic activity. Their findings suggest that HV creates a conductive environment for the spread of epileptiform activities but does not alone trigger seizures in focal epilepsy. This study underscores the role of HV in epilepsy diagnostics and the potential of cortical connectivity measures for understanding seizure propagation and developing targeted interventions.

3.2. Cognitive Neuroscience

The second research focus of the papers in this Special Issue is cognitive neuroscience, with three contributions (1, 7, and 8) exploring key themes: the sense of embodiment (contribution 1), perception (contribution 7), and emotion recognition (contribution 8).

Tomás et al. (**contribution 1**) reviewed 20 selected studies on BMIs that utilize multisensory feedback to support the sense of embodiment (SoE) in EEG-based applications. The sense of embodiment is fundamental to human perception, allowing individuals to perceive and control their own body parts. Their review indicates that factors such as immersive scenarios, human-like avatars, and coherent sensory feedback significantly enhance the embodiment experience. However, their analysis does not consistently support the idea that incorporating additional sensory modalities leads to stronger SoE or improved BMI performance. The authors underscore a critical gap in the literature: a lack of systematic experimental studies examining how different sensory modalities individually or cumulatively impact SoE and BMI outcomes. They emphasize the need for further empirical research to isolate and measure the contributions of each sensory modality to embodiment in BMIs.

Peña Serrano et al. (**contribution 7**) make a unique contribution to cognitive neuroscience by applying hypergraph theory to visual perception, marking the first use of hypergraphs in this domain. Hypergraphs are a sophisticated extension of graph theory with diverse applications across cognitive neuroscience and medicine [15]. Using MEG recordings, the authors constructed both traditional graphs and hypergraphs to capture connectivity patterns during the perception of a flickering image. Their analysis considered graph metrics such as degree centrality, betweenness centrality, eigenvector centrality, connected components, shortest-path distances, cycle counts, and node degrees. The hypergraph approach enabled them to capture individual differences across frequency bands, revealing dynamic insights into brain connectivity. The study identified key network features across delta, theta, alpha, beta, and gamma bands, with cortico-cortical interactions across the frontal, parietal, temporal, and occipital lobes. These findings highlight robust activation patterns in specific brain regions, supporting theories of lobe integration and multifunctionality and offering a deeper understanding of neural dynamics in visual perception.

Finally, Yao et al. (**contribution 8**) introduce a novel approach for constructing complex networks to enhance emotion recognition using EEG data. Unlike conventional methods, which typically rely on ordinal representations of time series as network nodes, their approach leverages dimension and delay to map time series data into phase space, enabling more nuanced network construction. To validate their method, they applied it to two test signals: random noise and Lorenz chaotic signals. Their approach achieved over 91% accuracy in emotion classification, surpassing existing techniques. This contribution offers a promising new pathway for high-accuracy emotion recognition models, with potential applications in affective computing and real-time emotion detection.

3.3. Machine Learning

Machine learning (ML), a transformative branch of Artificial Intelligence (AI), is rapidly advancing data science applications, including neuroscience. Reflecting the impact of ML, its pioneers, John J. Hopfield and Geoffrey E. Hinton, were awarded the Nobel Prize in Physics in 2024. ML has become indispensable in neuroscience, enhancing predictive accuracy in medical diagnostics, advancing BCIs, and serving as a powerful research tool. Five papers in this issue (contributions 2, 3, 8, 10, and 15) apply ML techniques to neuroscience, with two of these (contributions 2 and 8) discussed in previous sections. Here, we explore the remaining three studies (contributions 3, 10, and 15).

Kolodziej et al. (**contribution 3**) investigated the potential of CNNs to enhance the detection of steady-state visual evoked potentials (SSVEPs) in BCIs. SSVEPs are EEG signals elicited by visual stimuli at specific frequencies, often used in BCIs due to their simplicity and reliability. Typically, users observe flashing lights at designated frequencies, and SSVEPs are detected by analyzing power spectral density. Kolodziej et al. proposed a CNN model capable of classifying SSVEPs effectively, even with limited training data. Their findings indicate that CNNs significantly improve SSVEP-based BCI accuracy, with up to a 20% increase in performance over traditional methods. This improvement is attributed to the CNN classifier's resilience to artifacts in EEG signals, which often challenge conventional SSVEP detection techniques.

Chen et al. (**contribution 10**) presented an innovative approach to processing diffusion Magnetic Resonance Imaging (dMRI) data from macaque brains using a custom-designed primary–auxiliary dual GAN network (PadGAN). This end-to-end GAN model extracts latent space features from peak information maps to translate high-b-value images to lower-b-value images. In dMRI, the b-value determines the strength and timing of gradients, with higher b-values emphasizing diffusion effects. By translating these high-b-value images, PadGAN produces computed images that maintain a higher signal-to-noise ratio than directly acquired images [16]. This may enhance the quality and utility of dMRI data in brain connectivity studies.

Finally, Cedron et al. (**contribution 15**) developed a novel technique for optimizing multilayer perceptrons (MLPs), a form of ANN, to reduce memory usage and improve runtime. Their method involves pruning zero-weight elements from the ANN, creating a sparse matrix that proves advantageous with large datasets and dense networks. Their approach showed that the sparse matrix format is beneficial when non-zero data elements constitute around 10% of the matrix, particularly with data sets containing thousands of entries. This pruning technique prevents exponential memory consumption and shortens processing time, creating ANNs with enhanced efficiency for neuroscience applications. However, the authors noted that this method currently applies only to fully connected feedforward networks.

These papers highlight the growing role of machine learning in advancing neuroscience, offering methods that enhance data processing, analysis accuracy, and computational efficiency in various applications.

3.4. Statistical Methods

Statistical analysis is fundamental to neuroscience, as biological data are inherently noisy and nondeterministic [17]. This variability reflects differences in brain structure and function across individuals and populations, and effective statistical techniques help identify probable patterns and generalize findings from noisy data. However, noise can play a constructive role, as seen in phenomena like coherence resonance, where noise at an optimal level enhances signal coherence [18].

Petzold (**contribution 4**) introduces a simple yet powerful graphical method, the partial parallelism plot, to illustrate partial parallelism in data. Originally developed for laboratory tests, parallelism plots have been an essential tool for assessing similarity in test results. However, the experimental validation of parallelism remains challenging in bioanalytical method validation. While traditional methods, such as analysis of variance (ANOVA), are commonly applied to evaluate parallelism in linear data sets, they often fall short in identifying nuanced deviations from parallelism. Petzold's approach extends beyond traditional ANOVA limitations by offering a graphical assessment tool designed for cases where parallelism is only partially present. This method accommodates biomarker tests with subtle deviations, enhancing the evaluation of parallelism and addressing limitations within existing regulatory guidelines.

Gómez et al. (**contribution 12**) focus on the role of stochasticity in neuronal dynamics, particularly in the opening and closing of ion channels. Neuronal behavior is probabilistic, with neural noise influencing ion channel activity at the cellular level [19] and perceptual switching at the behavioral level [20]. This intrinsic process underscores the complexity of biological systems and highlights that purely random models, while insightful, are approximations. The inherent randomness is likely shaped by hidden or unknown deterministic factors influencing neuronal activity. By studying stochastic models of ion channel behavior, Gómez et al. contribute to a more comprehensive understanding of how noise impacts neural dynamics, shedding light on probabilistic mechanisms that may govern brain function at multiple scales.

These contributions underscore the critical role of statistical methods in neuroscience, providing tools to decipher complex, noisy biological data and elucidate patterns within inherently variable systems.

4. Conclusions

This Special Issue highlights the transformative role of computational and mathematical approaches in advancing neuroscience, showcasing a wide range of state-of-the-art methodologies, such as computational modeling, ML, network analysis, and BCIs, that have deepened our understanding of brain dynamics, network interactions, cognitive processes, and behavior. By addressing core challenges in data integration and model validation, the papers in this issue underscore the potential of these methods to drive breakthroughs with far-reaching implications across medicine, technology, and our understanding of the human mind.

Each contribution demonstrates not only cutting-edge technologies but also valuable applications that bring us closer to decoding the complexity of brain function. From applications in medical diagnostics to insights into cognitive neuroscience and innovative statistical frameworks, these works collectively enhance our capacity to model, predict, and interpret brain activity with increasing accuracy and reliability.

We extend our gratitude to the authors, reviewers, and editors whose dedication has culminated in this comprehensive volume. It is our hope that these collective efforts will enrich our understanding of neural function and inspire further exploration in neuroscience, pushing the field to new and exciting frontiers.

Conflicts of Interest: The author declares no conflict of interest.

List of Contributions:

1. Tomás, D.; Pais-Vieira, M.; Pais-Vieira, C. Sensorial Feedback Contribution to the Sense of Embodiment in Brain–Machine Interfaces: A Systematic Review. *Appl. Sci.* **2023**, *13*, 13011. https://doi.org/10.3390/app132413011.
2. Altwijri, O.; Alanazi, R.; Aleid, A.; Alhussaini, K.; Aloqalaa, Z.; Almijalli, M.; Saad, A. Novel Deep-Learning Approach for Automatic Diagnosis of Alzheimer's Disease from MRI. *Appl. Sci.* **2023**, *13*, 13051. https://doi.org/10.3390/app132413051.
3. Kołodziej, M.; Majkowski, A.; Rak, R.; Wiszniewski, P. Convolutional Neural Network-Based Classification of Steady-State Visually Evoked Potentials with Limited Training Data. *Appl. Sci.* **2023**, *13*, 13350. https://doi.org/10.3390/app132413350.
4. Petzold, A. Partial Parallelism Plots. *Appl. Sci.* **2024**, *14*, 602. https://doi.org/10.3390/app14020602.

5. Roy, S.; Ehrlich, S.; Lampe, R. Somatosensory Mismatch Response in Patients with Cerebral Palsy. *Appl. Sci.* **2024**, *14*, 1030. https://doi.org/10.3390/app14031030.
6. Rossi, D.; Aricò, P.; Di Flumeri, G.; Ronca, V.; Giorgi, A.; Vozzi, A.; Capotorto, R.; Inguscio, B.; Cartocci, G.; Babiloni, F.; et al. Analysis of Head Micromovements and Body Posture for Vigilance Decrement Assessment. *Appl. Sci.* **2024**, *14*, 1810. https://doi.org/10.3390/app14051810.
7. Peña Serrano, N.; Jaimes-Reátegui, R.; Pisarchik, A. N. Hypergraph of Functional Connectivity Based on Event-Related Coherence: Magnetoencephalography Data Analysis. *Appl. Sci.* **2024**, *14*, 2343. https://doi.org/10.3390/app14062343.
8. Yao, L.; Lu, Y.; Wang, M.; Qian, Y.; Li, H. Exploring EEG Emotion Recognition through Complex Networks: Insights from the Visibility Graph of Ordinal Patterns. *Appl. Sci.* **2024**, *14*, 2636 https://doi.org/10.3390/app14062636.
9. Koźmiński, P.; Gniazdowska, E. Design, Synthesis and Molecular Modeling Study of Radiotracers Based on Tacrine and Its Derivatives for Study on Alzheimer's Disease and Its Early Diagnosis. *Appl. Sci.* **2024**, *14*, 2827. https://doi.org/10.3390/app14072827.
10. Chen, Y.; Zhang, L.; Xue, X.; Lu, X.; Li, H.; Wang, Q. PadGAN: An End-to-End dMRI Data Augmentation Method for Macaque Brain. *Appl. Sci.* **2024**, *14*, 3229. https://doi.org/10.3390/app14083229.
11. Sait, A. A LeViT–EfficientNet-Based Feature Fusion Technique for Alzheimer's Disease Diagnosis. *Appl. Sci.* **2024**, *14*, 3879. https://doi.org/10.3390/app14093879.
12. Gómez, C.; Rodríguez-Martínez, E.; Altahona-Medina, M. Unavoidability and Functionality of Nervous System and Behavioral Randomness. *Appl. Sci.* **2024**, *14*, 4056. https://doi.org/10.3390/app14104056.
13. Billat, V.; Berthomier, C.; Clémençon, M.; Brandewinder, M.; Essid, S.; Damon, C.; Rigaud, F.; Bénichoux, A.; Maby, E.; Fornoni, L.; et al. Electroencephalography Response during an Incremental Test According to the $\dot{V}O_2$max Plateau Incidence. *Appl. Sci.* **2024**, *14*, 5411. https://doi.org/10.3390/app14135411.
14. Mattie, D.; Peña-Castillo, L.; Takahashi, E.; Levman, J. MRI Diffusion Connectomics-Based Characterization of Progression in Alzheimer's Disease. *Appl. Sci.* **2024**, *14*, 7001. https://doi.org/10.3390/app14167001.
15. Cedron, F.; Alvarez-Gonzalez, S.; Ribas-Rodriguez, A.; Rodriguez-Yañez, S.; Porto-Pazos, A. Efficient Implementation of Multilayer Perceptrons: Reducing Execution Time and Memory Consumption. *Appl. Sci.* **2024**, *14*, 8020. https://doi.org/10.3390/app14178020.
16. Ferri, L.; Mason, F.; Di Vito, L.; Pasini, E.; Michelucci, R.; Cardinale, F.; Mai, R.; Alvisi, L.; Zanuttini, L.; Martinoni, M.; et al. Cortical Connectivity Response to Hyperventilation in Focal Epilepsy: A Stereo-EEG Study. *Appl. Sci.* **2024**, *14*, 8494. https://doi.org/10.3390/app14188494.

References

1. Hodgkin, A.L.; Huxley, A.F. A quantitative description of membrane current and its application to conduction and excitation in nerve. *J. Physiol.* **1952**, *117*, 500–544. [CrossRef] [PubMed]
2. FitzHugh, R. Impulses and physiological states in theoretical models of nerve membrane. *Biophys. J.* **1961**, *1*, 445–466. [CrossRef] [PubMed]
3. Nagumo, J.; Arimoto, S.; Yoshizawa, S. An active pulse transmission line simulating nerve axon. *Proc. IRE* **1962**, *50*, 2061–2070. [CrossRef]
4. Hindmarsh, J.L.; Rose, R.M. A model of neuronal bursting using three coupled first order differential equations. *Proc. Roy. Soc. Lond.* **1984**, *221*, 87–102.
5. Wilson, H.R.; Cowan, J.D. Excitatory and inhibitory interactions in localized populations of model neurons. *Biophys. J.* **1972**, *12*, 1–24. [CrossRef] [PubMed]
6. Izhikevich, E.M. Simple model of spiking neurons. *IEEE Trans. Neural Netws.* **2003**, *14*, 1569–1572. [CrossRef] [PubMed]
7. Abbott, L.F. Lapicque's introduction of the integrate-and-fire model neuron (1907). *Brain Res. Bull.* **2003**, *50*, 303–304. [CrossRef] [PubMed]
8. Rulkov, N.F. Modeling of spiking-bursting neural behavior using two-dimensional map. *Phys. Rev. E* **2002**, *65*, 041922. [CrossRef] [PubMed]
9. Lytton, W.W. Computer modelling of epilepsy. *Nat. Rev. Neurosci.* **2008**, *9*, 626–637. [CrossRef] [PubMed]
10. Frolov, N.; Grubov, V.V.; Maksimenko, V.A.; Lüttjohann, A.; Makarov, V.V.; Pavlov, A.N.; Sitnikova, E.; Pisarchik, A.N.; Kurths, J.; Hramov, A.E. Statistical properties and predictability of extreme epileptic events. *Sci. Rep.* **2019**, *9*, 7243. [CrossRef] [PubMed]
11. Nicolas-Alonso, L.F.; Gomez-Gil, J. Brain computer interfaces, a review. *Sensors* **2012**, *12*, 1211–1279. [CrossRef] [PubMed]
12. Hramov, A.E.; Maksimenko, V.A.; Pisarchik, A.N. Physical principles of brain-computer interfaces and their applications for rehabilitation, robotics and control of human brain states. *Phys. Rep.* **2021**, *918*, 1–133. [CrossRef]
13. World Health Organization. Ageing and Health, 1 October 2024. Available online: https://www.who.int/news-room/fact-sheets/detail/ageing-and-health (accessed on 2 December 2024).
14. American Parkinson Disease Association. Parkinson's Disease, 2024. Available online: https://www.apdaparkinson.org/what-is-parkinsons (accessed on 2 December 2024).
15. Bretto, A. *Hypergraph Theory: An Introduction*; Springer: Cham, Switzerland, 2013.

16. Ogura, A.; Koyama, D.; Hayashi, N.; Hatano, I.; Osakabe, K.; Yamaguchi, N. Optimal b values for generation of computed high-b-value DW images. *AJR Am. J. Roentgenol.* **2016**, *206*, 713–718. [CrossRef] [PubMed]
17. Destexhe, A.; Rudolph-Lilith, M. *Neuronal Noise*; Springer: Berlin/Heidelberg, Germany, 2012.
18. Pisarchik, A.N.; Hramov, A.E. Coherence resonance in neural networks: Theory and experiments. *Phys. Rep.* **2023**, *1000*, 1–57. [CrossRef]
19. Jaimes-Reátegui, R.; Huerta-Cuellar, G.; García-López, J.H.; Pisarchik, A.N. Multistability and noise-induced transitions in the model of bidirectionally coupled neurons with electrical synaptic plasticity. *Eur. Phys. J. Spec. Top.* **2022**, *231*, 255–265. [CrossRef]
20. Pisarchik, A.N.; Hramov, A.E. *Multistability in Physical and Living Systems: Characterization and Applications*; Springer: Cham, Switzerland, 2022.

applied sciences

MDPI

Article

Hypergraph of Functional Connectivity Based on Event-Related Coherence: Magnetoencephalography Data Analysis

Natalia Peña Serrano [1,2], Rider Jaimes-Reátegui [2] and Alexander N. Pisarchik [3,*]

1 Escuela Nacional de Estudios Superiores Juriquilla, Universidad Nacional Autónoma de México, Boulevard Juriquilla 3001, Queretaro 76230, Queretaro, Mexico; natalia.pese@comunidad.unam.mx

2 Departamento de Ciencias Exactas y Tecnología, Centro Universitario de los Lagos, Universidad de Guadalajara, Enrique Díaz de León 1144, Colonia Paseos de la Monta na, Lagos de Moreno 47460, Jalisco, Mexico; rider.jaimes@academicos.udg.mx

3 Centro de Tecnología Biomédica, Universidad Politécnica de Madrid, Campus de Montegancedo, Pozuelo de Alarcón, 28223 Madrid, Spain

* Correspondence: alexander.pisarchik@ctb.upm.es

Abstract: We construct hypergraphs to analyze functional brain connectivity, leveraging event-related coherence in magnetoencephalography (MEG) data during the visual perception of a flickering image. Principal network characteristics are computed for the delta, theta, alpha, beta, and gamma frequency ranges. Employing a coherence measure, a statistical estimate of correlation between signal pairs across frequencies, we generate an edge time series, depicting how an edge evolves over time. This forms the basis for constructing an edge-to-edge functional connectivity network. We emphasize hyperedges as connected components in an absolute-valued functional connectivity network. Our coherence-based hypergraph construction specifically addresses functional connectivity among four brain lobes in both hemispheres: frontal, parietal, temporal, and occipital. This approach enables a nuanced exploration of individual differences within diverse frequency bands, providing insights into the dynamic nature of brain connectivity during visual perception tasks. The results furnish compelling evidence supporting the hypothesis of cortico–cortical interactions occurring across varying scales. The derived hypergraph illustrates robust activation patterns in specific brain regions, indicative of their engagement across diverse cognitive contexts and different frequency bands. Our findings suggest potential integration or multifunctionality within the examined lobes, contributing valuable perspectives to our understanding of brain dynamics during visual perception.

Keywords: brain; magnetoencephalography (MEG); network; hypergraph; coherence; visual perception

Citation: Peña Serrano, N.; Jaimes-Reátegui, R.; Pisarchik, A.N. Hypergraph of Functional Connectivity Based on Event-Related Coherence: Magnetoencephalography Data Analysis. *Appl. Sci.* **2024**, *14*, 2343. https://doi.org/10.3390/app14062343

Academic Editor: Vladislav Toronov

Received: 25 January 2024
Revised: 6 March 2024
Accepted: 8 March 2024
Published: 11 March 2024

1. Introduction

Understanding the intricacies of brain connectivity in response to diverse stimuli is crucial for unraveling the mechanisms underlying information processing and decision making within the brain. This study delves into three essential forms of brain connectivity: structural, functional, and efficient [1–4]. Structural connectivity entails the identification of anatomical neural networks, revealing potential pathways for neural communication [5,6]. On the other hand, functional connectivity explores active brain regions exhibiting correlated frequency, phase, and/or amplitude [7]. Finally, effective connectivity utilizes information from functional connectivity to discern the dynamic flow of information within the brain [8,9].

Measurement of effective and functional connectivity can be conducted in both the frequency domain, employing methods such as coherence [10], and in the time domain, utilizing approaches like Granger causality [4] or artificial-neural-network-based functional connectivity [11]. When a sufficiently large population of neurons synchronizes, their electrical and magnetic activities become detectable outside the skull through techniques like electroencephalography (EEG) and magnetoencephalography (MEG) [12]. While EEG measures return or bulk currents outside the neuron (secondary currents), MEG captures ionic currents inside the neuron (primary currents). Notably, MEG holds a distinct advantage over EEG due to its superior spatial resolution, rendering it an exceptional tool for investigating and characterizing interactions between distinct brain regions [13].

To study functional connectivity, some researchers use approaches borrowed from graph theory [14]. Using connections between the biorhythms of the brain in its various parts, a model of a complex network is recreated, in which parts of the brain are considered as nodes, and the connection forces between them are considered as links [15,16]. Using this approach makes it possible to identify not only individual cognitive differences between subjects [17], but also helps to diagnose some diseases at an early stage [18–20] and also monitor the aging process [21–24].

One of the important measures that can be used to quantify neuronal synchrony is event-related coherence [10,25,26]. It examines the frequency domain relationship between two signals, indicating the degree to which their spectral components are synchronized. Essentially, it is an assessment of the constancy of the relative amplitude and phase between two signals within a given frequency range. There is a linear mathematical method that creates a symmetrical matrix, devoid of any directional information. Identical signals produce a coherence value of 1, while the coherence value approaches 0 as the difference between the signals in question increases. Since then, coherence has been used in many brain connectivity studies with both patients and control individuals; these include but are not limited to studies of working memory [27], brain lesions [28], hemiparesis [29], resting state networks [30], schizophrenia [31,32], favorable responses to panic medications [33], and motor imagery [34]. As a result of the unique characteristics of human brains, distinct patterns of coherent neuronal activity were observed among different subjects. For instance, the presentation of flickering visual stimuli induces coherent responses in the visual cortex of subjects at both the flicker frequency and its harmonics, resulting in varied sizes of coherent neural networks [35,36].

In this work, we employ hypergraph analysis, a technique rooted in dynamic graph theory [37], to investigate functional connectivity. We analyze variations in functional connectivity networks using MEG data collected during the observation of flickering images. This approach is an extension of conventional graph theory methods. Specifically, we begin by defining a standard functional network that connects nodes across consecutive time segments. We then generate a set of edge time series, representing the fluctuation of edges over time. We process these edge time series similarly to node time series, creating a network of edge-to-edge functional connectivity. Within this framework, we focus on "hyperedges," which are the connected components of an absolute-valued end-to-end functional connectivity network.

Conventional graphs are limited in their ability to represent connections between pairs of nodes, while hypergraphs can depict relationships among sets of nodes, especially when complex interactions involve more than two elements. Using a hypergraph allows for a more intuitive interpretation of inter-element relationships, enhancing comprehension and result interpretation. It should be noted that Wang et al. [38] have already demonstrated the utility of hypergraphs in showcasing multiple relationships between vertices by employing Pearson correlation for interaction derivation. However, despite the fact that the hypergraph theory was applied several years ago, our paper introduces a significant advancement in hypergraph data representation by leveraging event-related coherence between brain lobes rather than conventional correlation metrics. We prefer coherence due to its ability to measure signal relationships based on relative phase, thereby capturing temporal synchronization between signals. By using a hypergraph, we can clearly and succinctly visualize the correlation between brain lobes and frequency bands, crucial for unraveling the intricate network of cerebral interactions. Moreover, our approach, centered on visualization with modulation, enriches our research with an additional layer of depth, offering an innovative and pertinent perspective in this burgeoning field.

2. Materials and Methods

2.1. Subjects

In this study, we analyze the MEG data of 15 control subjects (aged 17–64 years; 10 men and 5 women) obtained in the experiment based on a flickering image paradigm [39] at the Center for Biomedical Technology of the Universidad Politécnica de Madrid, Spain. The MEG data have been downloaded from https://zenodo.org/record/4408648#.X-72UdYo-Cc (accessed on 10 August 2023).

2.2. Experimental Paradigm

The experimental protocol, depicted in Figure 1, consists of two stages. During the first stage, subjects were presented with a static (unmodulated) black square with white lines for 120 s. They were instructed to fix their gaze on a red dot located at the center of the square. The MEG of the baseline neuronal activity (B-trial) was recorded during this stage After a brief break (40–390 s), the second stage of the experiment commenced. During this stage, the brightness of pixels on the square image was periodically modulated with a frequency of $f_m = 6.67$ Hz and a maximum amplitude of 50% of the RGB color model, oscillating between black (0) and gray (127). This frequency was chosen due to its ability to elicit a prominent spectral response in the visual cortex [35]. The flickering image was presented 2–5 times at intervals of 120 s, with a 30-second break between each presentation, and the MEG was recorded (F-trials). The averaged F-trials were then normalized to the B-trial for each subject and subsequently averaged across all subjects.

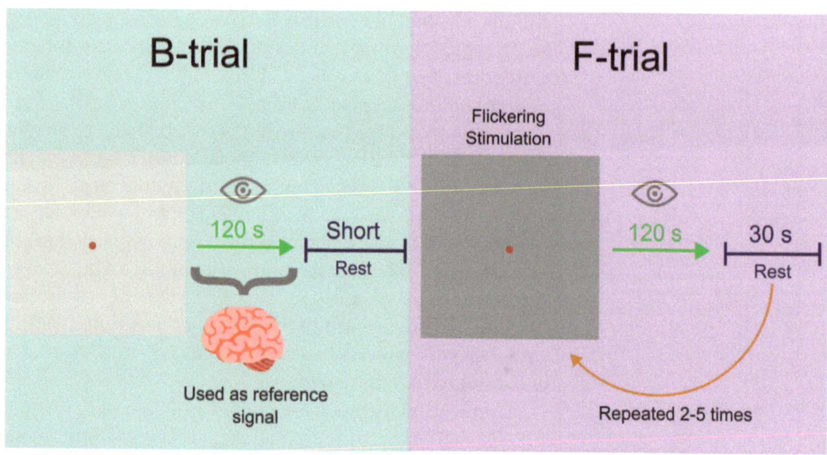

Figure 1. Experiment protocol. The B-trial corresponds to the baseline neuronal activity induced by the unmodulated visual stimulus and the F-trials correspond to the flickering image.

2.3. Signal Analysis in Brainstorm

The signal analysis was performed using the Brainstorm 3.231017 software, a collaborative, open-source application based on MATLAB R2022a that is dedicated to processing and analyzing brain recordings obtained through different brain imaging techniques [40]. The included tools, along with the interface, facilitated the creation of the scripts used in this article.

2.4. Head Model Adjustment

The default Brainstorm head model was adjusted to the head points recorded using a Polhemus Fastrak system, with 2% deformation and automatic refinement of head points.

2.5. Signal Processing

Signal analysis involved reading MEG data, and applying a Notch filter to eliminate 50 Hz power line frequencies and their harmonics. Artifacts from the electrooculogram (EOG) and electrocardiogram (ECG) signals were automatically identified and manually reviewed to ensure the inclusion of any potentially omitted artifacts. Signal–space projection (SSP) methods were applied to correct the artifacts by order.

2.6. Event Segmentation

The signals were segmented into 120 s epochs for two experimental phases: B-trial and F-trial (Figure 1). The signal recorded during the B section was used as a reference signal. These epochs were further divided into 3 s trials.

2.7. Source Reconstruction

Reconstruction of electrical activity in the brain from MEG measurements was carried out by creating a forward model and a lead field matrix. Brainstorm's overlapped spheres method was used, maintaining the recommended 15,000 cortical sources. The inverse solution was calculated using standardized low-resolution electromagnetic tomography (sLORETA).

2.8. Fourier Analysis

To analyze the recorded signal, a Fourier analysis technique using a hanning window was employed. First, a Hanning window was generated with an appropriate length corresponding to the duration of the recorded signal (3 s). Subsequently, this window was applied to the signal to mitigate edge effects and enhance frequency domain resolution. Once the frequency spectrum of the processed signals was obtained, the spectrum of the signal of interest was normalized to the reference signal. This was accomplished by computing the fast Fourier transform (FFT)—using MATLAB's FFT function—of each processed signal and subsequently dividing the frequency spectrum of the signal of interest by that of the reference (baseline) signal.

2.9. Signal Coherence

The brain is known to generate electromagnetic activity in a wide frequency range, from slow waves of 0.5 Hz to fast waves of 500 Hz and higher frequencies [41]. These rhythms are classified according to their frequencies and are assigned Greek letters. In this paper, we consider five frequency bands: delta (0.5–4 Hz), theta (4–7 Hz), alpha (8–12 Hz), beta (13–30 Hz), and gamma (31–90 Hz). Coherence measures the correlation within discrete frequency bands for selected epoch lengths and is independent of signal amplitude [42]. We constructed brain networks based on coherence, a mathematical measure quantifying synchronization patterns between spatially separated sensors or between brain areas [10].

After the Hanning window was applied to the signal, we estimated the strength of network interactions according to coherence between eight brain areas: frontal left—FL; frontal right—FR; occipital left—OL; occipital right—OR; parietal left—PL; parietal right—PR; temporal left—TL; temporal right—TR. The 15,000 brain sources were grouped into these eight lobes using Brainstorm's segmentation model, PALS-12 Lobes with ten structures, excluding the insula, resulting in eight structures or vertices.

The stored vertices were used to average signals within each lobe, reducing complexity to eight signals. The square magnitude coherence was then calculated between time series of each lobe with the rest, for both F and B trials (C_F and C_B, respectively). Subsequently, the absolute difference between the coherence values of F and B was obtained and normalized to the B activity, giving event-related coherence (ERC) as

$$ERC = \frac{|C_F - C_B|}{C_B}. \quad (1)$$

2.10. Visualization with BrainNet

The output consisted of a tensor with dimensions $8 \times 8 \times 15$. The average matrix of all subjects was calculated for each frequency band, and these matrices were saved in ".edge" text files for visualization with BrainNet Viewer. The latter is a tool that facilitates the visualization of structural and functional connectivity patterns in brain networks [43]. The surface template used was "BrainMesh_ICBM152_smoothed.nv", included in the "BrainNetViewer_20191031" folder when downloaded.

A "Node.node" text file was created with the format defined by BrainNet Viewer to set the position of nodes in the brain figure. The ".edge" files for each frequency band obtained earlier were used to display interactions.

2.11. Graph Construction

Coherence matrices showed coefficients between 0.1 and 19.16. A threshold σ_{th} was set for graph construction, with its value varying between 0.1 and 1.25. The analysis was conducted to assess how graph characteristics change when including interactions above σ_{th}. Various centrality measures were calculated such as degree centrality (number of edges [44]), betweenness centrality (fraction of shortest paths passing through a node [45]), and eigenvector centrality (importance of a node considering the importance of its neighbors [44]). Connected components of the graphs were explored [45]. The shortest path distances between all node pairs were calculated, and cycles in the graph were identified, and defined as connected graphs in which each vertex has degree 2 [46]. These metrics, with respect to σ_{th}, are presented in Appendix (Figures A1–A7).

To assess the connectivity of the graphs generated in this study, the 'conncomp' function in MATLAB was employed. This function facilitates the identification of connected components in an undirected graph, thus providing a measure of the level of connection within the network. This function assigns each node in the graph a connected component identifier, enabling the determination of the total number of connected components present in the graph. The graph's connectivity coefficient can be calculated as

$$\kappa = \frac{1}{n}, \tag{2}$$

where n is the number of connected components. This methodological approach allowed for the analysis of how the variation in the threshold σ_{th} affects the global connectivity of the graphs.

A threshold value of $\sigma_{th} = 0.5$ was decided upon, as lower values show coherence interactions that may include noise, and values higher than this completely lose one of our frequency bands.

With the specified threshold, we generated graphs explicitly depicting the mentioned measures using MATLAB. The calculation of the distances between node pairs was carried out without considering the weight of the edges, i.e., treating it as an unweighted graph. Betweenness centrality and eigenvector centrality measures can be found in Appendix A (Figures A6 and A7).

2.12. Hypergraph Construction

To contextualize the analysis of hypergraphs, we define the elements of graph theory used to construct hypergraphs formed by nodes and edges, where nodes denote brain regions, or groups of voxels, and edges denote correlations in activity between pairs of nodes over time. Significant correlation in activity between pairs of edges over time is denoted as links. In this context, we define a hyperedge as a group of links connecting two or more edges with significantly correlated temporal profiles. Finally, a set of hyperedges forms a hypergraph, H, defined as an ordered pair (V, ε). Here, V is a finite set of nodes or vertices $V = \{1, \ldots, N\}$ and $\varepsilon = \{h_1, h_2, \ldots, h_n\}$ is a family of nonempty subsets of elements of X, called hyperedges or hyperlinks, showing the interaction between elements of X [14].

The algorithm of [47] was used for part of the hypergraph visualization. The incidence matrix was given as input. The obtained representations included the hypergraph, the incidence matrix in linear form, and the star expansion. Hypergraph characteristics were obtained along with some matrix representations of the same.

2.12.1. Adjacency and Node Stars

Adjacency between vertices is established when at least one hyperedge contains two vertices [14]. This is illustrated by the star of the nodes, which is defined as the collection of hyperedges incident on node i: $\varepsilon(i) = \{h|i \in h\}$ is referred to as the star of i, where h is a hyperedge.

2.12.2. Degrees of Vertices and Hyperedges

The degree of vertices in the hypergraph is defined as $deg_H(i) = |\varepsilon(i)|$, which corresponds to the size of its stars, i.e., the number of hyperedges incident to i. The degree of a hyperedge, $h \in \varepsilon$, is the number of vertices it contains, denoted as $deg(h) = |h|$.

3. Results and Discussion

3.1. Coherence Brain Networks and Matrices

Figure 2 displays brain networks (left column) of the coherence and corresponding 8×8 coherence matrices (right column) for five frequency bands (delta, theta, alpha, beta, and gamma), presented using the BrainNet Viewer template. The coherence values shown in this figure were computed from the averaged F-trials normalized to the B-trials. One can observe that, in general, the coherence at low frequencies (delta and theta waves) is stronger than that at high frequencies (alpha, beta, and gamma waves). In particular, the strongest coherence at low frequencies occurred between the right and left temporal lobes (Figure 2a,b). At the same time, for alpha waves, the strongest coherence was observed between the right occipital and left parietal lobes (Figure 2c), while for beta waves, the strongest coherence occurred between the right frontal and left occipital lobes (Figure 2d). Finally, for gamma waves, the strongest coherence was observed between the left frontal and right occipital lobes (Figure 2e). The coherence values in both the brain and matrix representations are prominently depicted in the color bar, appearing as a distinct shade of dark red.

3.2. Frequency Spectra

Figure 3 displays the brain activity in the frequency domain. Here, we present the normalized power spectra in eight lobes, which correspond to the average fast Fourier transform (FFT) of the F-trials, normalized to the FFT of the B-trials.

The modulation frequency $f_m = 6.67$ Hz (M) is notably visible in the frequency spectrum of most lobes. However, its second harmonic, $2f_m = 13.34$ Hz (H), exhibits the highest amplitude across all lobes. Furthermore, the higher harmonics at $3f_m$ and $4f_m$ are discernible in the occipital lobes. This observation aligns cohesively with findings reported by Chholak et al. [39].

3.3. Network Characteristics

We computed various network characteristics. As depicted in Figure 2, the coherence coefficient ranges from 0 to 19.16 across different frequency bands, reaching a distinct maximum value for each band. Table 1 displays the maximum coherences observed within each frequency band, along with the lobes between which they were identified.

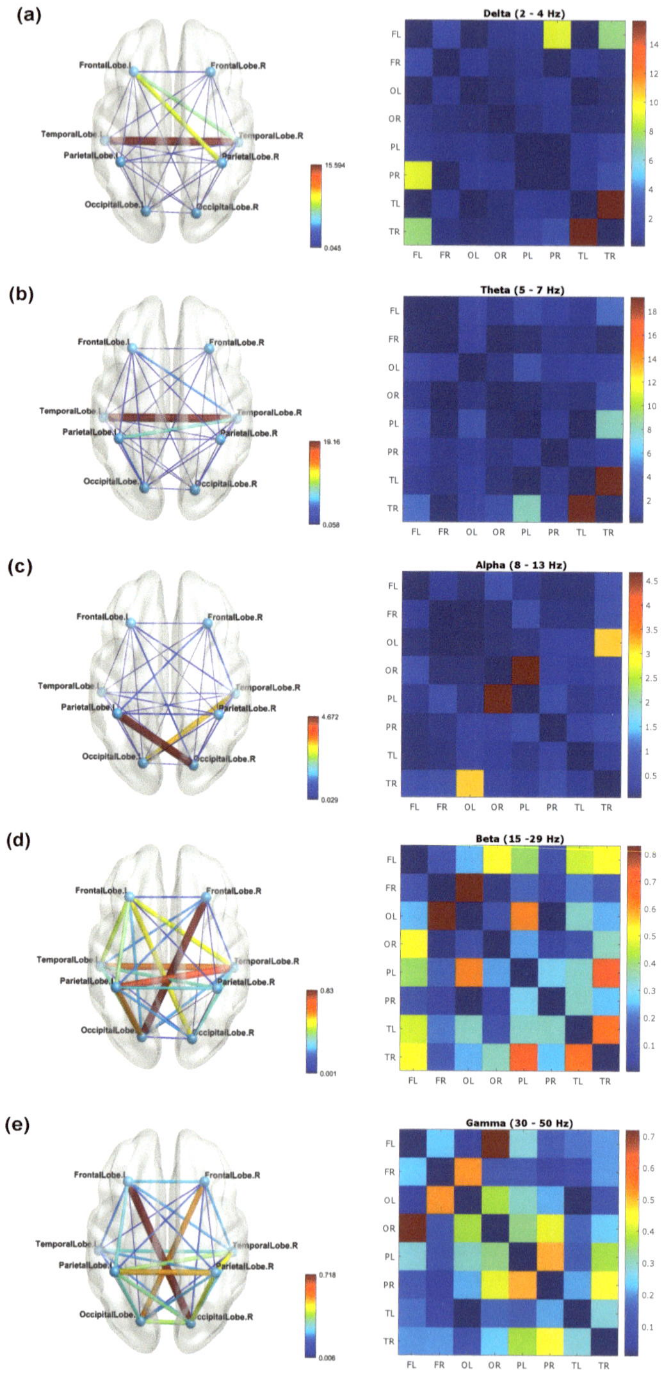

Figure 2. Coherence networks (**left column**) with their respective matrices (**right column**) for (**a**) delta, (**b**) theta, (**c**) alpha, (**d**) beta, and (**e**) gamma waves of the normalized F-trials. The stronger coherence is represented by wider lines.

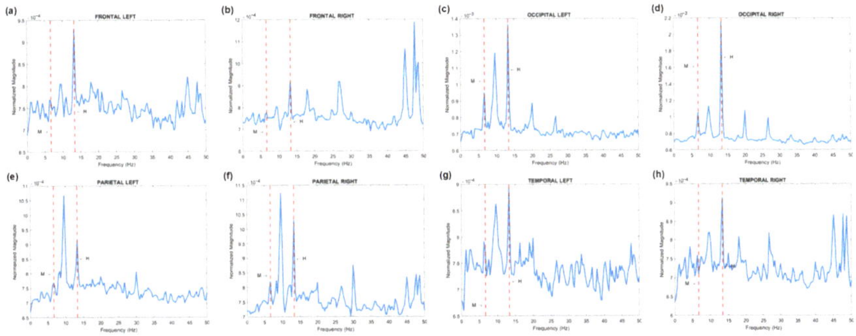

Figure 3. Power spectra of brain activity in different lobes: (**a**) left frontal, (**b**) right frontal, (**c**) left occipital, (**d**) right occipital, (**e**) left parietal, (**f**) right parietal, (**g**) left temporal, and (**h**) right temporal. The dotted lines marked by letters M and H denote the modulation frequency f_m and its second harmonic $2f_m$, respectively.

Table 1. Maximum coherence of each frequency band.

Frequency Band	Maximum Coherence	Lobes
Delta	15.5940	Temporal Left–Temporal Right
Theta	19.160	Temporal Left–Temporal Right
Alpha	4.6720	Parietal Left–Parietal Right
Beta	0.8300	Frontal Right–Occipital Left
Gamma	0.7180	Frontal Left–Occipital Right

In Figure 4, we plot the global connectivity coefficient κ found by Equation (2) for each frequency band with respect to the threshold value σ_{th}. A general decrease in the coherence coefficient between nodes was observed with increasing frequency.

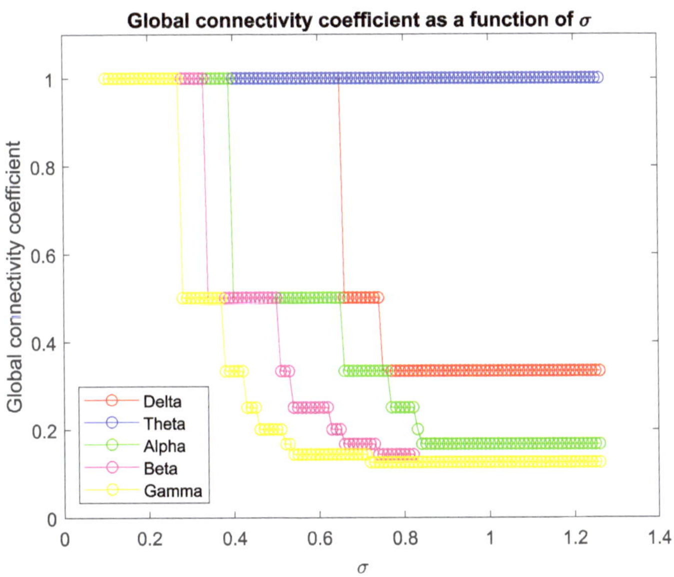

Figure 4. Global connectivity coefficient κ for different frequency bands as a function of the threshold value.

Additionally, it is noteworthy that our representation emphasizes the heightened connectivity of low-frequency components compared to high frequencies, aligning with previous observations by Salvador et al. [48]. This depiction offers a clearer insight into the variations in connectivity between lobes within each frequency band.

One can see that, for each frequency range, there is a coherence threshold value σ_{th} at which centrality measures, shortest-path distances, and degree of nodes undergo significant changes (Figures A1–A5). This threshold value depends on the wave frequency. Specifically, as seen in Figure 4, for delta waves $\sigma_{th} \approx 0.75$ and for theta waves, the global connectivity appears to remain constant at 1 at all these σ_{th} values: for alpha waves, this is $\sigma_{th} \approx 0.42$, for beta waves, this is $\sigma_{th} \approx 0.37$, and for gamma waves, this is $\sigma_{th} \approx 0.3$. This means that σ_{th} decreases as the wave frequency increases, i.e., the brain network of functional connectivity is more stable at low frequencies.

A connectivity threshold serves as a crucial parameter in delineating genuine connections within a functional network while filtering out spurious ones. This approach enables a focused examination of network properties. Analyzing a graph necessitates setting a connectivity threshold to discern valid connections among nodes and discard erroneous ones. The choice of threshold value is somewhat arbitrary, with increasing thresholds excluding weaker, potentially noisy connections. However, setting the threshold too high risks eliminating important frequency bands like beta and gamma, resulting in a connectivity coefficient of 0 for the resulting graph. Our chosen threshold value of $\sigma_{th} = 0.5$ ensures a minimum coherence of this magnitude between lobes' signals. Despite its strictness, this threshold preserves all connections across various frequency bands.

Figure 5 illustrates the results of the analysis of degree centrality in the brain network of the eight lobes for different frequency ranges. The node sizes indicate their importance as a function of edge weights. Centrality, as extensively documented in the electrophysiological literature, has consistently underscored the non-uniform distribution of coherence across frequencies [49]. It is well established that different systems of brain regions may exhibit varying levels of coherence at distinct frequencies [48].

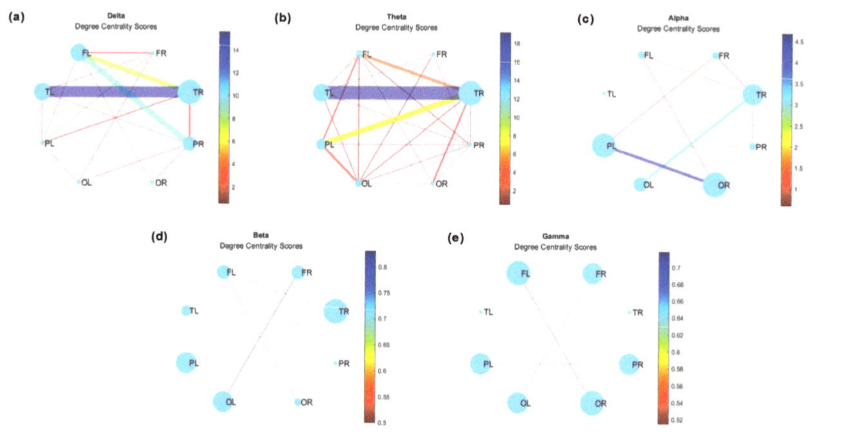

Figure 5. Degree centrality for (**a**) delta, (**b**) theta, (**c**) alpha, (**d**) beta, and (**e**) gamma waves at $\sigma = 0.5$.

The centrality patterns also demonstrate frequency-specific nuances. Specifically, at low frequencies, centrality predominantly manifests in temporal lobes, with noteworthy lateralization observed in delta and theta waves (Figure 5a,b). For alpha frequencies (Figure 5c), the coherence reveals a shift in centrality, now prominently observed in the right occipital and left parietal lobes, whereas in the case of beta and gamma waves (Figure 5d,e), the centrality is very weak and homogeneous between lobes. Contralateral coherences are observed between frontal and occipital lobes in the high-frequency bands.

Figure 6 presents another representation of the node degrees for the lobes. It is evident that the nodes exhibit larger coherence-related connections in the low-frequency bands (Figure 6a,b). However, in the higher-frequency bands (Figure 6c–e), the connections are less pronounced. Notably, the connections in the delta band (Figure 6a) are smaller compared to those in the theta network (Figure 6b), which shows nodes with degrees of 7. A more evident alteration is observed for alpha waves (Figure 6c), where the left temporal lobe is completely disconnected, while the other lobes experience a reduction in connections. In the beta graph (Figure 6d), the right parietal lobe ceases to participate entirely. Meanwhile, for gamma waves (Figure 6e), engagement diminishes for the temporal lobes, while the rest show a degree of 1.

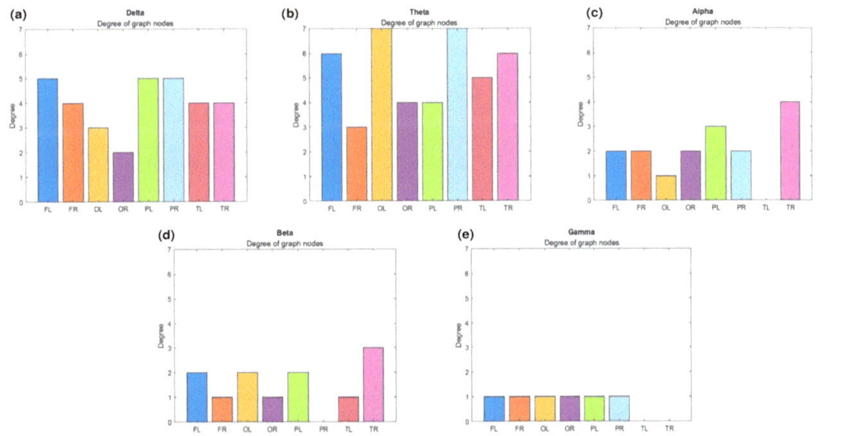

Figure 6. Node degrees for (**a**) delta, (**b**) theta, (**c**) alpha, (**d**) beta, and (**e**) gamma waves. FL—left frontal; FR—right frontal; OL—left occipital; OR—right occipital; PL—left parietal; PR—right parietal; TL—left temporal; TR—right temporal.

Figure 7 illustrates the connected graph components, which are the subset of network nodes where there is a path from each node in the subset to any other node in the same subset [45]. This representation allows us to observe the formation of groups as connections begin to dissolve. Identifying connected components within an undirected graph provides insight into the network's level of connectivity and aids in the extraction of coefficient κ (Equation (2)). Table 2 presents the global connectivity coefficients for $\sigma_{th} = 0.5$.

Figure 8 shows cycles formed in the networks for three frequency bands: delta, theta, and alpha. A cycle is a connected graph, in which each vertex has degree 2 [46]. The total number of cycles found for each graph is indicated in the figure. The theta graph (Figure 8b) displays the greatest connectivity and the most edges, resulting in a higher number of cycles within the network. In contrast, both the alpha and delta graphs (Figure 8a–c, respectively) exhibit fewer connections, with the alpha graph revealing only three cycles. Furthermore, no cycles were observed in the beta or gamma graphs. This observation is consistent with the broader literature, which suggests that slower rhythmic patterns typically exhibit a more widespread network configuration compared to faster ones [50,51].

Table 2. Connectivity coefficients for different frequency bands.

Frequency Band	κ
Delta	1
Theta	1
Alpha	0.5
Beta	0.5
Gamma	0.2

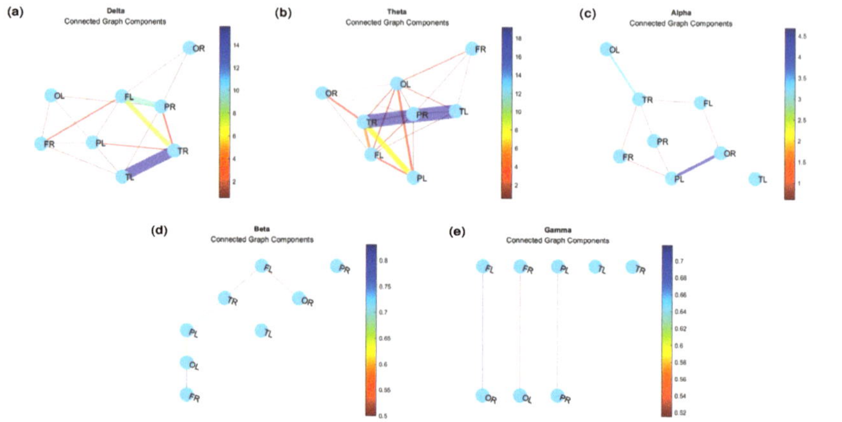

Figure 7. Connected graph components for (**a**) delta, (**b**) theta, (**c**) alpha, (**d**) beta, and (**e**) gamma bands.

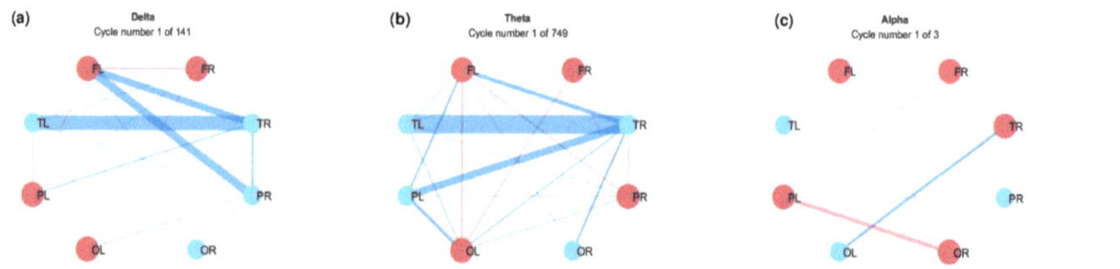

Figure 8. Cycles of the graphs for (**a**) delta, (**b**) theta, and (**c**) alpha bands.

Distances between nodes are represented as 8 × 8 matrices ($Delta_D$, $Theta_D$, $Alpha_D$, $Beta_D$, $Gamma_D$), showing the shortest path distances. When the nodes are not connected, the distance is infinite. The disconnection of the networks is also noticeable in these matrices. The largest value is a five-node distance, only seen in the beta wave.

$$Delta_D = \begin{bmatrix} 0 & 1 & 2 & 1 & 1 & 1 & 2 & 1 \\ 1 & 0 & 1 & 2 & 1 & 2 & 1 & 2 \\ 2 & 1 & 0 & 2 & 1 & 1 & 2 & 2 \\ 1 & 2 & 2 & 0 & 2 & 1 & 2 & 2 \\ 1 & 1 & 1 & 2 & 0 & 2 & 1 & 1 \\ 1 & 2 & 1 & 1 & 2 & 0 & 1 & 1 \\ 2 & 1 & 2 & 2 & 1 & 1 & 0 & 1 \\ 1 & 2 & 2 & 2 & 1 & 1 & 1 & 0 \end{bmatrix}$$

$$Theta_D = \begin{bmatrix} 0 & 2 & 1 & 1 & 1 & 1 & 1 & 1 \\ 2 & 0 & 1 & 2 & 2 & 1 & 1 & 2 \\ 1 & 1 & 0 & 1 & 1 & 1 & 1 & 1 \\ 1 & 2 & 1 & 0 & 2 & 1 & 2 & 1 \\ 1 & 2 & 1 & 2 & 0 & 1 & 2 & 1 \\ 1 & 1 & 1 & 1 & 1 & 0 & 1 & 1 \\ 1 & 1 & 1 & 2 & 2 & 1 & 0 & 1 \\ 1 & 2 & 1 & 1 & 1 & 1 & 1 & 0 \end{bmatrix}$$

$$Alpha_D = \begin{bmatrix} 0 & 2 & 2 & 1 & 2 & 2 & \infty & 1 \\ 2 & 0 & 2 & 2 & 1 & 2 & \infty & 1 \\ 2 & 2 & 0 & 3 & 3 & 2 & \infty & 1 \\ 1 & 2 & 3 & 0 & 1 & 2 & \infty & 2 \\ 2 & 1 & 3 & 1 & 0 & 1 & \infty & 2 \\ 2 & 2 & 2 & 2 & 1 & 0 & \infty & 1 \\ \infty & \infty & \infty & \infty & \infty & \infty & 0 & \infty \\ 1 & 1 & 1 & 2 & 2 & 1 & \infty & 0 \end{bmatrix}$$

$$Beta_D = \begin{bmatrix} 0 & 4 & 3 & 1 & 2 & \infty & 2 & 1 \\ 4 & 0 & 1 & 5 & 2 & \infty & 4 & 3 \\ 3 & 1 & 0 & 4 & 1 & \infty & 3 & 2 \\ 1 & 5 & 4 & 0 & 3 & \infty & 3 & 2 \\ 2 & 2 & 1 & 3 & 0 & \infty & 2 & 1 \\ \infty & \infty & \infty & \infty & \infty & 0 & \infty & \infty \\ 2 & 4 & 3 & 3 & 2 & \infty & 0 & 1 \\ 1 & 3 & 2 & 2 & 1 & \infty & 1 & 0 \end{bmatrix}$$

$$Gamma_D = \begin{bmatrix} 0 & \infty & \infty & 1 & \infty & \infty & \infty & \infty \\ \infty & 0 & 1 & \infty & \infty & \infty & \infty & \infty \\ \infty & 1 & 0 & \infty & \infty & \infty & 1 & \infty \\ 1 & \infty & \infty & 0 & \infty & \infty & \infty & \infty \\ \infty & \infty & \infty & \infty & 0 & 1 & \infty & \infty \\ \infty & \infty & \infty & \infty & 1 & 0 & \infty & \infty \\ \infty & \infty & \infty & \infty & \infty & \infty & 0 & \infty \\ \infty & \infty & \infty & \infty & \infty & \infty & \infty & 0 \end{bmatrix}$$

3.4. Hypergraphs

Figure 9 represents the hypergraph constructed based on a chosen threshold ($\sigma_{th} = 0.5$) in different forms with colors corresponding to different frequency bands. In particular, the hypergraph is shown as a network in Figure 9a, a star expansion in Figure 9b with connections for each node, and as its incidence matrix in Figure 9c. It is observed that all eight lobes are coupled in the low-frequency ranges of the delta and theta bands, seven lobes (FL, FR, OL, OR, TR, PL, and PR) are coupled for the alpha band, eight lobes (FL, FR, TL, TR, OL, OR, and PL) for the beta band, and six lobes (FL, FR, OL, OR, PL, and PR) for the gamma band.

The analysis was carried out following the basic properties of hypergraphs. The degrees of the vertices and hyperedges are given in Table 3, where $deg_H(i) = |\varepsilon(i)|$ is the number of vertices incident on i, and the degree of a hyperedge $deg(h) = |h|$ is the number of vertices it contains.

Table 3. Vertices and hyperedges degrees.

Vertice	$\|\varepsilon(i)\|$	Hyperedge	$\|e\|$
Frontal Left	5	Delta	8
Frontal Right	5	Theta	8
Occipital Left	5	Alpha	7
Occipital Right	5	Beta	7
Parietal Left	5	Gamma	6
Parietal Right	4		
Temporal Left	3		
Temporal Right	4		

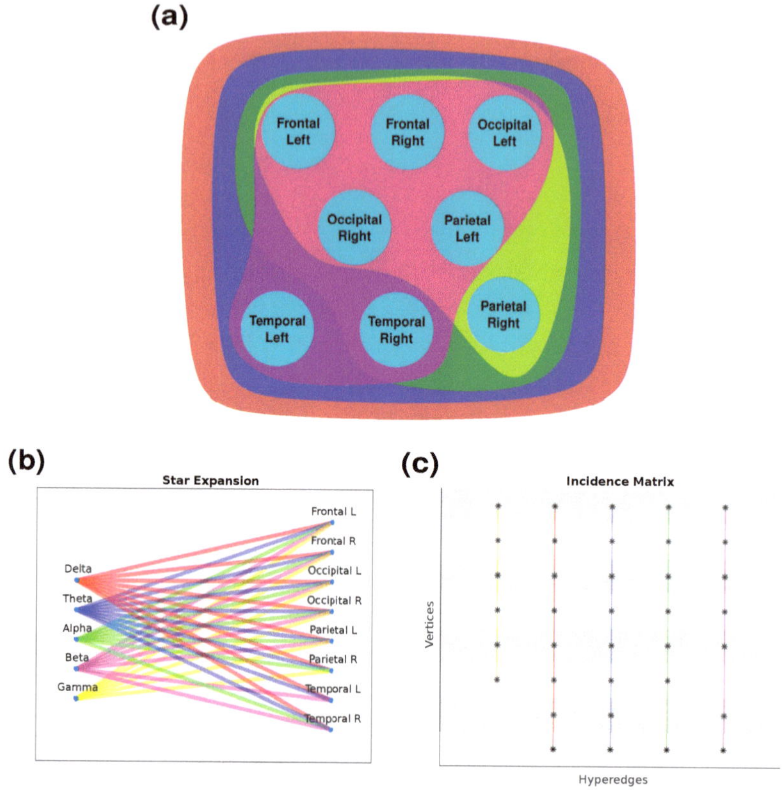

Figure 9. Hypergraph representation as (**a**) network, (**b**) star expansion, and (**c**) adjacency matrix. Colors are maintained between hypergraph representations. Delta (red), Theta (blue), Alpha (green), Beta (magenta), Gamma (yellow).

A star, as mentioned earlier, denotes a pattern where a central vertex—in this case, representing a cerebral lobe—is intricately connected to multiple peripheral vertices, symbolizing different brain frequency ranges. This graphical representation is valuable as it effectively portrays the intricate relationship between a specific cerebral lobe and its involvement across diverse frequency bands. The presence of stars within the graph signifies that the central cerebral lobe exhibits activation across various cognitive conditions or mental states, indicative of its multifunctional nature.

The stars of each lobe are the following:

- Left frontal lobe (node 1): Delta, theta, alpha, beta, and gamma.
- Right frontal lobe (node 2): Delta, theta, alpha, beta, and gamma.
- Left occipital lobe (node 3): Delta, theta, alpha, beta, and gamma.
- Right occipital lobe (node 4): Delta, theta, alpha, beta, and gamma.
- Left parietal lobe (node 5): Delta, theta, alpha, beta, and gamma.
- Right parietal lobe (node 6): Delta, theta, alpha, and gamma.
- Left temporal lobe (node 7): Delta, theta, and beta.
- Right temporal lobe (node 8): Delta, theta, alpha, and beta.

Although correlations between frequencies have been observed in previous studies [52], and biophysical models have been proposed to explain interactions among different frequency bands, such as theta and gamma [53], further research, similar to the current study, is necessary to elucidate the potential coupling between the mechanisms generating these distinct frequencies.

4. Conclusions

In this study, we conducted a comprehensive hypergraph analysis of functional connectivity using MEG data acquired from a modulated visual stimulus. By focusing on differences within distinct frequency bands, we constructed a coherence-based hypergraph to explore functional connectivity among the frontal, parietal, temporal, and occipital brain lobes. Our findings suggest that each frequency band has a specific coherence threshold; beyond this, significant changes occur in network characteristics, such as centrality, shortest-path distances, and node degree. Interestingly, this threshold is lower for higher-frequency bands, indicating stronger connectivity between lobes at lower frequencies and greater stability in the brain network of functional connectivity. Specifically, we observed strong coherence among all eight lobes in the delta and theta bands; this was only observed among six lobes in the gamma band.

Furthermore, we noted variations in global connectivity with coherence thresholds across different frequency bands. Strong coherence was observed between temporal right and left lobes for delta and theta waves, occipital right and parietal left lobes for alpha waves, frontal right and occipital left lobes for beta waves, and frontal and occipital lobes for gamma waves. These findings collectively contribute to our understanding of brain network dynamics across different frequency bands. Our results lend credence to the idea that cortico–cortical interactions can manifest at multiple levels. The resultant hypergraph exposes robust activation patterns in select brain regions across varied cognitive contexts associated with various frequency bands, hinting at possible integration or multifunctionality within these lobes.

While previous studies have noted correlations between frequencies and proposed biophysical models to explain interactions, our results highlight the importance of further research to unravel potential connections between the mechanisms generating different frequencies and with a modulated visual stimulus, using coherence instead of traditional correlation. Additionally, we argue that exploring hypergraph visualizations with a focused examination of finer-grained neural ensembles or smaller regions of interest (ROIs) can reveal compelling dynamics that merit further scholarly investigation.

Author Contributions: N.P.S.: investigation, formal analysis; R.J.-R.: resources, project administration, original draft preparation; A.N.P.: conceptualization, methodology, review and editing. All authors have read and agreed to the published version of the manuscript.

Funding: This research received no external funding.

Institutional Review Board Statement: The study was conducted according to the guidelines of the Declaration of Helsinki, and approved by the Ethics Committee of the Technical University of Madrid (protocol No. 19, 3 February 2020).

Informed Consent Statement: Informed consent was obtained from all subjects involved in the study.

Data Availability Statement: The original contributions presented in the study are included in the article, further inquiries can be directed to the corresponding author.

Conflicts of Interest: The authors declare no conflicts of interest. The funders had no role in the design of the study; in the collection, analyses, or interpretation of data; in the writing of the manuscript; or in the decision to publish the results.

Appendix A. Network Characteristics as a Function of Coherence Threshold

The dependencies of the main network characteristics on the coherence threshold σ for different brain waves are illustrated in Figures A1–A5 (delta in Figure A1; theta in Figure A2; alpha in Figure A3; beta in Figure A4; gamma in Figure A5). The lobes are represented with different colors. The shortest path distances and numbers of cycles depict averages in the case of distances and totals in the case of cycles.

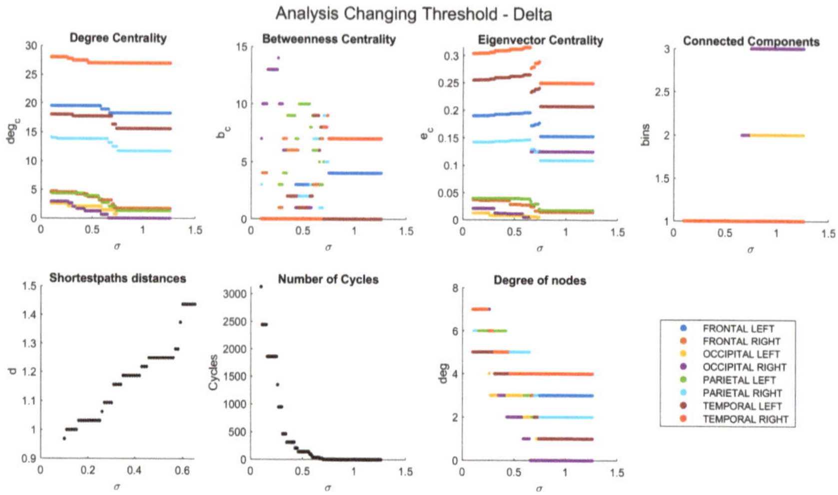

Figure A1. Network characteristics versus coherence threshold value σ_{th} for delta band: degree centrality (deg_C), betweenness centrality (b_C), eigenvector centrality (e_C), connected components (bins), shortest-path distances (d), number of cycles (Cycles), and degree of nodes (deg).

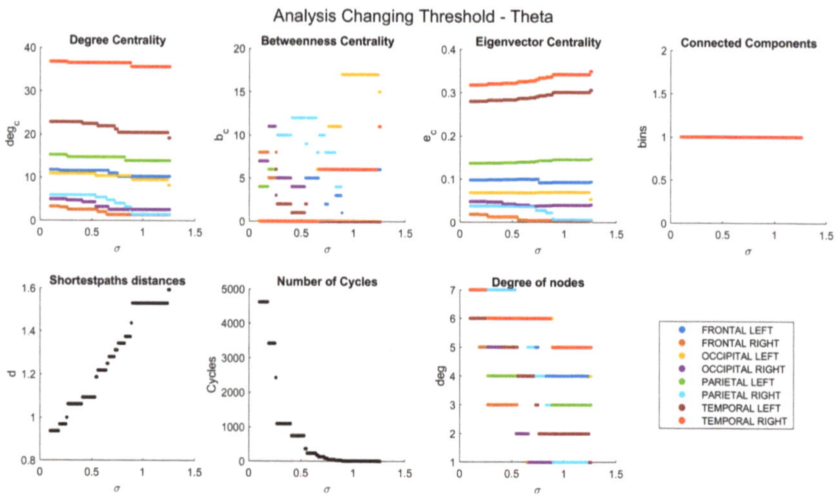

Figure A2. Network characteristics for theta band.

Figure A6 illustrates the results of the analysis of the betweenness centrality coefficient, which measures the importance of a node in terms of the number of shortest paths that pass through it. Once again, the connections are less pronounced at higher frequencies (Figure A6c,d) compared to lower frequencies (Figure A6a,b). Furthermore, the nodes with the highest betweenness centrality vary across different frequency bands. In the gamma band (Figure A6d), the nodes show a uniform and reduced size, whereas in the delta band (Figure A6a), the nodes with the highest betweenness centrality coefficient are the left parietal and left frontal lobes. In contrast, in the theta band (Figure A6b), a notable enlargement was observed in the right parietal lobe compared to other lobes, presenting a striking contrast to the sizes depicted in the degree centrality graph (Figure 5b), where it initially appeared to be one of the smallest.

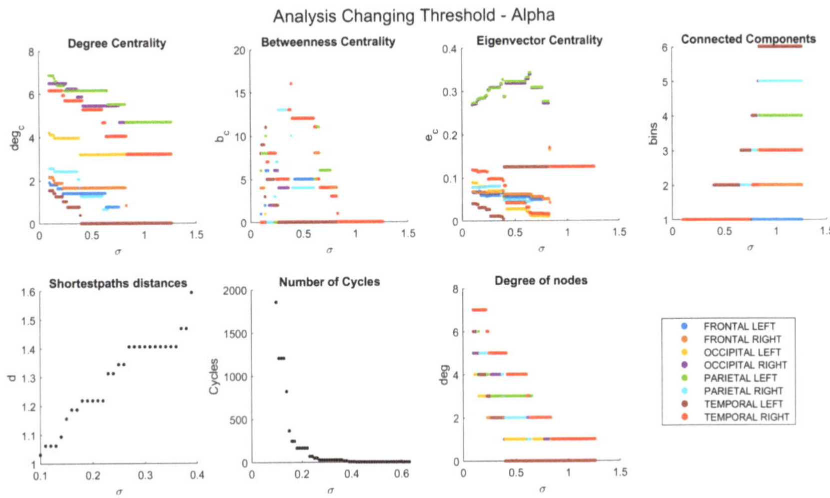

Figure A3. Network characteristics for alpha band.

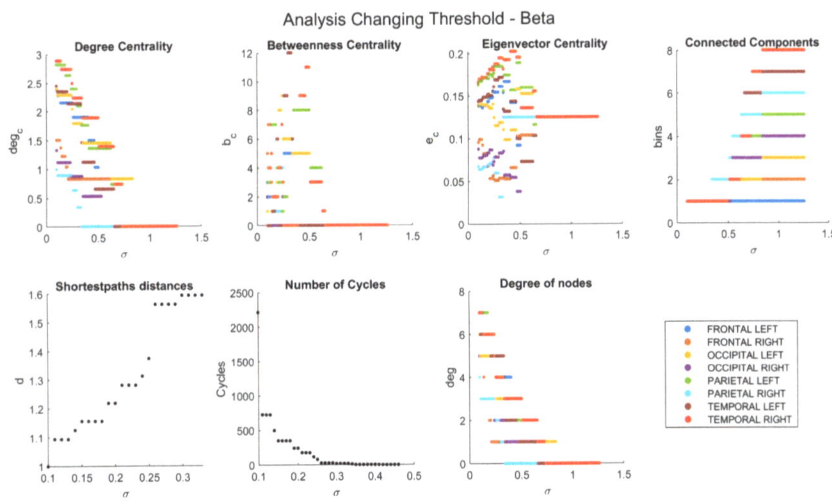

Figure A4. Network characteristics for beta band.

The eigenvector centrality, depicting the significance of a node based on the importance of its neighboring nodes, is illustrated in Figure A7. Node sizes reflect their respective importance, mirroring the patterns observed in Figure 5 across most graphs. However, in the case of gamma waves, node sizes remain consistent. Conversely, nodes associated with alpha waves, particularly in the right temporal region, exhibit notably diminished sizes compared to their prominence in degree centrality (Figure 5c).

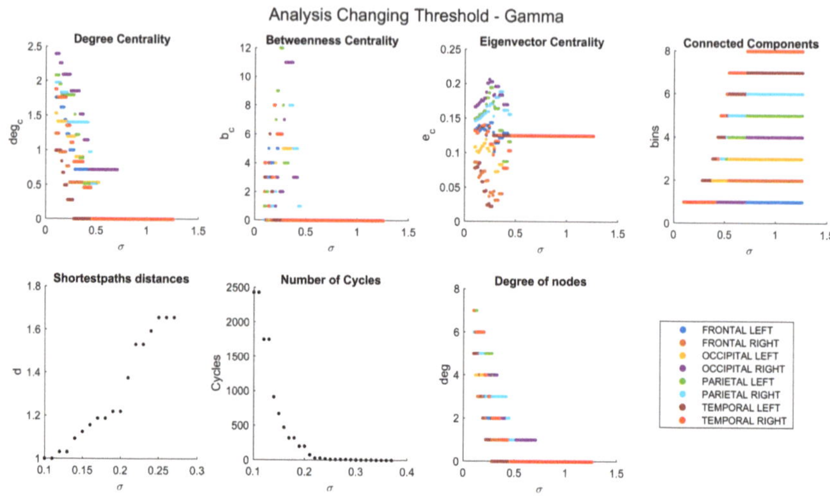

Figure A5. Network characteristics for gamma band.

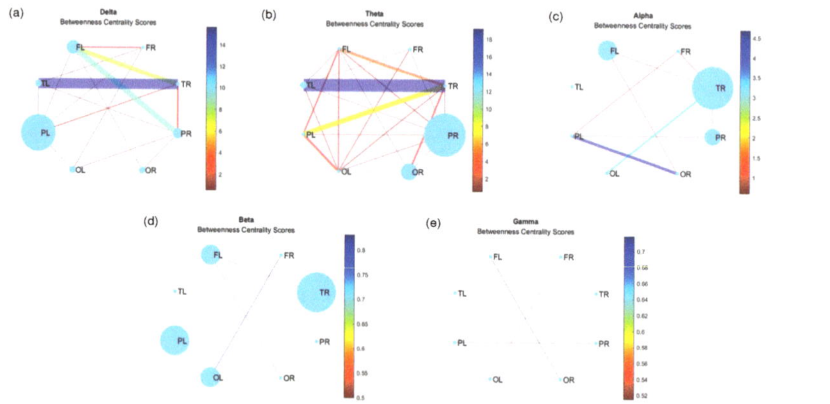

Figure A6. Betweenness centrality for (**a**) delta, (**b**) theta, (**c**) alpha, (**d**) beta, and (**e**) gamma bands.

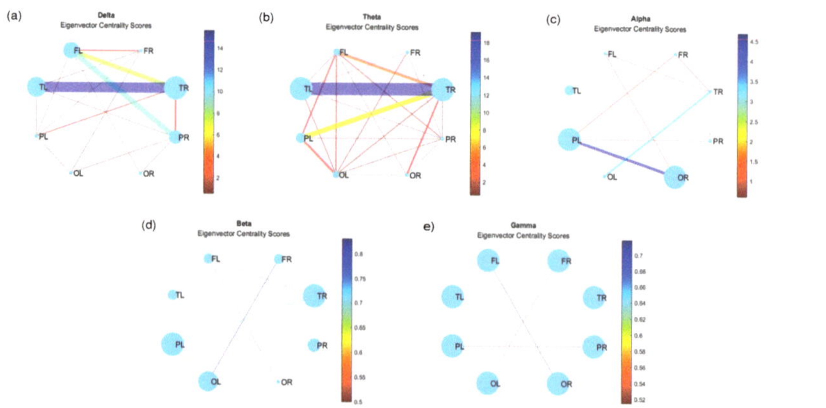

Figure A7. Eigenvector centrality for (**a**) delta, (**b**) theta, (**c**) alpha, (**d**) beta, and (**e**) gamma bands.

References

1. Friston, K.J.; Frith, C.D.; Liddle, P.F.; Frackowiak, R.S. Functional connectivity: the principal-component analysis of large (PET) data sets. *J. Cereb. Blood Flow Metab.* **1993**, *13*, 5–14. [CrossRef]
2. Greenblatt, R.E.; Pflieger, M.E.; Ossadtchi, A.E. Connectivity measures applied to human brain electrophysiological data. *J. Neurosci. Methods* **2012**, *207*, 1–16. [CrossRef] [PubMed]
3. Greenblatt, R.E.; Pflieger, M.E.; Ossadtchi, A.E. Review of advanced techniques for the estimation of brain connectivity measured with EEG/MEG. *Comput. Biol. Med.* **2011**, *41*, 1110–1117.
4. Hramov, A.E.; Frolov, N.S.; Maksimenko, V.A.; Kurkin, S.A.; Kazantsev, V.B.; Pisarchik, A.N. Functional networks of the brain: from connectivity restoration to dynamic integration. *Phys. Uspekhi* **2021**, *64*, 584–616. [CrossRef]
5. Le Bihan, D.; Mangin, J.F.; Poupon, C.; Clark, C.A.; Pappata, S.; Molko, N.; Chabriat, H. Diffusion tensor imaging: concepts and applications. *J. Magn. Reson. Imaging JMRI* **2001**, *13*, 534–546. [CrossRef] [PubMed]
6. Wedeen, V.J.; Wang, R.P.; Schmahmann, J.D.; Benner, T.; Tseng, W.Y.I.; Dai, G.; Pandya, D.N.; Hagmann, P.; D'Arceuil, H.; de Crespigny, A.J. Diffusion spectrum magnetic resonance imaging (DSI) tractography of crossing fibers. *NeuroImage* **2008**, *41*, 1267–1277. [CrossRef] [PubMed]
7. Towle, V.L.; Hunter, J.D.; Edgar, J.C.; Chkhenkeli, S.A.; Castelle, M.C.; Frim, D.M.; Kohrman, M.; Hecox, K. frequency domain analysis of human subdural recordings. *J. Clin. Neurophysiol.* **2007**, *24*, 205–213. [CrossRef] [PubMed]
8. Cabral, J.; Kringelbach, M.L.; Deco, G. Exploring the network dynamics underlying brain activity during rest. *Prog. Neurobiol.* **2014**, *114*, 102–131. [CrossRef] [PubMed]
9. Horwitz, B. The elusive concept of brain connectivity. *NeuroImage* **2003**, *19*, 466–470. [CrossRef]
10. Bowyer, S. Coherence a measure of the brain networks: Past and present. *Neuropsychiatr. Electrophysiol.* **2016**, *2*, 1. [CrossRef]
11. Frolov, N.; Maksimenko, V.; Lüttjohann, A.; Koronovskii, A.; Hramov, A. Feed-forward artificial neural network provides data-driven inference of functional connectivity. *Chaos* **2019**, *29*, 091101. [CrossRef]
12. Hämäläinen, M.; Hari, R.; Ilmoniemi, R.J.; Knuutila, J.; Lounasmaa, O.V. Magnetoencephalography—Theory, instrumentation, and applications to noninvasive studies of the working human brain. *Rev. Mod. Phys.* **1993**, *65*, 413. [CrossRef]
13. Burgess, R.C. Magnetoencephalography for localizing and characterizing the epileptic focus. *Handb. Clin. Neurol.* **2019**, *160*, 203–214.
14. Boccaletti, S.; De Lellis, P.; del Genio, C.; Alfaro-Bittner, K.; Criado, R.; Jalan, S.; Romance, M. The structure and dynamics of networks with higher order interactions. *Phys. Rep.* **2023**, *1018*, 1–64. [CrossRef]
15. Bullmore, E.; Sporns, O. Complex brain networks: graph theoretical analysis of structural and functional systems. *Nat. Rev, Neurosci.* **2009**, *10*, 186–198. [CrossRef] [PubMed]
16. Friston, K.J. Functional and effective connectivity: A review. *Brain Connectivity* **2011**, *1*, 13–36. [CrossRef] [PubMed]
17. Tavor, I.; Jones, O.P.; Mars, R.; Smith, S.; Behrens, T.; Jbabdi, S. Task-free MRI predicts individual differences in brain activity during task performance. *Science* **2016**, *352*, 216–220. [CrossRef] [PubMed]
18. Zhang, D.; Raichle, M.E. Disease and the brain's dark energy. *Nat. Rev. Neurol.* **2010**, *6*, 15–28. [CrossRef]
19. Greicius, M. Resting-state functional connectivity in neuropsychiatric disorders. *Curr. Opin. Neurol.* **2008**, *21*, 424–430. [CrossRef]
20. Dennis, E.L.; Thompson, P.M. Functional brain connectivity using fMRI in aging and Alzheimer's disease. *Neuropsychol. Rev.* **2014**, *24*, 49–62. [CrossRef] [PubMed]
21. Tomasi, D.; Volkow, N.D. Aging and functional brain networks. *Mol. Psychiatry* **2012**, *17*, 549–558. [CrossRef] [PubMed]
22. Contreras, J.A.; Goñi, J.; Risacher, S.L.; Sporns, O.; Saykin, A.J. The structural and functional connectome and prediction of risk for cognitive impairment in older adults. *Curr. Behav. Neurosci. Rep.* **2015**, *2*, 234–245. [CrossRef]
23. Sala-Llonch, R.; Bartrés-Faz, D.; Junqué, C. Reorganization of brain networks in aging: a review of functional connectivity studies. *Front. Psychol.* **2015**, *6*, 663. [CrossRef] [PubMed]
24. Davison, E.N.; Turner, B.O.; Schlesinger, K.J.; Miller, M.B.; Grafton, S.T.; Bassett, D.S.; Carlson, J.M. Individual differences in dynamic functional brain connectivity across the human lifespan. *PLoS Comput. Biol.* **2016**, *12*, e1005178. [CrossRef] [PubMed]
25. Andrew, C.; Pfurtscheller, G. Event-related coherence as a tool for studying dynamic interaction of brain regions. *Electroencephalogr. Clin. Neurophysiol.* **1996**, *98*, 144–148. [CrossRef]
26. Pisarchik, A.; Hramov, A. Coherence resonance in neural networks: Theory and experiments. *Phys. Rep.* **2023**, *1000*, 1–57. [CrossRef]
27. Gross, J.; Schmitz, F.; Schnitzler, I.; Kessler, K.; Shapiro, K.; Hommel, B.; Schnitzler, A. Modulation of long-range neural synchrony reflects temporal limitations of visual attention in humans. *Proc. Natl. Acad. Sci. USA* **2004**, *101*, 13050–13055. [CrossRef]
28. Guggisberg, A.G.; Honma, S.M.; Findlay, A.M.; Dalal, S.S.; Kirsch, H.E.; Berger, M.S.; Nagarajan, S.S. Mapping functional connectivity in patients with brain lesions. *Ann. Neurol.* **2008**, *63*, 193–203. [CrossRef]
29. Belardinelli, P.; Ciancetta, L.; Staudt, M.; Pizzella, V.; Londei, A.; Birbaumer, N.; Romani, G.L.; Braun, C. Cerebro-muscular and cerebro-cerebral coherence in patients with pre- and perinatally acquired unilateral brain lesions. *NeuroImage* **2007**, *37*, 1301–1314. [CrossRef]
30. de Pasquale, F.; Della Penna, S.; Snyder, A.Z.; Lewis, C.; Mantini, D.; Marzetti, L.; Belardinelli, P.; Ciancetta, L.; Pizzella, V.; Romani, G.L.; et al. Temporal dynamics of spontaneous MEG activity in brain networks. *Proc. Natl. Acad. Sci. USA* **2010**, *107*, 6040–6045. [CrossRef]

31. Kim, J.S.; Shin, K.S.; Jung, W.H.; Kim, S.N.; Kwon, J.S.; Chung, C.K. Power spectral aspects of the default mode network in schizophrenia: An MEG study. *BMC Neurosci.* **2014**, *15*, 104. [CrossRef]
32. Bowyer, S.M.; Gjini, K.; Zhu, X.; Kim, L.; Moran, J.E.; Rizvi, S.U.; Gumenyuk, N.T.; Tepley, N.; Boutros, N.N. Potential biomarkers of schizophrenia from MEG resting-state functional connectivity networks: Preliminary data. *J. Behav. Brain Sci.* **2015**, *5*, 1–11 [CrossRef]
33. Boutros, N.N.; Galloway, M.P.; Ghosh, S.; Gjini, K.; Bowyer, S.M. Abnormal coherence imaging in panic disorder: a magnetoencephalography investigation. *Neuroreport* **2013**, *24*, 487–491. [CrossRef] [PubMed]
34. Chholak, P.; Niso, G.; Maksimenko, V.A.; Kurkin, S.A.; Frolov, N.S.; Pitsik, E.N.; Hramov, A.E.; Pisarchik, A.N. Visual and kinesthetic modes affect motor imagery classification in untrained subjects. *Sci. Rep.* **2019**, *9*, 9838. [CrossRef] [PubMed]
35. Pisarchik, A.N.; Chholak, P.; Hramov, A.E. Brain noise estimation from MEG response to flickering visual stimulation. *Chaos Solitons Fractals X* **2019**, *1*, 100005. [CrossRef]
36. Chholak, P.; Maksimenko, V.A.; Hramov, A.E.; Pisarchik, A.N. Voluntary and involuntary attention in bistable visual perception: A MEG study. *Front. Hum. Neurosci.* **2020**, *14*, 555. [CrossRef] [PubMed]
37. Dai, Q.; Gao, Y. *Hypergraph Computation*; Springer: Singapore, 2023.
38. Wang, Z.; Liu, J.; Zhong, N.; Qin, Y.; Zhou, H.; Yang, J.; Li, K. A naive hypergraph model of brain networks. In Proceedings of the Brain Informatics: International Conference, BI 2012, Macau, China, 4–7 December 2012; Springer: Berlin/Heidelberg, Germany, 2012; pp. 119–129.
39. Chholak, P.; Kurkin, S.A.; Hramov, A.E.; Pisarchik, A.N. Event-related coherence in visual cortex and brain noise: An MEG study. *Appl. Sci.* **2021**, *11*, 375. [CrossRef]
40. Tadel, F.; Baillet, S.; Mosher, J.C.; Pantazis, D.; Leahy, R.M. Brainstorm: A user-friendly application for MEG/EEG analysis. *Comput. Intell. Neurosci.* **2011**, *2011*, 1–13. [CrossRef]
41. Bear, M.; Connors, B.; Paradiso, M.A. *Neuroscience: Exploring the Brain, Enhanced Edition: Exploring the Brain*; Jones & Bartlett Learning: Burlington, MA, USA, 2020.
42. French, C.C.; Beaumont, J.G. A critical review of EEG coherence studies of hemisphere function. *Int. J. Psychophysiol.* **1984**, *1*, 241–254. [CrossRef]
43. Xia, M.; Wang, J.; He, Y. BrainNet Viewer: A network visualization tool for human brain connectomics. *PLoS ONE* **2013**, *8*, e68910. [CrossRef]
44. Golbeck, J. *Analyzing the Social Web*; Morgan Kaufmann: Boston, MA, USA, 2013.
45. Zinoviev, D. *Complex Network Analysis in Python: Recognize-Construct-Visualize-Analyze-Interpret*; Pragmatic Bookshelf: Raleigh, NC, USA, 2018.
46. Voloshin, V.I. *Introduction to Graph and Hypergraph Theory*; Nova Science Publishers: Hauppauge, NY, USA, 2009.
47. Pickard, J.; Chen, C.; Salman, R.; Stansbury, C.; Kim, S.; Surana, A.; Bloch, A.; Rajapakse, I. HAT: Hypergraph analysis toolbox. *PLoS Comput. Biol.* **2023**, *19*, e1011190. [CrossRef] [PubMed]
48. Salvador, R.; Suckling, J.; Schwarzbauer, C.; Bullmore, E. Undirected graphs of frequency-dependent functional connectivity in whole brain networks. *Philos. Trans. R. Soc. B* **2005**, *360*, 937–946. [CrossRef] [PubMed]
49. Sun, F.T.; Miller, L.M.; D'esposito, M. Measuring interregional functional connectivity using coherence and partial coherence analyses of fMRI data. *Neuroimage* **2004**, *21*, 647–658. [CrossRef] [PubMed]
50. Von Stein, A.; Sarnthein, J. Different frequencies for different scales of cortical integration: From local gamma to long range alpha/theta synchronization. *Int. J. Psychophysiol.* **2000**, *38*, 301–313. [CrossRef]
51. Buzsaki, G.; Draguhn, A. Neuronal oscillations in cortical networks. *Science* **2004**, *304*, 1926–1929. [CrossRef]
52. Furl, N.; Coppola, R.; Averbeck, B.B.; Weinberger, D.R. Cross-frequency power coupling between hierarchically organized face-selective areas. *Cereb. Cortex* **2014**, *24*, 2409–2420. [CrossRef]
53. Pastoll, H.; Solanka, L.; van Rossum, M.C.; Nolan, M.F. Feedback inhibition enables theta-nested gamma oscillations and grid firing fields. *Neuron* **2013**, *77*, 141–154. [CrossRef]

applied sciences

MDPI

Article

Cortical Connectivity Response to Hyperventilation in Focal Epilepsy: A Stereo-EEG Study

Lorenzo Ferri [1,2,†], Federico Mason [2,3,†], Lidia Di Vito [1], Elena Pasini [1], Roberto Michelucci [1], Francesco Cardinale [4,5], Roberto Mai [4], Lara Alvisi [1,2], Luca Zanuttini [2], Matteo Martinoni [1,*] and Francesca Bisulli [1,2]

1 IRCCS Istituto delle Scienze Neurologiche di Bologna, Full Member of the European Reference Network EpiCARE, 40139 Bologna, Italy; lorenzo.ferri@ausl.bologna.it (L.F.); lidia.divito@ausl.bologna.it (L.D.V.); elena.pasini@isnb.it (E.P.); roberto.michelucci@isnb.it (R.M.); lara.alvisi@unibo.it (L.A.); francesca.bisulli@unibo.it (F.B.)
2 Department of Neuromotor and Biomedical Sciences, University of Bologna, 40127 Bologna, Italy; federico.mason@unipd.it (F.M.); luca.zanuttini@studio.unibo.it (L.Z.)
3 Department of Information Engineering, University of Padova, 35131 Padova, Italy
4 Centro Munari Chirurgia dell'Epilessia e del Parkinson, Niguarda Hospital, 20162 Milan, Italy; francesco.cardinale@ospedaleniguarda.it (F.C.); roberto.mai@ospedaleniguarda.it (R.M.)
5 Unit of Neuroscience, Department of Medicine and Surgery, University of Parma, 43124 Parma, Italy
* Correspondence: m.martinoni@isnb.it
† These authors contributed equally to this work.

Abstract: Hyperventilation (HV) is an activation technique performed during clinical practices to trigger epileptiform activities, supporting the neurophysiological evaluation of patients with epilepsy. Although the role of HV has often been questioned, especially in the case of focal epilepsy, no studies have ever assessed how cortical structures respond to such a maneuver via intracranial EEG recordings. This work aims to fill this gap by evaluating the HV effects on the Stereo-EEG (SEEG) signals from a cohort of 10 patients with drug-resistant focal epilepsy. We extracted multiple quantitative metrics from the SEEG signals and compared the results obtained during HV, awake status, non-REM sleep, and seizure onset. Our findings show that the cortical connectivity, estimated via the phase transfer entropy (PTE) algorithm, strongly increases during the HV maneuver, similar to non-REM sleep. The opposite effect is observed during seizure onset, as ictal transitions involve the desynchronization of the brain structures within the epileptogenic zone. We conclude that HV promotes a conductive environment that may facilitate the propagation of epileptiform activities but is not sufficient to trigger seizures in focal epilepsy.

Keywords: stereoelectroencephalography; focal epilepsy; hyperventilation; brain dynamics; network analysis; phase transfer entropy

Citation: Ferri, L.; Mason, F.; Di Vito, L.; Pasini, E.; Michelucci, R.; Cardinale, F.; Mai, R.; Alvisi, L.; Zanuttini, L.; Martinoni, M.; et al. Cortical Connectivity Response to Hyperventilation in Focal Epilepsy: A Stereo-EEG Study. *Appl. Sci.* **2024**, *14*, 8494. https://doi.org/10.3390/app14188494

Academic Editor: Alexander N. Pisarchik

Received: 30 August 2024
Revised: 16 September 2024
Accepted: 19 September 2024
Published: 20 September 2024

1. Introduction

Hyperventilation (HV) is a well-known activation technique performed during routine Electroencephalography (EEG) recording, recommended by the international guideline of the main clinical neuro-physiology and epilepsy societies [1]. In practice, HV involves deep and regular breathing at a rate of approximately 20 breaths per minute for a duration of 2 to 4 min. This technique was introduced in clinical practice by Otfrid Foerster who, in 1925, observed that HV could trigger *latent epilepsy* [2]. After the broad diffusion of EEG, HV became one of the most common procedures to elicit epileptic activities, offering valuable information for the medical management of epilepsy [3].

In the past years, several studies have demonstrated that the main HV effect is a physiological slowing of background EEG activity. Particularly, in people with epilepsy, it has been observed that HV could increase focal and generalized Interictal Epileptiform

Discharges (IEDs) [4]. A possible explanation for such a phenomenon relies on the vasoconstriction associated with cerebral hypoxia and the intracellular pH increment associated with respiratory hypocapnia. More specifically, it has been shown that cerebral hypoxia induces a negative DC shift variation over the EEG signal [5], while respiratory hypocapnia leads to higher excitatory postsynaptic potentials [6]. On the other hand, other studies have hypothesized that the cortical response to HV is due to sympathetic over-activation, which is notoriously considered a seizure trigger in temporal lobe epilepsy [7].

The effectiveness of HV is well established in generalized epilepsy: in the case of absence seizures, it has been demonstrated that HV generates a spike-and-wave complex in over 90% of the cases and triggers clinical seizures in around 50% of the cases [8]. On the other hand, there is limited agreement regarding the role of HV as an activation maneuver in focal epilepsy, especially in adults, who, notably, also present less significant autonomic abnormalities compared to the pediatric population [9]. A study carried out on a large patient cohort demonstrated that HV rarely triggers clinical seizures or increases the IED frequency [10]. However, other authors found that temporal lobe epilepsy might be more sensitive to HV than other types of epilepsy, suggesting its potential role in shortening presurgical evaluations [11].

A deeper comprehension of HV effects could be offered by using quantitative approaches for analyzing cortical or scalp EEG signals. In this context, the naive approach consists of analyzing the recorded signals in the time–frequency domain, evaluating the energy spectral density and the distribution of the signal phase. Other approaches aim to estimate the interdependency within different signals, considering connectivity-based metrics, which have proven to be extremely valuable for detecting epileptiform activities or discerning different populations [12]. Interestingly, most recent connectivity techniques focus on the analysis of the phase distribution of the EEG signals, using algorithms such as the Phase Locking Value (PLV), the Phase Lag Index (PLI), or the Phase Transfer Entropy (PTE) [13]. The latter algorithm combines the analysis of the signal instantaneous phase with the Granger causality and enables the estimation of the effective relations between signals generated by different brain structures [14].

A recent study concluded that HV increases the magnitude of the EEG power spectra, especially in the cingulate cortex, and demonstrated that different brain regions respond differently to respiratory hypocapnia [15]. The same study denoted how HV leads to a higher increase in cortical connectivity in people with epilepsy than in healthy individuals. Nevertheless, quantitative analysis of EEG traces during HV is scarce in the current literature, and it is still unclear how HV impacts the epileptogenic network. A better understanding of the relation between HV and seizure development could be offered by the analysis of intracranial EEG recording as allowed by Stereo-EEG (SEEG). This latter is an invasive surgical procedure that enables the recording of deep cortical signals, providing fundamental information for the accurate localization of the Epileptogenic Zone (EZ), i.e., the cortex area responsible for seizure generation [16].

To our knowledge, at the present time, there have been no research works investigating the HV mechanisms by exploiting the intracranial EEG signals. This study aimed to fill this gap and assess how the HV affects cortical brain structures in patients with focal epilepsy. Our fundamental hypothesis is that HV promotes an increase in cortical connectivity as occurs during the Non-REM sleep (N-REM) status, but such an effect is not directly associated with the outbreak of the seizure onset. To achieve this goal, we selected a cohort of focal epilepsy patients that underwent SEEG monitoring for pre-surgical evaluation. Then, we computed multiple quantitative metrics from the recorded signals, using the PTE algorithm to estimate cortical connectivity. To spotlight the HV effects, we compared the SEEG signals associated with HV against those associated with the awake status, N-REM, and ictal transition, i.e., the period during which the seizure starts to form.

2. Materials

This study considered a cohort of 10 *consecutive patients* that underwent SEEG monitoring at IRCCS Institute of Neurological Sciences of Bologna from January 2022 to June 2024. The study protocol was approved by the local ethics committee (protocol number 89-2021, committee code 20230), and written informed consent was obtained from each patient. The SEEG implants included multiple electrodes, each presenting 5–18 recording sites, named *contacts*; the number and location of the electrodes were patient tailored, depending on the EZ localization hypothesis [17]. Each contact was 2.2 mm in length and was separated by 1.5 mm from neighboring contacts (Microdeep Intracerebral Electrodes-D08, Dixi Medical, Besançon, France).

The SEEG implantation followed the workflow developed at Niguarda Hospital and involved the construction of a multimodal scene of the patient's brain. The multimodal scenes made it possible for clinicians to comprehensively evaluate all the anatomical information regarding the cortical area explored by each contact [18]. The SEEG signals were recorded using the Nihon Kohden EEG 2100 (Tokyo, Japan), using a maximum of 256 channels, and a sampling frequency of $f_{\text{sampling}} = 1000$ Hz. To correlate electrical and clinical features, a high-definition synchronized video of each patient was recorded for the whole duration of the SEEG monitoring (up to 20 days per patient).

All patients underwent a standardized 1-hour-recording protocol during the second day of SEEG monitoring, which included two activation maneuvers, namely, HV and intermittent photic stimulation. Particularly, HV sessions consisted of a sequence of deep breaths, at a rate of approximately 20 breaths/minute, for a total period of 4 min. To ensure the correct progress of the maneuver, expert clinical personnel, normally a neurologist and a neurophysiological technician, were in charge of explaining the procedure to the patient and assisting him/her for the exam duration.

In this work, we excluded all the SEEG contacts exploring the White Matter (WM) since they have only a propagator function [19], focusing only on contacts exploring the Grey Matter (GM). The discrimination between WM and GM, as well as the selection of the SEEG epochs, was performed by a board-certified neurophysiologist (L.F.). No other SEEG channel was excluded from the analysis, making our methodology agnostic to the specific SEEG implant and the clinical and demographic features associated with each patient. Hence, a subset of 26 SEEG epochs per patient was selected for the analysis, according to the following specifics:

- One epoch associated with the *ictal transition*, namely the period that included the 20 s preceding and following the first ictal change;
- Five epochs associated with the HV maneuver;
- Ten epochs associated with recording periods during which the patient was awake;
- Ten epochs associated with recording periods during which the patient was asleep (N-REM phase).

Each epoch lasted $T_{\text{epoch}} = 40$ seconds, resulting in a multidimensional signal, whose components, named *channels*, described the electrical activity generated by the cortical sites explored by the SEEG implant. An example of the epochs is reported in Figure 1.

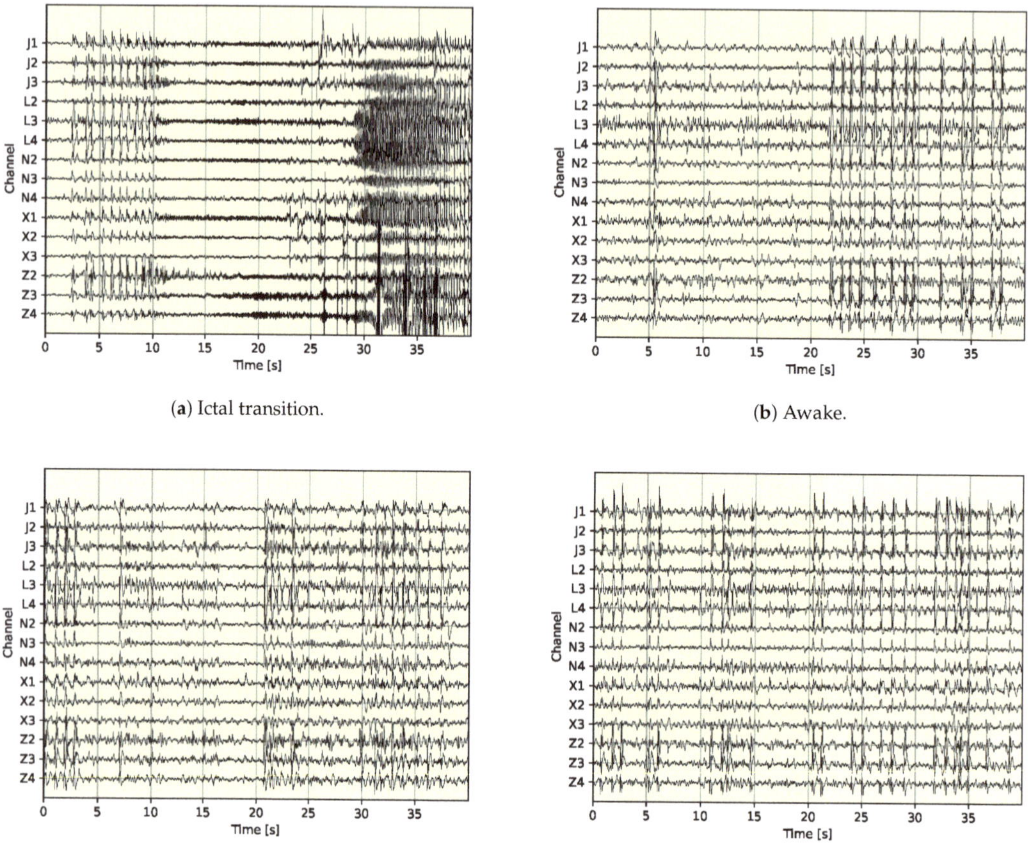

(**a**) Ictal transition.

(**b**) Awake.

(**c**) Non-REM sleep.

(**d**) Hyperventilation.

Figure 1. Example of SEEG epochs associated with one of the patients; the channels displayed are located within the EZ and are associated with marked epileptiform discharges; during the ictal transition, the epileptiform discharges evolve into low-voltage fast activities that trigger the epileptic seizure.

3. Methods

As a preliminary step, the SEEG epochs were processed by a *comb filter* to erase the powerline frequency f_{power} and its multiples $k \cdot f_{\text{power}}$, with $k \in \mathbb{Z}^+$, where $\mathbb{Z}^+$ represents the set of positive integers. Then, each channel was segmented into overlapping windows: the window duration was set to $T_{\text{window}} = 1.0$ s, and we inter-spaced consecutive windows by an interval $T_{\text{shift}} = 0.25$ s. Hence, each epoch was segmented in $W = \lfloor (T_{\text{signal}} - T_{\text{window}})/T_{\text{shift}} \rfloor = 156$ windows, where each window includes $n = T_{\text{window}} \cdot f_{\text{sampling}} = 1000$ samples. In the rest of the manuscript, we denote by $\mathcal{N}$ the set of channels within the same SEEG epoch and by $\mathcal{N}(t)$ the set of windows $x(t)$ lasting from time $t \cdot T_{\text{shift}}$ to time $t \cdot T_{\text{shift}} + T_{\text{window}}$, with $t \in \mathbb{Z}^+$. Therefore, a SEEG epoch including $N = |\mathcal{N}|$ channels was associated with a total of $N \cdot W$ windows.

3.1. Spectral Analysis

At first, we consider the Fourier Transform (FT) of the windows associated with each channel $x \in \mathcal{N}$ [20]. We write $\mathcal{X}(t)$ to indicate the FT of $x(t) \in \mathcal{N}(t)$: notably, $\mathcal{X}(t)$ includes $m = T_{\text{window}} \cdot f_{\text{sampling}}/2 = 500$ complex values, named *tones*, each associated with a positive frequency $f \in B_{\text{total}} = [0, f_{\text{sampling}}/2]$. In particular, we can obtain the total energy $E_x(t)$ of $x(t)$ as

$$E_x(t) = \int_{B_{\text{total}}} \|\mathcal{X}(t,f)\|^2 df, \tag{1}$$

where $\|\cdot\|$ is the norm function and $\mathcal{X}(t,f)$ is the tone associated with frequency $f \in B_{\text{total}}$.

Given the Fourier representation of the signal, we study the relation between different frequency bands. Specifically, considering $B_{\text{low}} = [4, 30]$ Hz and $B_{\text{high}} = [30, 250]$ Hz as target bands, we compute the *energy ratio* $ER_x(t)$ of $x(t)$ as

$$ER_x(t) = \frac{\int_{B_{\text{high}}} \|\mathcal{X}(t,f)\|^2 df}{\int_{B_{\text{low}}} \|\mathcal{X}(t,f)\|^2 df}. \tag{2}$$

The value of $ER_x(t)$ increases whenever the channel x starts exhibiting fast oscillations, a phenomenon commonly associated with the onset of epileptic discharges [21].

Besides considering the magnitude of the FT, we also analyze the phase $\theta_x(t,f)$ of the different signal tones. Specifically, $\theta_x(t,f)$ represents the relative distances, measured in radians, between the starting time of window $x(t)$ and the peak amplitude of the Fourier sinusoid associated with frequency $f \in [0, f_{\text{sampling}}/2]$. In this work, we model the phase distribution $\theta_x(t)$ of $x(t)$ as a histogram, whose range is within $-\pi$ and $+\pi$, and whose bin number is chosen according to the *Sturges rule* [22]. Each bin $\vartheta \in \Theta$ includes a phase range lasting $2\pi/(log_2(m)+1)$ radians, where $m = 500$ is the number of Fourier tones associated with each window. Hence, we compute the entropy of $\theta_x(t)$ as

$$H(\theta_x(t)) = -\sum_{\vartheta \in \Theta} p_\vartheta(\theta_x(t)) \log(p_\vartheta(\theta_x(t))), \tag{3}$$

where $p_\vartheta(\theta_x(t))$ represents the probability that $\theta_x(t)$ takes values in the phase range associated with ϑ. Appreciably, $H(\theta_x(t))$ denotes the tendency of the FT of $x(t)$ to assume a large variety of phase values and is maximized when $\theta_x(t)$ is uniform. As $x(t)$ becomes more complex, it includes more sinusoidal components, leading to a higher entropy value [23].

3.2. Connectivity Analysis

Afterwards, we compute the Hilbert Transform (HT), obtaining the analytic representation $\mathcal{X}_a(t) = x(t) + HT(x(t))$ of each window $x(t) \in \mathcal{N}(t)$. The values $\mathcal{X}_a(t)$ are associated with an *instantaneous phase* $\phi_x(t,\tau)$, where τ is the time index of the window samples [24]. Also in this case, we model the instantaneous phase distribution $\phi_x(t)$ as a histogram, choosing the bin number according to the Sturges rule. Practically, each bin $\varphi \in \Phi$ includes a phase range lasting $2\pi/(log_2(n)+1)$ radians, where $n = 1000$ is the number of signal samples associated with each window. Hence, we compute the entropy of $\phi_x(t)$ as

$$H(\phi_x(t)) = -\sum_{\varphi \in \Phi} p_\varphi(\phi_x(t)) \log(p_\varphi(\phi_x(t))), \tag{4}$$

where $p_\varphi(\phi_x(t))$ represents the probability that $\phi_x(t)$ takes values in the phase range associated with φ. Despite being computed in different domains, $H(\phi_x(t))$ has a similar meaning to $H(\theta_x(t))$ and tends to increase as $x(t)$ obtains more complex patterns.

The instantaneous phase distribution is also used for describing the relations within the different channels $x \in \mathcal{N}$ in time [25]. To this goal, we consider the PTE algorithm, which, given a couple of channels $x, y \in \mathcal{N}$, estimates the influence that each of the channels exerts on the network [14]. The algorithm takes as input the instantaneous phase distribution $\phi_x(t)$ and $\phi_y(t)$ of $x(t)$ and $y(t)$ and estimates the amount of information in $\phi_x(t)$ that can be used to predict the future evolution of $\phi_y(t)$. Specifically, the value of the PTE between $x(t)$ and $y(t)$, considering a lag δ, is obtained by

$$PTE_{x\to y}(t,\delta) = H(\phi_y(t),\phi_y(t+\tau))+ \\ H(\phi_x(t),\phi_y(t)) - H(\phi_x(t),\phi_y(t),\phi_y(t+\tau)) - H(\phi_y(t)), \tag{5}$$

where $H(\cdot)$ denotes both the entropy and the mutual entropy function.

We recall that $PTE_{x\to y}(t,\delta)$ is an effective connectivity measure, which means that, in general, $PTE_{x\to y}(t,\delta) \neq PTE_{y\to x}(t,\tau)$ [26]. The value $PTE_{x\to y}(t,\delta)$ depends on the lag δ, i.e., the time distance at which the information transfer is estimated. To remove the dependency from δ, we redefine the PTE between $x(t)$ and $y(t)$ as the maximum value of $PTE_{x\to y}(t,\tau)$ among multiple lags:

$$PTE_{x\to y}(t) = \max_{\delta\in\{0,\ldots,\delta_{max}\}} PTE_{x\to y}(t,\tau), \tag{6}$$

where we set $\delta_{max} = 100$ ms. By doing so, the magnitude of the effective connection exerted on the channel y by the channel x at time t is given by $PTE_{x\to y}(t)$, while the propagation delay associated with such a connection is

$$\delta_{x\to y}(t) = \arg\max_{\delta\in\{0,\ldots,\delta_{max}\}} PTE_{x\to y}(t,\delta). \tag{7}$$

3.3. *Statistical Analysis*

We observe that all the measures are window-dependent and, thus, a series of $W = 156$ multiple measures is obtained for each epoch and measure. Hence, before performing the statistical analysis, we perform two additional steps. First, we normalize each measure by the average value observed during the epochs associated with the awake conditions. In this way, we implicitly assume that the awake epochs constitute a baseline condition, enabling a fair comparison between patients with different characteristics. Then, we compute the median ($Med[\cdot]$) and the interquartile range ($IQR[\cdot]$) of the measures obtained from each epoch, and consider such values to be the input information.

To assess if the SEEG epochs present significant differences according to the median or the interquartile range of any of the measures, we consider the *Welch's t-test*. The latter enables the comparison of the means of two data groups in case the variances are unknown and not equal [27]. Specifically, we perform a one-tailed test, checking whether the mean of a certain group of SEEG epochs is higher than another group, setting $\alpha = 0.05$ as the significance level. We recall that, by comparing the medians, we verify if a certain measure takes higher or lower values in a specific SEEG group with respect to another. Instead, by comparing the interquartile ranges, we assess if a measure takes more or less variable outcomes.

4. Results

The study included a total of 10 patients (6 Males), with a mean age of 37 years at the time of SEEG implantation. As shown in Table 1, it was possible to delineate an EZ that was located in the temporal lobe in seven patients, temporal–occipital cortex in two patients, and frontal in one patient. All the SEEG implants were unilateral (7 right, 3 left) and included 15 electrodes on average. In particular, four patients underwent temporo-parieto-occipital exploration, three fronto-temporal, two fronto-temporo-parietal, and one temporal. At the time of the study, six patients underwent epilepsy surgery, and the histopathological analysis revealed hippocampal sclerosis in four cases, focal cortical dysplasia in one case, and aspecific findings in the remaining one.

Table 1. Clinical and demographic characteristics of the studied population; F is for Frontal, T for Temporal, O for Occipital, THC for Thermocoagulation, ATL for Anterior Temporal Lobectomy, HS for Hippocampus Sclerosis, and FCD for Focal Cortical Dysplasia.

	Patient 1	Patient 2	Patient 3	Patient 4	Patient 5	Patient 6	Patient 7	Patient 8	Patient 9	Patient 10
Age (SEEG)	24	39	50	39	36	52	36	24	25	48
Sex	Male	Male	Female	Female	Male	Female	Male	Male	Female	Male
EZ localization	Left F	Left T	Left T	Right T-O	Right T-O	Right T	Left T	Right T	Right T	Right T
SEEG implant	Left F-T	Left F-T-P	Left F-T-P	Right T-P-O	Right T-P-O	Right F-T	Left T	Right T-P-O	Right T-P-O	Right F-T
Surgery	THC	THC + Left ATL	THC + Left ATL	THC	THC	THC + Right ATL	THC + Left ATL	THC	THC + Right ATL	THC + Right ATL
Histopathology	N/A	HS 1	HS 1	N/A	N/A	HS 1	FCD	N/A	Aspecific	HS 1

Our analysis focused on the following metrics: the energy ratio (ER), the entropy ($H(\theta)$) of the spectral phase distribution, and the Phase Transfer Entropy (PTE) computed from the instantaneous phase distribution. As shown in Table 2, both the ictal and awake periods presented higher energy ratio values than N-REM and awake periods. The energy ratio proved to be an effective biomarker for discerning HV activities from the N-REM conditions since it was sensibly lower when patients performed the HV maneuver. Taking the interquartile range into account, the differences are the same as those expressed in terms of median: in other words, the epochs presenting higher energy values are also associated with higher energy variability. Notably, there was slight evidence (p-value 0.09) that ictal transitions present more variable energy ratios than awake periods, although the latter showed higher median values for this metric (0.0672 vs. 0.0600).

Table 2. Comparison of the median and inter-quartile range of the energy ratio (ER) computed in different groups of SEEG epochs.

Metric	Group 1		Group 2		p-Value	
	Label	Mean ± CI	Label	Mean ± CI	Greater	Less
$Med[ER]$	Ictal	0.0600 ± 0.0126	Awake	0.0672 ± 0.0207	0.2534	0.7466
	Ictal	0.0600 ± 0.0126	N-REM	0.0463 ± 0.0185	0.0587	0.9413
	Ictal	0.0600 ± 0.0126	HV	0.0277 ± 0.0077	0.0092	0.9908
	Awake	0.0672 ± 0.0207	N-REM	0.0463 ± 0.0185	0.0034	0.9966
	Awake	0.0672 ± 0.0207	HV	0.0277 ± 0.0077	<0.0001	>0.9999
	N-REM	0.0463 ± 0.0185	HV	0.0277 ± 0.0077	0.0066	0.9934
$IQR[ER]$	Ictal	0.0988 ± 0.0177	Awake	0.0849 ± 0.0192	0.0874	0.9126
	Ictal	0.0988 ± 0.0177	N-REM	0.0556 ± 0.0194	0.0185	0.9995
	Ictal	0.0988 ± 0.0177	HV	0.0382 ± 0.0115	0.0059	0.9941
	Awake	0.0849 ± 0.0192	N-REM	0.0556 ± 0.0194	0.0005	0.9995
	Awake	0.0849 ± 0.0192	HV	0.0382 ± 0.0115	<0.0001	>0.9999
	N-REM	0.0556 ± 0.0194	HV	0.0382 ± 0.0115	0.0181	0.9819

The entropy measures allowed us to differentiate between the ictal and the awake epochs from the N-REM and HV epochs (Table 3). At the same time, no significant difference was observed when comparing the ictal transition with the awake status, as well as the HV maneuver with the N-REM phase. Taking the interquartile range into account, the results followed an opposite trend compared to that before since higher entropy values were associated with lower entropy variability. The entropy during the ictal transition was slightly more variable (p-value of 0.063) than the one measured during the awake periods. In general, the variability was maximized during the ictal transition (with a value of 0.0988) and minimized during the HV (with a value of 0.0383).

Table 3. Comparison of the median and inter-quartile range of the spectral entropy ($H(\theta)$) computed in different groups of SEEG epochs.

Metric	Group 1		Group 2		p-Value	
	Label	Mean ± CI	Label	Mean ± CI	Greater	Less
$Med[H(\theta)]$	Ictal	1.8208 ± 0.0482	Awake	1.8350 ± 0.0445	0.6280	0.3720
	Ictal	1.8208 ± 0.0482	N-REM	1.7071 ± 0.0460	0.0038	0.9962
	Ictal	1.8208 ± 0.0482	HV	1.6903 ± 0.0307	0.0028	0.9972
	Awake	1.8350 ± 0.0445	N-REM	1.7071 ± 0.0460	<0.0001	>0.9999
	Awake	1.8350 ± 0.0445	HV	1.6903 ± 0.0307	0.0002	0.9998
	N-REM	1.7071 ± 0.0460	HV	1.6903 ± 0.0307	0.3287	0.6713
$IQR[H(\theta)]$	Ictal	0.5149 ± 0.0414	Awake	0.4726 ± 0.0464	0.0630	0.9370
	Ictal	0.5149 ± 0.0414	N-REM	0.5623 ± 0.0386	0.8126	0.1874
	Ictal	0.5149 ± 0.0414	HV	0.5975 ± 0.0125	0.9131	0.0869
	Awake	0.4726 ± 0.0464	N-REM	0.5623 ± 0.0386	0.9995	0.0005
	Awake	0.4726 ± 0.0464	HV	0.5975 ± 0.0125	0.9905	0.0095
	N-REM	0.5623 ± 0.0386	HV	0.5975 ± 0.0125	0.7999	0.2001

The cortical connectivity, estimated via the PTE algorithm, enabled the most significant discrimination between all the epochs. As shown in Table 4, the PTE was minimized during the ictal transition (0.8692), took higher values during the awake status (0.9063), was even higher during the N-REM period (0.9239), and was maximized during the HV maneuver (0.9399). In this case, the statistical test led to significant results even with a significance level of $\alpha < 0.01$. Besides showing the lowest median connectivity, the ictal transition reported the highest values in terms of the interquartile range. On the other hand, the HV and the N-REM epochs presented very similar interquartile ranges ($\approx$0.152), lower than the ictal transition (0.1773) but higher than the awake periods (0.1305). Hence, the connectivity increment during the N-REM and HV phases was also associated with a higher variability, which goes against what was observed during the seizure onset. A visual representation of the overall results is given in Figure 2

Table 4. Comparison of the median and inter-quartile range of the Phase Transfer Entropy (PTE) computed in different groups of SEEG epochs.

Metric	Group 1		Group 2		p-Value	
	Label	Mean ± CI	Label	Mean ± CI	Greater	Less
$Med[PTE]$	Ictal	0.8692 ± 0.01976	Awake	0.9063 ± 0.0056	0.9922	0.0078
	Ictal	0.8692 ± 0.01976	N-REM	0.9239 ± 0.0057	0.9992	0.0008
	Ictal	0.8692 ± 0.01976	HV	0.9399 ± 0.0045	0.9999	0.0001
	Awake	0.9063 ± 0.0056	N-REM	0.9239 ± 0.0057	>0.9999	<0.0001
	Awake	0.9063 ± 0.0056	HV	0.9399 ± 0.0045	>0.9999	<0.0001
	N-REM	0.9239 ± 0.0057	HV	0.9399 ± 0.0045	>0.9999	<0.0001
$IQR[PTE]$	Ictal	0.1773 ± 0.0245	Awake	0.1305 ± 0.0046	0.0018	0.9982
	Ictal	0.1773 ± 0.0245	N-REM	0.1527 ± 0.0047	0.0461	0.9539
	Ictal	0.1773 ± 0.0245	HV	0.1522 ± 0.0053	0.0426	0.9574
	Awake	0.1305 ± 0.0046	N-REM	0.1527 ± 0.0047	>0.9999	<0.0001
	Awake	0.1305 ± 0.0046	HV	0.1522 ± 0.0053	0.9999	0.0001
	N-REM	0.1527 ± 0.0047	HV	0.1522 ± 0.0053	0.4360	0.5640

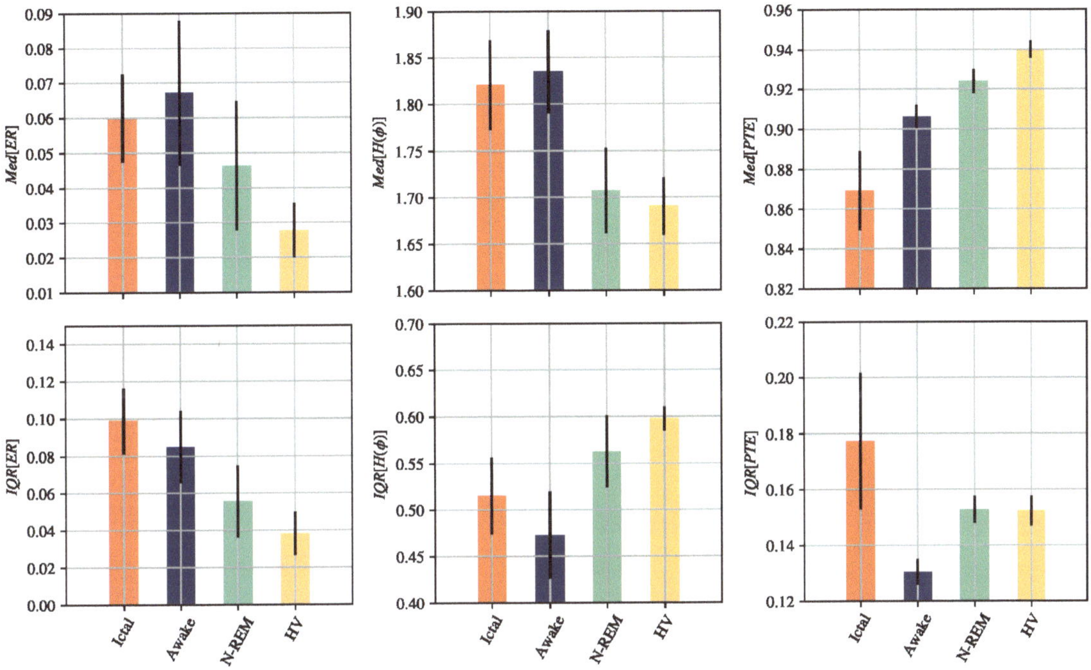

Figure 2. Comparison between the median and interquartile range of the analyzed metrics for the different groups of SEEG epochs; the top of each bar denotes the expected values, while the black vertical lines denote the confidence intervals.

5. Discussion

This is the first study to investigate HV using SEEG intracranial electrodes, offering novel insights into the impact of such a maneuver on the cortical brain signals. The results denote how HV shares similar characteristics in terms of energy ratio and phase entropy to N-REM sleep, and the same occurs for the awake status and ictal transition, where the latter includes the 20 s preceding and following the seizure onset. Specifically, ER and $H(\theta)$ are higher when the patient is conscious and during the ictal transition; in contrast, N-REM and HV present slower electrical activities, characterized by lower energy in high-frequency bands and more regular signal patterns. Interestingly, HV presents a reduced ER than awake status and, thus, seems to mitigate the rise in fast oscillations: this is in apparent contradiction to the scope of such an activation maneuver.

The low effectiveness of the energy ratio and the phase entropy in discerning HV from the N-REM status may be explained by the fact that such metrics do capture the interdependency between different cortical sites but analyze each signal as an independent element. More insights into HV are obtained by looking at the PTE, which, instead, enables full discrimination between all the SEEG groups. Particularly, the PTE algorithm captures the characteristics of intrasignal relationships across multiple frequency bands. This is very beneficial for analyzing epileptiform activities, which, notably, are not confined to specific frequency ranges but affect both fast and slow oscillations [28]. Our results show that HV increases the PTE, even more strongly than N-REM, while ictal transition behaves in the opposite fashion. This is in agreement with past studies that proved that the epileptogenic network presents reduced connectivity during the early ictal phase while being characterized by higher synchronization during seizure propagation. The ictal connectivity pattern may be explained by observing that the epileptogenic area adopts a pathological behavior and, thus, results in being desynchronized by the rest of the network [29].

At first glance, our study denotes that HV is on the opposite side with respect to ictal transition, both in terms of the cortical connectivity and energy ratio, making its role in triggering seizures questionable. These findings are in line with other clinical studies on adults with focal epilepsy, suggesting that HV rarely triggers either clinical seizures or increases epileptiform discharges [10]. The connectivity increase during HV may be related to the specific characteristics of slow oscillations, which has been identified as a primary driver of interictal activity during N-REM sleep [30–32]. The cortical hypersynchronization has also been called into question to explain the IED diffusion that usually is observed in both N-REM sleep and HV, in opposite fashion with respect to the REM phase in which the IEDs becomes more focused [32,33]. The fact that HV replicates the N-REM patterns explains the effectiveness of this activation maneuver in generalized epilepsy, such as absence epilepsy, where IED exploits the burst-firing mode of the wide-projecting corticothalamic system as is well documented during N-REM sleep [30].

The fact that cortical brain regions during HV replicate the conditions observed during the N-REM status has been hypothesized by some previous studies but, to our knowledge, was never shown via intracranial EEG recordings. This may steer against the current clinical practices, which recommend the use ofHV as an activation maneuver for both generalized and focal epilepsy. One could speculate that HV promotes a conductive environment that facilitates seizure propagation in focal epilepsy, but such an effect is not sufficient to trigger the seizure onset. In our experience during long-term EEG and SEEG monitoring, we have rarely observed seizures during such a procedure in the case of focal epilepsy: in some cases, seizures developed only after several HV sessions or minutes after the activation maneuver was carried out. This may suggest that the transition from a state of increased connectivity to a state with relatively decreased connectivity facilitates the focal desynchronization of epileptogenic sites and the consecutive rise in ictal discharges.

It is noteworthy that the EZ, namely, the cortex area responsible for seizure generation, responds differently to the HV maneuver compared to healthy cortical and subcortical regions. A past study demonstrated that, during HV, the mean decreases in cerebral blood flow were 20.9% and 10.8%, in epileptic and non-epileptic temporal cortical regions, respectively [5]. The authors found a linear dependency between the blood flow reduction and the interval between seizures, concluding that the EZ is more susceptible to hypoperfusion and particularly vulnerable to ischemia. Even if HV exerts a global connectivity effect, different pathological and physiological areas may respond in a personalized fashion. Therefore, studying how HV affects the cortical sites within the EZ may enable the definition of new biomarkers for better EZ localization.

Our study has several limitations. First, the sample size was limited and heterogeneous in terms of etiology and explored cortical areas, though most patients had a temporal EZ and exploration focused mainly on temporal structures. Further validation of our findings over a broader population with uniform characteristics represents an essential step for continuing this research and its effectiveness from a clinical point of view. Secondly, we did not systematically assess the relative increase or decrease in IED frequency during N-REM sleep, wakefulness, and HV periods, which could be either a cause or effect of the observed connectivity changes. Additionally, due to the small sample size, we did not separately analyze patients in whom HV induced IED versus those in whom the maneuver had no effect. From a methodological point of view, we considered only the PTE algorithm for estimating cortical connectivity, while extending the analysis by considering frequency-dependent metrics may enable discerning connectivity patterns according to specific frequency bands. We did not assess connectivity changes within the EZ compared to propagation zones, unaffected zones, and contralateral healthy control areas. Lastly, SEEG explorations are only conducted in people with drug-resistant epilepsy, as the invasive nature of the procedure precludes the inclusion of healthy control subjects.

6. Conclusions

In this work, we studied the effects of HV on cortical brain structures of 10 patients with drug-resistant focal epilepsy who underwent SEEG monitoring. We exploited different quantitative metrics to analyze the intracranial signals, considering the PTE algorithm to estimate the effective connectivity between the cortical sites explored by the SEEG implants. We observed that HV strongly increases cortical connectivity in focal fronto-temporal epilepsy, similar to what occurs during N-REM sleep. While HV may induce a conductive environment that facilitates the propagation of epileptiform activity, it seems insufficient to trigger seizure development, contrary to what occurs in generalized epilepsy. Hence, our findings suggest that HV should be considered a facilitating maneuver rather than an activation procedure. At the same time, analyzing the specific connectivity behavior of epileptogenic versus healthy areas during the HV maneuver could provide useful information for identifying new biomarkers that characterize the EZ.

Author Contributions: Conceptualization, L.F., F.M., M.M. and F.B.; Formal analysis, L.F. and F.M.; Investigation, L.F., F.M., L.D.V., E.P., R.M. (Roberto Michelucci), M.M. and F.B.; Methodology, L.F. and F.M.; Supervision, M.M. and F.B.; Writing—original draft, L.F. and F.M.; Writing—review and editing, L.F., F.M., L.D.V., E.P., R.M. (Roberto Michelucci), F.C., R.M. (Roberto Mai), L.A., L.Z., M.M. and F.B. All authors have read and agreed to the published version of the manuscript.

Funding: This research received no external funding.

Institutional Review Board Statement: The study was conducted in accordance with the Declaration of Helsinki. The study received approval from the Central Emilia Wide Area Ethical Committee of the Emilia-Romagna Region (protocol number 89-2021; committee code 20230)

Informed Consent Statement: Written informed consent was obtained from each patient participating in the study.

Data Availability Statement: The datasets presented in this article cannot be shared according to the approved Istitutional Review Board. Requests to access the datasets should be directed to the corresponding author.

Conflicts of Interest: The authors declare no conflicts of interest.

References

1. Kane, N.; Grocott, L.; Kandler, R.; Lawrence, S.; Pang, C. Hyperventilation during electroencephalography: Safety and efficacy. *Seizure* **2014**, *23*, 129–134. [CrossRef] [PubMed]
2. Foerster, O. Zur Pathogenese und chirurgischen Behandlung der Epilepsie. *Zentralbl Chir* **1925**, *52*, 531–549.
3. Gibbs, F.A.; Lennox, W.G.; Gibbs, E.L. The electro-encephalogram in diagnosis and in localization of epileptic seizures. *AMA Arch. Neurol. Psychiatry* **1936**, *36*, 1225–1235. [CrossRef]
4. Siddiqui, S.R.; Zafar, A.; Khan, F.S.; Shaheen, M. Effect of hyperventilation on electroencephalographic activity. *J. Pak. Med. Assoc.* **2011**, *61*, 850–852. [PubMed]
5. Weinand, M.E.; Carter, L.P.; Oommen, K.J.; Hutzler, R.; Labiner, D.M.; Talwar, D.; El-Saadany, W.; Ahern, G.L. Response of human epileptic temporal lobe cortical blood flow to hyperventilation. *Epilepsy Res.* **1995**, *21*, 221–226. [CrossRef] [PubMed]
6. John, W.; Wang, S. Response of medullary respiratory neurons to hypercapnia and isocapnic hypoxia. *J. Appl. Physiol.* **1977**, *43*, 812–821. [CrossRef]
7. Assenza, G.; Mecarelli, O.; Tombini, M.; Pulitano, P.; Pellegrino, G.; Benvenga, A.; Assenza, F.; Campana, C.; Di Pino, G.; Di Lazzaro, V. Hyperventilation induces sympathetic overactivation in mesial temporal epilepsy. *Epilepsy Res.* **2015**, *110*, 221–227. [CrossRef]
8. Rana, M.; Steenari, M.; Shrey, D. Hyperventilation and Seizures: Not a New sense: A Literature review. *Neuropediatrics* **2023**, *54*, 359–364. [CrossRef]
9. Mason, F.; Scarabello, A.; Taruffi, L.; Pasini, E.; Calandra-Buonaura, G.; Vignatelli, L.; Bisulli, F. Heart Rate Variability as a Tool for Seizure Prediction: A Scoping Review. *J. Clin. Med.* **2024**, *13*, 747. [CrossRef]
10. Holmes, M.D.; Dewaraja, A.S.; Vanhatalo, S. Does hyperventilation elicit epileptic seizures? *Epilepsia* **2004**, *45*, 618–620. [CrossRef]
11. Guaranha, M.S.; Garzon, E.; Buchpiguel, C.A.; Tazima, S.; Yacubian, E.M.; Sakamoto, A.C. Hyperventilation revisited: Physiological effects and efficacy on focal seizure activation in the era of video-EEG monitoring. *Epilepsia* **2005**, *46*, 69–75. [CrossRef] [PubMed]
12. Li, Z.; Hwang, K.; Li, K.; Wu, J.; Ji, T. Graph-generative neural network for EEG-based epileptic seizure detection via discovery of dynamic brain functional connectivity. *Sci. Rep.* **2022**, *12*, 18998. [CrossRef] [PubMed]

13. Sakkalis, V. Review of advanced techniques for the estimation of brain connectivity measured with EEG/MEG. *Comput. Biol. Med.* **2011**, *41*, 1110–1117. [CrossRef]
14. Lobier, M.; Siebenhühner, F.; Palva, S.; Palva, J.M. Phase transfer entropy: A novel phase-based measure for directed connectivity in networks coupled by oscillatory interactions. *Neuroimage* **2014**, *85*, 853–872. [CrossRef]
15. Mazzucchi, E.; Vollono, C.; Losurdo, A.; Testani, E.; Gnoni, V.; Di Blasi, C.; Giannantoni, N.M.; Lapenta, L.; Brunetti, V. Della Marca, G. Hyperventilation in Patients With Focal Epilepsy: Electromagnetic Tomography, Functional Connectivity and Graph Theory—A Possible Tool in Epilepsy Diagnosis? *J. Clin. Neurophysiol.* **2017**, *34*, 92–99. [CrossRef]
16. Mercier, M.R.; Dubarry, A.S.; Tadel, F.; Avanzini, P.; Axmacher, N.; Cellier, D.; Del Vecchio, M.; Hamilton, L.S.; Hermes, D. Kahana, M.J.; et al. Advances in human intracranial electroencephalography research, guidelines and good practices. *Neuroimage* **2022**, *260*, 119438. [CrossRef] [PubMed]
17. Talairach, J.; Bancaud, J. Lesion, "irritative" zone and epileptogenic focus. *Ster. Funct. Neurosurg.* **1966**, *27*, 91–94. [CrossRef]
18. Cardinale, F.; Cossu, M.; Castana, L.; Casaceli, G.; Schiariti, M.P.; Miserocchi, A.; Fuschillo, D.; Moscato, A.; Caborni, C. Arnulfo, G.; et al. Stereoelectroencephalography: Surgical methodology, safety, and stereotactic application accuracy in 500 procedures. *Neurosurgery* **2013**, *72*, 353–366. [CrossRef]
19. Greene, P.; Li, A.; González-Martínez, J.; Sarma, S.V. Classification of stereo-EEG contacts in white matter vs. Gray matter using recorded activity. *Front. Neurol.* **2021**, *11*, 605696. [CrossRef]
20. Pigeau, R.; Hoffmann, R.; Moffitt, A. A multivariate comparison between two EEG analysis techniques: Period analysis and fast Fourier transform. *Electroencephalogr. Clin. Neurophysiol.* **1981**, *52*, 656–658. [CrossRef]
21. Bartolomei, F.; Chauvel, P.; Wendling, F. Epileptogenicity of brain structures in human temporal lobe epilepsy: A quantified study from intracerebral EEG. *Brain* **2008**, *131*, 1818–1830. [CrossRef] [PubMed]
22. Scott, D.W. Sturges' rule. *Wiley Interdiscip. Rev. Comput. Stat.* **2009**, *1*, 303–306. [CrossRef]
23. Chua, K.; Chandran, V.; Rajendra Acharya, U.; Lim, C. Analysis of epileptic EEG signals using higher order spectra. *J. Med. Eng. Technol.* **2009**, *33*, 42–50. [CrossRef] [PubMed]
24. De Clercq, W.; Lemmerling, P.; Van Paesschen, W.; Van Huffel, S. Characterization of interictal and ictal scalp EEG signals with the Hilbert transform. In Proceedings of the 25th Annual International Conference of the IEEE Engineering in Medicine and Biology Society, Cancún, Mexico, 17–21 September 2003; Volume 3, pp. 2459–2462.
25. Chiarion, G.; Sparacino, L.; Antonacci, Y.; Faes, L.; Mesin, L. Connectivity analysis in EEG data: A tutorial review of the state of the art and emerging trends. *Bioengineering* **2023**, *10*, 372. [CrossRef] [PubMed]
26. Friston, K.J. Functional and effective connectivity in neuroimaging: A synthesis. *Hum. Brain Mapp.* **1994**, *2*, 56–78. [CrossRef]
27. West, R.M. Best practice in statistics: Use the Welch t-test when testing the difference between two groups. *Ann. Clin. Biochem.* **2021**, *58*, 267–269. [CrossRef]
28. Edakawa, K.; Yanagisawa, T.; Kishima, H.; Fukuma, R.; Oshino, S.; Khoo, H.M.; Kobayashi, M.; Tanaka, M.; Yoshimine, T. Detection of epileptic seizures using phase–amplitude coupling in intracranial electroencephalography. *Sci. Rep.* **2016**, *6*, 25422. [CrossRef]
29. Mason, F.; Ferri, L.; Vito, L.D.; Alvisi, L.; Zanuttini, L.; Martinoni, M.; Mai, R.; Cardinale, F.; Tinuper, P.; Michelucci, R.; et al. Desynchronization Index: A New Approach for Exploring Complex Epileptogenic Networks in Stereoelectroencephalography, *arXiv* **2024**, arXiv: 2408.16347.
30. Nobili, L.; Frauscher, B.; Eriksson, S.; Gibbs, S.A.; Halasz, P.; Lambert, I.; Manni, R.; Peter-Derex, L.; Proserpio, P.; Provini, F.; et al. Sleep and epilepsy: A snapshot of knowledge and future research lines. *J. Sleep Res.* **2022**, *31*, e13622. [CrossRef]
31. Steriade, M. Grouping of brain rhythms in corticothalamic systems. *Neuroscience* **2006**, *137*, 1087–1106. [CrossRef]
32. Frauscher, B.; von Ellenrieder, N.; Ferrari-Marinho, T.; Avoli, M.; Dubeau, F.; Gotman, J. Facilitation of epileptic activity during sleep is mediated by high amplitude slow waves. *Brain* **2015**, *138*, 1629–1641. [CrossRef] [PubMed]
33. Sammaritano, M.; Gigli, G.L.; Gotman, J. Interictal spiking during wakefulness and sleep and the localization of foci in temporal lobe epilepsy. *Neurology* **1991**, *41*, 290–290. [CrossRef] [PubMed]

applied sciences

MDPI

Article

Efficient Implementation of Multilayer Perceptrons: Reducing Execution Time and Memory Consumption

Francisco Cedron [1,*], Sara Alvarez-Gonzalez [1], Ana Ribas-Rodriguez [1], Santiago Rodriguez-Yañez [2] and Ana Belen Porto-Pazos [1]

1 Research Center on Information and Communication Technologies, Computer Science and Information Technology Department, University of A Coruña, 15071 A Coruña, Spain; sara.alvarezg@udc.es (S.A.-G.); ana.ribas@udc.es (A.R.-R.); ana.portop@udc.es (A.B.P.-P.)
2 Center for Technological Innovation in Construction and Civil Engineering, Computer Science and Information Technology Department, University of A Coruña, 15071 A Coruña, Spain; santiago.rodriguez@udc.es
* Correspondence: francisco.cedron@udc.es

Abstract: A technique is presented that reduces the required memory of neural networks through improving weight storage. In contrast to traditional methods, which have an exponential memory overhead with the increase in network size, the proposed method stores only the number of connections between neurons. The proposed method is evaluated on feedforward networks and demonstrates memory saving capabilities of up to almost 80% while also being more efficient, especially with larger architectures.

Keywords: neural networks; multilayer perceptron; compressed weight matrix; weight density; sparsity

Citation: Cedron, F.; Alvarez-Gonzalez, S.; Ribas-Rodriguez, A.; Rodriguez-Yañez, S.; Porto-Pazos, A.B. Efficient Implementation of Multilayer Perceptrons: Reducing Execution Time and Memory Consumption. *Appl. Sci.* **2024**, *14*, 8020. https://doi.org/10.3390/app14178020

Academic Editor: Alexander N. Pisarchik

Received: 30 July 2024
Revised: 25 August 2024
Accepted: 6 September 2024
Published: 7 September 2024

1. Introduction

Artificial neural network (ANN) is an established machine learning technique that is widely used due to the flexibility it provides. Several problems have been solved using ANNs and they are currently used in different commercial products [1]. Researchers have been interested in creating different types of ANNs to solve specific problems [2]. In the present work, no new type of network is proposed. Instead, a very versatile type of network has been selected and a new way of implementation is proposed for its use. This type of network is known as a multilayer perceptron (MLP), where the processing elements (PEs) are fully connected to all the PEs of the following layer [3]. MLP is one of the most employed models in neural network applications, where its main characteristic is the use of the backpropagation training algorithm, and achieving an implementation that may reduce execution times and memory consumption when using such an algorithm [4,5]. Figure 1 shows an example of MLP.

This work focuses mainly on the connectivity of the network, known as weights, since it is the most performance-consuming part of the network. The conventional way of storing the weights of a network is by using a two-dimensional array, such as the one displayed in Figure 2a. The reason why this method has been used is because the array indexes map the network connections. That is, the position *i*,*j* of the array indicates the connection w_{ij} that goes from neuron *i* to neuron *j*, bearing resemblance to what is known as a sparse matrix [6].

However, as can be seen in Figure 2a, the network connections do not occupy the entire matrix and are not dispersed. It can be seen that they follow a pattern and that small submatrices are formed. The proposed method (Figure 2c) will save solely these submatrices despite saving the general matrix. One of the challenges we face regarding the indexing of submatrices is finding the proper method, but this is solved by generating new indexes from a function.

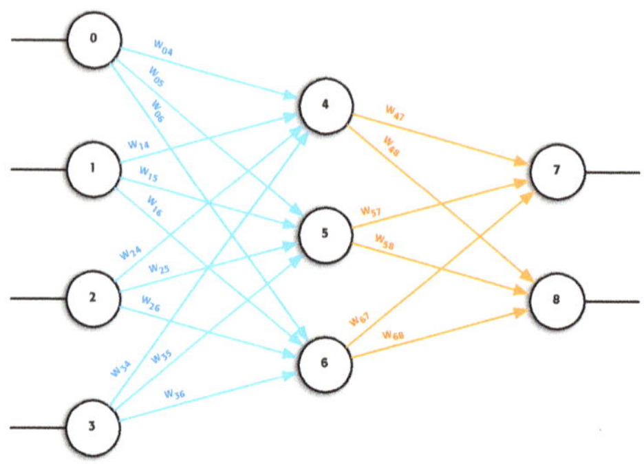

Figure 1. Graphic representation of an MLP. It can be observed that in its architecture, it has an input layer formed by neurons 0 to 3, another hidden layer with neurons 4, 5 and 6, and finally, the last layer, the output layer is formed by neurons 7 and 8. It can be observed that each neuron of a layer is connected with all the neurons of the next layer.

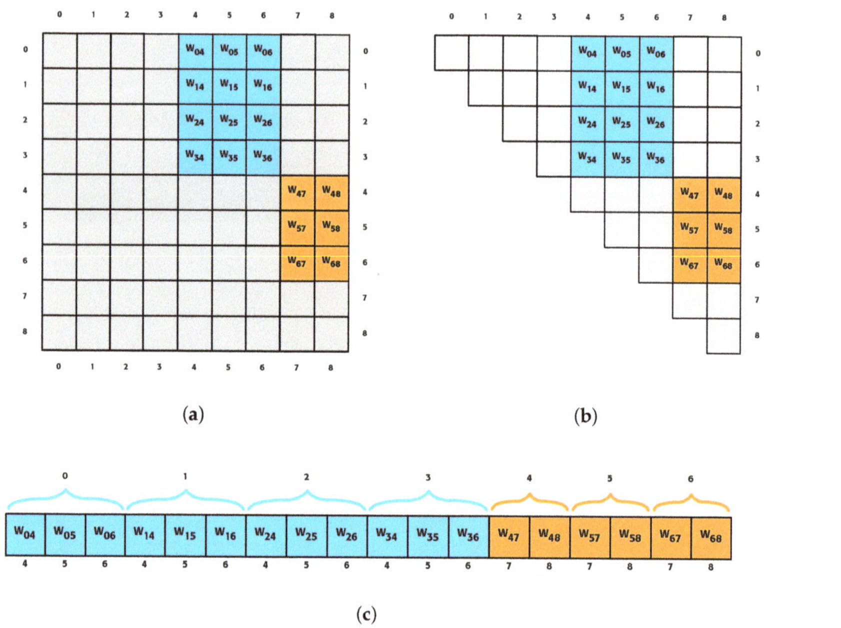

Figure 2. Different ways of storing the weights of a network. Specifically, the example network in Figure 1 is being represented. The gray boxes indicate positions that occupy memory but have no relevant value; the blue boxes represent connections from the first layer to the second layer, while the orange boxes are the connections of neurons from the second layer to the last layer. (**a**) The traditional way in which the weights of a network are stored by using a two-dimensional array. (**b**) Storing the weights of the network in an upper triangular array. (**c**) Proposed way of storing the network weights, where they are all stored consecutively.

The reason for focusing on network connectivity lies in Amdahl's Law, which states that the best option for improving a system is the one that has the greatest impact [7]. Thus, this paper proposes a method to reduce the memory needed to store a network by focusing on the connectivity of an MLP.

To conclude, after demonstrating the saving of the used memory by the proposed method to store the network information, it is shown that this method does not negatively affect the speed at which the network runs, in fact, the proposed method improves performance by making the ANN run even faster.

2. Materials and Methods

2.1. Proposed Approach

In the event of saving the neural network configuration, it is necessary to emphasize how to store the weights of the connections, because this is what occupies most space and the other data hardly take up any [8]. The first option we find is to save the weights in a square matrix where one of the dimensions would be the input neurons and the other the output dimensions. With this technique, we can keep all the weights without any problem, the only thing we have are several positions that are at zero, pointing out that there are no connections between them.

A way of not consuming so much memory would be to eliminate those values that are useless. By not storing values that are never actually used, we will manage to avoid having so many cache failures and we will be able to take better advantage of the location of the data [9,10]. To achieve this, we can look at the literature where we would find out that this is an issue that occurs sometimes, and the matrices with values which do not serve us are called the sparse matrix [11]. What is performed with this is to eliminate all the positions that do not work for us by storing the matrix otherwise. Although there are different implementations of using the sparse matrix, all of them have something in common, that is, they are only profitable to use if you have a lot of unusable values and massive data. Generally, useful data should be around 10% at most, and you should have several thousands of records. If we analyze the weight matrices of our neural networks, the weights occupy approximately 30% of the total matrix, and we also do not meet the condition of several thousands of records since we are not close to that number. This fact will be discussed in detail in later sections.

We should therefore analyze the feedforward networks, as, like it says in the literature, these networks can be stored in an upper triangular matrix, as shown in Figure 2b [12,13].

Through this, an important memory saving method is achieved, since half of the storage is used, with the triangular matrix, although there are still many values that are not employed. This is due to the fact that the connectivity of an MLP is feedforward from layer to layer, implying that those neurons that are in the same layer, or more than one layer away, will not be linked by a weight.

Furthermore, if we look at how these networks are stored, it is possible to observe that they follow a common pattern (see Figure 2). These are collected in small subarrays, which are defined by the PEs in one layer and in the next. Since it is feasible to know how many weights each of these networks has, we can store these submatrices, a method which is based on the present paper.

As described above, the proposed solution will involve storing only the values that are useful in an array. If, for instance, we use a network with 4 neurons in the input layer, 2 in the output layer and 3 in the only hidden layer, as per the example in Figure 2, it is shown how much memory is used to store the network. In Table 1, it is also possible to see how much space is used.

Table 1. Amount of memory used to store the weights of the different approximations used in Figure 2.

Approach	Cells	Saving
Matrix	81	-
Superior triangular matrix	45	44.4%
Proposed approach	18	77.7%

2.2. Implementation

In the previous section, the proposed solution stores all the weights consecutively but one very important thing is lost, i.e., the matrix indices, which serve to reference the positions of the weights in the neural network. In order to know which position of the array of weights is related to the weight in the network, some additional information needs to be stored. This section explains how the entire neural network is held in memory.

2.2.1. Required Variables

- num_layer: Number of layers in the network (including input, output and hidden layers).
- num_neurons: Number of neurons in the network.
- Layer (array): The i position indicates the number of processing elements in the i layer.
- Position (array): This indicates which layer the processing element i belongs to.
- Index (array): This indicates the position that the processing element occupies within the layer.
- num_weights: Number of weights that the network has.
- Weight (array): Value of the network connections.
- Stride (array): The i position indicates where the outgoing weights of the i neuron start.

2.2.2. Length of Arrays

- Layer: The size of this array is given by the value of num_layer, that is, the number of layers.
- Index and position: The size of this array is given by the value of num_neurons, that is, the number of neurons in the network.
- Stride: The size of this array is given by the number of neurons that have outgoing connections, i.e., all the neurons except the output layer.

$$size(stride) = num\_neurons - layer[num\_layer - 1] \tag{1}$$

- Weight: The size of this array is determined by the number of outgoing connections that can exist, which is determined by the num_weight variable:

$$num\_weights = \sum_{i=0}^{num\_layers-1} layer[i] * layer[i+1] \tag{2}$$

Once we have gathered all the necessary information, we still need to know how to access a weight by referencing it with the input and output neurons. We can perform this using the formula below:

$$w_{ij} = weight[stride[i] + index[j]] \tag{3}$$

Figure 3 shows the values of the different arrays used in the proposed method for a neural network.

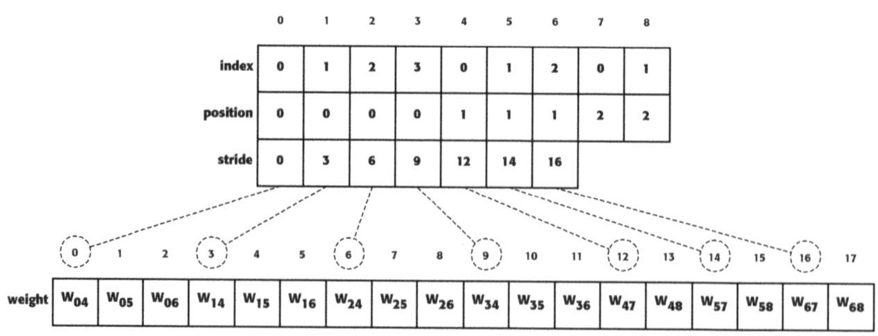

Figure 3. Array values using the proposed method for the example in Figure 1. The index of the arrays "index" and "position" and "stride" refer to the number of neurons to which it refers. The values

of the "index" array indicate the position that a neuron occupies within a layer. The array "position" values indicate the layer number to which a neuron belongs. The array "weights" values are the values of the network connections. The "stride" values indicate at which position in the array weight the outgoing weights of a neuron start.

Finally, it remains to be checked if all this auxiliary information that we added improves the memory consumption used in the code. To know the amount that is consumed, they are used in the equations shown below (the *sizeof(int)* and *sizeof(float)* simply imply the value of the size of an integer and a float, respectively):

$$\begin{aligned} Traditional = {} & (2 + num\_neurons) * sizeof(int) \\ & + num\_neurons^2 * sizeof(float) \end{aligned} \tag{4}$$

$$\begin{aligned} Proposed = {} & (3 + 3 * num\_neurons + num\_layers + \\ & layer[num\_layers - 1]) * sizeof(int) \\ & + num\_weights * sizeof(float) \end{aligned} \tag{5}$$

In light of these equations, the comparative calculation to estimate the difference in consumption between the traditional and the proposed method is not a trivial issue. While in the former it is only necessary to know the number of neurons (*num_neurons*), our method incorporates another series of variables that depend directly on the architecture of the MLP network under analysis.

2.2.3. Pseudocode

The pseudocode of the proposed solution is shown in Listing 1. It is important to highlight that the variables *NUM_NEURONS* and *NUM_LAYERS* are predefined by the network to be simulated, while *getArrayInt* is a function that returns an array of integers of the size indicated by the parameter.

Listing 1. Pseudocode of the proposed solution.

```
POSITION = getArrayInt(NUM_NEURONS);
INDEX    = getArrayInt(NUM_NEURONS);
for (l = i = 0; i < NUM_LAYERS; i++) {
for (j = 0; j < LAYER[i]; j++, l++) {
POSITION[l] = i;
INDEX[l] = j;
}
}

TMP = NUM_NEURONS - LAYER[NUM_LAYERS - 1];
STRIDE = getArrayInt(TMP);
STRIDE[0] = 0;
for (i = 1; i < TMP; i++) {
STRIDE[i] = STRIDE[i - 1] +
LAYER[POSITION[i - 1] + 1];
}

NUM_WEIGHTS = 0;
for (i = 0; i < NUM_LAYERS - 1; i++) {
NUM_WEIGHT += LAYER[i] * LAYER[i + 1];
}
```

Listing 2 shows how the output of the network would be obtained. The *NET* is an array that stores the output of each neuron, while *ACTIVATION* is an array containing the activation functions of each neuron.

Listing 2. Pseudocode to obtain the output network.

```
first_pe = in_pe = 0;
for (l = 1; l < NUM_LAYERS; l++) {
first_pe += LAYER[l - 1];
for (in = 0; in < LAYER[l - 1];
in++, in_pe++) {
pe = first_pe;
for (out = 0; out < LAYER[l];
out++, pe++) {
NET[pe] += NET[pe] *
WEIGHT[out+STRIDE[in_pe]];
}
}
for (tmp = first_pe; tmp < pe; tmp++) {
NET[tmp] = ACTIVATION[tmp](NET[TMP]);
}
}
```

3. Results

It is important to emphasize that the proposed method does not look to improve the network results, i.e., it does not obtain better metrics in accuracy, F1 score or others [14]. The proposed method aims to reduce the memory consumption needed to store a network.

We have tested the method described for networks whose structures are designed to solve the problems found in the UCI repository, specifically for iris [15], cancer [16] and ionosphere [17]. The networks that have been tested have 1, 2 and 3 hidden layers for all three problems. The network structures used can be seen in Table 2. These datasets have been chosen since they are classical problems used in classification problems, based on MLP structures used in previous works, and which have been found to achieve good performance [18].

Table 2. Topology of the networks used for the experimental results. The numbers indicate the amount of neurons on each layer. The first and last values represent the input and output layers, respectively, and the middle numbers are the hidden layers.

Dataset	One Hidden Layer	Two Hidden Layers	Three Hidden Layers
Iris	4, 5, 3	4, 5, 7, 3	4, 4, 5, 5, 3
Cancer	9, 7, 1	9, 7, 3, 1	9, 12, 8, 4, 1
Ionosphere	34, 9, 1	34, 9, 4, 1	34, 12, 8, 4, 1

3.1. Memory Consumption

The algorithm presented in this work has a different memory consumption than the traditional method. Next, you can see in Figure 4 the memory saving process when using the proposed method against the traditional one. In addition, Table 3 shows the memory used for all the cases employed.

The minimum improvement in memory consumption is over 50%, although most comparisons range from over 60% to almost 80%. This significant improvement in the bytes required for storing the network structure and weights shows a clear significance that is especially beneficial as the MLP grows in complexity.

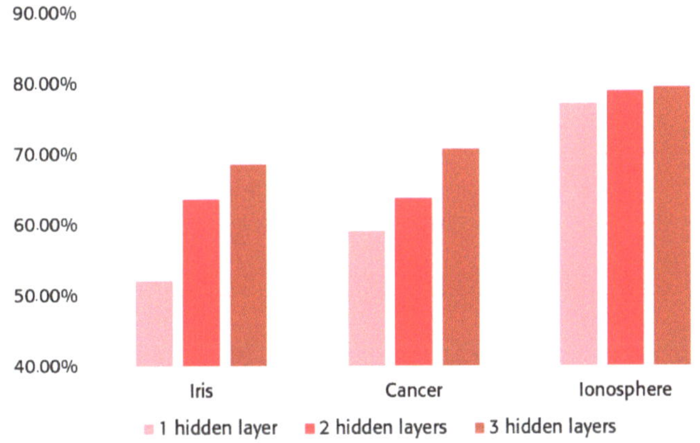

Figure 4. Percentage of saved memory using the proposed method instead of the traditional one for the networks specified in Table 2.

Table 3. Memory in bytes used to store ANNs. The second column indicates the topology of the network. The third and fourth columns represent the memory used in bytes to store the network using the traditional and proposed methods, respectively. The last column indicates the amount of memory saved by using the proposed method instead of the traditional method.

Dataset	Network Topology	Traditional Approach	Proposed Approach	Memory Saving
Iris	4, 5, 3	644 B	308 B	52.17%
	4, 5, 7, 3	1544 B	560 B	63.73%
	4, 4, 5, 5, 3	1876 B	588 B	68.66%
Cancer	9, 7, 1	1244 B	508 B	59.16%
	9, 7, 3, 1	1704 B	616 B	63.85%
	9, 12, 8, 4, 1	4788 B	1400 B	70.76%
Ionosphere	34, 9, 1	7940 B	1812 B	77.18%
	34, 9, 4, 1	9432 B	1988 B	78.92%
	34, 12, 8, 4, 1	14188 B	2900 B	79.56%

3.2. *Operation Time*

Achieving such a noticeable improvement when storing a network is a great accomplishment, but it is necessary to check how it affects the speed of execution. This is necessary since the scenario where the proposed method can be used may vary depending on the performance of the system.

With the aim of achieving representative and reliable results, all the networks run a million patterns to measure the time. The way we have measured the times has been to run each test 10 times and then use the average of those runs as a result. Furthermore, in order to avoid equipment bias, times are taken both on a laptop and on a server, in addition to measuring the times with and without using the optimization options [19,20].

3.2.1. Personal Computer

All tests were conducted on a 2017 macbook pro with i5-7360U CPU @ 2.30 GHz (Apple, Cupertino, CA, USA) and with the clang compiler in the clang-900.0.39.2 version [21].

These first times have been taken without using any of the optimization options provided by the compiler. The results can be checked in Table 4, and Figure 5 shows the increased speed. The reduction in time can be observed in all tested runs, showing a pattern of our proposed model of performance improvement at a higher network complexity.

The times obtained by using compiler optimization options have also been analyzed Specifically, the following parameters were used: "-O3-ffast-math-funroll-loops-ftree-vectorize-march=native". The results can be checked in Table 4, and Figure 5 shows the increased speed. Once again, the results obtained in the execution of the problems on a server will follow the pattern observed in our proposed model. It is worth noting that the improvement achieved in execution on a personal laptop using a compiler optimization versus not using the optimization is still significantly better on average.

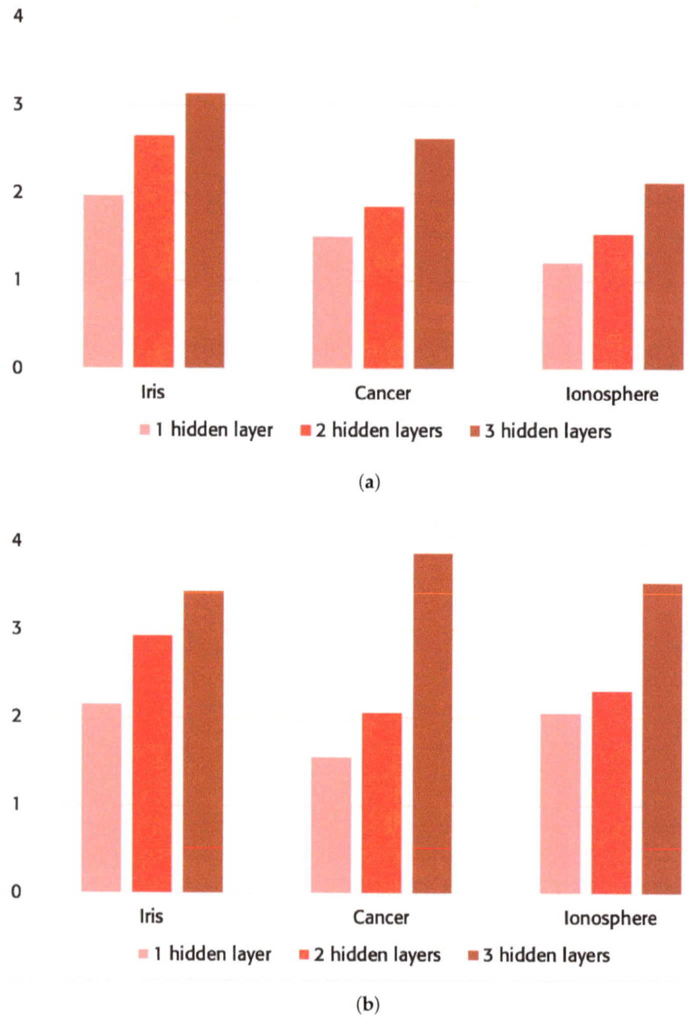

Figure 5. Sped-up improvement of the proposed method vs. the traditional method in network execution on a personal computer. (**a**) Without using the compiler optimization flags. (**b**) Using the compiler optimization flags "-O3-ffast-math-funroll-loops-ftree-vectorize-march=native".

Table 4. Performance in network execution on a personal computer without using compiler optimization flags and using "-O3-ffast-math-funroll-loops-ftree-vectorize-march=native".

		No Optimization Flags			Using Optimization Flags		
	Network Topology	**Traditional Approach**	**Proposed Approach**	**Speed Up**	**Traditional Approach**	**Proposed Approach**	**Speed Up**
Iris	4, 5, 3	0.4108 s	0.2084 s	1.9712x	0.1295 s	0.0604 s	2.1440x ↑
	4, 5, 7, 3	1.1077 s	0.4185 s	2.6468x	0.3152 s	0.1075 s	2.9320x ↑
	4, 4, 5, 5, 3	1.3989 s	0.4470 s	3.1295x	0.3942 s	0.1147 s	3.4367x ↑
Cancer	9, 7, 1	0.5568 s	0.3690 s	1.5089x	0.1459 s	0.0942 s	1.5488x ↑
	9, 7, 3, 1	0.8677 s	0.4695 s	1.8481x	0.2295 s	0.1117 s	2.0546x ↑
	9, 12, 8, 4, 1	2.9614 s	1.1292 s	2.6225x	0.8527 s	0.2204 s	3.8688x ↑
Iono...	34, 9, 1	1.6810 s	1.3907 s	1.2087x	0.5276 s	0.2569 s	2.0537x ↑
	34, 9, 4, 1	2.3799 s	1.5479 s	1.5375x	0.6511 s	0.2823 s	2.3064x ↑
	34, 12, 8, 4, 1	4.9259 s	2.3174 s	2.1256x	1.4572 s	0.4120 s	3.5368x ↑

3.2.2. Computer Server

All these tests were carried out at CESGA in the HPC finisterrae 2 using the thin-node partition, which has an Intel(R) Xeon(R) CPU E5-2680 v3 @ 2.50 GHz (Intel, Santa Clara, CA, USA), and using the GCC compiler in version 7.2.0 [22,23].

These times have been taken without using any of the optimization options provided by the compiler. The results can be checked in Table 5, and Figure 6 shows the increased speed.

The times obtained by using compiler optimization options have also been analyzed. Specifically, the following parameters were used: "-O3-ffast-math-funroll-loops-ftree-vectorize-march=native". The results can be checked in Table 5, and Figure 6 shows the increased speed.

Table 5. Performance in network execution on a server using compiler optimization flags and using "-O3 -ffast-math -funroll-loops-ftree-vectorize-march=native".

		No Optimization Flags			Using Optimization Flags		
	Network Topology	**Traditional Approach**	**Proposed Approach**	**Speed Up**	**Traditional Approach**	**Proposed Approach**	**Speed Up**
Iris	4, 5, 3	0.3791 s	0.2221 s	1.7068x	0.1301 s	0.0641 s	2.0296x ↑
	4, 5, 7, 3	1.0686 s	0.4412 s	2.4220x	0.3036 s	0.1171 s	2.5926x ↑
	4, 4, 5, 5, 3	1.2403 s	0.4587 s	2.7039x	0.3659 s	0.1280 s	2.8585x ↑
Cancer	9, 7, 1	0.5180 s	0.3852 s	1.3447x	0.1389 s	0.0941 s	1.4760x ↑
	9, 7, 3, 1	0.8367 s	0.4926 s	1.6985x	0.2103 s	0.1203 s	1.7481x ↑
	9, 12, 8, 4, 1	2.9419 s	1.1932 s	2.4655x	0.7555 s	0.2593 s	2.9136x ↑
Iono...	34, 9, 1	1.6926 s	1.5160 s	1.1164x	0.3911 s	0.2678 s	1.4604x ↑
	34, 9, 4, 1	2.3677 s	1.6797 s	1.4095x	0.5905 s	0.3029 s	1.9494x ↑
	34, 12, 8, 4, 1	4.9747 s	2.5451 s	1.9546x	1.6443 s	0.4714 s	3.4881x ↑

As was the case for the laptop runs, the server will follow a very similar behavior. With a minimal improvement of 130% that can be seen in the simplest case for the cancer problem in Table 5, the improvements in computational speed are impressively increased with higher network complexity and with the use of optimization in the compiler.

It is interesting to remark that, contrary to what one might think, a greater improvement in times resulted when the executions were carried out on the laptop.

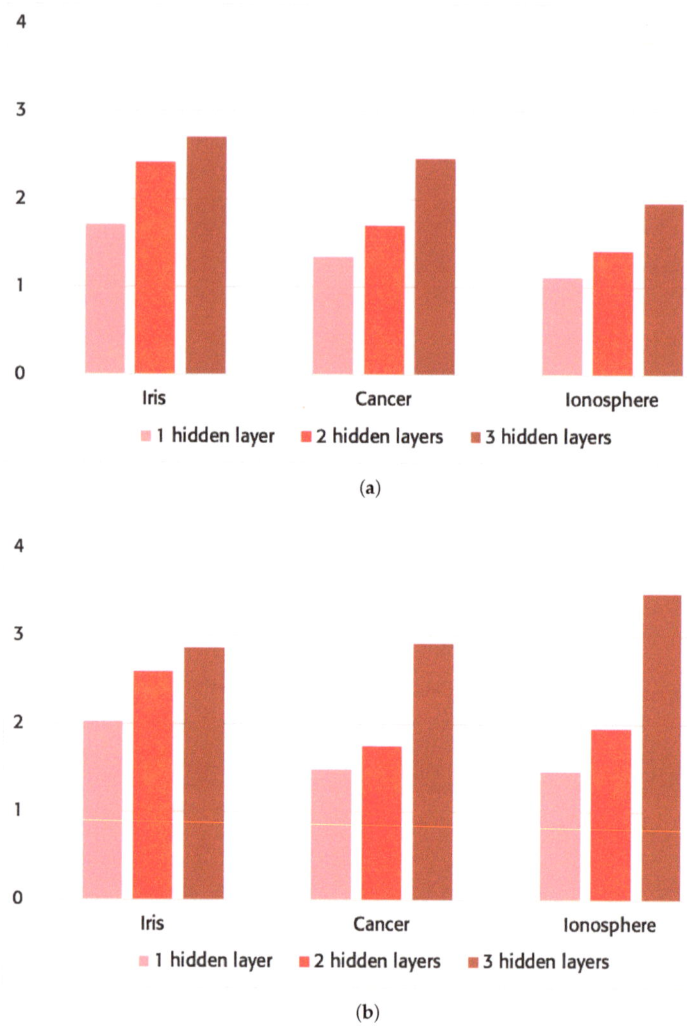

Figure 6. Sped-up up improvement of the proposed method vs. the traditional method in network execution on a server. (**a**) Without using the compiler optimization flags; (**b**) using the compiler optimization flags "-O3 -ffast-math -funroll-loops -ftree-vectorize -march=native".

4. Conclusions

It should be stressed that the proposed method really does save memory and also improves runtime. Equation (4) shows that the traditional method stores more memory as the number of neurons increases. This is because the traditional method has an exponential growth of consumption that grows very fast. However, with the proposed method, this does not happen, since the important thing is not the number of neurons, but the number of connections between them, so that the memory consumption is not so abusive (see Table 3).

The proposed method avoids the exponential memory consumption, and although it also improves the execution time, the improvement is not so drastic. In Tables 4 and 5, it can be observed that the algorithm works better when it has more layers and these, in turn, have many neurons. Moreover, as can be seen in the tables previously mentioned, the proposed method takes much better advantage of the compiler optimizations.

Also, it should be pointed out that the proposed method assumes that everything put into it is correct. In the example of Figure 3, where neurons 0 and 3 do not have a weight, using the w_{ij} formula (Equation (3)) will give you a value for the weight w_{03}, when in fact, this does not exist. Thus, in this method, we only need to enter the correct values for a feedforward network.

It is important to note that the method proposed in this paper only works for fully forward connected networks. Other types such as recurrent networks would not be able to store all the weights. For instance, for a network with four neurons in the input layer, three in the hidden layer and two in the output layer, the w_{ij} formula (Equation (3)) would return the same value when retrieving the values of w_{65} and w_{68}.

Although the proposed method is created for feedforward networks, its internal structure allows for the use of techniques that are commonly used in networks such as MLP; an example is the dropout technique where connections can be removed from a network. To carry out this technique of dropping connections, it would simply be necessary to set the connection value to zero [24].

Finally, it is important to mention that the proposed method can involve the use of networks in IoT or ubiquitous computing devices using modest hardware [25,26]. Devices such as the Arduino Mega have only 256Kb of memory, making it impossible to use networks using the traditional method [27].

Author Contributions: Conceptualization, F.C.; Methodology S.R.-Y., S.A.-G. and F.C.; Software, F.C.; Validation, A.B.P.-P and F.C.; Writing—draft preparation, S.A.-G. and F.C.; Writing—review and editing A.R.-R. and F.C.; Visualization, A.B.P.-P. and F.C.; Funding acquisition S.R.-Y. All authors have read and agreed to the published version of the manuscript.

Funding: This research received no external funding.

Institutional Review Board Statement: Not applicable.

Informed Consent Statement: Not applicable.

Data Availability Statement: Data is contained within the article.

Acknowledgments: We are particularly grateful to CESGA (Galician Supercomputing Centre) for providing access to their infrastructure.

Conflicts of Interest: The authors declare no conflicts of interest.

Abbreviations

The following abbreviations are used in this manuscript.

ANN	Artificial neural network
MLP	Multilayer perceptron
PEs	Process elements

References

1. Misra, J. ; Saha, I. Artificial neural networks in hardware: A survey of two decades of progress. *Neurocomputing* **2010**, *74*, 239–255. [CrossRef]
2. Abiodun, O.I.; Jantan, A.; Omolara, A.E.; Dada, K.V.; Mohamed, N.A.; Arshad, H. State-of-the-art in artificial neural network applications: A survey. *Heliyon* **2018**, *4*, e00938. [CrossRef] [PubMed]
3. Gardner, M.W.; Dorling, S. Artificial neural networks (the multilayer perceptron)—A review of applications in the atmospheric sciences. *Atmos. Environ.* **1998**, *32*, 2627–2636. [CrossRef]
4. Hecht-Nielsen, R. Theory of the backpropagation neural network. In *Neural Networks for Perception*; Elsevier: Amsterdam, The Netherlands, 1992; pp. 65–93.
5. Popescu, M.C.; Balas, V.E.; Perescu-Popescu, L.; Mastorakis, N. Multilayer perceptron and neural networks. *WSEAS Trans. Circuits Syst.* **2009**, *8*, 579–588.
6. Yan, D.; Wu, T.; Liu, Y.; Gao, Y. An efficient sparse-dense matrix multiplication on a multicore system. In Proceedings of the 2017 IEEE 17th International Conference on Communication Technology (ICCT), Chengdu, China, 27–30 October 2017; pp. 1880–1883.
7. Amdahl, G.M. Computer architecture and amdahl's law. *Computer* **2013**, *46*, 38–46. [CrossRef]

8. Brunel, N.; Hakim, V.; Isope, P.; Nadal, J.P.; Barbour, B. Optimal information storage and the distribution of synaptic weights Perceptron versus Purkinje cell. *Neuron* **2004**, *43*, 745–757. [PubMed]
9. Nishtala, R.; Vuduc, R.W.; Demmel, J.W.; Yelick, K.A. When cache blocking of sparse matrix vector multiply works and why *Appl. Algebra Eng. Commun. Comput.* **2007**, *18*, 297–311. [CrossRef]
10. Blanco Heras, D.; Blanco Pérez, V.; Carlos Cabaleiro Domínguez, J.; Fernández Rivera, F. Modeling and improving locality for irregular problems: Sparse matrix-Vector product on cache memories as a case study. In *Proceedings of the High-Performance Computing and Networking*; Sloot, P., Bubak, M., Hoekstra, A., Hertzberger, B., Eds.; Springer: Berlin/Heidelberg, Germany, 1999 pp. 201–210.
11. Buluc, A.; Gilbert, J.R. Challenges and advances in parallel sparse matrix-matrix multiplication. In Proceedings of the 2008 37th International Conference on Parallel Processing, Portland, OR, USA, 9–12 September 2008; pp. 503–510.
12. Vincent, K.; Tauskela, J.; Thivierge, J.P. Extracting functionally feedforward networks from a population of spiking neurons *Front. Comput. Neurosci.* **2012**, *6*, 86. [CrossRef] [PubMed]
13. Bilski, J.; Rutkowski, L. Numerically robust learning algorithms for feed forward neural networks. In *Neural Networks and Soft Computing*; Springer: Zakopane, Poland, 2003; pp. 149–154.
14. Caruana, R.; Niculescu-Mizil, A. Data Mining in Metric Space: An Empirical Analysis of Supervised Learning Performance Criteria. In Proceedings of the KDD'04: 10th ACM SIGKDD International Conference on Knowledge Discovery and Data Mining, New York, NY, USA, 22–25 August 2004; pp. 69–78. [CrossRef]
15. Fisher, R. UCI Machine Learning Repository Iris Data Set. 1936. Available online: https://archive.ics.uci.edu/ml/datasets/Iris (accessed on 19 January 2022).
16. Zwitter, M.; Soklic, M. UCI Machine Learning Repository Breast Cancer Data Set. 1988. Available online: https://archive.ics.uci.edu/ml/datasets/breast+cancer (accessed on 19 January 2022).
17. Sigillito, V. UCI Machine Learning Repository Ionosphere Data Set. 1989. Available online: https://archive.ics.uci.edu/ml/datasets/Ionosphere (accessed on 19 January 2022).
18. Porto-Pazos, A.B.; Veiguela, N.; Mesejo, P.; Navarrete, M.; Alvarellos, A.; Ibáñez, O.; Pazos, A.; Araque, A. Artificial astrocytes improve neural network performance. *PLoS ONE* **2011**, *6*, e19109. [CrossRef] [PubMed]
19. Haneda, M.; Knijnenburg, P.M.W.; Wijshoff, H.A.G. Optimizing general purpose compiler optimization. In Proceedings of the CF'05: 2ND Conference on Computing Frontiers, New York, NY, USA, 4–6 May 2005. [CrossRef]
20. Dong, S.; Olivo, O.; Zhang, L.; Khurshid, S. Studying the influence of standard compiler optimizations on symbolic execution. In Proceedings of the 2015 IEEE 26th International Symposium on Software Reliability Engineering (ISSRE), Washington, DC, USA, 2–5 November 2015. [CrossRef]
21. Intel Core i5 7360U Processor 4M Cache up to 3.60 Ghz Product Specifications. Available online: https://ark.intel.com/content/www/us/en/ark/products/97535/intel-core-i57360u-processor-4m-cache-up-to-3-60-ghz.html (accessed on 23 January 2022).
22. CESGA—Centro de Supercomputación de Galicia. Available online: https://www.cesga.es/ (accessed on 19 January 2022).
23. Intel Xeon Processor E5 2680 v3 30 M Cache 2.50 Ghz Product Specifications. Available online: https://ark.intel.com/content/www/us/en/ark/products/81908/intel-xeon-processor-e52680-v3-30m-cache-2-50-ghz.html (accessed on 23 January 2022).
24. Tan, S.Z.K.; Du, R.; Perucho, J.A.U.; Chopra, S.S.; Vardhanabhuti, V.; Lim, L.W. Dropout in Neural Networks Simulates the Paradoxical Effects of Deep Brain Stimulation on Memory. *Front. Aging Neurosci.* **2020**, *12*, 273 [CrossRef] [PubMed]
25. Madakam, S.; Ramaswamy, R.; Tripathi, S. Internet of Things (IoT): A Literature Review. *J. Comput. Commun.* **2015**, *3*, 164–173. [CrossRef]
26. Raman Kumar, S.P. *Applications in Ubiquitous Computing*; Springer: Cham, Switzerland, 2021.
27. Arduino Board Mega 2560. Available online: https://www.arduino.cc/en/Main/ArduinoBoardMega2560 (accessed on 23 January 2022).

applied sciences

MDPI

Article

MRI Diffusion Connectomics-Based Characterization of Progression in Alzheimer's Disease

David Mattie [1,2,3], Lourdes Peña-Castillo [1], Emi Takahashi [4,5] and Jacob Levman [2,5,6,*]

1 Department of Computer Science, Memorial University of Newfoundland, St. John's, NL A1C 5S7, Canada; dmattie@stfx.ca (D.M.); lourdes@mun.ca (L.P.-C.)
2 Department of Computer Science, St. Francis Xavier University, Antigonish, NS B2G 2W5, Canada
3 Department of Marketing and Enterprise Systems, St. Francis Xavier University, Antigonish, NS B2G 2W5, Canada
4 Department of Radiology, Harvard Medical School, Boston, MA 02115, USA; emi.oki@mgh.harvard.edu
5 Athinoula A. Martinos Center for Biomedical Imaging, Massachusetts General Hospital, Charlestown, MA 02129, USA
6 Nova Scotia Health Authority—Research, Innovation and Discovery Center for Clinical Research, Halifax, NS B3J 0EB, Canada
* Correspondence: jlevman@stfx.ca

Abstract: Characterizing Alzheimer's disease (AD) progression remains a significant clinical challenge. The initial stages of AD are marked by the accumulation of amyloid-beta plaques and Tau tangles, with cognitive functions often appearing normal, and clinical symptoms may not manifest until up to 20 years after the prodromal period begins. Comprehensive longitudinal studies analyzing brain-wide structural connectomics in the early stages of AD, especially those with large sample sizes, are scarce. In this study, we investigated a longitudinal diffusion-weighted imaging dataset of 264 subjects to assess the predictive potential of diffusion data for AD. Our findings indicate the potential of a simple prognostic biomarker for disease progression based on the hemispheric lateralization of mean tract volume for tracts originating from the supramarginal and paracentral regions, achieving an accuracy of 86%, a sensitivity of 86%, and a specificity of 93% when combined with other clinical indicators. However, diffusion-weighted imaging measurements alone did not provide strong predictive accuracy for clinical variables, disease classification, or disease conversion. By conducting a comprehensive tract-by-tract analysis of diffusion-weighted characteristics contributing to the characterization of AD and its progression, our research elucidates the potential of diffusion MRI as a tool for the early detection and monitoring of neurodegenerative diseases and emphasizes the importance of integrating multi-modal data for enhanced predictive analytics.

Keywords: Alzheimer's disease; diffusion tensor imaging; whole-brain tractography; biomarkers

Citation: Mattie, D.; Peña-Castillo, L.; Takahashi, E.; Levman, J. MRI Diffusion Connectomics-Based Characterization of Progression in Alzheimer's Disease. *Appl. Sci.* **2024**, *14*, 7001. https://doi.org/10.3390/app14167001

Academic Editor: Alexander N. Pisarchik

Received: 20 May 2024
Revised: 30 July 2024
Accepted: 31 July 2024
Published: 9 August 2024

1. Introduction

An estimated 35 million people worldwide suffered from Alzheimer's Disease in 2022, with 7 million new cases every year [1]. The percentage of people with Alzheimer's disease, the most common form of dementia, increases with age, where 5.0% of people aged 65 to 74, 13.1% of people aged 75 to 84, and 33.2% of people aged 85 and older have Alzheimer's [2].

1.1. Research in Context

1.1.1. Research before This Study

Comprehensive whole-brain analyses on longitudinal data remain relatively rare. There is mixed support for the utility of diffusion-weighted imaging data in developing biomarkers. Few studies establish prognostic biomarkers to indicate disease conversion. Numerous studies conduct correlational analysis to understand the effect size of the diffusion characteristics between disease stages, and many studies develop classification models with high accuracy, often relying on clinical indicators as predictors. However, little research has been conducted to assess the ability of diffusion characteristics in isolation to characterize disease stages or conversion.

1.1.2. Added Value of This Study

To our knowledge, this study is the first to systematically evaluate the potential of diffusion-weighted imaging (DWI) metrics, in isolation, to predict a range of neuropsychological and neurobiological indicators commonly used in staging Alzheimer's disease with an aim to develop objective, non-invasive cognitive assessment methods that could complement or potentially reduce reliance on traditional clinical testing procedures.

This study integrates tractography metrics with rich phenotypic and neuropsychological data across a substantial longitudinal dataset focused on Alzheimer's research. This integration allows for a thorough analysis of neurodevelopmental changes over time as subjects progress in Alzheimer's disease with the aim of identifying early-stage prognostic biomarkers. The interaction between the hemispheric lateralization of tract volumes connected to the supramarginal gyrus and paracentral regions offers a potential prognostic biomarker of Alzheimer's disease progression, with AUROC of 74% and AUPRC of 75% being important findings of this study. While this region has been implicated in previous studies, our analysis is novel in that it uses tractography metrics to establish this connection. To the best of our knowledge, this is the first study to implicate the supramarginal gyrus in AD progression by using detailed tractography measurements.

1.1.3. Implications

Diffusion-weighted imaging measurements alone did not provide strong predictive accuracy for clinical variables, disease classification, or disease conversion. Our analysis of white matter tract features revealed moderate but notable associations with neurobiological and neuropsychological markers, opening the door to future potential models, likely based on multi-modal data capable of predicting these clinical indicators. Our findings also demonstrate the potential for a simple prognostic biomarker of disease conversion when combined with other clinical indicators.

A brief overview of this study is provided in Table 1.

Table 1. Study overview.

Variable	Measurement
Participants	264 (434 sessions)
Whole-brain tractography measurements	12 diffusion measurements per tract
Disease stages	CN, MCI, and AD
Neuropsychological measurements	9
Neurobiological measurements	7
Features used	Whole-brain tractography measurements
Feature reduction	ElasticNet
Evaluation metrics	RMSE, AUROC, and AUPRC
Performance	AUROC of 74% and AUPRC of 75% based exclusively on diffusion measurements

A definitive diagnosis of Alzheimer's disease (AD) can only be confirmed by a histological examination of brain tissue post-mortem [3]. The initial stages of AD are characterized by the accumulation of plaques of the protein amyloid-beta ($A\beta$) in the medial parietal cortex. In this prodromal stage, cognitive function appears normal, and patients may not exhibit clinical symptoms up to 20 years after the start of the prodromal stage [4]. As the condition advances, other signs of neurodegeneration, such as neuronal death [5], atrophy (depending on the subtype) [6], and gliosis [7], become discernible after a variable period of latency. These changes correlate with clinical cognitive evaluations taken over multiple years and align with a suite of biomarkers, including hippocampal volume and heightened concentrations of Aβ and Tau proteins, which were pinpointed as indicators in the timeline of AD progression as described in the established literature [8].

The prevailing consensus is that early-stage therapeutic interventions could offer the greatest potential to improve health outcomes before irreversible neuronal loss and damage to brain tissue occur. Estimates suggest that providing treatment during the disease's preclinical phase could significantly curtail its progression. In fact, some projections indicate that "a delay of 10 years would result in virtual disappearance of the disease" [9]. The first neurons damaged are those responsible for memory, language, and cognition. However, the pathophysiological processes that cause this damage are thought to begin 20 years before symptoms are reported [1,10]. Since Alzheimer's disease is a gradual and progressive neurodegenerative disorder, understanding the potential of biomarkers to characterize the disease's pathology and its long-term development is a key motivation behind this study. These biomarkers may support the identification of individuals who could benefit from treatment, potentially improving health outcomes for patients with the disease.

The current gold standard for the diagnosis of Alzheimer's disease is biopsy or autopsy [11]. Recent studies have invested considerable effort and resources in the early detection of AD in the prodromal stages of mild cognitive impairment (MCI). Altered brain asymmetry of subcortical structures, reduction in cortical thickness, and hippocampal, entorhinal, fusiform and medial temporal lobe volumes are all proposed biomarkers of AD [11]. However, the net improvement in AD diagnostic accuracy from structural MRI tests following clinical neurocognitive memory assessments has been shown to be low, +1.1% (95% CI 0.1 to 3.9) [12]. In contrast, diffusion-weighted imaging (DWI) techniques have shown promise [13]. DWI was designed to study white matter (WM) structure [14], a tissue to which AD has been associated [15–17]. This modality is particularly useful, given that AD exhibits degeneration of cellular barriers of neurons and fiber tracts as a result of the buildup of Tau proteins [18]. In recent years, a large body of research has focused on leveraging DWI to classify AD stages and predict disease progression [10,13,19–24].

Despite growing evidence that diffusion-weighted imaging correlates with disease severity [25], we have been unable to find a comprehensive analysis of whether diffusion-weighted imaging can be used to predict established biomarkers used for staging AD. In this study, we investigate the potential of tractography metrics to predict neuropsychological and neurobiological test results in the context of Alzheimer's disease progression while also aiming to identify early prognostic biomarkers of AD. Our analysis addresses four key questions: (Q1) Do phenotypic characteristics predict cognitive decline within our current dataset? (Q2) Are WM tract features predictive of neuropsychological and neurobiological indicators? (Q3) Does baseline tract volume change with cognitive decline? (Q4) To what extent can tractography metrics predict cognitive decline? We present our methodology for data acquisition and analysis, followed by results corresponding to each research question. Finally, we discuss the implications of our findings for both clinical practice and future research directions in neuroimaging and Alzheimer's disease.

2. Materials and Methods

2.1. Participants

In this study, two hundred and sixty-four participants across multiple exams totaling 434 sessions were obtained from the Alzheimer's Disease Neuroimaging Initiative (ADNI) database (adni.loni.usc.edu). The ADNI was launched in 2003 as a public–private partnership, led by Principal Investigator Michael W. Weiner, MD. The primary goal of the ADNI is to test whether serial magnetic resonance imaging (MRI), positron emission tomography (PET), other biological markers, and clinical and neuropsychological assessment could be combined to measure the progression of mild cognitive impairment (MCI) and early Alzheimer's disease (AD). The images obtained are associated with a clinical diagnosis and a specific time point. As such, individuals may span one or more clinical diagnoses, such as progression from cognitive normal (CN) $\rightarrow$ MCI $\rightarrow$ AD, and therefore can have images at each diagnosis. Table 2 shows the number of participants by clinical diagnosis at imaging and clinical progression.

Table 2. Stratification of ADNI images and associated demographic and clinical details.

Diagnosis at Imaging	Progression Profile	Subjects (F:M)	Mean Age (F:M)	SD (F:M)
CN	CN	126:92	71:78	7.4:7.7
CN	CN → MCI	5:4	74:79	10.8:8.6
CN	CN → AD	1:0	88: -	-:-
MCI	MCI	60:75	76:76	7.7:6.9
MCI	CN → MCI	4:5	73:82	13.4:5.8
MCI	MCI → AD	2:1	82:83	8.1:-
AD	AD	28:27	78:76	7.9:8.4
AD	CN → AD	1:0	87: -	-:-
AD	MCI → AD	2:1	83:85	7.6:7.7

Note: CN = cognitively normal, MCI = mild cognitive impairment, and AD = Alzheimer's disease.

2.2. MRI Acquisition

T1-weighted (T1w) images were acquired at 3T, 208 × 240 × 256 voxels of size 1 mm^3. Diffusion MRI data were acquired by using diffusion-weighted single-shot spin-echo echo-planar imaging. Fifty-six slices of 2 mm in thickness, yielding 2 mm isotropic voxels, were obtained. Forty-nine diffusion-weighted measurements (b = 1000 s/mm^2) and seven non-diffusion-weighted measurements (b = 0 s/mm^2) were acquired with TR = 7200 ms, TE = 56 ms, and field of view = 232 mm × 232 mm.

The imaging sequences and parameters of the anatomical scans followed the Alzheimer's Disease Neuroimaging Initiative 3 protocols (https://adni.loni.usc.edu/wp-content/themes/freshnews-dev-v2/documents/mri/ADNI3-MRI-protocols.pdf, accessed on 16 June 2022) and were collected across 57 imaging centers. The data used in our research included all subjects in the ADNI3 cohort with at least one T1-weighted MP-RAGE and a corresponding diffusion-weighted image (DWI).

2.3. Preprocessing of MRI Data and Estimation of Structural Networks

We developed a pipeline management software package [26] offering a robust, fault-tolerant, and extensible platform to execute the processing workflows of the ADNI data. The pipeline code was containerized by using Apptainer [27] to facilitate reproducibility as well as environment management during pipeline execution on each of the super-computing clusters in which software was executed. We sought to process all ADNI3 participant sessions that contained both a T1-weighted image and a diffusion-weighted image. A total of 961 participants, totaling 1873 exam sessions, were considered for our study. After excluding scans without DWI images or bval/bvec files, 264 participants represented in 434 sessions were processed by our pipeline.

The preprocessing pipeline to extract tractography metrics from the ADNI dataset has been described elsewhere [28,29]. For the ADNI dataset, T1w MRI images were segmented into sub-regions by using the FreeSurfer software package, version 7.2 [30], with cortical [31] and subcortical [30] labeling pipelines. The white matter volume generated by Freesurfer's recon-all was further separated into 181 regions of interest (ROIs) by using the Freesurfer program mri_extract_label. Labels were extracted by using the Desikan--Killiany atlas [31]. The diffusion-weighted image was registered to the T1w image before orientation distribution function (ODF) estimation was performed. ODF maps were created from the preprocessed DWI images by using the Diffusion Toolkit (DTK v0.6.4) [32] software package. The HARDI/Q-ball imaging model [33] with a fiber orientation distribution function was estimated at each voxel. The Fiber Assignment by Continuous (FACT)-alike tracking algorithm [32,34] was employed for deterministic fiber tracking. Seed points for tractography were generated throughout the entire brain volume where valid diffusion data existed, using a 35° angle threshold for stopping criteria. This whole-brain tract file was constructed by using the odf_recon and odf_tracker utilities from DTK.

The generated tract file and individual white matter ROI masks were postprocessed for the extraction of 32,580 tracts (cartesian product of ROIs), with multiple measures extracted for each, including mean tract length, tract volume, mean fractional anisotropy (FA), FA standard deviation, mean diffusivity (MD) calculated by using the mean of the three eigenvalues, MD standard deviation, and the corresponding left–right asymmetries of each of these measurements (12 measurements in total). We derived these measurements for tracts that started or terminated in our ROIs as well as tracts that passed through our ROIs, for a total of 781,920 measurements per scan session. White matter tract identification was assessed by using q-ball imaging [33,35]. Although a variety of structural connectomics analytics technologies have been developed for the analysis of diffusion-enabled brain MRI examinations [34,36–39], our approach includes a thorough assessment of features with characterization potential, such as the variability (as measured with the standard deviation) in FA and MD along each given fiber pathway, as well as hemispheric asymmetry measurements for all aforementioned features.

2.4. Statistics

We applied whole-brain deterministic tractography techniques to generate a high-dimensional dataset across 434 exams. Each subject had approximately 1.3 million feature measurements on average being evaluated for potential as a biomarker for characterizing AD, depending on the number of visits a subject participated in. The ADNI3 dataset includes detailed clinical biomarkers for most participants at each imaging session, providing tremendous benefit, as they offer potential indicators of early decline in the preclinical stages or for those with mild cognitive impairment (MCI). We targeted 16 neuropsychological and neurobiological phenotypic characteristics (see Table 3) as response variables in our analysis of white matter tract features.

Table 3. Table of ADNI biomarkers available per scan/session.

Phenotypic Characteristic	Range	Description
ADAS-11 (neuropsychological)	0–70	A rating scale to assess the severity of cognitive and non-cognitive dysfunction from mild to severe AD. ADAS-11 assesses the cognitive domains of memory, language, and praxis. Specific tasks include Word Recall, Naming Objects and Fingers, Commands, Constructional Praxis, Ideational Praxis, Orientation, Word Recognition, and Language
ADAS-13 (neuropsychological)	0–85	A rating scale to assess cognitive domains hypothesized to be important treatment targets of antidementia drugs that are not assessed by the ADAS-11: attention and concentration, planning and executive function, verbal memory, nonverbal memory, and praxis.
AV45 (PET image analysis)		An imaging biomarker for amyloid plaque accumulation in subjects with cognitive impairment that may be attributed to the presence of Alzheimer's disease. Average AV45 regional standardized uptake values of frontal, anterior cingulate, precuneus, and parietal cortex relative to the cerebellum.
BRAIN VOLUME (MR image analysis)		Volume (mm^3) of brain
Diagnosis	[CN,MCI,AD]	Diagnosis classification at scan
ENTORHINAL VOLUME (MR image analysis)		Volume (mm^3) of entorhinal cortex
FDG (PET scan information)		Average fluorodeoxyglucose PET of angular, temporal, and posterior cingulate. It reflects loss of neuropil, loss of synapse, and functional impairment of neurons. Lower FDG-PET was regarded as a signal of neuronal hypometabolism due to neurodegeneration.
FUSIFORM VOLUME (MR image analysis)		Volume (mm^3) of fusiform
HIPPOCAMPUS VOLUME (MR image analysis)		Volume (mm^3) of hippocampus
ICV (MR image analysis)		Intracranial volume. In patients with dementia, but not in MCI, severity of cognitive impairment and ICV were moderately correlated. The effect of ICV on cognition was not mediated by hippocampal atrophy.
MIDTEMPORAL VOLUME (MR image analysis)		Volume (mm^3) of mid temporal
VENTRICLE VOLUME (MR image analysis)		Volume (mm^3) of ventricles
CDR SB (neuropsychological)	0–36	Clinical Dementia Rating scale Sum of Boxes (CDR-SB) score. This score has been used to accurately stage severity of Alzheimer dementia and mild cognitive impairment (MCI).

Table 3. *Cont.*

Phenotypic Characteristic	Range	Description
FAQ (neuropsychological)	0–30	The Functional Activities Questionnaire (FAQ) measures instrumental activities of daily living such as preparing balanced meals and preparing finances. A cut-point of 9 (dependent in 3 or more activities) is recommended to indicate impaired function and possible cognitive impairment.
MMSE (neuropsychological)	0–30	Mini Mental State Examination. It is an 11-question measure that tests five areas of cognitive function: orientation, registration, attention and calculation, recall, and language. The maximum score is 30. A score of 23 or lower is indicative of cognitive impairment.
MoCA (neuropsychological)	0–30	Montreal Cognitive Assessment Test for Dementia. This test is a 30-item test of language, memory, visual and spatial thinking, reasoning, and orientation skills. A score of 26 or above is considered normal.
RAVLT Immediate (neuropsychological)		Rey Auditory Verbal Learning Test evaluating short-term memory, working memory, and long-term memory. RAVLT Immediate is the sum of scores from the 5 first trials.

We performed a univariate statistical analysis on each feature derived within our dataset to understand range, central tendency, standard deviation, and correlation to each of the 16 selected phenotypic characteristics provided by the ADNI as they related to cognitive decline. Cognitive decline is presented in ADNI data as a categorical variable representing diagnosis. Using our dataframe containing diffusion measurements derived from dMRI and T1-weighted images, we sought to answer the following research questions.

Q1. Do phenotypic characteristics predict cognitive decline within our current dataset?

To investigate potential associations between phenotypic biomarkers (Table 3) and diagnostic categories in our study, we employed the Kruskal–Wallis test, a non-parametric test suitable for comparing distributions across multiple groups. The Kruskal–Wallis test enabled an overall assessment of whether statistically significant differences exist in the distribution of biomarker values across diagnostic categories (CN, MCI, and AD). This approach was chosen due to its robustness in handling non-normally distributed data and its ability to discern variations in central tendencies across multiple groups. All statistical analyses were performed by using R (version 4.1.3) with a significance threshold set at $p < 0.05$ after Benjamini–Hochberg (BH) correction.

Following the establishment of the relationship between these phenotypic characteristics and diagnosis, our next step involved exploring diffusion-weighted imaging (DWI) measurements known to be strong predictors of these biomarkers. This exploration aims to identify additional features that could enhance subsequent predictive models.

Q2. Are WM tract features predictive of neuropsychological and neurobiological indicators?

We divided our data into training and test subsets with an 80–20 split that sought to ensure a balance of our response variables across the training/test datasets. Given the longitudinal nature of our dataset, comprising multiple MRI sessions per patient over time, we employed a stratification approach that involved ensuring that all sessions pertaining to a single patient were exclusively included in either the training or the test set, but not both. This was achieved through a randomized allocation process until no participants were found in either the training and test sets to avoid data leakage. Due to the high dimensionality of our data, it was necessary to perform aggressive feature reduction to include only the features exhibiting the top 10% highest variance. We used mean imputation across the remaining features to make it possible to employ regularization techniques for further reduction. We normalized our data; then, for each independent neuropsychological and neurobiological response variable, we employed ElasticNet [40,41] regularization with alpha ranging from 0.5 to 0.8 to reduce the number of features. The final features expected to offer the most discriminatory power for our response variables were scaled and re-imputed by using k-NN nearest neighbour imputation [42]. Imputation using k-NN was initially unable to perform complete imputation with such a wide dataset, forcing us to initially use mean imputation until we could perform ElasticNet regularization. Subsequently, for each neuropsychological and neurobiological response variable, we employed a repeated 10-fold cross-validation method to ensure robustness with various models, including support vector machines [42,43], decision trees [44], random forest [45], multi-layer perceptron [46], and gradient boosting [47]. The repeated cross-validation approach minimized variability in performance metrics due to random partitioning. We evaluated our results by using RMSE, MAE, and R2 metrics. This multi-step process is depicted in Figure 1.

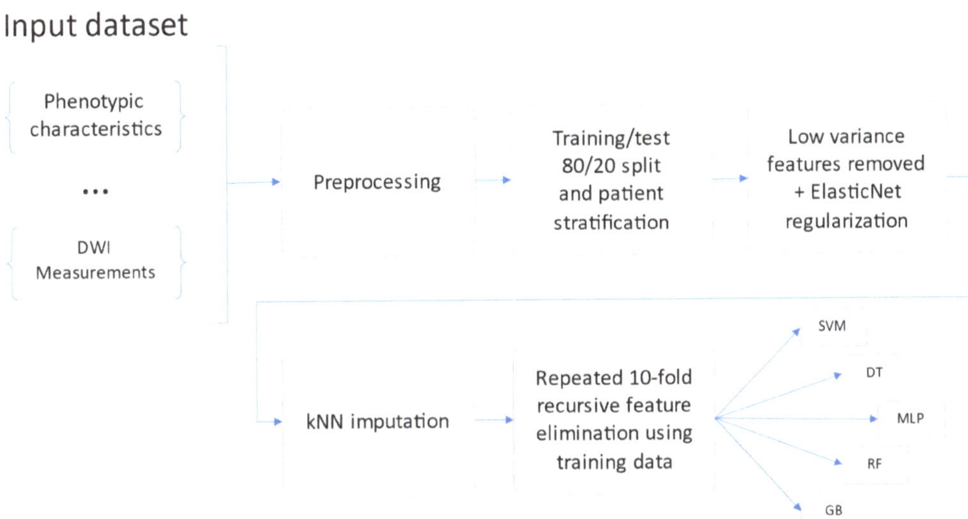

Figure 1. Processing steps of the feature selection framework and the subsequent regression of the neuropsychological and neurobiological response variables. After removing features with low variance, ElasticNet was used for feature selection.

Q3. Does baseline tract volume change with cognitive decline?

Brain atrophy is a major symptom of AD observed in vivo [48–50]. There is compelling evidence to suggest that Aβ facilitates the spread of Tau neurofibrillary tangles, which may then drive neurodegeneration, atrophy, and subsequent dementia [20]. While almost all aged brains show characteristic changes linked to neurodegeneration [51], Alzheimer's disease has different neurodegenerative processes compared with normal ageing, with distinctive neuron loss profiles [6]. This atrophy is understood to begin in the entorhinal cortex, progressing then to the hippocampus, temporal, frontal, and parietal areas, before spreading to the entire cerebral cortex [19,22].

DWI tractography enables the reconstruction of white matter tract bundles by estimating the principal directions of diffusion within a voxel, thereby enabling the segmenting of tracts and providing tract-specific measures such as volume, MD, and FA. Since DWI does not directly measure neurons themselves, but rather assesses the diffusion characteristics of water in tissue, these metrics represent water, not the neurons. We investigated the changes in tract volume, MD, and FA across different stages of cognitive impairment. Our analyses aimed to elucidate the relationship between these neuroimaging biomarkers and the progression from cognitively normal (CN) status to mild cognitive impairment (MCI) and Alzheimer's disease.

Linear mixed effects models were employed to assess the effects of diagnosis and age on tract volume, MD, and FA, accounting for random effects due to individual differences. Disease-related changes were examined by using linear mixed effects models to understand the interaction of disease stages, age, tract volume, MD, FA, and tract length. We employed the lmer function from the R package lme4 [52]. Specifically, the model was formulated as $y_{ij} = \beta_0 + \beta_1 \times x_{1ij} + \beta_2 \times x_{2ij} + u_j + \epsilon_{ij}$, where y_{ij} represents the tractography measurement (tract volume, MD, FA, and length were each considered separately) for the ith observation within the jth subject. The predictors x_{1ij}, x_{2ij} represent the fixed effects disease stage and age. The term u_j is the random intercept for subject j, assumed to follow a normal distribution $u_j \sim N(O, \sigma_u^2)$. The residual error term ϵ_{ij} is also assumed to be normally distributed with $\epsilon_{ij} \sim N(O, \sigma^2)$. This modeling approach allowed us to examine the effects of the predictors while accounting for the hierarchical structure of the data and the within-subject variability.

Q4. To what extent can tractography metrics predict cognitive decline?

Pairwise Wilcoxon Rank Sum Tests were conducted to identify significant differences in diffusion metrics across these groups, with the Benjamini–Hochberg correction [53] applied to control the false discovery rate, acknowledging the heightened risk of type I errors due to multiple comparisons. Features exhibiting a false discovery rate (FDR) corrected p-value lower than 0.05 were considered for further investigation as potential predictors in our classification models. To explore the predictive capacity of tractography metrics for cognitive decline, we segmented our dataset by current disease sub-stage as determined by physicians based on established clinical criteria. We created a cumulative distribution plot to understand the disease stages in which features exhibit significant differences. We further elucidated our understanding of which features were exhibiting significant differences by ranking the frequency of occurrence in a simple bar chart. Our expectation is that features exhibiting significant differences between disease classifications may offer predictive potential as classifiers, as we seek to develop models that indicate the potential for disease conversion towards AD.

In our total sample population of 264 participants across 434 sessions, only 14 participants converted to the next disease stage within 3 years (these participants are henceforth referred to as Converters), and only 15 converted within 10 years. Given our significant class imbalance, we divided our data into training and test subsets with a 60–40 split that sought to ensure the balance of our response variables across the training/test datasets, leaving 5 postitive classes (Converters) in our test data. We employed Adaptive Synthetic Sampling Approach for Imbalanced Learning (ADASYN), version 1.3.1, from the smotefamily package (version 1.3.1) [54] to compensate for our significant class imbalance. Subsequently, we applied a rigorous machine learning model development process involving the use of the ElasticNet grid search strategy to perform feature reduction, using conversion to the next disease stage within three years as a response variable; we varied α between 0 and 1.0 to identify the most parsimonious feature set for subsequent model development.

We employed a repeated 10-fold cross-validation strategy to ensure robustness with various models, including support vector machines, decision trees, random forest, and eXtreme gradient boosting. We used a grid search technique to identify the best classification model (Normal vs. Converter) and hyper-parameters predictive of cognitive decline and used Kappa as an evaluation metric for model comparison.

3. Results

Q1. Do phenotypic characteristics predict cognitive decline within our dataset?

In our analysis of various neuropsychological and neurobiological indicators, the Kruskal–Wallis test revealed statistically significant differences across the three Alzheimer's disease stages: cognitively normal (CN), mild cognitive impairment (MCI), and Alzheimer's disease (AD). Significant differences were observed for all of our neuropsychological indicators and many of our neurobiological indicators, as shown in Table 4.

As expected [55], the Clinical Dementia Rating-Sum of Boxes (CDR-SB) scores, a comprehensive measure of dementia severity, exhibited very high chi-squared values ($\chi^2 = 205.51,\ p < 0.001$), indicating pronounced differences across disease stages, thus supporting CDR-SB's ability to stage severity of Alzheimer dementia and mild cognitive impairment. The observation of a high chi-squared value for the CDR-SB test along with an observed long tail depicted in the density plot found in Figure 2 suggests that the MCI and AD disease stages have more variability and more extreme values than CN, signalling a potential sensitivity for early detection of conversion from CN.

Table 4. Differences in the distribution of neuropsychological and neurobiological indicator values across diagnostic categories (CN, MCI, and AD).

Indicator	CN	MCI	AD	Mean	IQR	χ^2 (Kruskal–Wallis)	*p*-adj
			Neurophysiological Indicators				
ADAS-11	139	80	40	8.9	7.2	115.39	1.05×10^{-24}
ADAS-13	139	80	38	13.9	10.3	131.18	4.58×10^{-28}
AV45	92	42	25	1.2	0.4	24.96	2.29×10^{-5}
CDR SB	140	78	41	1.4	1.5	205.51	3.78×10^{-44}
FAQ	138	75	40	3.3	3.0	149.02	6.56×10^{-32}
FDG	5	57	30	1.2	0.1	33.05	5.34×10^{-7}
MMSE	138	82	40	27.5	3.0	108.51	2.74×10^{-23}
MOCA	259	82	38	24.1	6.0	128.39	1.72×10^{-27}
RAVLT	139	80	38	40.2	20.0	115.06	1.14×10^{-24}
			Neurobiological Indicators				
Brain vol	132	76	35	1,031,622.0	140,524.0	8.49	2.86×10^{-2}
Entorhinal vol	130	77	34	3940.6	1031.0	29.08	3.39×10^{-6}
Fusiform vol	131	77	33	17,953.6	3295.0	22.52	5.16×10^{-5}
Hippocampus vol	133	76	32	7060.3	1531.6	60.00	8.41×10^{-13}
ICV	136	73	37	1,468,892.6	233,267.5	1.43	4.90×10^{-1}
Middle temporal vol	131	77	33	20,163.5	3979.0	23.70	3.56×10^{-5}
Ventricle vol	135	75	35	40,103.4	27,051.5	21.45	6.60×10^{-5}

p-adj refers to Benjamini–Hochberg-corrected *p*-values.

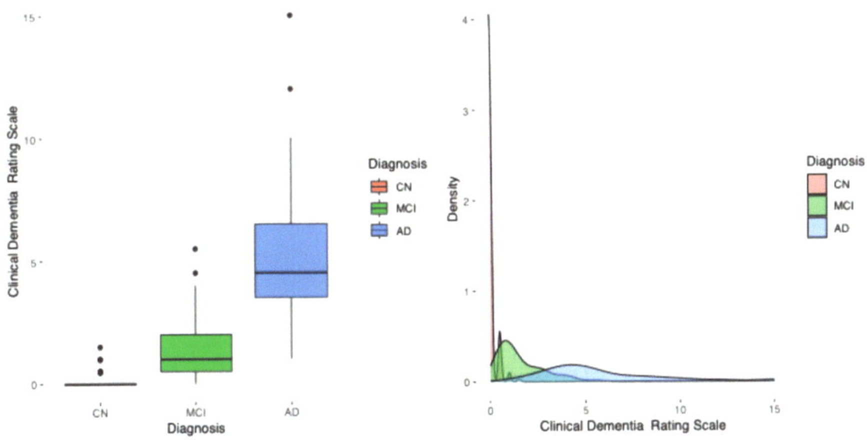

Figure 2. CDR-SB density plot exhibiting differences across diagnosis classification (n = 140 for CN, n = 78 for MCI, and n = 41 for AD). Outliers appear as points beyond the whiskers.

Other neuropsychological assessments, such as the Alzheimer's Disease Assessment Scale (ADAS-11 and ADAS-13), Mini-Mental State Examination (MMSE), and the Montreal Cognitive Assessment (MOCA), showed substantial discriminative power (ADAS-11: ($\chi^2 = 115.39$); ADAS-13: ($\chi^2 = 131.18$); MMSE: ($\chi^2 = 108.51$); MOCA: ($\chi^2 = 128.39$); all have $p < 0.001$). These results highlight the efficacy of these tests in differentiating among the stages of Alzheimer's disease.

The Rey Auditory Verbal Learning Test (RAVLT) immediate recall scores exhibited significant differences among the diagnostic groups (CN, MCI, and AD), as evidenced by the Kruskal–Wallis test ($\chi^2 = 115.06$, $p < 0.001$). Figure 3 demonstrates how the RAVLT scores progressively decrease from cognitively normal individuals to those with MCI and further to individuals with AD.

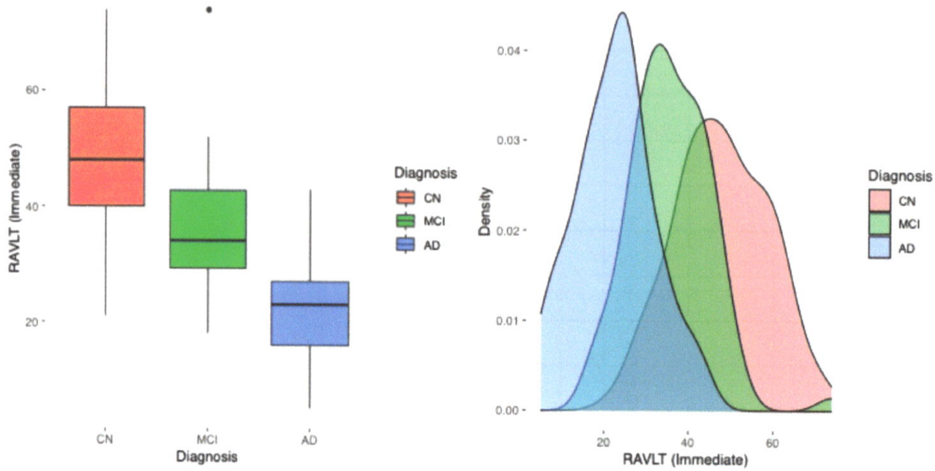

Figure 3. RAVLT density plot exhibiting differences across diagnosis classification (n = 139 for CN, n = 80 for MCI, and n = 38 for AD). Outliers appear as points beyond the whiskers.

In terms of neurobiological markers, volumes of key brain regions, such as the hippocampus, entorhinal cortex, fusiform gyrus, and the middle temporal gyrus, showed significant differences across the disease stages (hippocampus: $\chi^2 = 60.00$, $p < 0.001$; entorhinal: $\chi^2 = 29.08$, $p < 0.001$; fusiform: $\chi^2 = 22.52$, $p < 0.001$; middle temporal: $\chi^2 = 23.70$, $p < 0.001$).

Q2. Are WM tract features predictive of neuropsychological and neurobiological indicators?

The feature reduction strategy employed by this study substantially streamlined the initial high-dimensional feature space of DWI tractography measurements to a manageable subset. From an initial 1.3 million features, we identified high variance features to reduce our dataset to 134,340 explanatory variables that were imputed and further refined by using ElasticNet regularization. The ElasticNet model's α parameter, which balances L2 and L2 penalties, was optimized independently for each response variable, with values ranging from 0.5 to 0.8. This approach aimed to balance feature retention and model complexity. The final feature count for each response variable is depicted in Table 5.

Table 5. Feature reduction profile of DWI tractography data.

Response	Features after Elasticnet Regularization		
	α=0.5	α=0.75	α=0.8
ADAS-11	507	295	352
ADAS-13	175	203	164
CDR-SB	353	155	40
FAQ	77	85	101
FDG	129	45	27
MMSE	12	6	9
MOCA	36	89	96
RAVLT	181	217	197
Entorhinal Vol	69	19	94
Fusiform Vol	360	267	392
Hippocampus Vol	521	319	128
Mid Temp Vol	154	202	168

We used repeated 10-fold cross-validation (10 iterations) to train our models for each of the neuropsychological and neurobiological indicators identified (Table 4) as being discriminative of diagnosis classification. Despite the rigorous model development process, our predictive models showed varying levels of predictive performance across different indicators (Table 6). For AV45, an imaging biomarker for amyloid plaque accumulation (mean: 1.2; IQR: 0.4), the model yielded error metrics (RMSE: 0.2; MAE: 0.15) that were lower than the sample mean and IQR. Hippocampal volume (mean: 7060.3; IQR: 1531.6) predictions resulted in error metrics of RMSE of 885.20 and MAE of 702.23, and the model for MoCA scores (mean: 24.1; IQR: 6.0) produced error metrics of RMSE of 4.09 and MAE of 3.12, suggesting a 13% error on average in assessing a patient's MoCA scores based on DWI MRI analysis alone. Most of our models had R^2 values below 0.4, indicating limited explanatory power. The model for FDG, a marker for glucose metabolism, captured the most variability in the data, with an R^2 of 0.66.

Table 6. Best prediction models of neuropsychological and neurobiological indicators.

Indicator	Mean	IQR	Model	α	RMSE	MAE	R^2
Neuropsychological Indicators							
ADAS-11	8.9	7.2	Random forest	0.75	5.91	4.22	0.18
ADAS-13	13.9	10.3	Random forest	0.5	8.25	6.20	0.30
AV45	1.2	0.4	Decision trees	0.5	0.20	0.15	0.15
CDR-SB	1.4	1.5	SVM Radial	0.8	2.32	1.61	0.19
FAQ	3.3	3.0	Random forest	0.8	3.53	2.76	0.66
FDG	1.2	0.1	Multilayer Perceptron	0.5	0.09	0.07	0.08
MMSE	27.5	3.0	Linear	0.8	3.72	2.83	0.12
MOCA	24.1	6.0	Gradient boosting	0.75	4.09	3.12	0.25
RAVLT	40.2	20.0	Decision trees	0.5	11.9	9.70	0.29
Neurobiological Indicators							
Entorhinal Vol	3940.6	1031.0	Decision trees	0.8	864.96	659.14	0.11
Fusiform Vol	17,953.6	3295.0	SVM Radial	0.8	2352.72	1841.73	0.17
Hippocampus Vol	7060.3	1531.6	Random forest	0.8	885.20	702.23	0.46
Mid Temporal Vol	26,163.5	3979.0	Gradient boosting	0.75	2622.42	2151.96	0.23

While we found that there is lack of comprehensive analysis using DWI-only metrics to predict these indictors, there is alignment with several studies.

A study by Patil et. al. [56] found that no strong correlation was observed for any DWI measurements in any region with respect to MMSE. This was supported by Jokinen et al. [57], who determined that white matter ADC was not predictive of poor cognitive outcomes.

Correlational analysis is the most common approach to presenting the association between DWI and clinical scores. A recent study by Saito et al. [58] consistently reported low correlations between DWI and many of the indicators we evaluated in this study.

Q3. Does baseline WM tract volume change with cognitive decline?

Our findings (Table 7) show significant decreases in both MD and FA across both the MCI and AD stages compared with CN, even after adjusting for age, with tract volume analyses revealing significant increases with the progression of cognitive decline towards the later stages of the disease (AD), after adjusting for age. There was not a significant difference in tract volume between the CN and MCI stages.

Table 7. Estimated effects of diagnosis and age on MD, FA, and tract volume using linear mixed effects models.

Variable	Estimate	Std. Error	df	t-Value	p-adj
		Mean diffusivity			
Intercept	1.41×10^{-3}	3.39×10^{-5}	2.83×10^{2}	41.55	2.08×10^{-122}
Diagnosis (MCI)	-3.36×10^{-5}	8.16×10^{-7}	1.12×10^{7}	−41.19	2.00×10^{-16}
Diagnosis (AD)	-3.90×10^{-5}	1.29×10^{-6}	1.12×10^{7}	−30.17	8.52×10^{-200}
Age	-6.13×10^{-6}	9.42×10^{-8}	1.10×10^{7}	−65.05	2.00×10^{-16}
		Fractional anisotropy			
Intercept	3.94×10^{-1}	2.79×10^{-3}	1.30×10^{3}	141.16	2.00×10^{-16}
Diagnosis (MCI)	-2.64×10^{-3}	2.44×10^{-4}	7.20×10^{6}	−10.80	6.75×10^{-27}
Diagnosis (AD)	-1.97×10^{-3}	3.86×10^{-4}	5.35×10^{6}	−5.08	3.69×10^{-7}
Age	-2.93×10^{-4}	2.80×10^{-5}	1.05×10^{6}	−10.47	1.60×10^{-25}
		Tract volume			
Intercept	2.73×10^{3}	2.19×10^{2}	5.69×10^{3}	12.48	1.07×10^{-34}
Diagnosis (MCI)	2.78×10^{1}	2.38×10^{1}	1.00×10^{6}	1.17	2.43×10^{-1}
Diagnosis (AD)	-1.22×10^{2}	3.76×10^{1}	5.35×10^{5}	−3.24	1.60×10^{-3}
Age	2.51×10^{1}	0.26×10^{1}	6.76×10^{4}	9.37	1.57×10^{-20}
		Tract length			
Intercept	1.62×10^{1}	0.19×10^{1}	1.78×10^{3}	8.41	8.36×10^{-17}
Diagnosis (MCI)	0.20×10^{1}	1.79×10^{-1}	5.54×10^{6}	10.87	2.27×10^{-27}
Diagnosis (AD)	-0.36×10^{1}	2.83×10^{-1}	3.72×10^{6}	−12.98	3.23×10^{-38}
Age	6.35×10^{-1}	2.05×10^{-2}	6.01×10^{5}	30.97	27.21×10^{-210}

p-adj: linear mixed effects model *p*-values were estimated based on Satterthwaite's approximation, and subsequently FDR-corrected.

The MCI group is associated with an increase in mean tract length compared with the CN group, holding age constant.

Q4. To what extent can tractography metrics predict cognitive decline?

Among the 1.3 million tract measurements assessed, 5394 tract measurements (0.3%) exhibited statistically significant differences among groups after performing pairwise Wilcoxon Rank Sum Tests with Benjamini–Hochberg correction [53] ($p < 0.05$). In the cumulative distribution shown in Figure 4, we note the curve deviation of the statistics from the null distribution increases modestly as participants transition into later stages of cognitive impairment, suggesting tract anomalies may be more pronounced in later stages of the disease. The curve deviations of the comparison of mild cognitive impairment (MCI) to late mild cognitive impairment (LMCI) (blue) are larger than other groups, suggesting potentially higher effect sizes [59].

Figure 5 presents a frequency plot of measurement types identified as significant during the transition across the MCI → LMCI → AD cognitive impairment stages. The plot ranks measurement types by their frequency of occurrence, emphasizing which measurements are most prevalent in highlighting tract anomalies associated with cognitive decline. The higher-frequency measurements relate to the morphometrics of detected tracts (length and volume) as well as differences in MD as expected [60].

We employed four machine learning algorithms, including support vector machine, random forest, XGBoost, and MARS, to predict progression from a normal disease stage to a "Converter" status, indicative of advancement to a more severe disease stage within three years of imaging. These models were evaluated by using a repeated 10-fold cross-validation technique to ensure the reliability and stability of our predictions.

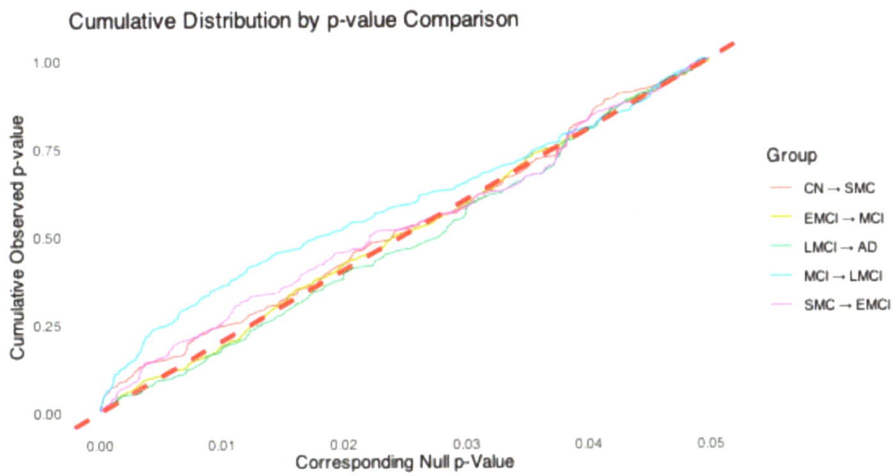

Figure 4. Distribution of tract measurement anomalies across cognitive impairment stages. This figure illustrates the deviation of tract measurement statistics from the null distribution across different stages of cognitive impairment. The curve deviations increase during transitions to later stages of cognitive impairment. The comparisons between mild cognitive impairment (MCI) and late mild cognitive impairment (LMCI) (shown in blue) exhibit larger deviations compared with other groups, suggesting more pronounced tract anomalies in these later stages of the disease.

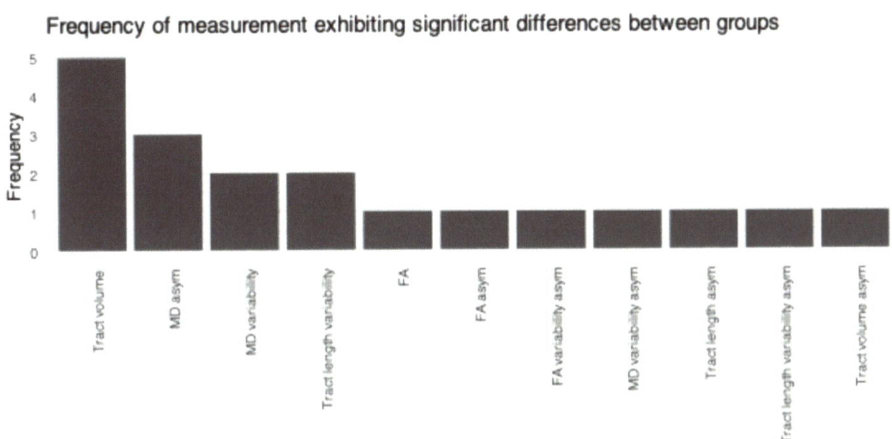

Figure 5. The frequency of measurement type for those tracts that exhibited significant differences among groups suggests that tracts may deteriorate quickly. Tract volume dominates the anomalies detected and is more likely to characterize differences between late mild cognitive impairment and Alzheimer's disease.

All four models exhibited nearly identical performance metrics across the evaluation scheme. The accuracy for each model was observed to be between 0.4836 and 0.4985, with a 95% confidence interval ranging from 42.89% to 55.33%.

3.1. Comparison with Similar Studies

Despite many studies claiming that diffusion metrics offer potential for prognostic biomarkers of AD (Table 8), many of these studies highlight significant effect sizes but do not actually attempt prediction. Most studies that classify current disease stages involve the consolidation of clinical indicators, health record data, and data from multiple imaging modalities to achieve maximum accuracy. Our analysis has determined that clinical variables alone are sufficient to achieve an averaged balanced accuracy of 88%, a specificity of 92%, and a sensitivity of 84% with our current dataset. The addition of individual tract-specific diffusion data contributed very little to our models (accuracy of 88%, specificity of 93%, and sensitivity of 85%). When tract-specific measurements are used in isolation, DWI data appear to offer relatively weak performance when classifying disease stages of AD.

Table 8. Comparative analysis of our proposed model with other models applied to the ADNI dataset.

Authors	Highlights	Participants	Performance
Classification of current disease stage			
Chen (2023) [61]	The study investigated white matter alterations in the Alzheimer's continuum by using diffusion tensor imaging, finding widespread changes correlated with Tau pathology, particularly in the cingulum, which may serve as a promising biomarker for preclinical Alzheimer's disease.	236 ADNI3 subjects (176 CN, 36 MCI, and 24 AD)	74% Acc, 69% AUC, 58% Sens, and 78% Spec
Chen (2023) [62]	A model that enhances multi-modal AD diagnosis by using orthogonal latent space learning, feature weighting, and graph learning to improve discriminative information retention and relationship encoding among samples.	757 ADNI2 subjects (283 CN, 330 MCI, and 144)	67% Acc, 69% Sens, 64% Spe, and 71% AUC
Deng (2023) [63]	The Fully Connected Multi-Kernel Convolutional Neural Network model accurately diagnoses Alzheimer's disease and mild cognitive impairment from diffusion tensor imaging (DTI) data while also generating fiber probability maps to assist in clinical diagnosis.	413 subjects (162 CN, 130 MCI, and 121 AD)	96% Acc, 97% Sens, 100% Spec, and 98% Auc
Khan (2022) [64]	Developed a 3-tiered cognitive hybrid machine learning algorithm for disease prediction.	818 ADNI1 subjects (229 CN, 396 MCI, and 193 AD)	95% Acc, 95% Sens 97% Spe, and 99% Auc
Razzak (2022) [65]	Proposes an integrative deep ensemble learning framework called PartialNet, tailored for Alzheimer's detection using brain MRIs, demonstrating improved predictive performance and efficiency compared with DenseNet, with notable gains in both multiclass and binary class AD detection on benchmark datasets.	350 subjects (95 AD, 146 MCI, and 95 CN)	98% Acc (mean of CN, MCI, and AD)
Hazarika (2022) [66]	The study discusses various deep learning models for Alzheimer's disease classification, highlighting DenseNet-121's strong performance and computational inefficiency, and proposes a modified DenseNet-121 with depth-wise convolutions.	210 (70 CN, 70 MCI, and 70 AD)	98% Acc (mean of CN, MCI, and AD)
Prediction of future conversion			
Stone (2021) [67]	This study identified diffusivity measures from specific white matter tracts, particularly axial diffusivity, by using only DTI data.	87 subjects: 34 Converted and 53 Not converted	72% Acc and 67% AUC
Velazquez (2022) [68]	Prediction of conversion from mild cognitive impairment (MCI) to AD using DTI data with clinical variables from health records.	384 subjects: 49 Converted and 335 Not converted	98% Acc and AUC 99%

3.2. Adaptions to Methodology

After reflecting on the results obtained while answering Questions 1–4 in our research methodology, it was important to consider alternative mechanisms for understanding the probability with which a participant will exhibit further disease progression. We found limited evidence that tractography metrics could be useful to predict established neuropsychological and neurobiological biomarkers (Table 6), and we found considerable evidence by using linear mixed effects modeling that tractography metrics exhibit an effect on disease stages while accounting for age (Table 7).

Given the high dimensionality of our data and probable loss of meaning resulting from aggressive feature selection, we considered an approach whereby we condensed our features into aggregate measures at the seed level to provide a representative feature as an alternative to feature reduction strategies. Formula (1) represents the Z-score for a given region measurement of a participant's session.

$$Z_{ijkl} = \frac{X_{ijkl} - \mu_k}{\sigma_k} \quad (1)$$

where x_{ijkl} is the kth measurement (e.g., those identified in Figure 5) for the ith participant in the jth session from the lth seed region targeting a specific region. μ_{kl} is the mean of the kth measurement, and σ_k represents the standard deviation for the kth measurement.

After scaling the measurements, we determined the mean Z-score for each participant, session, and ROI (Formula (2)). A value at this aggregated level gives insights into how a participant's ROI may be different from the same region in other participants. For example, mean tract length of all tracts connected to the paracentral cortex for a given participant's session.

$$\overline{Z}_{ijk} = \frac{1}{M}\sum_{l=1}^{M} Z_{ijkl} \quad (2)$$

With a more condensed set of features to move forward with, we continued to rely on an ElasticNet grid search to perform feature selection, resulting in a reduction from 2994 features to 2, relying on an α value of 0.6. These two features identified were the features representing the hemispheric lateralization of mean tract volume for tracts originating from the supramarginal and paracentral regions.

Relying on a dataset with only two features representing these diffusion-derived anatomical measurements, we applied several machine learning models to predict the classification of individuals into two groups: Normal (no conversion) and those who would convert to Alzheimer's within three years. The models tested included eXtreme gradient boosting (XGB), multivariate adaptive regression splines (MARS), support vector machine (SVM), and random forest (RF), with results presented in Table 9.

Table 9. Best prediction models of Normal vs. Converter within 3 years.

Model	AUROC	AUPRC	F1	Accuracy	Sensitivity	Specificity	Kappa	*p*-adj
RF [1]	0.74	0.75	0.64	0.71	0.52	0.90	0.42	1.11×10^{-14}
XGBTree [2]	0.77	0.72	0.57	0.67	0.44	0.90	0.34	3.78×10^{-10}
MARS [3]	0.64	0.56	0.50	0.61	0.48	0.74	0.26	7.20×10^{-5}
SVM [4]	0.54	0.50	0.24	0.50	0.16	0.84	−0.00	0.05×10^{1}

[1] Random forest mtry = c(2, floor(sqrt(num_features)), floor(num_features/3)) [2] eXtreme gradient boosting nrounds = seq(from = 25, to = 100, by = 25); max_depth = seq(from = 5, to = 35, by = 10); eta = seq (from = 0.2, to = 1, by = 0.2); gamma = seq(from = 1, to = 10, by = 1); colsample_bytree = seq(from = 0.6, to = 1, by = 0.2); min_child_weight = seq(from = 2, to = 5, by = 1); subsample = 1 [3] Multivariate adaptive regression splines; degree = seq(from = 1, to = 3, by = 1); nprune = seq(from = 1, to = 10, by = 1) [4] Support vector machine; sigma = (0.001,0.01,0.1,1,10,100); C = (0.001,0.01,0.1,1,10,100).

Overall, the random forest model performed the best, achieving an Area Under the Receiver Operating Characteristic (AUROC) of 0.74, indicating a good ability to differentiate between "Converter" and "Normal" classes. The Area Under the Precision–Recall Curve (AUPRC) was 0.75, reflecting a strong performance in capturing the "Converter" class, which is particularly important given the class imbalance of our original data (see Figure 6). The F1-score was 0.64, providing a harmonic mean of precision and recall, and the overall accuracy of predicting the correct class was 71%. The model's ability to detect "Converters" (true positive) has room for improvement, with a sensitivity of 0.52. The specificity was much better, 0.90, suggesting a strong ability to identify Normal subjects. This was expected given the large class imbalance. The model exhibited a Cohen's Kappa of 0.42, indicating that there is moderate agreement between the predicted classifications and the actual classifications. These results are consistent with other studies who attempted to predict future disease conversion by using DWI-only data (Table 8).

To enhance the predictive power of this model, we integrated these two diffusion metrics with traditional clinical variables, including MMSE, MoCA, RAVLT, CDR-SB, FAQ, hippocampal volume, entorhinal volume, and Aβ and Tau indicators. This hybrid model achieved a significant improvement, yielding an accuracy of 86%, a sensitivity of 86%, and a specificity of 93%. These results surpass the diagnostic performance of current clinical assessments, where the sensitivity ranges from 70.9% to 87.3% and the specificity from 44.3% to 70.8% [69]. Our findings emphasize the value of a hybrid machine learning approach that combines advanced neuroimaging techniques with conventional clinical assessments.

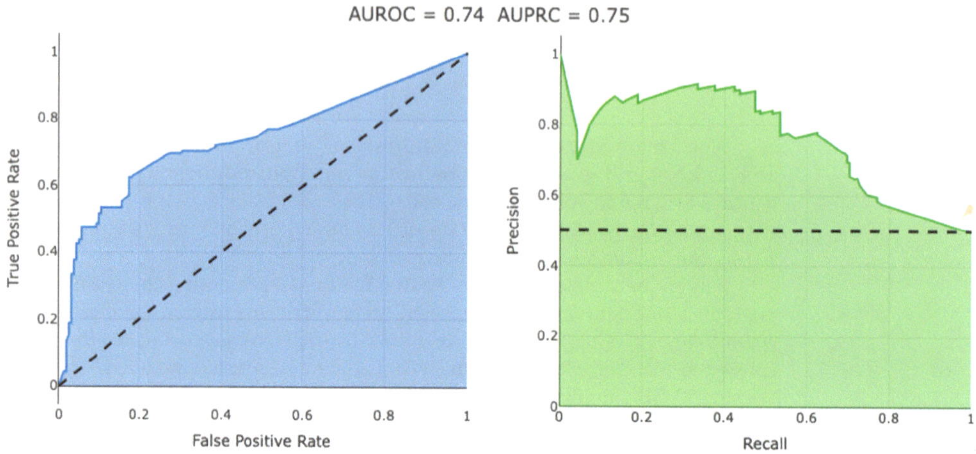

Figure 6. Receiver operating characteristic plot and corresponding precision recall plot for eXtreme gradient boosting model predicting subjects who exhibit worsening disease stages within three years (Converters).

4. Discussion

Alzheimer's disease is a complex and widespread [20] neurodegenerative disease that manifests itself in multiple ways [6,70], including atrophy of the hippocampus, entorhinal region, and middle temporal regions, as well as accumulation of Aβ and Tau proteins. Tractography measurements derived from diffusion-weighted images appear to show limited potential as a single scan test capable of offering predictions for many of the traditional neuropsychlogical and neurobiological assessment metrics used today; however, they may still contribute as an important tool to a comprehensive approach to understanding the complexity of AD and characterization of disease staging for some of these clinical variables.

Assessing and diagnosing Alzheimer's disease remain complex and challenging due to the lack of a complete model that can identify the disease in any stage. Currently, diagnosis often relies heavily on subjective assessments derived from neuropsychological and neurobiological tests carried out in primary care settings. These tests, while valuable, can be influenced by various factors, such as the examiner's expertise, the time of day when the test is administered, the testing environment, and the patient's physical and emotional state at the time of testing [71].

The inherent variability and subjectivity in these evaluations can lead to inconsistent diagnoses, particularly in the early or preclinical stages of Alzheimer's disease, where symptoms may be subtle or overlap with other conditions, such as ageing. Clinic pathological studies have shown that the diagnostic sensitivity of clinicians is between 70.9% and 87.3% and the specificity is between 44.3% and 70.8% [69]. Additionally, traditional diagnostic methods may fail to capture the full spectrum of neuropathological changes associated with Alzheimer's, limiting their effectiveness in early detection and intervention. In light of this, the aim of our study was to elucidate the contribution that diffusion-weighted imaging can make to improving model development towards the early detection of Alzheimer's. Machine learning models hold significant promise for improving the evaluation of Alzheimer's disease by offering more objective, accurate, and scalable diagnostic tools. Machine learning can uncover subtle and complex relationships within the data that may not be apparent through traditional methods, potentially leading to earlier and more accurate diagnoses, a better monitoring of disease progression, and personalized treatment plans.

4.1. Novel Contributions

Our study identified the hemispheric lateralization of tract volumes connected to the supramarginal gyrus and paracentral regions as a potential prognostic biomarker of AD disease. The supramarginal gyrus is part of the parietal lobe and plays a role in language perception and processing [72], spatial orientation and tool use [73], emotion recognition [74], writing and word recognition [75], and the integration of sensory information [76]. This region's association with AD and dementia has been reported in the literature [77–80], typically in later stages of the disease. In the most recent study referenced, the authors used magnetoencephalography to reveal that decreased beta-band intensity in the left supramarginal gyrus is associated with decreased neuropsychological assessment scores and increased clinical severity of cognitive impairment, suggesting its importance in assessing cognitive status. Our findings support this observation by providing complimentary evidence that changes in volume asymmetries of tracts connecting the supramarginal gyrus may be associated with cognitive impairment and dementia. The deterioration of white matter tracts connecting the supramarginal gyrus may lead to reduced efficiency of the default mode network, to which the supramarginal gyrus belongs.

While our final model for identifying "Converters" leaves some room for improvement, overall, the model demonstrates commendable ability in distinguishing Converter from Normal subjects, particularly by achieving good AUROC and AUPRC scores based on only two features. However, the moderate sensitivity and low specificity suggest that there is more work to do in terms of enhanced feature selection, alternative model development, or the fine tuning of the current models. This analysis could be expanded by integrating DWI and resting state functional MRI data around the regions implicated in our model. Existing resting state studies indicate that significant differences in signal intensity exist for the same regions our models use [81,82]. Graph theoretical metrics could also be included in this analysis to understand if changes in nodal efficiency across different disease stages could also strengthen our model. There do appear to be imaging data available for the ADNI3 cohort that calculate the network failure quotient from resting state functional MRI images, which may encompass these two potential enhancements to our analysis.

In addition, the application of proven machine learning algorithms with a consolidated dataset of whole-brain tractography, phenotypic, and neuropsychological data for early biomarker identification in Alzheimer's disease (AD) represents a thorough and integrative approach. While tractography-focused predictive analytics has been widely used in neuroscience research, comprehensive whole-brain analyses on longitudinal data remain relatively rare. Our study reinforces the utility of this approach by demonstrating its application in a comprehensive longitudinal dataset. Utilizing well-established machine learning techniques in combination with exhaustive tractography and neuropsychological data provides a robust methodology for investigating early biomarkers of AD. Our findings add to the existing body of evidence elucidating the potential of diffusion MRI as a tool for the early detection and monitoring of neurodegenerative diseases, highlighting the importance of integrating multi-modal data for enhanced predictive analytics.

Our findings confirm the significant predictive value of existing neurobiological and neuropsychlogical biomarkers in detecting Alzheimer's disease. The biomarkers identified in Table 8 demonstrate that while some biomarkers may be more effective in different stages of the disease, they collectively provide a robust toolbox for disease detection. The RAVLT is of particular interest, given that it is one of the earliest indicators of conversion from CN to MCI [8]. Our findings highlight the pronounced impact of Alzheimer's disease on memory function. Notably, the RAVLT is recognized as a critical diagnostic tool, particularly due to its sensitivity in detecting early memory deficits that often signify the transition from mild cognitive impairment (MCI) to Alzheimer's disease (AD) [8]. Furthermore, Figure 3 provides a visual depiction of these differences, illustrating a clear distinction among the CN, MCI, and AD groups. This separation is indicative of the progressive nature of memory impairment in AD pathology.

Our analysis of white matter tract features revealed moderate but notable associations with neurobiological and neuropsychological markers. Our results based on predicting cognitive test scores indicate some potential for relying on DWI-based MRI to non-subjectively assess cognitive progression in AD. The results demonstrate an MAE of 3.12 for the MoCA, a test which is on a scale of 0 to 30, implying a 13% error on average in assessing a patient's cognitive outcomes based on DWI MRI analysis alone. This is an interesting finding, which implies that one day, we may be able to create predictive technologies informed by MRI that may be able to accurately predict a patient's cognitive test scores. Many patients with AD despise taking cognitive tests, implying that technologies developed to monitor their disease progression may be a welcome development in AD patient management.

A significant finding of our study is the relationship between tract characteristics and cognitive decline (Table 7). Our results suggest that the brain might be undergoing specific microstructural changes that both restrict diffusion (lower MD) and disrupt the coherence of white matter tracts (lower FA), while at the same time, the increases in tract volume during later disease stages might reflect underlying processes, such as the cellular proliferation of astrocytes [70,83], or changes in the extracellular matrix [84], potentially confounding volume measurement, although further studies are needed to explore this. Conceivably, an increase in volume may be symptomatic of inflammation leading to edema, which could increase the extracellular space. A corresponding increase in MD would have supported this hypothesis [85]; however, that was not observed in our data.

Our results highlight that changes in tract length may offer a useful biomarker for disease staging. A significant increase ($p = 2.266 \times 10^{-27}$) suggests that on average, MCI diagnosis is associated with a longer tract length than observed in CN participants, whereas the AD group is associated with a decrease in a mean tract length while holding age constant, which is consistent with expectations of neurodegeneration leading to tract deterioration. There is support in the existing literature of increased tract length with age [86], and future research may explore the potential that early or mild stages of cognitive decline could trigger compensatory mechanisms [87] in the brain, potentially leading to an increase in tract length as the brain attempts to maintain connectivity.

Our attempts to leverage tract-specific measurements from diffusion-weighted images that were correlated with existing neurobiological and neuropsychlogical biomarkers as a means to identify cognitive impairment were initially unsuccessful. Our model accuracy aligned with the No Information Rating, indicating that our models' predictions were no better than random chance. Further, the Cohen's Kappa statistic for each model was 0, reflecting the absence of agreement beyond chance between the predicted outcomes and the actual disease progression status.

There are several practical implications from these findings. The identification of tract volume asymmetries in the supramarginal gyrus and paracentral regions offers a nascent but promising potential prognostic biomarker as a non-invasive method for the detection and monitoring of disease progression. If sufficiently advanced, this approach could be integrated into routine clinical practice, providing clinicians with a valuable tool to assess disease progression and inform treatment plans.

4.2. Limitations

Limited sample sizes in neuroimaging studies can compromise the reliability and validity of the findings reported [88]. Small sample sizes reduce the statistical power, increasing the likelihood of Type 1 (false positive) and Type 2 (false negative) errors. While the original ADNI-3 cohort is larger than many studies (960 subjects across 6050 scans), many subjects lacked diffusion-weighted imaging data. Our study included 264 participants with a limited number of scan sessions per participant (between one and five).

We acknowledge the potential source of error resulting from the smoothing effects of interpolation as a result of registering DWI images to T1. Our pipeline strategy was initially developed for a large dataset of noisy clinical data [26,29], where it was determined after many approaches that registration of DWI to T1-weighted images before ODF reconstruction offered the most reliable alignment approach with the highest number of successful registrations. It was felt that lower rates of successful registration were highly undesirable and could potentially skew the results of analyses more so than the error associated with the simple smoothing that results from the interpolation process.

We relied on single-shell DWI, where diffusion measurements are acquired with a single b-value. Single-shell DWI has been shown to underperform in resolving complex fiber configurations within a voxel compared with multi-shell DWI [89]. The limitations of single-shell DWI mean that our study might not fully capture the complexity of white matter architecture, particularly in regions where multiple fiber pathways intersect [90]. This can potentially lead to the mischaracterization of fiber tract integrity and connectivity, especially with respect to the tract length measurement we considered in our study. Consequently, our findings regarding tract length might be less reliable than if we had access to more sensitive multi-shell images that offer opportunities for improved delineation of crossing fibers.

Given the breadth of our data, we relied on aggressive feature reduction strategies to make machine learning feasible. This included only considering the features with the highest variability (top 10%) after removing features with low variance or highly correlated redundant features. This may have resulted in the exclusion of potentially important data that were never introduced during model training. As a consequence, there may be significant characteristics within our data that could have enhanced model performance but were excluded early in the process.

The results of our adapted machine learning strategy are promising, though they offer room for improvement. While it is encouraging to achieve this classification accuracy with only two measures based on the hemispheric lateralization of mean tract volume for tracts originating from the supramarginal and paracentral regions, there are specific limitations that should be addressed. Our highly imbalanced proportion of Converter to Normal participants (14/434) necessitated a reliance on synthetic data to better balance for reliable predictions. This deficiency is likely a contributing factor to our low sensitivity scores, as synthetic data do not perfectly capture the complexity and variability of real-world data. The limited sample size restricts the statistical power and generalizability of our findings. More data would enhance the model training process of our ML models, allowing for better feature learning and reducing the risk of overfitting. In particular, increasing the number of individuals who exhibit progressive disease pathology would help provide a more balanced dataset.

4.3. Conclusions

Overall, our results align with the existing literature on the neurodegenerative patterns characteristic of Alzheimer's disease [5,91]. The observed microstructural changes and their impact on cognitive function highlight the importance of integrating advanced neuroimaging techniques with traditional neuropsychological assessments. Future research should focus on refining these predictive models, exploring additional biomarkers, and validating our findings in larger, more balanced cohorts to enhance the robustness and generalizability of our conclusions.

Author Contributions: Conceptualization, D.M. and J.L.; methodology, D.M.; software, D.M.; validation, D.M., J.L., and L.P.-C.; formal analysis, D.M.; investigation, D.M.; resources, J.L.; data curation, D.M.; writing—original draft preparation, D.M.; writing—review and editing, J.L., L.P.-C., and E.T.; visualization, D.M.; supervision, J.L. and L.P.-C.; funding acquisition, J.L. All authors have read and agreed to the published version of the manuscript.

Funding: This study was financially supported by a Canada Foundation for Innovation grant, a Nova Scotia Research and Innovation Trust grant, a St. Francis Xavier University research startup grant to J.L., and a Compute Canada Resource Allocation grant to J.L. Data collection and sharing for the Alzheimer's Disease Neuroimaging Initiative (ADNI) are funded by the National Institute on Aging (National Institutes of Health grant U19 AG024904). The grantee organization is the Northern California Institute for Research and Education. In the past, the ADNI has also received funding from the National Institute of Biomedical Imaging and Bioengineering, the Canadian Institutes of

Health Research, and private sector contributions through the Foundation for the National Institutes of Health (FNIH), including generous contributions from the following: AbbVie, Alzheimer's Association; Alzheimer's Drug Discovery Foundation; Araclon Biotech; BioClinica, Inc.; Biogen Bristol-Myers Squibb Company; CereSpir, Inc.; Cogstate; Eisai, Inc.; Elan Pharmaceuticals, Inc.; Eli Lilly and Company; EuroImmun; F. Hoffmann-La Roche, Ltd., and its affiliated company Genentech, Inc.; Fujirebio; GE Healthcare; IXICO, Ltd.; Janssen Alzheimer Immunotherapy Research & Development, LLC.; Johnson & Johnson Pharmaceutical Research & Development, LLC.; Lumosity; Lundbeck; Merck & Co., Inc.; Meso Scale Diagnostics, LLC.; NeuroRx Research; Neurotrack Technologies; Novartis Pharmaceuticals Corporation; Pfizer, Inc.; Piramal Imaging; Servier; Takeda Pharmaceutical Company; and Transition Therapeutics.

Institutional Review Board Statement: Not applicable.

Informed Consent Statement: Not applicable.

Data Availability Statement: The data used in the preparation of this article were obtained from the Alzheimer's Disease Neuroimaging Initiative (ADNI) database (adni.loni.usc.edu). The ADNI was launched in 2003 as a public–private partnership, led by Principal Investigator Michael W Weiner, MD. The primary goal of ADNI is to test whether serial magnetic resonance imaging (MRI), positron emission tomography (PET), other biological markers, and clinical and neuropsychological assessment can be combined to measure the progression of mild cognitive impairment (MCI) and early Alzheimer's disease (AD). For up-to-date information, see www.adni-info.org. The code used for data preprocessing, analysis, and figure generation in this study is available on GitHub (https://github.com/dmattie/pacs-adni-eab, accessed on 1 August 2024) and archived on Zenodo [92].

Conflicts of Interest: J. Levman is the founder of Time Will Tell Technologies, Inc. The authors declare no competing financial interests.

References

1. Alzheimer's Disease Facts and Figures. Alzheimer's Association. Available online: https://www.alz.org/media/Documents/alzheimers-facts-and-figures.pdf (accessed on 8 August 2023).
2. Rajan, K.B.; Weuve, J.; Barnes, L.L.; McAninch, E.A.; Wilson, R.S.; Evans, D.A. Population estimate of people with clinical Alzheimer's disease and mild cognitive impairment in the United States (2020–2060). *Alzheimer's Dement.* **2021**, *17*, 1966–1975. [CrossRef] [PubMed]
3. Ashton, N.J.; Schöll, M.; Heurling, K.; Gkanatsiou, E.; Portelius, E.; Höglund, K.; Brinkmalm, G.; Hye, A.; Blennow, K.; Zetterberg, H. Update on biomarkers for amyloid pathology in Alzheimer's disease. *Biomark. Med.* **2018**, *12*, 799–812. [CrossRef] [PubMed]
4. Masters, C.L.; Bateman, R.; Blennow, K.; Rowe, C.C.; Sperling, R.A.; Cummings, J.L. Alzheimer's disease. *Nat. Rev. Dis. Prim.* **2015**, *1*, 15056. [CrossRef] [PubMed]
5. Blennow, K.; de Leon, M.J.; Zetterberg, H. Alzheimer's disease. *Lancet* **2006**, *368*, 387–403. [CrossRef] [PubMed]
6. Ferreira, D.; Verhagen, C.; Hernández-Cabrera, J.A.; Cavallin, L.; Guo, C.J.; Ekman, U.; Muehlboeck, J.-S.; Simmons, A.; Barroso, J.; Wahlund, L.-O.; et al. Distinct subtypes of Alzheimer's disease based on patterns of brain atrophy: Longitudinal trajectories and clinical applications. *Sci. Rep.* **2017**, *7*, 46263. [CrossRef] [PubMed]
7. Bronzuoli, M.R.; Iacomino, A.; Steardo, L.; Scuderi, C. Targeting neuroinflammation in Alzheimer's disease. *J. Inflamm. Res.* **2016**, *9*, 199–208. [CrossRef] [PubMed]
8. Jedynak, B.M.; Lang, A.; Liu, B.; Katz, E.; Zhang, Y.; Wyman, B.T.; Raunig, D.; Jedynak, C.P.; Caffo, B.; Prince, J.L. A computational neurodegenerative disease progression score: Method and results with the Alzheimer's disease neuroimaging initiative cohort. *NeuroImage* **2012**, *63*, 1478–1486. [CrossRef] [PubMed]
9. DeKosky, S.T.; Marek, K. Looking Backward to Move Forward: Early Detection of Neurodegenerative Disorders. *Science* **2003**, *302*, 830–834. [CrossRef]
10. Peraza, L.R.; Díaz-Parra, A.; Kennion, O.; Moratal, D.; Taylor, J.; Kaiser, M.; Bauer, R.; Initiative, A.D.N. Structural connectivity centrality changes mark the path toward Alzheimer's disease. *Alzheimer's Dement. Diagn. Assess. Dis. Monit.* **2019**, *11*, 98–107. [CrossRef]
11. Lombardi, G.; Crescioli, G.; Cavedo, E.; Lucenteforte, E.; Casazza, G.; Bellatorre, A.-G.; Lista, C.; Costantino, G.; Frisoni, G.; Virgili, G.; et al. Structural magnetic resonance imaging for the early diagnosis of dementia due to Alzheimer's disease in people with mild cognitive impairment. *Cochrane Database Syst. Rev.* **2020**, *3*, CD009628. [CrossRef]
12. Richard, E.; Schmand, B.A.; Eikelenboom, P.; Van Gool, W.A. MRI and cerebrospinal fluid biomarkers for predicting progression to Alzheimer's disease in patients with mild cognitive impairment: A diagnostic accuracy study. *BMJ Open* **2013**, *3*, e002541. [CrossRef] [PubMed]
13. Kamagata, K.; Andica, C.; Hatano, T.; Ogawa, T.; Takeshige-Amano, H.; Ogaki, K.; Akashi, T.; Hagiwara, A.; Fujita, S.; Aoki, S. Advanced diffusion magnetic resonance imaging in patients with Alzheimer's and Parkinson's diseases. *Neural Regen. Res.* **2020**, *15*, 1590–1600. [CrossRef] [PubMed]

14. Basser, P.J.; Jones, D.K. Diffusion-tensor MRI: Theory, experimental design and data analysis—A technical review. *NMR Biomed.* **2002**, *15*, 456–467. [CrossRef] [PubMed]
15. Bartzokis, G.; Cummings, J.L.; Sultzer, D.; Henderson, V.W.; Nuechterlein, K.H.; Mintz, J. White Matter Structural Integrity in Healthy Aging Adults and Patients With Alzheimer Disease. *Arch. Neurol.* **2003**, *60*, 393–398. [CrossRef]
16. Fieremans, E.; Benitez, A.; Jensen, J.; Falangola, M.; Tabesh, A.; Deardorff, R.; Spampinato, M.; Babb, J.; Novikov, D.; Ferris, S.; et al. Novel White Matter Tract Integrity Metrics Sensitive to Alzheimer Disease Progression. *Am. J. Neuroradiol.* **2013**, *34*, 2105–2112. [CrossRef]
17. Rose, S.E.; Chen, F.; Chalk, J.B.; O Zelaya, F.; E Strugnell, W.; Benson, M.; Semple, J.; Doddrell, D.M. Loss of connectivity in Alzheimer's disease: An evaluation of white matter tract integrity with colour coded MR diffusion tensor imaging. *J. Neurol. Neurosurg. Psychiatry* **2000**, *69*, 528–530. [CrossRef]
18. Ballatore, C.; Lee, V.M.-Y.; Trojanowski, J.Q. Tau-mediated neurodegeneration in Alzheimer's disease and related disorders. *Nat. Rev. Neurosci.* **2007**, *8*, 663–672. [CrossRef]
19. Salawu, F.K.; Umar, J.T.; Olokoba, A.B. Alzheimer's disease: A review of recent developments. *Ann. Afr. Med.* **2011**, *10*, 73–79. [CrossRef]
20. Jagust, W. Imaging the evolution and pathophysiology of Alzheimer disease. *Nat. Rev. Neurosci.* **2018**, *19*, 687–700. [CrossRef]
21. de Jong, L.W.; van der Hiele, K.; Veer, I.M.; Houwing, J.J.; Westendorp, R.G.J.; Bollen, E.L.E.M.; de Bruin, P.W.; Middelkoop, H.A.M.; van Buchem, M.A.; van der Grond, J. Strongly reduced volumes of putamen and thalamus in Alzheimer's disease: An MRI study. *Brain* **2008**, *131*, 3277–3285. [CrossRef] [PubMed]
22. Casanova, R.; Whitlow, C.T.; Wagner, B.; Williamson, J.; Shumaker, S.A.; Maldjian, J.A.; Espeland, M.A. High Dimensional Classification of Structural MRI Alzheimer?s Disease Data Based on Large Scale Regularization. *Front. Neurosci.* **2011**, *5*, 22. [CrossRef] [PubMed]
23. Teipel, S.; Drzezga, A.; Grothe, M.J.; Barthel, H.; Chételat, G.; Schuff, N.; Skudlarski, P.; Cavedo, E.; Frisoni, G.B.; Hoffmann, W.; et al. Multimodal imaging in Alzheimer's disease: Validity and usefulness for early detection. *Lancet Neurol.* **2015**, *14*, 1037–1053. [CrossRef] [PubMed]
24. Wen, Q.; Mustafi, S.M.; Li, J.; Risacher, S.L.; Tallman, E.; Brown, S.A.; West, J.D.; Harezlak, J.; Farlow, M.R.; Unverzagt, F.W.; et al. White matter alterations in early-stage Alzheimer's disease: A tract-specific study. *Alzheimer's Dement. Diagn. Assess. Dis. Monit.* **2019**, *11*, 576–587. [CrossRef] [PubMed]
25. Horgusluoglu-Moloch, E.; Xiao, G.; Wang, M.; Wang, Q.; Zhou, X.; Nho, K.; Saykin, A.J.; Schadt, E.; Zhang, B. Systems modeling of white matter microstructural abnormalities in Alzheimer's disease. *Neuroimage Clin.* **2020**, *26*, 102203. [CrossRef]
26. Mattie, D. A Generalized Tool for Deriving Connectomes in Support of Computational Neuroscience. In Proceedings of the SMRM Annual Conference, Vancouver, BC, Canada, 15–20 May 2021.
27. Kurtzer, G.M.; Sochat, V.; Bauer, M.W. Singularity: Scientific containers for mobility of compute. *PLoS ONE* **2017**, *12*, e0177459. [CrossRef] [PubMed]
28. Mattie, D.; Fang, Z.; Takahashi, E.; Castillo, L.P.; Levman, J. Baseline Structural Connectomics Data of Healthy Brain Development Assessed with Multi-Modal Magnetic Resonance Imaging. *Information* **2024**, *15*, 66. [CrossRef]
29. Levman, J.; Fang, Z.; Zumwalt, K.; Cogger, L.; Vasung, L.; MacDonald, P.; Lim, A.R.; Takahashi, E. Asymmetric Insular Connectomics Revealed by Diffusion Magnetic Resonance Imaging Analysis of Healthy Brain Development. *Brain Connect.* **2019**, *9*, 2–12. [CrossRef] [PubMed]
30. Fischl, B. *Freesurfer*, Software Version 7.2.0; Athinoula A; Martinos Center for Biomedical Imaging: Boston, MA, USA. Available online: https://surfer.nmr.mgh.harvard.edu/ (accessed on 5 April 2022).
31. Desikan, R.S.; Ségonne, F.; Fischl, B.; Quinn, B.T.; Dickerson, B.C.; Blacker, D.; Buckner, R.L.; Dale, A.M.; Maguire, R.P.; Hyman, B.T.; et al. An automated labeling system for subdividing the human cerebral cortex on MRI scans into gyral based regions of interest. *NeuroImage* **2006**, *31*, 968–980. [CrossRef] [PubMed]
32. Wang, R.; Benner, T.; Sorensen, A.G.; Wedeen, V.J. Diffusion Toolkit: A Software Package for Diffusion Imaging Data Processing and Tractography. *Proc. Intl. Soc. Mag. Reson. Med.* **2007**, *15*, 3720.
33. Hess, C.P.; Mukherjee, P.; Han, E.T.; Xu, D.; Vigneron, D.B. Q-ball reconstruction of multimodal fiber orientations using the spherical harmonic basis. *Magn. Reson. Med.* **2006**, *56*, 104–117. [CrossRef]
34. Mori, S.; Crain, B.J.; Chacko, V.P.; Van Zijl, P.C.M. Three-dimensional tracking of axonal projections in the brain by magnetic resonance imaging. *Ann. Neurol.* **1999**, *45*, 265–269. [CrossRef]
35. Tuch, D.S. Q-ball imaging. *Magn. Reson. Med.* **2004**, *52*, 1358–1372. [CrossRef] [PubMed]
36. Basser, P.J.; Pajevic, S.; Pierpaoli, C.; Duda, J.; Aldroubi, A. In vivo fiber tractography using DT-MRI data. *Magn. Reason. Med.* **2000**, *44*, 625–632. [CrossRef]
37. Lazar, M.; Weinstein, D.M.; Tsuruda, J.S.; Hasan, K.M.; Arfanakis, K.; Meyerand, M.E.; Badie, B.; Rowley, H.A.; Haughton, V.; Field, A.; et al. White matter tractography using diffusion tensor deflection. *Hum. Brain Mapp.* **2003**, *18*, 306–321. [CrossRef] [PubMed]
38. Behrens, T.E.J.; Berg, H.J.; Jbabdi, S.; Rushworth, M.F.S.; Woolrich, M.W. Probabilistic diffusion tractography with multiple fibre orientations: What can we gain? *NeuroImage* **2007**, *34*, 144–155. [CrossRef] [PubMed]

39. Conturo, T.E.; Lori, N.F.; Cull, T.S.; Akbudak, E.; Snyder, A.Z.; Shimony, J.S.; McKinstry, R.C.; Burton, H.; Raichle, M.E. Tracking neuronal fiber pathways in the living human brain. *Proc. Natl. Acad. Sci. USA* **1999**, *96*, 10422–10427. [CrossRef]
40. Zou, H.; Hastie, T. Regularization and Variable Selection Via the Elastic Net. *J. R. Stat. Soc. Stat. Methodol. Ser. B* **2005**, *67*, 301–320. [CrossRef]
41. Friedman, J.; Hastie, T.; Tibshirani, R. Regularization Paths for Generalized Linear Models via Coordinate Descent. *J. Stat. Softw.* **2010**, *33*, 1–22. [CrossRef]
42. Kuhn, M. Building Predictive Models in R Using the caret Package. *J. Stat. Softw.* **2008**, *28*, 1–26. [CrossRef]
43. Boser, B.E.; Guyon, I.M.; Vapnik, V.N. A training algorithm for optimal margin classifiers. In Proceedings of the Fifth Annual Workshop on Computational Learning Theory, Pittsburgh, PA, USA, 27–29 July 1992; pp. 144–152. [CrossRef]
44. Breiman, L.; Friedman, J.H.; Olshen, R.A.; Stone, C.J. *Classification and Regression Trees*; Routledge: Boca Raton, FL, USA, 2017. [CrossRef]
45. Breiman, L. Random Forests. *Mach. Learn.* **2001**, *45*, 5–32. [CrossRef]
46. Bishop, C.M. *Neural Networks for Pattern Recognition*; Oxford University Press: Oxford, UK, 1995.
47. Friedman, J.H. Greedy Function Approximation: A Gradient Boosting Machine. *Ann. Stat.* **2001**, *29*, 1189–1232. [CrossRef]
48. Fjell, A.M.; Walhovd, K.B.; Fennema-Notestine, C.; McEvoy, L.K.; Hagler, D.J.; Holland, D.; Brewer, J.B.; Dale, A.M. One-Year Brain Atrophy Evident in Healthy Aging. *J. Neurosci.* **2009**, *29*, 15223–15231. [CrossRef] [PubMed]
49. Driscoll, I.; Davatzikos, C.; An, Y.; Wu, X.; Shen, D.; Kraut, M.; Resnick, S.M. Longitudinal pattern of regional brain volume change differentiates normal aging from MCI. *Neurology* **2009**, *72*, 1906–1913. [CrossRef] [PubMed]
50. Raz, N.; Lindenberger, U.; Rodrigue, K.M.; Kennedy, K.M.; Head, D.; Williamson, A.; Dahle, C.; Gerstorf, D.; Acker, J.D. Regional Brain Changes in Aging Healthy Adults: General Trends, Individual Differences and Modifiers. *Cereb. Cortex* **2005**, *15*, 1676–1689. [CrossRef] [PubMed]
51. Wyss-Coray, T. Ageing, neurodegeneration and brain rejuvenation. *Nature* **2016**, *539*, 180–186. [CrossRef] [PubMed]
52. Bates, D.; Mächler, M.; Bolker, B.; Walker, S. Fitting Linear Mixed-Effects Models using lme4. *J. Stat. Softw.* **2014**, *67*, 1–48. [CrossRef]
53. Benjamini, Y.; Hochberg, Y. Controlling the False Discovery Rate: A Practical and Powerful Approach to Multiple Testing. *J. R. Stat. Soc. Ser. B Methodol.* **1995**, *57*, 289–300. Available online: http://www.jstor.org/stable/2346101 (accessed on 1 August 2024). [CrossRef]
54. Siriseriwan, W. Smotefamily: A Collection of Oversampling Techniques for Class Imbalance Problem Based on SMOTE. 2019. Available online: https://CRAN.R-project.org/package=smotefamily (accessed on 1 August 2024).
55. O'Bryant, S.E.; Lacritz, L.H.; Hall, J.; Waring, S.C.; Chan, W.; Khodr, Z.G.; Massman, P.J.; Hobson, V.; Cullum, C.M. Validation of the New Interpretive Guidelines for the Clinical Dementia Rating Scale Sum of Boxes Score in the National Alzheimer's Coordinating Center Database. *Arch. Neurol.* **2010**, *67*, 746–749. [CrossRef] [PubMed]
56. Patil, R.B.; Ramakrishnan, S. Analysis of sub-anatomic diffusion tensor imaging indices in white matter regions of Alzheimer with MMSE score. *Comput. Methods Programs Biomed.* **2014**, *117*, 13–19. [CrossRef]
57. Jokinen, H.; Schmidt, R.; Ropele, S.; Fazekas, F.; Gouw, A.A.; Barkhof, F.; Scheltens, P.; Madureira, S.; Verdelho, A.; Ferro, J.M.; et al. Diffusion changes predict cognitive and functional outcome: The LADIS study. *Ann. Neurol.* **2012**, *73*, 576–583. [CrossRef]
58. Saito, Y.; Kamagata, K.; Andica, C.; Taoka, T.; Tuerxun, R.; Uchida, W.; Takabayashi, K.; Owaki, M.; Yoshida, S.; Yamazaki, K.; et al. Multisite harmonization of diffusion tensor image analysis along the perivascular space using the COMBined Association Test. *Jpn. J. Radiol.* **2023**, *41*, 1072–1083. [CrossRef]
59. Dong, Q.; Zhang, W.; Wu, J.; Li, B.; Schron, E.H.; McMahon, T.; Shi, J.; Gutman, B.A.; Chen, K.; Baxter, L.C.; et al. Applying surface-based hippocampal morphometry to study APOE-E4 allele dose effects in cognitively unimpaired subjects. *NeuroImage Clin.* **2019**, *22*, 101744. [CrossRef]
60. Douaud, G.; Menke, R.A.L.; Gass, A.; Monsch, A.U.; Rao, A.; Whitcher, B.; Zamboni, G.; Matthews, P.M.; Sollberger, M.; Smith, S. Brain Microstructure Reveals Early Abnormalities more than Two Years prior to Clinical Progression from Mild Cognitive Impairment to Alzheimer's Disease. *J. Neurosci.* **2013**, *33*, 2147–2155. [CrossRef]
61. Chen, Q.; Abrigo, J.; Deng, M.; Shi, L.; Wang, Y.-X.; Chu, W.C.W. Diffusion Changes in Hippocampal Cingulum in Early Biologically Defined Alzheimer's Disease. *J. Alzheimer's Dis.* **2023**, *91*, 1007–1017. [CrossRef]
62. Chen, Z.; Liu, Y.; Zhang, Y.; Li, Q. Orthogonal latent space learning with feature weighting and graph learning for multimodal Alzheimer's disease diagnosis. *Med. Image Anal.* **2023**, *84*, 102698. [CrossRef]
63. Deng, L.; Wang, Y. Fully Connected Multi-Kernel Convolutional Neural Network Based on Alzheimer's Disease Diagnosis. *J. Alzheimer's Dis.* **2023**, *92*, 209–228. [CrossRef]
64. Khan, A.; Zubair, S. Development of a three tiered cognitive hybrid machine learning algorithm for effective diagnosis of Alzheimer's disease. *J. King Saud Univ. Comput. Inf. Sci.* **2022**, *34*, 8000–8018. [CrossRef]
65. Razzak, I.; Naz, S.; Ashraf, A.; Khalifa, F.; Bouadjenek, M.R.; Mumtaz, S. Mutliresolutional ensemble PartialNet for Alzheimer detection using magnetic resonance imaging data. *Int. J. Intell. Syst.* **2022**, *37*, 6613–6630. [CrossRef]
66. Hazarika, R.A.; Kandar, D.; Maji, A.K. An experimental analysis of different Deep Learning based Models for Alzheimer's Disease classification using Brain Magnetic Resonance Images. *J. King Saud Univ. Comput. Inf. Sci.* **2021**, *34*, 8576–8598. [CrossRef]

67. Stone, D.B.; Ryman, S.G.; Hartman, A.P.; Wertz, C.J.; Vakhtin, A.A. Specific White Matter Tracts and Diffusion Properties Predict Conversion From Mild Cognitive Impairment to Alzheimer's Disease. *Front. Aging Neurosci.* **2021**, *13*, 711579. [CrossRef]
68. Velazquez, M.; Lee, Y. Multimodal ensemble model for Alzheimer's disease conversion prediction from Early Mild Cognitive Impairment subjects. *Comput. Biol. Med.* **2022**, *151*, 106201. [CrossRef]
69. Beach, T.G.; Monsell, S.E.; Phillips, L.E.; Kukull, W. Accuracy of the Clinical Diagnosis of Alzheimer Disease at National Institute on Aging Alzheimer Disease Centers, 2005–2010. *J. Neuropathol. Exp. Neurol.* **2012**, *71*, 266–273. [CrossRef]
70. Nam, M.-H.; Ko, H.Y.; Kim, D.; Lee, S.; Park, Y.M.; Hyeon, S.J.; Won, W.; Chung, J.-I.; Kim, S.Y.; Jo, H.H.; et al. Visualizing reactive astrocyte-neuron interaction in Alzheimer's disease using 11C-acetate and 18F-FDG. *Brain* **2023**, *146*, 2957–2974. [CrossRef]
71. Feeney, J.; Savva, G.M.; O'Regan, C.; King-Kallimanis, B.; Cronin, H.; Kenny, R.A. Measurement Error, Reliability, and Minimum Detectable Change in the Mini-Mental State Examination, Montreal Cognitive Assessment, and Color Trails Test among Community Living Middle-Aged and Older Adults. *J. Alzheimer's Dis.* **2016**, *53*, 1107–1114. [CrossRef]
72. Deschamps, I.; Baum, S.R.; Gracco, V.L. On the role of the supramarginal gyrus in phonological processing and verbal working memory: Evidence from rTMS studies. *Neuropsychologia* **2014**, *53*, 39–46. [CrossRef]
73. Reynaud, E.; Lesourd, M.; Navarro, J.; Osiurak, F. On the neurocognitive origins of human tool use: A critical review of neuroimaging data. *Neurosci. Biobehav. Rev.* **2016**, *64*, 421–437. [CrossRef]
74. Wada, S.; Honma, M.; Masaoka, Y.; Yoshida, M.; Koiwa, N.; Sugiyama, H.; Iizuka, N.; Kubota, S.; Kokudai, Y.; Yoshikawa, A.; et al. Volume of the right supramarginal gyrus is associated with a maintenance of emotion recognition ability. *PLoS ONE* **2021**, *16*, e0254623. [CrossRef]
75. Stoeckel, C.; Gough, P.M.; Watkins, K.E.; Devlin, J.T. Supramarginal gyrus involvement in visual word recognition. *Cortex* **2009**, *45*, 1091–1096. [CrossRef]
76. Lee, D.H.; Chung, C.K.; Kim, J.S.; Ryun, S. Unraveling tactile categorization and decision-making in the subregions of supramarginal gyrus via direct cortical stimulation. *Clin. Neurophysiol.* **2024**, *158*, 16–26. [CrossRef]
77. Hoshi, H.; Kobayashi, M.; Hirata, Y.; Fukasawa, K.; Ichikawa, S.; Shigihara, Y. Decreased beta-band activity in left supramarginal gyrus reflects cognitive decline: Evidence from a large clinical dataset in patients with dementia. *Hum. Brain Mapp.* **2023**, *44*, 6214–6226. [CrossRef]
78. Desikan, R.S.; Cabral, H.J.; Hess, C.P.; Dillon, W.P.; Glastonbury, C.M.; Weiner, M.W.; Schmansky, N.J.; Greve, D.N.; Salat, D.H.; Buckner, R.L.; et al. Automated MRI measures identify individuals with mild cognitive impairment and Alzheimer's disease. *Brain* **2009**, *132*, 2048–2057. [CrossRef] [PubMed]
79. Grignon, Y.; Duyckaerts, C.; Bennecib, M.; Hauw, J.-J. Cytoarchitectonic alterations in the supramarginal gyrus of late onset Alzheimer's disease. *Acta Neuropathol.* **1998**, *95*, 395–406. [CrossRef] [PubMed]
80. Penniello, M.-J.; Lambert, J.; Eustache, F.; Petit-Taboué, M.C.; Barré, L.; Viader, F.; Morin, P.; Lechevalier, B.; Baron, J.-C. A PET study of the functional neuroanatomy of writing impairment in Alzheimer's disease The role of the left supramarginal and left angular gyri. *Brain* **1995**, *118*, 697–706. [CrossRef]
81. Hafkemeijer, A.; Möller, C.; Dopper, E.G.; Jiskoot, L.C.; Berg-Huysmans, A.A.v.D.; van Swieten, J.C.; van der Flier, W.M.; Vrenken, H.; Pijnenburg, Y.A.; Barkhof, F.; et al. A Longitudinal Study on Resting State Functional Connectivity in Behavioral Variant Frontotemporal Dementia and Alzheimer's Disease. *J. Alzheimer's Dis.* **2016**, *55*, 521–537. [CrossRef] [PubMed]
82. Cai, S.; Chong, T.; Zhang, Y.; Li, J.; von Deneen, K.M.; Ren, J.; Dong, M.; Huang, L. Altered Functional Connectivity of Fusiform Gyrus in Subjects with Amnestic Mild Cognitive Impairment: A Resting-State fMRI Study. *Front. Hum. Neurosci.* **2015**, *9*, 471. [CrossRef] [PubMed]
83. Cercignani, M.; Wheeler-Kingshott, C.G. From micro- to macro-structures in multiple sclerosis: What is the added value of diffusion imaging. *NMR Biomed.* **2018**, *32*, e3888. [CrossRef] [PubMed]
84. Wiese, S.; Karus, M.; Faissner, A. Astrocytes as a Source for Extracellular Matrix Molecules and Cytokines. *Front. Pharmacol.* **2012**, *3*, 120. [CrossRef] [PubMed]
85. Kim, E.; Figueiredo, I.C.; Simmons, C.; Randall, K.; Gonzalez, L.R.; Wood, T.; Ranieri, B.; Sureda-Gibert, P.; Howes, O.; Pariante, C.; et al. Mapping acute neuroinflammation in vivo with diffusion-MRI in rats given a systemic lipopolysaccharide challenge. *Brain Behav. Immun.* **2023**, *113*, 289–301. [CrossRef]
86. Liu, X.; Gao, X.; Zhang, L.; Yuan, Z.; Zhang, C.; Lu, W.; Cui, D.; Zheng, F.; Qiu, J.; Xie, J. Age-related changes in fiber tracts in healthy adult brains: A generalized q-sampling and connectometry study. *J. Magn. Reson. Imaging* **2018**, *48*, 369–381. [CrossRef]
87. Lin, L.; Jin, Y.; Xiong, M.; Wu, S.; Sun, S. The Protective Power of Cognitive Reserve: Examining White Matter Integrity and Cognitive Function in the Aging Brain for Sustainable Cognitive Health. *Sustainability* **2023**, *15*, 11336. [CrossRef]
88. Baker, M. 1500 scientists lift the lid on reproducibility. *Nature* **2016**, *533*, 452–454. [CrossRef] [PubMed]
89. Tur, C.; Grussu, F.; Prados, F.; Charalambous, T.; Collorone, S.; Kanber, B.; Cawley, N.; Altmann, D.R.; Ourselin, S.; Barkhof, F.; et al. A multi-shell multi-tissue diffusion study of brain connectivity in early multiple sclerosis. *Mult. Scler. J.* **2019**, *26*, 774–785. [CrossRef] [PubMed]
90. Jeurissen, B.; Leemans, A.; Tournier, J.; Jones, D.K.; Sijbers, J. Investigating the prevalence of complex fiber configurations in white matter tissue with diffusion magnetic resonance imaging. *Hum. Brain Mapp.* **2012**, *34*, 2747–2766. [CrossRef] [PubMed]

91. Busatto, G.F.; Garrido, G.E.; Almeida, O.P.; Castro, C.C.; Camargo, C.H.; Cid, C.G.; A Buchpiguel, C.; Furuie, S.; Bottino, C.M A voxel-based morphometry study of temporal lobe gray matter reductions in Alzheimer's disease. *Neurobiol. Aging* **2002**, 24 221–231. [CrossRef]
92. Mattie, D. *Dmattie/pacs-adni-eab: 2024-07-09*, Version 2024-07-09; Zenodo: Boston, MA, USA, 2024. [CrossRef]

applied sciences

MDPI

Article

Electroencephalography Response during an Incremental Test According to the $\dot{V}O_2$max Plateau Incidence

Véronique Billat [1,2,*], Christian Berthomier [3], Michel Clémençon [4,5], Marie Brandewinder [3], Slim Essid [6], Cécilia Damon [6], François Rigaud [6], Alexis Bénichoux [6], Emmanuel Maby [7], Lesly Fornoni [7], Patrick Bouchet [7], Pascal Van Beers [8], Bertrand Massot [9], Patrice Revol [10], Luc Poinsard [1], Thomas Creveaux [4], Christian Collet [4], Jérémie Mattout [7] and Vincent Pialoux [4,11]

1 Movement, Balance, Performance, and Health Laboratory (EA 4445), Université de Pau et des Pays de l'Adour, 65000 Tarbes, France; lpoinsard@univ-pau.fr
2 Faculté des Sciences du Sport, Université Paris-Saclay, Univ Evry, 91000 Evry-Courcouronnes, France
3 Physip, 6 Rue Gobert, 75011 Paris, France; c.berthomier@physip.fr (C.B.); marie.brandewinder@ensam.eu (M.B.)
4 Inter-University Laboratory of Human Movement Biology (EA 7424), Claude Bernard University of Lyon 1, 69100 Villeurbanne, France; michel.clemencon@univ-rouen.fr (M.C.); thomas.creveaux@gmail.com (T.C.); christian.collet@univ-lyon1.fr (C.C.); vincent.pialoux@univ-lyon1.fr (V.P.)
5 Centre d'Etudes des Transformations des Activités Physiques et Sportives, Normandie Université, UNIROUEN, CETAPS, 76000 Rouen, France
6 Institut Telecom Paris, CNRS-LTCI, 91120 Palaiseau, France; slim.essid@telecom-paristech.fr (S.E.); ceciliadamon@gmail.com (C.D.); francois.rigaud@telecom-paristech.fr (F.R.); abenicho@gmail.com (A.B.)
7 Lyon Neuroscience Research Center, Brain Dynamics and Cognition Team, INSERM UMRS 1028, CNRS UMR 5292, Université Claude Bernard Lyon 1, 69100 Villeurbanne, France; manu.maby@inserm.fr (E.M.); lesly.fornoni@inserm.fr (L.F.); patrick.bouchet@inserm.fr (P.B.); jeremie.mattout@inserm.fr (J.M.)
8 Unité Fatigue et Vigilance, Institut de Recherche Biomédicale des Armées (IRBA), 91220 Brétigny-sur-Orge, France; pascal.vanbeers@intradef.gouv.fr
9 INSA Lyon, Ecole Centrale de Lyon, CNRS, Université Claude Bernard Lyon 1, CPE Lyon, INL, UMR5270, 69621 Villeurbanne, France; bertrand.massot@insa-lyon.fr
10 Service de Médecine Physique et Réadaptation, Hôpital Henry Gabrielle, Hospices Civils de Lyon, 69230 Saint-Genis-Laval, France; patrice.revol@inserm.fr
11 Institut Universitaire de France, 75231 Paris, France
* Correspondence: veronique.billat@univ-evry.fr

Citation: Billat, V.; Berthomier, C.; Clémençon, M.; Brandewinder, M.; Essid, S.; Damon, C.; Rigaud, F.; Bénichoux, A.; Maby, E.; Fornoni, L.; et al. Electroencephalography Response during an Incremental Test According to the $\dot{V}O_2$max Plateau Incidence. *Appl. Sci.* **2024**, *14*, 5411. https://doi.org/10.3390/app14135411

Academic Editor: Alexander N. Pisarchik

Received: 13 May 2024
Revised: 17 June 2024
Accepted: 18 June 2024
Published: 21 June 2024

Abstract: $\dot{V}O_2$max is recognized as a key measure in exercise physiology and sports medicine. However, only 20–50% of maximal incremental exercise tests (IET) result in a plateau of $\dot{V}O_2$ ($\dot{V}O_2$pl). To our knowledge, no study has yet examined the possible difference in brain activity during an IET, in $\dot{V}O_2$pl and non-plateau athletes with the same $\dot{V}O_2$max and age. This study aimed to shed light on the central governor hypothesis, namely that the inability to reach a $\dot{V}O_2$pl may be dictated by the brain rather than by a peripheral physical limit. This hypothesis can now be explored using electroencephalography (EEG) during IET, measuring concomitant power in specific frequency bands. Forty-two athletes were divided into two groups: those who practiced endurance sports and those who did not, and were asked to perform an IET. EEG signals and gas exchange were recorded. A $\dot{V}O_2$pl was observed in twenty-two subjects (52%). EEG power increased in all subjects during IET, except in the alpha band, which showed variability, but not significantly (64% increase, 34% decrease, $p = 0.07$). No differences were found between endurance athletes and non-endurance athletes, except for $\dot{V}O_2$max (60.10 ± 6.16 vs. 51.77 ± 6.41, $p < 0.001$). However, the baseline-corrected ratio of EEG power to $\dot{V}O_2$ was found to decrease in all subjects during IET, in the alpha, beta and theta bands. In conclusion, the presence or absence of a $\dot{V}O_2$pl is not related to the type of EEG response during an IET. Nevertheless, the decline in brain and $\dot{V}O_2$ powers/ratios in all frequency bands suggests that aerobic power may be constrained by brain mobilization.

Keywords: EEG; exhausting exercise; maximal oxygen consumption; fatigue; central governor; endurance; cycling

1. Introduction

The fundamental tenet of maximal oxygen consumption ($\dot{V}O_2max$), positing a threshold in speed or work rate beyond which no further increases in $\dot{V}O_2$ is observed, traces its origins back in the seminal work of Hill and Lupton in 1923 [1]. They observed a limit to the body's ability to use oxygen during exercise, beyond which oxygen consumption reaches a plateau despite increasing exercise intensity. Currently, $\dot{V}O_2max$ is recognized as a key measure in exercise physiology and sports medicine [2,3]. However, it should be noted that the certainty of reaching $\dot{V}O_2max$ requires the observation of a $\dot{V}O_2$ plateau ($\dot{V}O_2pl$), defined as the point at which $\dot{V}O_2$ remains relatively constant despite increasing work rate. Therefore, by definition, diagnosis of $\dot{V}O_2max$ requires a $\dot{V}O_2pl$, whether it occurs at the end of a continuous incremental test, between the final stages of a discontinuous test, or between an incremental test and a subsequent verification test, as Poole and Jones pointed out [4].

The observation that only a subset of participants in an incremental test have a $\dot{V}O_2pl$ highlights the variability inherent in such assessments. Reported incidences vary considerably, ranging from 17 to 94%, even in studies involving more than 50 participants, particularly depending on the definition of $\dot{V}O_2pl$ used [5]. This variability has led to an extensive debate on the conceptualization and diagnosis of $\dot{V}O_2max$ over the last three decades [2,4]. Only 20–50% of $\dot{V}O_2max$ tests result in a plateau according to this definition [6–14]. The low frequency of the $\dot{V}O_2pl$ has been reported by [15].

However, behind the discussion about the absolute necessity of observing a $\dot{V}O_2pl$ to assess true "$\dot{V}O_2max$", and not just a "$\dot{V}O_2peak$" [16–19], lies the fundamental question of the limiting factors of $\dot{V}O_2$. Reasons for the presence of $\dot{V}O_2pl$ in some, but not all, athletes are not yet clear [18]. Lower anaerobic power or capacity has been suggested as a cause of the absence of plateau [19–21]. Furthermore, it has been shown that faster oxygen kinetics, which minimizes the anaerobic contribution to metabolism in the severe intensity range between the respiratory compensation threshold (RCP) and $\dot{V}O_2max$ (or $\dot{V}O_2peak$) during an incremental exercise test (IET), increases the chances of a plateau occurring at $\dot{V}O_2max$ [5].

However, the role of the brain in regulating exercise intensity to exhaustion has been hotly debated [22–25]. More specifically, the brain has been given greater consideration in models describing the factors responsible for continuing or stopping ongoing exercise. Noakes has suggested that unpleasant sensations of fatigue provide useful feedback to the central nervous system and are used as key regulators to stop exhausting exercise before there is a risk of physiological damage [26]. This hypothesis underlines the importance of recording brain activity alongside exercise characteristics and environmental conditions, particularly under conditions of exhaustion. A better understanding of brain responses to exhaustive exercise up to $\dot{V}O_2max$ would rely on a multimodal approach that could combine neurological, physiological, and biomechanical data.

Regarding neurophysiological aspects, two methods are less restrictive and allow body and head movements: near-infrared spectroscopy (NIRS) and electroencephalogram (EEG). Indeed, brain activity has already been studied during high-intensity exercise, notably when performing an IET on an ergobicycle until exhaustion using NIRS [27–30] or EEG [31–33]. EEG provides robust information on changes in cortical potentials, particularly rhythmic activity and frequency of synaptic processes [33–35]. EEG is one of the most pragmatic means of monitoring changes in brain activity in humans during exercise, certainly because it is less intrusive [31,34,36]. EEG is used to measure cortical brain activity, which is categorized into distinct frequency ranges such as alpha (α: 8–12 Hz) and beta (β: 12–30 Hz), each related to different cognitive functions regulated by the brain. Alpha activity, characterized by quite low-frequency oscillations, is associated with perceptual awareness and inhibition of non-essential processing, which facilitates task performance [37,38]. Conversely, beta activity, characterized by high-frequency oscillations, is associated with voluntary contractions, alertness, and arousal, enhancing the perception of stimuli [39–41]. Spectral analysis is a common means of quantifying the frequency content of these bands, the main result being the power spectral density (PSD), which indicates the strength or energy of variations between frequencies. An increase in oscillations in cortical regions yields to an increase in the EEG's spectral power.

Several studies have coupled EEG recording with a breath-by-breath expired gas system [31,33,34]. EEG activity has generally been recorded during pedaling tests at a constant load without exhaustion [42–45]. It has also been recorded during exercise to exhaustion in an incremental test [31–33,46]. While the role of the prefrontal cortex in exercise tolerance and termination has been investigated [47–50] showing, for example, that the alpha/beta ratio decreased after the subject experienced exercise as "hard" in reference to the Rate of Perception of Exhaustion [51] and then above the RCP [33,47,48], no study has yet examined the relationship between the occurrence of $\dot{V}O_2pl$ with EEG response in well-trained athletes.

To our knowledge, no study has yet examined the possible difference in brain activity during maximal incremental testing in $\dot{V}O_2pl$ and non-plateau athletes with the same $\dot{V}O_2max$ and age. This study aimed to shed light on the central governor hypothesis, namely that the inability to reach a $\dot{V}O_2pl$ may be dictated by the brain rather than by a peripheral physical limit. This hypothesis can now be explored using EEG during IET, measuring concomitant power in specific frequency bands. Forty-two athletes were divided into two groups: those who practiced endurance sports and those who did not, and were asked to perform an IET.

We hypothesize that the difference between athletes who reached the plateau of their $\dot{V}O_2max$ and those who did not could be present in different EEG characteristics and in EEG/$\dot{V}O_2$ ratios. Specifically, we believe that athletes who do not achieve a $\dot{V}O_2pl$ will show a more pronounced decrease in the ratio between EEG power and $\dot{V}O_2$ as metabolic demand increases. To test this hypothesis, we compared, in the $\dot{V}O_2pl$ and non-plateau groups of athletes, the relationship between increasing metabolic demand and EEG response by examining the ratios between the α, β, θ (theta) bands of the EEG and $\dot{V}O_2$. The aim was to tackle the possible issue of a decrease in the ratio between metabolic demand and brain activity during physical exercise.

2. Materials and Methods

2.1. Participants Recruitment and Ethical Approval

Forty-two volunteers participated in the experimentation (Table 1).

Table 1. Subjects' characteristics for the incremental exercise test (n = 42).

Subjects	Mean	Standard Deviation (SD)
Age (years)	25.81	4.92
Height (cm)	180.55	6.78
Weight (kg)	73.90	10.71
Body mass index (kg/m^2)	22.59	2.27
$\dot{V}O_2max$ (mL·min^{-1}·kg^{-1})	55.74	7.51

The population consisted of active men aged between 18 and 35 years old with no declared neurological or motor deficits. Participants were recruited through posters and communication among well-trained physical students and multisport practitioners. Volunteers were included in the experimental group and remunerated for their participation. This study was approved by the Léon Bérard Centre's Research and Ethics Committee under the number A 13–160.

In addition, to take into account the types of sports practiced by the subjects, we divided them into two categories: sports with endurance (END) characteristics (triathlon, running, cycling and trail running) and those without (climbing, volleyball, basketball, judo, water polo, etc.). We had 22 athletes in the NONEND group and 20 in the END group.

2.2. Participants Recruitment and Ethical Approval

Each participant completed an IET. The purpose of the IET was to assess maximal aerobic power (MAP) and maximal oxygen consumption ($\dot{V}O_2max$).

Prior to the test, a standardized warm-up allowed participants to familiarize themselves with the protocol and equipment. Each participant was asked to pedal constantly for 8 min: 2 min and 45 s at 50 W, and 5 min and 15 s, including six stages of 15 s each from 90 to 240 W (increasing by 30 W per stage), interspersed with 45 s intervals at 50 W basal power output. During the 5 min recovery period that followed, the participants were fully equipped before beginning the IET. This test was preceded by a 30 s period at 50 W before performing an incremental test until exhaustion. The test itself began at the power of 90 W and consisted of a series of two-minute increments of 30 W, with participants pedaling at their own frequency. No information on power or time was given during the test.

No verbal encouragement was given during the test. The experimental session is presented in Figure 1.

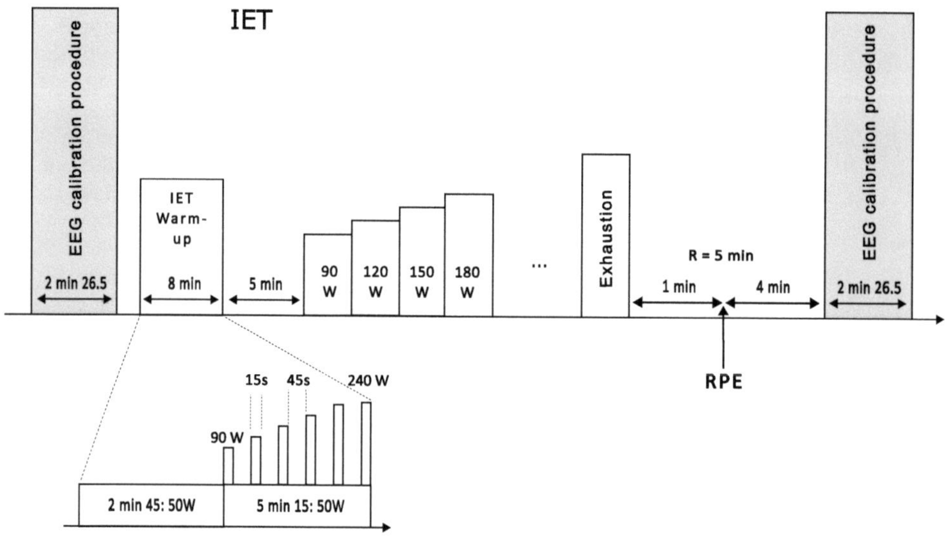

Figure 1. Summary of experimental session. The session began with a brief EEG calibration, followed by an 8 min warm-up. After a 5 min recovery period, the IET was performed until exhaustion. RPE test was then performed during the next 5 min of recovery, before the final EEG calibration procedure. EEG: electroencephalogram, IET: incremental exercise test, and RPE: rate of perceived exertion.

2.3. Experimental Design

Participants were seated in a chair. Using an abrasive cream (Nuprep®, Weaver and Company, Aurora, CO, USA) and a cotton swab, the skin was rubbed at the location of the reference and ground electrodes. This operation removed dead cells, impurities, and excess sebum to improve conductivity. Additionally, the forehead, scalp, and hair were degreased with a compress soaked in 70% alcohol [52]. They were then fitted with a Polar® heart rate belt, a gas mask, and an EEG headset. The EEG electrodes were positioned with conductive gel between the scalp and the electrodes to improve contact between the skin and the sensor. Subjects were then seated in a semi-recumbent position on a cyclo-ergometer, as shown in Figure 2.

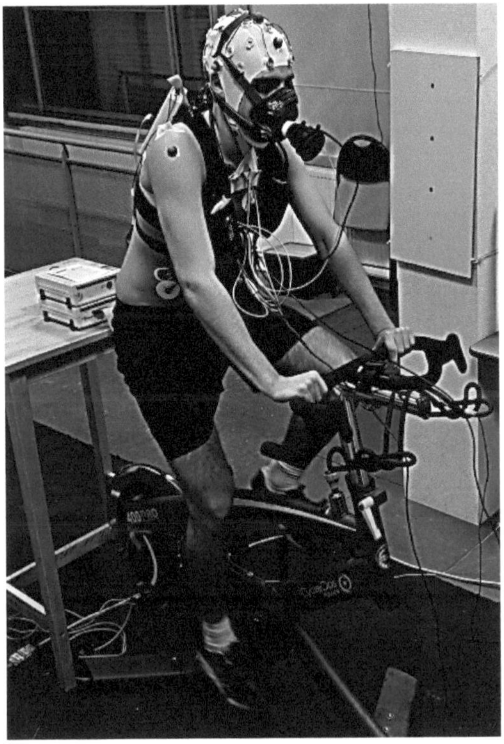

Figure 2. Positioning of sensors and participants on the ergocycle.

In accordance with traditional recommendations [53,54], the electrodermal sensors were placed on the second phalanx of the second and third fingers of the non-dominant hand (or on the third phalanx, in case of frequent potential contact between the sensors and the environment). A conductive gel (Teca, ref. 822-201210) was applied between the sensors and the skin to improve contact after cleansing the skin with a mixture of alcohol and ether. The electrodes were then firmly strapped to the fingers using hypoallergenic adhesive tape. Finally, six 3D kinematic markers were placed on the head and shoulder (see Figure 2).

The impedance of the EEG electrodes was then checked, allowing the conductive EEG gel time to warm up to body temperature. The target impedance values were between 1 and 5 kΩ. If the impedance values were higher than 5 kΩ, conductive gel was added until the impedance reached the required values. At this point, the EEG calibration procedure began, as shown in Figure 3, and proceeded in the following sequence: 30 s with eyes open, 30 s with eyes closed, 15 s of eye blink, 15 s of eye movement, and 12.5 s of head movement alone in the four directions (left, right, up, and down). A 7.5 s pause was observed between each test sequence. The experimental EEG calibration procedure was carried out using Presentation® software (version 18.1, www.neurobs.com (accessed on 19 June 2024)) and presented on a computer in front of the subject. Participants were asked to look straight ahead and to remain motionless during the procedure, except for what they were asked to do.

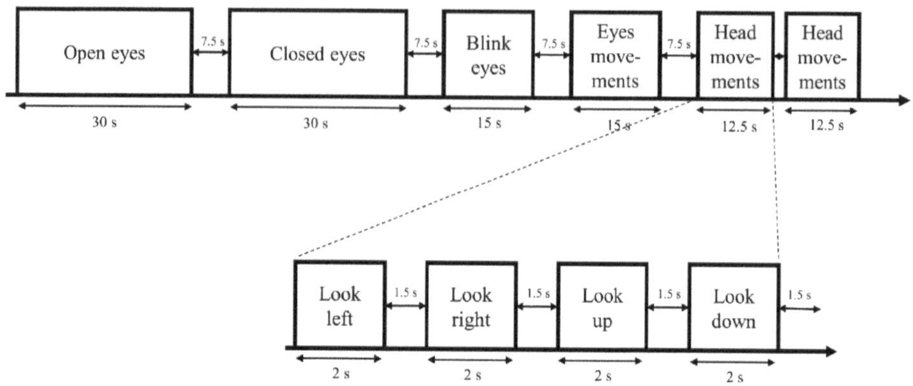

Figure 3. EEG calibration procedure. The whole procedure takes approximately 2′30″. After 30 s of open eyes followed by 30 s of closed eyes, the procedure requires 15 s of blink eyes and 15 s of eyes movement before 2 sets of 12.5 s of demanding head movements.

After remaining static during the EEG calibration sequence, each participant warmed up for 8 min (see Figure 1). They were then fitted with a nafion/permapure sampling tube connected to a turbine for measurements of pulmonary gas exchange. Each subject then performed the physical test described above. The participants were asked to remain seated on the saddle in order to limit head and upper body movements as much as possible. After each test and one minute's rest, each participant was asked to complete a perceived exertion rating scale (see below). After a 5 min rest, the EEG calibration procedure (Figure 3) was again performed, along with the impedance check.

Rating of Perceived Exertion (RPE) Scale

We asked each participant to rate the perceived exertion on a scale of 6 to 20 at the end of the IET. A score of 6 corresponds to rest and is closely correlated to a resting heart rate value, while 20/20 corresponds to maximal effort with the highest heart rate values [55].

2.4. Measurements

During the test, six apparatus recorded measurements throughout the exercise. Descriptions of each device are presented here, along with the relevant data recorded.

2.4.1. Ergocycle Data

We used the CycleOps 400 Pro Indoor Cycle (Saris Cycling Group, Inc., 5253 Verona Road, Madison, WI 53711, USA) with the CycleOps Joule 3.0 (Saris Cycling Group, Inc.) computer. The Joule 3.0 CPU uses ANT+ technology to communicate wirelessly with the bike's sensors (i.e., the PowerTap power meter in the rear flywheel, speed sensor, cadence sensor, and heart rate strap sensor). The warm-up and the IET have been programmed in advance. This ergocycle makes it possible to adjust the power output regardless of the pedaling cadence. All data were stored on the Joule 3.0 CPU and downloaded to a computer running Power agent software (version 7.8.28) (Saris Cycling Group, Inc.). Power (W), torque (Nm), speed (km/h), cadence (rpm), and heart rate (bpm) were recorded and stored at a sampling rate of 1 Hz.

2.4.2. Electroencephalography Measurements

EEG was recorded using a 32-channel ActiCap system (Brain Products, Gilching, Germany), which combines active electrodes based on high-quality Ag/AgCl sensors with the application of a conductive gel to lower impedances. Ten sensors were used for the measurements. We used the following sites from the extended 10–10 system: Fp1, Fp2, Fz, C3, Cz, C4, Pz, O1, Oz, O2. The ground electrode was placed on the lateral third of the right scapula spine. All electrodes were referenced to an electrode placed on the right mastoid, and impedances were kept below 5 kΩ for all sensors. Analog signals were amplified (analog band-pass filter 0.016 Hz–1000 Hz) with a BrainAmp amplifier (Brain Products, Germany) and digitized at a frequency of 5000 Hz. EEG data were downsampled to 1000 Hz (with a 400 Hz anti-aliasing filter) and recorded using the Brain Vision Recorder software (version 1.20.0601, Brain Products, Gilching, Germany).

2.4.3. Heart Rate (HR) Measurements

Heart rate was measured in two different ways. Firstly, it was measured using a heart rate belt (Polar®) and synchronized with the gas measurements. Secondly, it was measured using a three-channel electrocardiogram (ECG) and synchronized with the electrodermal apparatus. The time of occurrence of the R-waves could thus be accurately determined. The D2 derivation signal (the interval between two consecutive ECG R-waves) was electronically processed and delivered in the form of instantaneous heart rate. In the case of missed ECG R-waves or false detection due to artifacts, HR could be estimated offline using an algorithm that iteratively replaced changes in IHR above a threshold of 10 beat per minute (bpm) with interpolated values. The interpolation was calculated between pairs of values below the threshold. The new signal, free of artifact, was then resampled. The smallest appreciable variation was 0.5 bpm, and the calibrated scale ranged from zero to 200 bpm. The IHR signal was then extracted directly from the ECG at the sensors. The IHR was therefore treated as an analog signal. Data acquisition was performed at 10 Hz on this analog signal.

2.4.4. Gas Measurements

O_2 and CO_2 concentrations during the test were measured using a Metamax® 3B mobile gas analyzer (Cortex Biophysik GmbH, Leipzig, Germany). Breath-by-breath data on respiratory volume and gas concentrations were sent in real-time by telemetry to a PC. Metasoft® software (version 3.9.9 SR5) calculated ventilation rate ($\dot{V}E$), oxygen consumption ($\dot{V}O_2$), carbon dioxide output ($\dot{V}CO_2$), and synchronized all gas data with heart rate. The system ran for at least 30 min and was calibrated before each test in accordance with the manufacturer's recommendations.

As the ergocycle used in this study could not receive any analog or digital signal, we used its pedaling signal, as well as the pedaling signal provided by the trigger, to synchronize the ergocycle's data with the others (Figure 4).

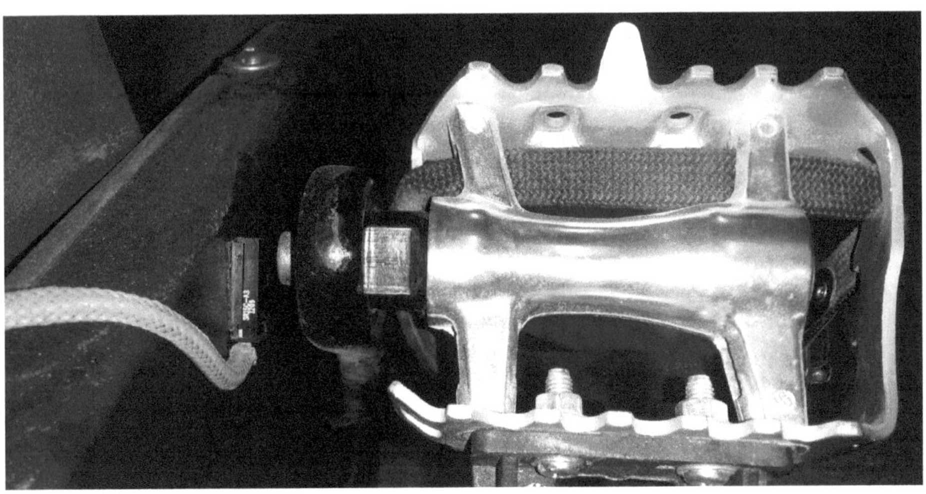

Figure 4. Magnet attached to the pedal and Hall-effect sensor attached to the ergocycle frame.

Similarly, as the gas analyzer used in this study had no possible signal input, we used its ability to record heart rate to synchronize it with the electrodermal activity (EDA) device which also provided a heart rate measurement.

In summary, Figure 5 shows the different devices and how they are synchronized with each other.

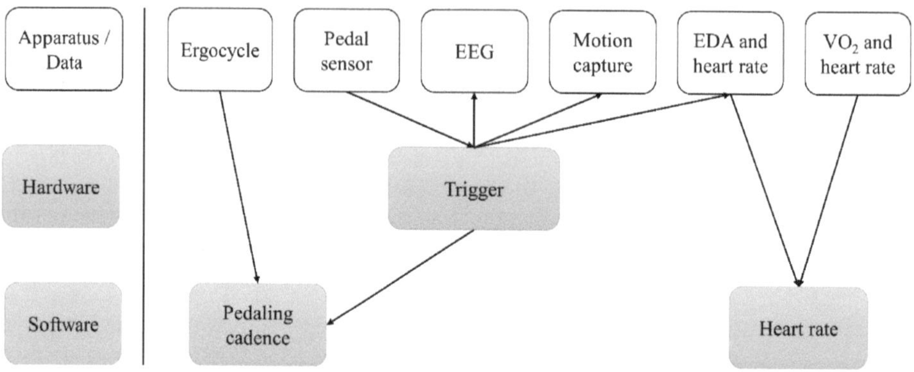

Figure 5. Device and data synchronization strategies (hardware and software). EEG: electroencephalogram, EDA: electrodermal activity, and $\dot{V}O_2$: oxygen consumption.

2.4.5. Determination of the $\dot{V}O_2$max Plateau

Following the recommendations of Niemeyer and colleagues in the major review of the oxygen uptake plateau as a "frequently misunderstood phenomenon" [5], we used fairly wide sampling intervals (i.e., the $\dot{V}O_2$pl was determined from more than the final 30 W or 60 s) and a cut-off that was set at approximately half the expected increase in $\dot{V}O_2$ in the submaximal intensity range. Additionally, as recommended by Poole and colleagues [14], we did not use the "so-called" secondary criteria as a maximal value of respiratory exchange ratio (RER) $\geq$ 1.1, maximum HR value (HRmax) greater than 95% of the maximum value predicted for age (220-age), and the end-exercise blood lactate criterion greater than 8.0 mmol/L, although we only checked that all subjects had achieved these at the end of the IET at the same time as we asked them for their RPE value. If subjects did not satisfy the $\dot{V}O_2$ criteria but satisfied the secondary ones, we considered the higher $\dot{V}O_2$ value was a $\dot{V}O_2$peak. The workload eliciting the RCP was determined using the criteria of an increase in both the VE/$\dot{V}O_2$ and VE/$\dot{V}O_2$ and a decrease in end-tidal carbon dioxide pressure ($PetCO_2$) [56]. To determine RCP, the values of the gas-exchange variables were averaged for every 1 min period and plotted against workload. For statistical comparisons between groups (see below), RCP was expressed as the mean value of %$\dot{V}O_2$max for the corresponding 1 min interval. Two experienced independent observers detected RCP. In case of disagreement, we sought the opinion of a third investigator.

2.5. *Signal Processing*

EEG: Data Reduction Procedures for Artifact Correction and Removal

We paid particular attention to preventing and/or limiting artifacts due to muscle contraction, eye, or body movement [57]. The use of active gel EEG electrodes was likely to overcome this problem thanks to an integrated noise subtraction circuit, thus improving the reliability of data collection [32]. We rejected abrupt variations in the signal by analyzing it using a 2 s time window. The resulting EEG signal shows a noise level (Figure 6) that:

- Increases systematically towards the end of the recording and the highest effort.
- Is higher on the occipital derivation, probably due to muscle contraction.
- Is kept to an acceptable level (approximately 20 dB at 50 Hz).

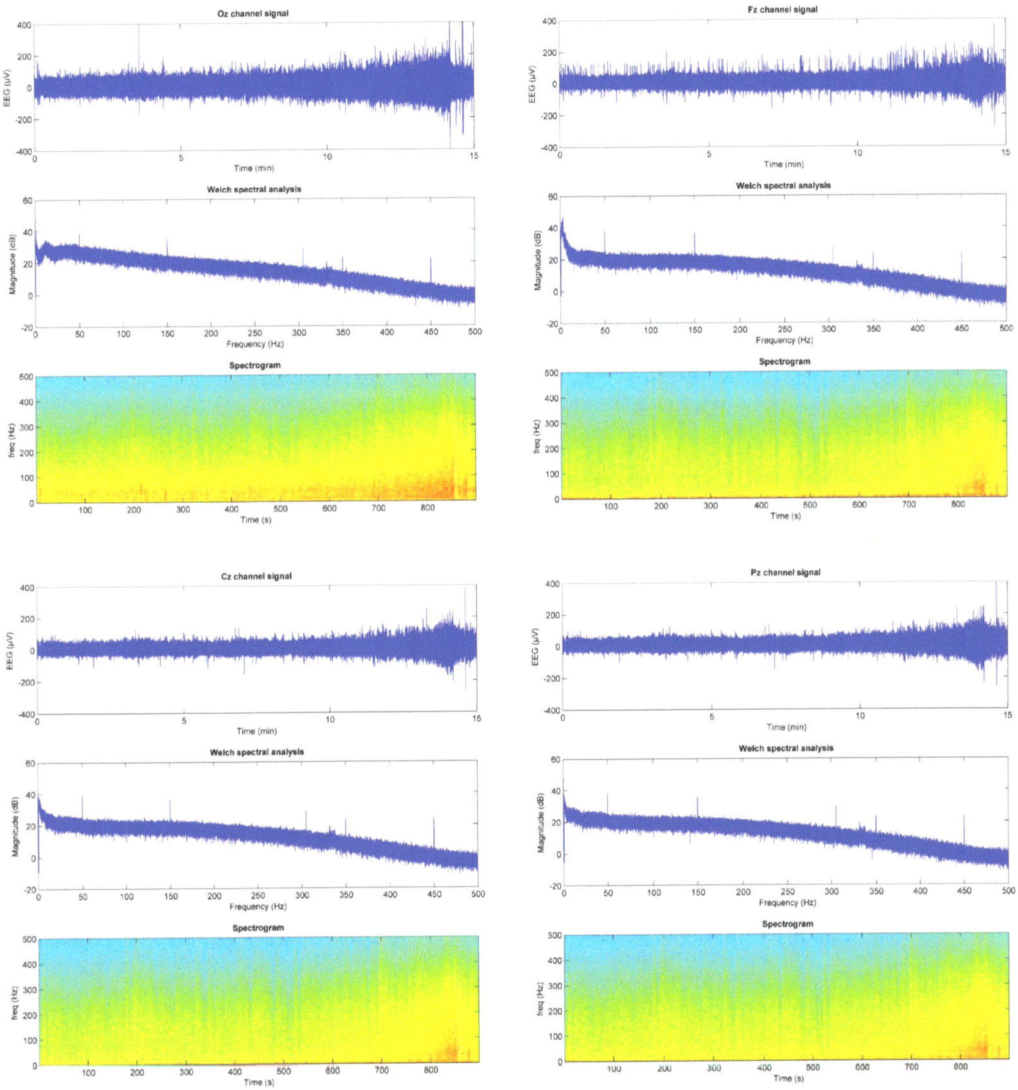

Figure 6. Typical raw traces obtained during the incremental exercise test for Oz, Fz, Cz and Pz channel signals with the raw signal, Welch spectral analysis and spectrogram for each channel. In the spectrogram, cooler colors (blue and green) represent low power and warmer colors (white, red and yellow) represent high power. EEG: electroencephalogram; freq: frequency.

It also shows the presence of residual artifacts. Eye blinks are visible in the temporal signal at the frontal site. Some artifacts are found in the lower frequency bands where EEG activity is expected to occur, but at an amplitude that does not generate signal saturation. These artifacts will present a challenge to subsequent EEG analysis, but can be subject to artifact rejection procedures. A specific artifact rejection method has been developed [58].

2.6. Statistics

A two-way ANOVA was used to analyze changes in the mean of a quantitative variable as a function of the levels of two categorical variables (END and NONEND, PLAT and NONPLAT). More specifically, we investigated the main effect A: the average difference in physiological and neurophysiological responses due to variations in the first independent variable, i.e., PLAT vs. NONPLAT factor, and for the main effect B: the average difference in the dependent variable due to variations in the second independent variable, i.e., the END vs. NONEND sports practice. We also examined the interaction effect in order to determine whether the effect of PLAT compared with NONPLAT depended on the level of the END vs. NONEND variable. In other words, we wanted to check whether the combined effect of both variables was different from what we expected on the basis of their individual effects. We therefore measured the impact of achieving $\dot{V}O_2max$ NONPLAT and PLAT on the ratio of EEG power to $\dot{V}O_2$ profile during the IET for θ, α, β, α/β ratio, RPE, $\dot{V}O_2max$, HRmax, and RCP (%$\dot{V}O_2max$). We also proceeded in the same manner for the NONEND and END factors. To achieve this, we used a two-way ANOVA (XLSTAT 2023.2.0, Paris, France).

In addition, we used the Pearson product-moment correlation coefficient to measure the strength of the linear relationship between the EEG response (α, β, α/β ratio) and $\dot{V}O_2$. If the relationship between the variables is not linear, the correlation coefficient does not correctly represent the strength of the relationship between the variables. We have considered the value alpha < 0.05 for the significance level, i.e., the probability that you will make the mistake of rejecting the null hypothesis when it is true. If the p value is greater than alpha, you accept the null hypothesis. If it is less than alpha, you reject the null hypothesis.

3. Results

3.1. Occurrence of $\dot{V}O_2pl$ in the Whole Group (n = 42)

All subjects completed the test. The results are presented in Table 2.

A $\dot{V}O_2pl$ was observed in twenty-two subjects (52%) (Figure 7, example of a $\dot{V}O_2pl$), while the remaining twenty subjects achieved a $\dot{V}O_2peak$ and met the secondary $\dot{V}O_2max$ criterion. It should be noted that the RPE did not reach maximal values corresponding to very hard (17/20), since the average value for the whole group was only 16.8 ± 1.5 (Table 2). On the other hand, there was no difference between the NONPLAT and PLAT groups or NONEND and END groups (Table 3).

Table 2. Physiological, mechanical and RPE results for the IET.

Variable	Mean	SD	Range
$\dot{V}O_2max$ (mL·min^{-1}·kg^{-1})	55.74	7.51	40.0
Maximal Aerobic Power (W)	273.78	42.49	180.0
Relative Maximal Aerobic Power (W/kg)	3.74	0.67	3.4
HRmax (bpm)	188.52	9.83	45.0
Time (s)	967.44	162.52	713.0
RPE	16.79	1.50	5.0

SD: standard deviation, RPE: rating of perceived exertion, and $\dot{V}O_2max$: maximal oxygen consumption.

Table 3. Effects of NONPLAT or PLAT and NONEND or END groups on characteristics and physiological variables.

	n	$\dot{V}O_2max$ (mL·min^{-1}·kg^{-1})	Age (Years)	Height (cm)	Training/Week (Hour)	Final RPE	RCP (%$\dot{V}O_2max$)
NONEND	n = 22	51.77 ± 6.41	24.23 ± 4.34	181.95 ± 8.00	6.95 ± 4.74	16.55 ± 1.72	73.97 ± 11.89
END	n = 20	60.10 ± 6.16	27.55 ± 5.04	179.00 ± 4.87	8.05 ± 4.75	17.08 ± 3.99	77.02 ± 7.52
F value		−4.10	−2.25	1.08	−0.91	−1.04	−0.557
p value		**<0.001**	**0.03**	0.285	0.368	0.305	0.581
NONPLAT	n = 20	55.00 ± 5.64	25.85 ± 5.17	181.75 ± 6.84	8.15 ± 5.82	16.66 ± 4.02	73.88 ± 11.08
PLAT	n = 22	56.41 ± 8.96	25.77 ± 4.81	179.45 ± 6.70	6.86 ± 3.45	16.91 ± 1.47	76.82 ± 9.05
F value		0.353	0.684	0.665	1.149	−0.271	−1.11
p value		0.72	0.498	0.510	0.258	0.788	0.273

Values in bold indicate a statistically significant p-value (α = 0.05). RCP: respiratory compensation threshold, RPE: rating of perceived exertion, $\dot{V}O_2max$: maximal oxygen consumption, NONEND: non-endurance group, and END: endurance group.

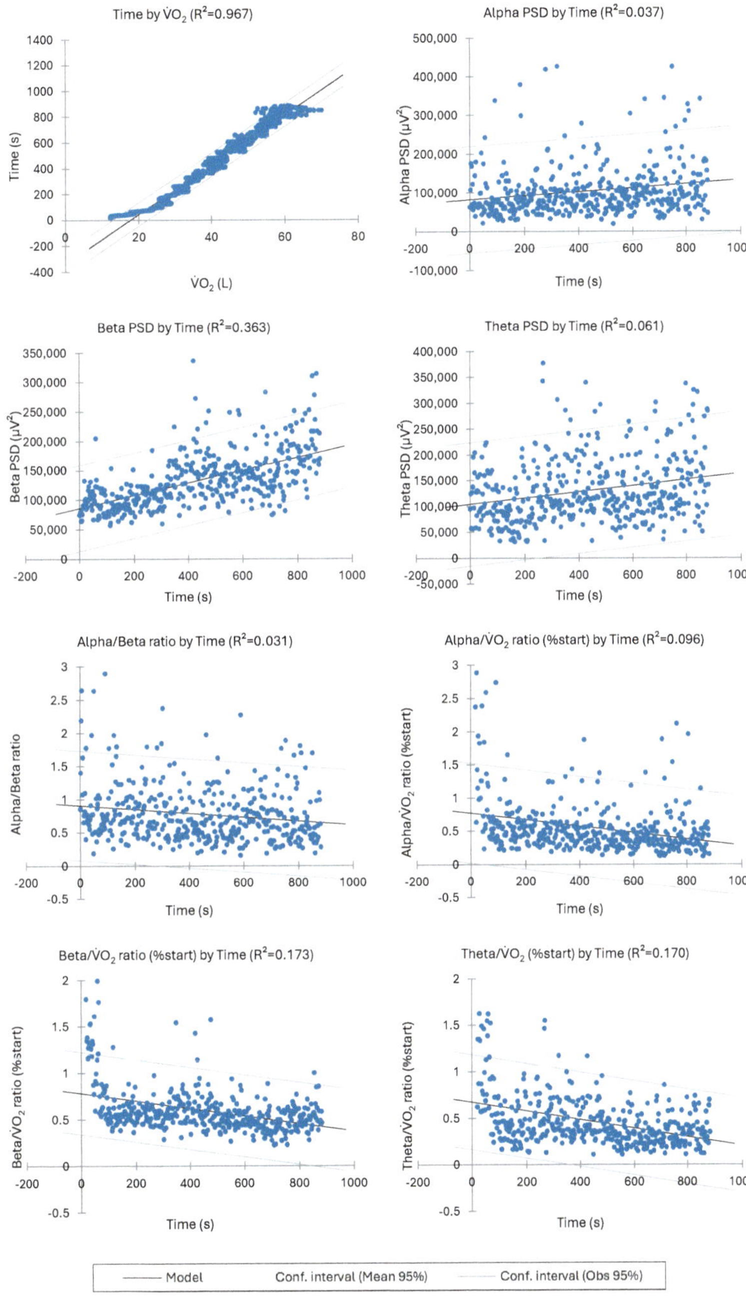

Figure 7. Linear regression of various physiological parameters by time for the subject 940 reaching a $\dot{V}O_2$pl. Subfigures include $\dot{V}O_2$, theta power, alpha power, beta power, alpha/beta power ratio, alpha/$\dot{V}O_2$ as a percentage of starting alpha/$\dot{V}O_2$ power, beta/$\dot{V}O_2$ as a percentage of starting beta/$\dot{V}O_2$ power, and theta/$\dot{V}O_2$ power as a percentage of starting theta/$\dot{V}O_2$ power. PSD: power spectral density.

3.2. Difference in Physiological and Training Parameters between the NONPLAT and PLAT Groups or NONEND and END Ones

Table 3 reports the non-significant values of $\dot{V}O_2max$, training volume, height, final RPE, and fractional use of $\dot{V}O_2max$ at the RCP. The only significant difference between plateau or endurance practice criteria was that the endurance group had a higher $\dot{V}O_2max$ ($p = 0.0002$) and was older ($p = 0.03$). Training volume did not differ significantly between the groups.

3.3. EEG Responses vs. $\dot{V}O_2$ Increase, between the NONPLAT and PLAT Groups or NONEND and END Ones during the IET

In all subjects, we observed a decrease in the α/β ratio and an increase in theta and beta (presented as percentage of the start value in Table 4).

Table 4. Effects of NONPLAT or PLAT and NONEND or END groups on EEG variables. Theta and beta values are expressed as a percentage of the starting value of the IET.

		Alpha/Beta Ratio	Theta Increase (%start)	Beta Increase (%start)
NONEND	$n = 22$	−20.30 ± 12.93	20.39 ± 16.43	40.06 ± 19.97
END	$n = 20$	−15.74 ± 11.93	20.24 ± 16.43	47.00 ± 16.14
F value		1.36	0.01	1.33
p value		0.250	0.907	0.256
NONPLAT	$n = 20$	−18.58 ± 12.71	18.29 ± 13.31	44.26 ± 18.59
PLAT	$n = 22$	−17.93 ± 12.67	22.25 ± 18.65	42.34 ± 18.93
F value		<0.001	0.74	0.08
p value		0.988	0.395	0.778

NONEND: non-endurance group; END: endurance group.

Figure 7 (and Table 5) shows an example of the EEG response during the IET in one subject (id 940) who was in the PLAT and NONEND groups. These EEG responses were the same for all subjects, with the exception of the alpha response, which exhibited some variability. Specifically, it increased in 27 of 42 subjects and decreased in the remaining 15. However, the frequency distribution of the alpha value tendency did not differ significantly from the expected frequencies (Table 6).

Table 5. Summary table of analysis of variance for $\dot{V}O_2$, theta power, alpha power, beta power, alpha power/beta power ratio, alpha/$\dot{V}O_2$ as a percentage of initial alpha/$\dot{V}O_2$ power, beta/$\dot{V}O_2$ as a percentage of initial beta/$\dot{V}O_2$ power, and theta/$\dot{V}O_2$ power as a percentage of initial theta/$\dot{V}O_2$ power.

	$\dot{V}O_2$	Alpha PSD	Beta PSD	Theta PSD	Alpha/Beta Ratio	Alpha/$\dot{V}O_2$	Beta/$\dot{V}O_2$	Theta/$\dot{V}O_2$
R^2	0.967	0.037	0.363	0.061	0.031	0.096	0.173	0.170
F	25,011.7	16.0	238.1	27.0	13.1	42.7	85.6	84.0
Pr > F	**<0.001**	**<0.001**	**<0.001**	**<0.001**	**<0.001**	**<0.001**	**<0.001**	**<0.001**

PSD: power spectral density; $\dot{V}O_2$: maximal oxygen consumption.

Nevertheless, although not significant, given the alpha threshold of 5% against which we compare p values ($p = 0.07$), fewer END subjects show a decrease in alpha power compared to the NONEND group. However, when we focused on the ratio of alpha power expressed as a percentage of the starting value to $\dot{V}O_2$ as a percentage of the starting value, it decreases in all subjects, indicating a "depletion" of alpha power compared to the metabolic demand, as was observed in all subjects for the other EEG power bands.

In addition, there was also no impact of the distribution of these different tendencies between groups for NONPLAT and PLAT subjects (Table 7) and the increase or decrease tendency of cadence, EEG $\theta/\dot{V}O_2$, EEG $\alpha/\dot{V}O_2$, EEG $\beta/\dot{V}O_2$, and α/β ratio (Table 8).

Table 6. Frequency distribution of alpha value trend in subjects.

	Positive Alpha Tendency	Negative Alpha Tendency	*p* Value
NONEND	12	10	0.07
END	15	5	
NOPLAT	14	6	0.75
PLAT	12	10	

NONEND: non-endurance group, END: endurance group, NOPLAT: non-plateau group, and PLAT: plateau group.

Table 7. Effects of NONPLAT or PLAT and NONEND or END groups on alpha power expressed as a percentage of alpha power at the start of the IET.

Chi-Square (Observed Value)	Chi-Square (Critical Value)	DF	*p* Value	Alpha
3.663	7.815	3	0.300	0.05

Table 8. Chi-square test for cadence, EEG $\beta/\dot{V}O_2$, EEG $\theta/\dot{V}O_2$, EEG $\alpha/\dot{V}O_2$ and α/β ratio between the NONPLAT (n = 22) and PLAT (n = 20) groups.

	Cadence	EEG $\theta/\dot{V}O_2$	EEG $\alpha/\dot{V}O_2$	EEG $\beta/\dot{V}O_2$	α/β Ratio	All
Chi-square (Observed value)	0.005	0.225	0.431	0.558	0.288	7.61
Chi-square (Critical value)	3.84	3.85	3.86	3.84	3.87	16.9
p value	0.945	0.636	0.512	0.455	0.591	0.57

EEG: electroencephalogram, α: alpha frequency band, β: beta frequency band, and θ: theta frequency band.

4. Discussion

The main aim of this study was to compare the EEG responses during an incremental exercise test in two groups of subjects who had achieved a $\dot{V}O_2$pl or not.

The debate around the $\dot{V}O_2$pl and its cause remains a long story of exercise physiology for nearly half a century, and as Noakes [59] states, "it is time to move beyond a brainless exercise physiology". We hypothesize that the difference between athletes who reached the plateau of their $\dot{V}O_2$max and those who did not could be present in different EEG characteristics and in EEG/$\dot{V}O_2$ ratios. Our main results showed no incidence of the $\dot{V}O_2$pl or not, nor of endurance or non-endurance practice, on EEG activity during IET. Regardless of group, all subjects showed an increase in beta and theta band power, while alpha band power was less uniform (half increasing and half decreasing, independent of group membership).

However, when we plotted the EEG band power against oxygen power demand, both relative to the start value, all subjects showed a decrease in alpha, beta, and theta/$\dot{V}O_2$ ratios during IET. Therefore, even if the brain is not the limiting factor of $\dot{V}O_2$max or responsible for the attainment of $\dot{V}O_2$pl or not, these results, given the relationship between EEG responses and oxygen demand, could constitute a new marker of a progressive "depletion" of EEG power capacity. Considering each of these elements, it may be recalled that at rest, alpha waves are generally more observed when a person is in a relaxed mental state [60]. During exercise, an increase in alpha activity may indicate increased attention or alertness, as the brain focuses on the task at hand [61]. However, an excessive reduction in alpha activity during exercise may reflect a large effort, yielding mental fatigue or decrease cognitive performance [62]. Beta waves are associated with active concentration, problem solving, and alertness [63]. During exercise, increased beta wave activity may reflect heightened cognitive engagement and concentration, particularly during tasks requiring attention and coordination [64,65]. Theta waves are linked to deep relaxation, meditation, and drowsiness [66]. During exercise, a decrease in theta wave activity can occur when a person is overexerted, which can lead to difficulties in maintaining precise sensorimotor control and monitoring sensory inputs due to reduced executive functions or decreased attention [67]. Overall, beyond EEG changes in the alpha, beta, and theta bands during exercise, the increases observed relative to that of $\dot{V}O_2$ highlight the importance of including central measures in our physical activity studies. This was the goal of this interdisciplinary approach using physiology, neurophysiology and biomechanics synchronizing all measurements during exercise performed at free cadence.

- Difference in maximal value of $\dot{V}O_2$, heart rate, RPE, and training parameters between the NONPLAT and PLAT groups or NONEND and END ones.

The occurrence of $\dot{V}O_2$pl was not dependent on specialty and therefore on maximum $\dot{V}O_2$, given that, unsurprisingly, the END group had a higher $\dot{V}O_2$max than the NONEND group. Thus, given that the END and NONEND athletes were equally distributed in the NONPLAT and PLAT groups, we can consider that specialty does not influence the occurrence of $\dot{V}O_2$pl. Since this factor (specialty) also has no impact on RCP, we can assume that they did not have contract a significant oxygen deficit before reaching $\dot{V}O_2$max (with or without plateau).

- The occurrence of $\dot{V}O_2$pl

The "plateau phenomenon", described by Mitchell and Blomqvist in 1971 [68], was observed during 3 consecutive workloads. Before them, in 1955, Taylor and colleagues introduced a criterion for identifying a $\dot{V}O_2$pl during exercise [69]. This threshold was set at $\leq$150 mL/min (or $\leq$2.1 mL$\cdot$min$^{-1}\cdot$kg^{-1}) based on the average increase in subjects' $\dot{V}O_2$ over incremental exercise increments. However, this method has been criticized [5,11]. Moreover, according to the positioning of eminent colleagues [4] on this issue, we did not consider the so-called "secondary" criteria (heart rate (HR) $\leq$10 beats/min or $\leq$5% of the maximum predicted by age (220-age), blood lactate concentration $\geq$8 mM, or respiratory exchange ratio (RER) >1.00, 1.10, or 1.15). Thus, when we obtain these criteria without an individual reaching a plateau, the "$\dot{V}O_2$peak" referred to the highest $\dot{V}O_2$ reached during a graded exercise test, as has been done in most research on maximal exercise tests in children for whom there was some difficulty in demonstrating a $\dot{V}O_2$pl [70]. Thus, in this study, to compare the EEG response during IET as a function of the occurrence of a $\dot{V}O_2$pl, this was defined as less than half the expected increase. According to this definition, we observed a $\dot{V}O_2$pl occurrence in 52.6% of our subjects, in accordance with previous studies [4,5,9,11,18,19,71,72].

- RPE response between the groups

Hill's model emphasizes the concept of a "critical metabolic state" in which metabolite accumulation reaches a threshold level, triggering fatigue and necessitating a reduction in exercise intensity to avoid metabolic imbalance and potential damage to muscle tissue. In essence, Hill proposed that fatigue is a protective mechanism that prevents excessive muscular strain and maintains physiological homeostasis [1,73,74]. The central governor theory challenged (or we might say, achieved) Hill's peripheral model of fatigue by emphasizing the role of the brain as a central regulator of exercise performance. The founders and protagonists of this "central governor theory" argue that while peripheral factors, such as muscle fatigue and metabolic stress, undoubtedly contribute to fatigue, the brain's role in pacing and regulating exercise intensity is paramount [75,76]. They suggest that the brain integrates sensory feedback from the muscles, cardiovascular system, and other physiological systems to modulate exercise effort and prevent catastrophic failure [59,71]. Here, we used perceived exertion to measure the subjective experience of intensity, stress, discomfort, and/or fatigue during physical activity. The RPE scale, first introduced in 1970 and subsequently refined by Borg [77], remains the predominant tool for assessing perceived exertion in adults [78]. Some studies suggest an even stronger correlation between RPE and $\dot{V}O_2$ [79,80]. The widespread adoption of RPE to assess exercise intensity may be explained by the ability of human beings to perceive effort, which derives from continuous use of a well-developed sensory system. During exercise, individuals are aware of their overall effort and can discern the location of the effort, allowing them to gauge the intensity of the exercise and estimate their endurance at a given level [81,82]. The experience accumulated with different intensities of exercise enables individuals to numerically evaluate or at least classify the intensity of exercise via the RPE scale. Whatever the group (END or NONEND; PLAT or NONPLAT), the subjects rated their RPE just below 17/20, i.e., the beginning of what they perceived as "very hard". We can therefore consider that the RPE does not play a role in the difference in potential EEG response during the IET. However, Samuele Marcora's theory [75] postulates that the perception of effort during exercise may be independent of afferent feedback from the muscles, heart, and lungs. He argues that the prevailing notion attributing a substantial influence to afferent feedback on the perception of effort may result of an overly broad framework. Marcora emphasizes the role of hedonicity in overall perception and argues for a narrower definition of exertion, aligned with the descriptors of Borg's RPE scale. He suggests that while sensory mechanisms may influence the perception of effort, experimental studies have dissociated the perception of effort from metabolic stress, indicating that the brain primarily generates the sensation of effort, with limited influence from afferent feedback.

- EEG responses between the NONPLAT and PLAT groups or NONEND and END ones during the IET

Therefore, there is a consensus that an individual's exercise performance is modulated by feedback from various physiological systems under the control of the brain. This is a different debate to that of the factors limiting $\dot{V}O_2max$ and its definition and here, by examining the EEG response as a function of the occurrence of $\dot{V}O_2pl$, we showed that the EEG response was affected neither by plateau nor by sports specialty and hence, the value of $\dot{V}O_2max$. Independently of the occurrence of a $\dot{V}O_2pl$, we found an increase in theta and beta power, while alpha power decreased or increased. These results regarding the increase in beta and theta frequency band and alpha/beta ratio, are in accordance with those of previous studies [83]. In their examination on EEG during exercise, Hosang and colleagues observed a predominant increase in alpha and beta activity following high-intensity exercise compared to low or moderate intensity exercise sessions [83]. Previous studies have noted that heat stress induced by exhaustive exercise can increase cortical activity, which could explain the link between exercise intensity and changes in alpha, beta, and delta activity [84–87]. Another plausible explanation for the increase in beta activity, particularly in fronto–central regions, is its association with high levels of psychomotor arousal [84] and increased cortical activation during voluntary movements [88–90]. Increased theta activity has also been associated with the processing of novel information, suggesting that the theta results observed may be related to the control and regulation of attentional resources [91,92].

Emerging evidence suggests that the onset of fatigue leading to exercise cessation is associated with afferent feedback, a neural factor regulated and interpreted by the brain [93]. This feedback is linked to sensory information detecting unpleasant stimuli such as lactate accumulation in active muscles, fatigue of peripheral locomotor muscle or an increase in central temperature, leading to a cerebral response. Afferent information from the periphery is transmitted to the prefrontal cortex, where it is interpreted and influences the decision to stop exercise. EEG has been proposed as a practical, non-invasive approach to gathering valuable information about changes in brain activity during rest and exercise [34]. This study attempted to provide information on brain regulations to exhaustion as a function of their respective limiting factors [31,33,34]. In both cases, our multimodal device may allow fatigue to be monitored by several methods, to study the interactions between the central nervous system, the autonomic nervous system, and respiratory exchanges during acute exercise, which may provide information on exercise tolerance and regulation [25,26,49,50]. Here, we attempted a multimodal approach providing different indicators of cortical activity changes that were synchronized with autonomic nervous system, gas exchanges, heart rate and behavioral performance while controlling body movements during an IET and comparing the EEG/$\dot{V}O_2$ profile according to the occurrence or non-occurrence of $\dot{V}O_2$pl.

- Impact of the cadence on the EEG responses between the NONPLAT and PLAT groups or NONEND and END ones during the IET

As cadence has an impact on EEG and response during exercise [43,94,95], we allowed the participant to freely choose the pedaling frequency associated with a fixed power output. This should facilitate exercise, particularly when participants were not accustomed to pedaling on a cyclo-ergometer. In addition, changes in pedaling frequency can indicate the onset of fatigue or disengagement and can be objectively confirmed by physiological measures. Recording body movements should help to control the quality of EEG data by providing information that helps to distinguish the signal from movement noise. Thanks to the quality of the recordings, and even with a reduced number of electrodes (10 compared with the usual 20 to 64), a wide range of brain areas was covered. We were thus able to analyze the four frequency bands usually studied in sports science in the frontal, parietal and central regions of the brain: theta (4–8 Hz), alpha (8–12 Hz), beta (12–30 Hz), and low gamma (30–50 Hz) [31,33]. As Bailey and colleagues suggest, EEG activity from previous research is difficult to compare due to specific differences in exercise protocol and apparatus [31]. For example, the site-specific density of electrical activity in the motor cortex was found to increase with exercise intensity until exhaustion [32]. Previously reported changes in EEG response to exercise showed that EEG activity varied during incremental or constant load exercise tests across all bandwidths [31,33,34]. Furthermore, it has been found that alpha activity in the motor cortex is not fully activated and is maintained during exercise exhaustion [50]. In contrast, alpha activity in the frontal cortex decreases from the second ventilatory threshold until exhaustion [33]. The frontal cortex could therefore play a central role in the cessation of exercise because of the changes it undergoes during high levels of physical exertion [50]. The precise cause of the effects of high-intensity exercise on cortical activity is difficult to determine because of concomitant physiological responses such as elevated body temperature and increased blood flow, which may influence oscillatory activity [84,85]. Conversely, the effects of moderate-intensity exercise are less likely influenced by such factors because the increase in body temperature is not as pronounced [96]. The results of studies on moderate-intensity exercise have mainly revealed an increase in alpha and beta activity in different regions of the brain, which can be explained by the fact that alpha activity is linked to arousal [97] and beta activity plays a role in maintaining exercise at a steady state [98]. Analysis of the effects of low-intensity exercise on cortical activity has not revealed any significant trends, the most frequent observation being the absence of significant changes in the oscillatory bands, exception of the alpha band. The effects on alpha band activity vary, with some studies reporting an increase [32,99,100], others a decrease [101–103], and still others showing no change [104]. Like studies on moderate-intensity exercise, research on low-intensity exercise has focused mainly on the activity of the alpha band, limiting discussion of the other oscillatory bands.

- Application of the EEG/$\dot{V}O_2$ ratio approach in exercise physiology and medicine.

The application of the EEG/$\dot{V}O_2$ ratio approach in exercise physiology and medicine offers a novel method for understanding the complex relationship between neural activity and metabolic demand during physical exertion. This innovative approach integrates EEG to monitor brain activity and oxygen consumption measurements to assess metabolic function, providing a comprehensive picture of how the brain and body respond to exercise. By analyzing the EEG/$\dot{V}O_2$ ratio, researchers and clinicians can better understand the cognitive and neural mechanisms underlying physical performance, fatigue, and recovery. This method has significant implications for optimizing athletic training, as it makes it possible to identify the mental states that correlate with peak performance and periods of reduced efficiency. For example, understanding how brain activity changes at different $\dot{V}O_2$max levels can help to design more effective training regimes that improve both physical and cognitive endurance. Furthermore, in the medical field, the EEG/$\dot{V}O_2$ ratio can be used to adapt rehabilitation programs for patients suffering from cardiovascular or neurological disorders, ensuring that both the cognitive and physical aspects of recovery are taken into account. This dual approach can improve patient outcomes by enhancing not only physical capacity, but also mental resilience and cognitive function. This approach also makes it possible to study the effects of environmental stress factors, such as high altitude, extreme temperatures, or hypoxic conditions, on brain function and overall performance. By monitoring how the brain adapts to these challenging conditions, it is possible to develop strategies to mitigate their adverse effects, thereby improving safety and performance in extreme environments. In this way, the EEG/$\dot{V}O_2$ ratio represents a powerful tool for advancing our understanding of the dynamic interaction between the brain and body in both health and disease, paving the way for more integrated and effective approaches in exercise science and clinical practice.

5. Conclusions

In this study, we mainly demonstrated that, regardless of the occurrence of $\dot{V}O_2$pl, a decline was observed for all bandwidths according to the EEG/$\dot{V}O_2$ decline throughout the test. Therefore, we suggest the existence of a "EEG reserve depletion" while alpha activity in motor cortex is preferentially maintained. The EEG responses to fatigue in this study can be associated with other variables to determine brain behavior during exercise, both before and at the end of exercise. Information on heart rate, ventilatory thresholds and maximum oxygen consumption associated with EEG data could emerge from this integrated data analysis. We hypothesized that overall EEG activity would change as fatigue developed throughout both exercise durations and might depend on the sport experience profile. In addition, we hypothesized that EEG analysis could provide a predictive index of exercise exhaustion.

Research into changes in cortical brain activity during an incremental exercise test has produced mixed results. For example, Bailey and colleagues reported increases in alpha and beta frequencies throughout an incremental exercise test [31], whereas Robertson and Marino [33] observed increases in these frequency bands only until RCP, after which EEG activity decrease until the end of exercise.

Therefore, in order to go further on the debate on the impact of exercise intensification on the EEG response, we proposed here to have a systemic approach to exercise limitation, especially applied to the concept of $\dot{V}O_2$max, by merging the fields of neuroscience, biomechanics, and physiology. This could open up the black box of the "central governor" that has been at the heart of the debates on factors limiting $\dot{V}O_2$max.

In conclusion, the combination of all the scientific fields in this study could help delineate the signature of exhaustive exercise. We might then be able to identify the evolution of multiple signals during exercise and potentially anticipate the decision to stop this exercise by looking for a minimum EEG/$\dot{V}O_2$ threshold or a threshold for continuing exercise beyond a specific EEG/$\dot{V}O_2$ threshold.

6. Limitations

This study was conducted using an "open-loop" control mode, since the power was imposed by the experimenters. Consequently, the movements were pre-programmed or predetermined without continuous feedback from the sensory inputs. In real performance conditions, such as running, the subject is in a closed-loop condition and the EEG response could also be used as a pace controller [48]. This closed-loop configuration is probably the most appropriate exercise model for understanding individual critical threshold [105] and subjective experiences of fatigue, such as in a marathon race [106], and for understanding the limits of exercise tolerance with an integrative model [107,108]. However, our experimental study is limited by the small sample size, which restricts the statistical power and may affect the reliability and generalizability of the results. A smaller sample size increases the risk of type II errors, where true effects may not be detected, and reduces confidence in effect sizes estimates. This limitation may also lead to greater variability and margin of error in the results, which could impact the robustness and reproducibility of our findings Therefore, although our results provide valuable preliminary information, further research with larger and more diverse samples is essential to validate and extend these findings.

Author Contributions: Conceptualization, C.B., E.M., M.B. and J.M.; methodology, V.B., C.B., M.C., E.M., M.B., S.E., P.B., P.V.B., V.P., J.M., P.R., B.M. and C.C.; software, C.B., E.M., S.E., C.D., F.R. and A.B.; formal analysis, V.B., C.B., L.P., E.M. and J.M.; data curation, E.M. and T.C.; writing, V.B., C.B and M.C.; review and editing, L.P.; project administration, C.B., M.B., S.E., L.F., V.P., J.M. and C.C.; funding acquisition, C.B., M.B. and J.M. All authors have read and agreed to the published version of the manuscript.

Funding: This study was conducted with the financial support of the French Defense Procurement Agency (DGA/AID) through a RAPID Project named MEEGAPERF, grant number 112906033.

Institutional Review Board Statement: This study was conducted in accordance with the Declaration of Helsinki, and approved by the Research and Ethics Committee of the Léon Bérard center under the ref A 13-160 number 2013-A00503-42 for studies involving humans.

Informed Consent Statement: Informed consent was obtained from all subjects involved in this study.

Data Availability Statement: The data presented in this study are available on request from the corresponding author. The data are not publicly available due to privacy.

Conflicts of Interest: The authors declare no conflicts of interest.

References

1. Hill, A.V.; Lupton, H. Muscular Exercise, Lactic Acid, and the Supply and Utilization of Oxygen. *QJM Int. J. Med.* **1923**, *16*, 135–171. [CrossRef]
2. Bassett, D.R., Jr.; Howley, E.T. Limiting factors for maximum oxygen uptake and determinants of endurance performance. *Med. Sci. Sports Exerc.* **2000**, *32*, 70–84. [CrossRef] [PubMed]
3. Ross, R.; Blair, S.N.; Arena, R.; Church, T.S.; Despres, J.P.; Franklin, B.A.; Haskell, W.L.; Kaminsky, L.A.; Levine, B.D.; Lavie, C.J.; et al. Importance of Assessing Cardiorespiratory Fitness in Clinical Practice: A Case for Fitness as a Clinical Vital Sign: A Scientific Statement from the American Heart Association. *Circulation* **2016**, *134*, e653–e699. [CrossRef] [PubMed]
4. Poole, D.C.; Jones, A.M. Measurement of the maximum oxygen uptake VO_{2max}: Vo_{2peak} is no longer acceptable. *J. Appl. Physiol.* **2017**, *122*, 997–1002. [CrossRef] [PubMed]
5. Niemeyer, M.; Knaier, R.; Beneke, R. The Oxygen Uptake Plateau—A Critical Review of the Frequently Misunderstood Phenomenon. *Sports Med.* **2021**, *51*, 1815–1834. [CrossRef] [PubMed] [PubMed Central]
6. Vella, C.A.; Marks, D.; Robergs, R.A. Oxygen cost of ventilation during incremental exercise to VO_2 max. *Respirology* **2006**, *11*, 175–181. [CrossRef] [PubMed]
7. Bassett, D.R., Jr. Scientific contributions of A. V. Hill: Exercise physiology pioneer. *J. Appl. Physiol.* **2002**, *93*, 1567–1582. [CrossRef] [PubMed]
8. Astorino, T.A.; Willey, J.; Kinnahan, J.; Larsson, S.M.; Welch, H.; Dalleck, L.C.; Shephard, R.J. Elucidating determinants of the plateau in oxygen consumption at VO_2max. *Br. J. Sports Med.* **2005**, *39*, 655–660. [CrossRef] [PubMed]
9. Day, J.R.; Rossiter, H.B.; Coats, E.M.; Skasick, A.; Whipp, B.J. The maximally attainable Vo_2during exercise in humans: The peak vs. maximum issue. *J. Appl. Physiol.* **2003**, *95*, 1901–1907. [CrossRef] [PubMed]
10. Duncan, G.E.; Howley, E.T.; Johnson, B.N. Applicability of VO_2max criteria: Discontinuous versus continuous protocols. *Med. Sci. Sports Exerc.* **1997**, *29*, 273–278. [CrossRef] [PubMed]
11. Rossiter, H.B.; Kowalchuk, J.M.; Whipp, B.J. A test to establish maximum O_2uptake despite no plateau in the O_2uptake response to ramp incremental exercise. *J. Appl. Physiol.* **2006**, *100*, 764–770. [CrossRef] [PubMed]
12. Astorino, T.A. Alterations in VO_2max and the VO_2 plateau with manipulation of sampling interval. *Clin. Physiol. Funct. Imaging* **2009**, *29*, 60–67. [CrossRef] [PubMed]
13. Astorino, T.A.; White, A.C.; Dalleck, L.C. Supramaximal testing to confirm attainment of VO_2max in sedentary men and women. *Int. J. Sports Med.* **2009**, *30*, 279–284. [CrossRef] [PubMed]
14. Poole, D.C.; Wilkerson, D.P.; Jones, A.M. Validity of criteria for establishing maximal O_2 uptake during ramp exercise tests. *Eur. J. Appl. Physiol.* **2008**, *102*, 403–410. [CrossRef] [PubMed]
15. Doherty, M.; Nobbs, L.; Noakes, T.D. Low frequency of the "plateau phenomenon" during maximal exercise in elite British athletes. *Eur. J. Appl. Physiol.* **2003**, *89*, 619–623. [CrossRef] [PubMed]
16. Green, S.; Askew, C.D. VO_{2peak} is an acceptable estimate of cardiorespiratory fitness but not VO_{2max}. *J. Appl. Physiol.* **2018**, *125*, 229–232. [CrossRef] [PubMed]

17. Azevedo, P.; Bhammar, D.M.; Babb, T.G.; Bowen, T.S.; Witte, K.K.; Rossiter, H.B.; Brugniaux, J.V.; Perry, B.D.; Dantas de Lucas, R.; Turnes, T.; et al. Commentaries on Viewpoint: VO_2peak is an acceptable estimate of cardiorespiratory fitness but not VO_2max. *J. Appl. Physiol.* **2018**, *125*, 970. [CrossRef] [PubMed]
18. Niemeyer, M.; Leithaeuser, R.; Beneke, R. Oxygen uptake plateau occurrence depends on oxygen kinetics and oxygen deficit accumulation. *Scand. J. Med. Sci. Sports* **2019**, *29*, 1466–1472. [CrossRef] [PubMed]
19. Niemeyer, M.; Bergmann, T.G.J.; Beneke, R. Oxygen uptake plateau: Calculation artifact or physiological reality? *Eur. J. Appl. Physiol.* **2020**, *120*, 231–242. [CrossRef] [PubMed]
20. Gordon, D.; Hopkins, S.; King, C.; Keiller, D.; Barnes, R.J. Incidence of the plateau at VO_2max is dependent on the anaerobic capacity. *Int. J. Sports Med.* **2011**, *32*, 1–6. [CrossRef] [PubMed]
21. Demarle, A.P.; Slawinski, J.J.; Laffite, L.P.; Bocquet, V.G.; Koralsztein, J.P.; Billat, V.L.; Cruz, R.S.d.O.; de Aguiar, R.A.; Turnes, T.; Pereira, K.L.; et al. Decrease of O_2 deficit is a potential factor in increased time to exhaustion after specific endurance training. *J. Appl. Physiol.* **2001**, *90*, 947–953. [CrossRef] [PubMed]
22. Esteve-Lanao, J.; Lucia, A.; Dekoning, J.J.; Foster, C. How Do Humans Control Physiological Strain during Strenuous Endurance Exercise? *PLoS ONE* **2008**, *3*, e2943. [CrossRef] [PubMed] [PubMed Central]
23. Marcora, S.M. Do we really need a central governor to explain brain regulation of exercise performance? *Eur. J. Appl. Physiol.* **2008**, *104*, 929–931. [CrossRef] [PubMed]
24. Tucker, R.; Noakes, T.D. The physiological regulation of pacing strategy during exercise: A critical review. *Br. J. Sports Med.* **2009**, *43*, e1. [CrossRef] [PubMed]
25. Weir, J.P.; Beck, T.W.; Cramer, J.T.; Housh, T.J. Is fatigue all in your head? A critical review of the central governor model. *Br. J. Sports Med.* **2006**, *40*, 573–586. [CrossRef] [PubMed] [PubMed Central]
26. Noakes, T.D. Fatigue is a Brain-Derived Emotion that Regulates the Exercise Behavior to Ensure the Protection of Whole Body Homeostasis. *Front. Physiol.* **2012**, *3*, 82. [CrossRef] [PubMed] [PubMed Central]
27. Ganesan, G.; Leu, S.-Y.; Cerussi, A.; Tromberg, B.; Cooper, D.M.; Galassetti, P. Cerebral and Muscle Tissue Oxygenation During Incremental Cycling in Male Adolescents Measured by Time-Resolved Near-Infrared Spectroscopy. *Pediatr. Exerc. Sci.* **2016**, *28*, 275–285. [CrossRef] [PubMed] [PubMed Central]
28. Jung, R.; Moser, M.; Baucsek, S.; Dern, S.; Schneider, S. Activation patterns of different brain areas during incremental exercise measured by near-infrared spectroscopy. *Exp. Brain Res.* **2015**, *233*, 1175–1180. [CrossRef] [PubMed]
29. Racinais, S.; Buchheit, M.; Girard, O. Breakpoints in ventilation, cerebral and muscle oxygenation, and muscle activity during an incremental cycling exercise. *Front. Physiol.* **2014**, *5*, 142. [CrossRef] [PubMed] [PubMed Central]
30. Rooks, C.R.; Thom, N.J.; McCully, K.K.; Dishman, R.K. Effects of incremental exercise on cerebral oxygenation measured by near-infrared spectroscopy: A systematic review. *Prog. Neurobiol.* **2010**, *92*, 134–150. [CrossRef] [PubMed]
31. Bailey, S.P.; E Hall, E.; E Folger, S.; Miller, P.C. Changes in EEG during graded exercise on a recumbent cycle ergometer. *J. Sports Sci. Med.* **2008**, *7*, 505–511. [PubMed]
32. Brümmer, V.; Schneider, S.; Strüder, H.; Askew, C. Primary motor cortex activity is elevated with incremental exercise intensity. *Neuroscience* **2011**, *181*, 150–162. [CrossRef] [PubMed]
33. Robertson, C.V.; Marino, F.E. Prefrontal and motor cortex EEG responses and their relationship to ventilatory thresholds during exhaustive incremental exercise. *Eur. J. Appl. Physiol.* **2015**, *115*, 1939–1948. [CrossRef] [PubMed]
34. Enders, H.; Nigg, B.M. Measuring human locomotor control using EMG and EEG: Current knowledge, limitations and future considerations. *Eur. J. Sport Sci.* **2016**, *16*, 416–426. [CrossRef] [PubMed]
35. Thompson, T.; Steffert, T.; Ros, T.; Leach, J.; Gruzelier, J. EEG applications for sport and performance. *Methods* **2008**, *45*, 279–288. [CrossRef] [PubMed]
36. Park, J.L.; Fairweather, M.M.; Donaldson, D.I. Making the case for mobile cognition: EEG and sports performance. *Neurosci. Biobehav. Rev.* **2015**, *52*, 117–130. [CrossRef] [PubMed]
37. Klimesch, W.; Doppelmayr, M.; Schimke, H.; Pachinger, T. Alpha frequency, reaction time, and the speed of processing information. *J. Clin. Neurophysiol.* **1996**, *13*, 511–518. [CrossRef] [PubMed]
38. Samaha, J.; Bauer, P.; Cimaroli, S.; Postle, B.R. Top-down control of the phase of alpha-band oscillations as a mechanism for temporal prediction. *Proc. Natl. Acad. Sci. USA* **2015**, *112*, 8439–8444. [CrossRef] [PubMed] [PubMed Central]
39. De, A.; Mondal, S. Yoga and Brain Wave Coherence: A Systematic Review for Brain Function Improvement. *Heart. Mind* **2020**, *4*, 33–39. [CrossRef]
40. Kropotov, J.D. Beta and gamma rhythms. In *Functional Neuromarkers for Psychiatry: Applications for Diagnosis and Treatment*; Academic Press: Cambridge, MA, USA, 2016; pp. 107–119. [CrossRef]
41. Abhang, P.A.; Gawali, B.W.; Mehrotra, S.C. *Introduction to EEG- and Speech-Based Emotion Recognition*; Elsevier: Amsterdam, The Netherlands, 2016.
42. Hilty, L.; Langer, N.; Pascual-Marqui, R.; Boutellier, U.; Lutz, K. Fatigue-induced increase in intracortical communication between mid/anterior insular and motor cortex during cycling exercise. *Eur. J. Neurosci.* **2011**, *34*, 2035–2042. [CrossRef] [PubMed]
43. Ludyga, S.; Gronwald, T.; Hottenrott, K. Effects of high vs. low cadence training on cyclists' brain cortical activity during exercise. *J. Sci. Med. Sport* **2016**, *19*, 342–347. [CrossRef] [PubMed]
44. Pontifex, M.B.; Hillman, C.H. Neuroelectric and behavioral indices of interference control during acute cycling. *Clin. Neurophysiol.* **2007**, *118*, 570–580. [CrossRef] [PubMed]

45. Schneider, S.; Rouffet, D.M.; Billaut, F.; Strüder, H.K. Cortical current density oscillations in the motor cortex are correlated with muscular activity during pedaling exercise. *Neuroscience* **2013**, *228*, 309–314. [CrossRef] [PubMed]
46. Brümmer, V.; Schneider, S.; Abel, T.; Vogt, T.; Strüder, H.K. Brain cortical activity is influenced by exercise mode and intensity. *Med. Sci. Sports Exerc.* **2011**, *43*, 1863–1872. [CrossRef] [PubMed]
47. Mechau, D.; Liesen, H.; Mücke, S.; Weiß, M. Effect of increasing running velocity on electroencephalogram in a field test. *Eur. J Appl. Physiol.* **1998**, *78*, 340–345. [CrossRef] [PubMed]
48. Dykstra, R.M.; Hanson, N.J.; Miller, M.G. Brain activity during self-paced vs. fixed protocols in graded exercise testing. *Exp. Brain Res.* **2019**, *237*, 3273–3279. [CrossRef] [PubMed]
49. Robertson, C.V.; Marino, F.E. A role for the prefrontal cortex in exercise tolerance and termination. *J. Appl. Physiol.* **2016**, *120* 464–466. [CrossRef] [PubMed]
50. Robertson, C.V.; Marino, F.E. Last Word on Viewpoint: A role for the prefrontal cortex in exercise tolerance and termination. *J Appl. Physiol.* **2016**, *120*, 470. [CrossRef] [PubMed]
51. Borg, G.A. Psychophysical bases of perceived exertion. *Med. Sci. Sports Exerc.* **1982**, *14*, 377–381. [CrossRef] [PubMed]
52. Maby, E. Practical Guide to Performing an EEG Experiment. In *Brain–Computer Interfaces 2: Technology and Applications*; Clerc, M., Bougrain, L., Lotte, F., Eds.; John Wiley & Sons, Inc.: Hoboken, NJ, USA, 2016; pp. 163–177.
53. Boucsein, W.; Fowles, D.C.; Grimnes, S.; Ben-Shakhar, G.; Roth, W.T.; Dawson, M.E.; Filion, D.L. Society for Psychophysiological Research Ad Hoc Committee on Electrodermal Measures Publication recommendations for electrodermal measurements *Psychophysiology* **2012**, *49*, 1017–1034. [CrossRef] [PubMed]
54. Fowles, D.C.; Christie, M.J.; Edelberg, R.; Grings, W.W.; Lykken, D.T.; Venables, P.H. Committee report. Publication recommendations for electrodermal measurements. *Psychophysiology* **1981**, *18*, 232–239. [CrossRef] [PubMed]
55. Borg, G.; Hassmén, P.; Lagerström, M. Perceived exertion related to heart rate and blood lactate during arm and leg exercise. *Eur. J. Appl. Physiol. Occup. Physiol.* **1987**, *56*, 679–685. [CrossRef] [PubMed]
56. Midgley, A.W.; McNaughton, L.R.; Carroll, S. Time at VO_{2max} during intermittent treadmill running: Test protocol dependent or methodological artefact? *Int. J. Sports Med.* **2007**, *28*, 934–939. [CrossRef] [PubMed]
57. Pontifex, M.B.; Hillman, C.H. Neuroelectric measurement of cognition during aerobic exercise. *Methods* **2008**, *45*, 271–278. [CrossRef] [PubMed]
58. Damon, C.; Liutkus, A.; Gramfort, A.; Essid, S. Non-negative Tensor Factorization for single-channel EEG artifact rejection. In Proceedings of the 2013 IEEE International Workshop on Machine Learning for Signal Processing (MLSP), Southampton, UK, 22–25 September 2013; pp. 1–6.
59. Noakes, T.D. Time to move beyond a brainless exercise physiology: The evidence for complex regulation of human exercise performance. *Appl. Physiol. Nutr. Metab.* **2011**, *36*, 23–35. [CrossRef] [PubMed]
60. Desai, R.; Tailor, A.; Bhatt, T. Effects of yoga on brain waves and structural activation: A review. *Complement. Ther. Clin. Pract.* **2015**, *21*, 112–118. [CrossRef] [PubMed]
61. Klimesch, W.; Schimke, H.; Pfurtscheller, G. Alpha frequency, cognitive load and memory performance. *Brain Topogr.* **1993**, *5*, 241–251. [CrossRef] [PubMed]
62. Klimesch, W. EEG-alpha rhythms and memory processes. *Int. J. Psychophysiol.* **1997**, *26*, 319–340. [CrossRef] [PubMed]
63. Neuper, C.; Pfurtscheller, G. Event-related dynamics of cortical rhythms: Frequency-specific features and functional correlates. *Int. J. Psychophysiol.* **2001**, *43*, 41–58. [CrossRef] [PubMed]
64. Sauseng, P.; Klimesch, W. What does phase information of oscillatory brain activity tell us about cognitive processes? *Neurosci. Biobehav. Rev.* **2008**, *32*, 1001–1013. [CrossRef] [PubMed]
65. Gola, M.; Magnuski, M.; Szumska, I.; Wróbel, A. EEG beta band activity is related to attention and attentional deficits in the visual performance of elderly subjects. *Int. J. Psychophysiol.* **2013**, *89*, 334–341. [CrossRef] [PubMed]
66. Sowndhararajan, K.; Kim, S. Influence of Fragrances on Human Psychophysiological Activity: With Special Reference to Human Electroencephalographic Response. *Sci. Pharm.* **2016**, *84*, 724–751. [CrossRef] [PubMed] [PubMed Central]
67. Baumeister, J.; Reinecke, K.; Schubert, M.; Schade, J.; Weiss, M. Effects of induced fatigue on brain activity during sensorimotor control. *Eur. J. Appl. Physiol.* **2012**, *112*, 2475–2482. [CrossRef] [PubMed]
68. Mitchell, J.H.; Blomqvist, G. Maximal Oxygen Uptake. *N. Engl. J. Med.* **1971**, *284*, 1018–1022. [CrossRef] [PubMed]
69. Taylor, H.L.; Buskirk, E.; Henschel, A.; Poole, D.C.; Jones, A.M.; Périard, J.D.; Racinais, S.; Joyner, M.J.; Casey, D.P.; Tipton, C.M.; et al. Maximal Oxygen Intake as an Objective Measure of Cardio-Respiratory Performance. *J. Appl. Physiol.* **1955**, *8*, 73–80. [CrossRef] [PubMed]
70. Eisenmann, J.C.; Malina, R.M. Secular trend in peak oxygen consumption among United States youth in the 20th century. *Am. J. Hum. Biol.* **2002**, *14*, 699–706. [CrossRef] [PubMed]
71. Noakes, T.D.; St Clair Gibson, A. Logical limitations to the "catastrophe" models of fatigue during exercise in humans. *Br. J. Sports Med.* **2004**, *38*, 648–649. [CrossRef] [PubMed] [PubMed Central]
72. Lacour, J.-R.; Messonnier, L.; Bourdin, M. The leveling-off of oxygen uptake is related to blood lactate accumulation. Retrospective study of 94 elite rowers. *Eur. J. Appl. Physiol.* **2007**, *101*, 241–247. [CrossRef] [PubMed]
73. Noakes, T.D. Maximal oxygen uptake: "classical" versus "contemporary" viewpoints: A rebuttal. *Med. Sci. Sports Exerc.* **1998**, *30*, 1381–1398. [CrossRef] [PubMed]

74. Bassett, D.R., Jr.; Howley, E.T. Maximal oxygen uptake: "classical" versus "contemporary" viewpoints. *Med. Sci. Sports Exerc.* **1997**, *29*, 591–603. [CrossRef] [PubMed]
75. Marcora, S. Perception of effort during exercise is independent of afferent feedback from skeletal muscles, heart, and lungs. *J. Appl. Physiol.* **2009**, *106*, 2060–2062. [CrossRef] [PubMed]
76. Noakes, T.D.; Marino, F.E. Does a central governor regulate maximal exercise during combined arm and leg exercise? A rebuttal. *Eur. J. Appl. Physiol.* **2008**, *104*, 757–759. [CrossRef] [PubMed]
77. Borg, G. Perceived exertion as an indicator of somatic stress. *Scand J. Rehabil. Med.* **1970**, *2*, 92–98. [CrossRef] [PubMed]
78. Borg, G. *Borg's Perceived Exertion and Pain Scales*; Human Kinetics: Champaign, IL, USA, 1998; p. 104.
79. Chen, M.J.; Fan, X.; Moe, S.T. Criterion-related validity of the Borg ratings of perceived exertion scale in healthy individuals: A meta-analysis. *J. Sports Sci.* **2002**, *20*, 873–899. [CrossRef] [PubMed]
80. Coquart, J.B.; Legrand, R.; Robin, S.; Duhamel, A.; Matran, R.; Garcin, M. Influence of successive bouts of fatiguing exercise on perceptual and physiological markers during an incremental exercise test. *Psychophysiology* **2009**, *46*, 209–216. [CrossRef] [PubMed]
81. Eston, R.G.; Davies, B.L.; Williams, J.G. Use of perceived effort ratings to control exercise intensity in young healthy adults. *Eur. J. Appl. Physiol.* **1987**, *56*, 222–224. [CrossRef] [PubMed]
82. Eston, R.G.; Faulkner, J.A.; Mason, E.A.; Parfitt, G. The validity of predicting maximal oxygen uptake from perceptually regulated graded exercise tests of different durations. *Eur. J. Appl. Physiol.* **2006**, *97*, 535–541. [CrossRef] [PubMed]
83. Hosang, L.; Mouchlianitis, E.; Guérin, S.M.R.; Karageorghis, C.I. Effects of exercise on electroencephalography-recorded neural oscillations: A systematic review. *Int. Rev. Sport Exerc. Psychol.* **2022**, 1–54. [CrossRef]
84. Nielsen, B.; Nybo, L. Cerebral changes during exercise in the heat. *Sports Med.* **2003**, *33*, 1–11. [CrossRef] [PubMed]
85. Nybo, L.; Nielsen, B.; Sato, K.; Oue, A.; Yoneya, M.; Sadamoto, T.; Ogoh, S.; De Pauw, K.; Roelands, B.; Marušič, U.; et al. Perceived exertion is associated with an altered brain activity during exercise with progressive hyperthermia. *J. Appl. Physiol.* **2001**, *91*, 2017–2023. [CrossRef] [PubMed]
86. Nakata, H.; Oshiro, M.; Namba, M.; Shibasaki, M. Effects of aerobic exercise under different thermal conditions on human somatosensory processing. *Am. J. Physiol. Regul. Integr. Comp. Physiol.* **2016**, *311*, R629–R636. [CrossRef] [PubMed]
87. Shibasaki, M.; Namba, M.; Oshiro, M.; Kakigi, R.; Nakata, H. Suppression of cognitive function in hyperthermia; From the viewpoint of executive and inhibitive cognitive processing. *Sci. Rep.* **2017**, *7*, srep43528. [CrossRef] [PubMed] [PubMed Central]
88. Pogosyan, A.; Gaynor, L.D.; Eusebio, A.; Brown, P. Boosting cortical activity at Beta-band frequencies slows movement in humans. *Curr. Biol.* **2009**, *19*, 1637–1641. [CrossRef] [PubMed] [PubMed Central]
89. Chung, J.W.; Ofori, E.; Misra, G.; Hess, C.W.; Vaillancourt, D.E. Beta-band activity and connectivity in sensorimotor and parietal cortex are important for accurate motor performance. *NeuroImage* **2017**, *144*, 164–173. [CrossRef] [PubMed] [PubMed Central]
90. Jenkinson, N.; Brown, P. New insights into the relationship between dopamine, beta oscillations and motor function. *Trends Neurosci.* **2011**, *34*, 611–618. [CrossRef] [PubMed]
91. Demiralp, T.; Ademoglu, A.; Istefanopulos, Y.; Başar-Eroglu, C.; Başar, E. Wavelet analysis of oddball P300. *Int. J. Psychophysiol.* **2001**, *39*, 221–227. [CrossRef] [PubMed]
92. Pesonen, M.; Hämäläinen, H.; Krause, C.M. Brain oscillatory 4–30 Hz responses during a visual n-back memory task with varying memory load. *Brain Res.* **2007**, *1138*, 171–177. [CrossRef] [PubMed]
93. Taylor, J.L.; Amann, M.; Duchateau, J.; Meeusen, R.; Rice, C.L. Neural Contributions to Muscle Fatigue: From the Brain to the Muscle and Back Again. *Med. Sci. Sports Exerc.* **2016**, *48*, 2294–2306. [CrossRef] [PubMed] [PubMed Central]
94. Hottenrott, K.; Taubert, M.; Gronwald, T. Cortical Brain Activity is Influenced by Cadence in Cyclists. *Open Sports Sci. J.* **2013**, *6*, 9–14. [CrossRef]
95. Ludyga, S.; Hottenrott, K.; Gronwald, T. Four weeks of high cadence training alter brain cortical activity in cyclists. *J. Sports Sci.* **2017**, *35*, 1–6. [CrossRef] [PubMed]
96. Takeda, R.; Okazaki, K. Body Temperature Regulation During Exercise and Hyperthermia in Diabetics. In *Diabetes and Its Complications*; IntechOpen: London, UK, 2018; pp. 88–109. [CrossRef]
97. Cantero, J.L.; Atienza, M.; Salas, R.M. Human alpha oscillations in wakefulness, drowsiness period, and REM sleep: Different electroencephalographic phenomena within the alpha band. *Neurophysiol. Clin.* **2002**, *32*, 54–71. [CrossRef] [PubMed]
98. Plattner, K.; Lambert, M.I.; Tam, N.; Lamberts, R.P.; Baumeister, J. Changes in cortical beta activity related to a biceps brachii movement task while experiencing exercise induced muscle damage. *Physiol. Behav.* **2014**, *123*, 1–10. [CrossRef] [PubMed]
99. Büchel, D.; Sandbakk, Ø.; Baumeister, J. Exploring intensity-dependent modulations in EEG resting-state network efficiency induced by exercise. *Eur. J. Appl. Physiol.* **2021**, *121*, 2423–2435. [CrossRef] [PubMed] [PubMed Central]
100. Chaire, A.; Becke, A.; Düzel, E. Effects of Physical Exercise on Working Memory and Attention-Related Neural Oscillations. *Front. Neurosci.* **2020**, *14*, 239. [CrossRef] [PubMed] [PubMed Central]
101. Kao, S.-C.; Wang, C.-H.; Hillman, C.H. Acute effects of aerobic exercise on response variability and neuroelectric indices during a serial n-back task. *Brain Cogn.* **2020**, *138*, 105508. [CrossRef] [PubMed]
102. Kubitz, K.A.; Mott, A.A. EEG power spectral densities during and after cycle ergometer exercise. *Res. Q. Exerc. Sport* **1996**, *67*, 91–96. [CrossRef] [PubMed]
103. Mierau, A.; Schneider, S.; Abel, T.; Askew, C.; Werner, S.; Strüder, H.K. Improved sensorimotor adaptation after exhaustive exercise is accompanied by altered brain activity. *Physiol. Behav.* **2009**, *96*, 115–121. [CrossRef] [PubMed]

104. Gutmann, B.; Zimmer, P.; Hülsdünker, T.; Lefebvre, J.; Binnebößel, S.; Oberste, M.; Bloch, W.; Strüder, H.; Mierau, A. The effects of exercise intensity and post-exercise recovery time on cortical activation as revealed by EEG alpha peak frequency. *Neurosci. Lett.* **2018**, *668*, 159–163. [CrossRef] [PubMed]
105. Amann, M. Central and peripheral fatigue: Interaction during cycling exercise in humans. *Med. Sci. Sports Exerc.* **2011**, *43*, 2039–2045. [CrossRef] [PubMed]
106. Doherty, C.; Keogh, A.; Davenport, J.; Lawlor, A.; Smyth, B.; Caulfield, B. An evaluation of the training determinants of marathon performance: A meta-analysis with meta-regression. *J. Sci. Med. Sport* **2020**, *23*, 182–188. [CrossRef] [PubMed]
107. Marcora, S.M.; Staiano, W. The limit to exercise tolerance in humans: Mind over muscle? *Eur. J. Appl. Physiol.* **2010**, *109*, 763–770. [CrossRef] [PubMed]
108. Meyniel, F.; Sergent, C.; Rigoux, L.; Daunizeau, J.; Pessiglione, M. Neurocomputational account of how the human brain decides when to have a break. *Proc. Natl. Acad. Sci. USA* **2013**, *110*, 2641–2646. [CrossRef] [PubMed] [PubMed Central]

applied sciences

MDPI

Article

Unavoidability and Functionality of Nervous System and Behavioral Randomness

Carlos M. Gómez *, Elena I. Rodríguez-Martínez and María A. Altahona-Medina

Department of Experimental Psychology, University of Seville, 41018 Seville, Spain; elisroma@us.es (E.I.R.-M.); maraltmed@alum.us.es (M.A.A.-M.)
* Correspondence: cgomez@us.es

Abstract: The basic functioning of the central nervous system is based on the opening and closing of ionic channels in the membranes of neurons. The behavior of ionic channels is considered to be a random process with an exponential probability distribution function. The central limit theorem implies that the mean of the sum of random variables generates a distribution in which the new variable tends to be normally distributed. The theorem implicitly implies that randomness can be embedded in a certain probability distribution but does not disappear. The present report will explore the possible implications for the functioning of nervous system and behavior of the constituent neural randomness. The possible functionality of "noise" to increase the exploratory space of nervous and behavioral systems will be considered.

Keywords: ionic channels; random process; psychometric function; geometric distribution; Gaussian distribution; threshold

Citation: Gómez, C.M.; Rodríguez-Martínez, E.I.; Altahona-Medina, M.A. Unavoidability and Functionality of Nervous System and Behavioral Randomness. *Appl. Sci.* **2024**, *14*, 4056. https://doi.org/10.3390/app14104056

Academic Editor: Alexander N. Pisarchik

Received: 26 March 2024
Revised: 2 May 2024
Accepted: 8 May 2024
Published: 10 May 2024

1. Introduction

Behavioral and cognitive activities depend on electrophysiological neural activity based on ionic currents of chemical, mechanical and voltage membrane channels. Opening and closing of these channels are considered as probabilistic random processes and are modeled via an exponential probability distribution [1]. Open dwelling times follow a simple exponential model, but to model the closed channel interval, a sum of many exponential functions is needed. The quantal nature of synaptic vesicle liberation is another source of neural activity variability [2,3]. The stochastic activity of ionic channels would be an intrinsic source of randomness in the nervous system, while the random structure of stimulation would be another source of variability, i.e., in the case of light perception, the quantal nature of photon emission. The large number of ionic channels cannot override this intrinsic and extrinsic variability, and experimental evidence demonstrates that membrane potential presents fluctuations that affect the generation of spontaneous action potentials [4,5]. In situations in which neurons receive steady inputs, the inter-spike interval distribution shows broad variability [6,7]. During constant inputs, the exact time at which a given spike is generated reflects the internal noise more than the inputs that the neuron receives [8].

Given that behavioral and cognitive activities depend on the activity of neural networks activity, and given that those are dependent on the activity of individual neurons that, at a very basic level, depend on the probabilistic behavior of ionic channels, it can be inferred that high-order levels of brain function as cognitive and behavioral activities should show a stochastic component, given that operating with a combination of random variables preserves the random nature of the outcome. One important consequence of that is that the sum of the random variables is another random variable, which converges to the Gaussian distribution when N tends toward infinity: the so-called central limit theorem [9].

The present report tries to demonstrate that there are traces of random processes at the organismic level, as well as that a possible interpretation of that would be the successive alternation of geometric and Gaussian distribution across levels of integration in the central nervous system. This would be achieved by reviewing the four different approaches followed by our group:

The lever press response of rats during variable interval schedules of reinforcement (VI) [10,11]; the fluctuations of the macroscopic activity of the abducens nerve [6], giving place to the ocular tremor; the perceptual changes during the observation of images in eye rivalry and ambiguous images [12]; and the psychometric (also called psychophysics) function [13]. The integration of unavoidable levels of neural randomness at the organismic level would facilitate flexibility in behavior [14].

2. The Inter-Response Time (IRT) Intervals in Variable Interval Schedules of Reinforcement

Animal behavior, although predictable in many instances, could also be considered to be unpredictable and modeled through stochastic processes. In fact, the reinforcement of certain responses (the increases in certain motor responses when generating positive outcomes in the subjects) in animal behavior would be a selective process on a pool of behavioral variability, in a similar manner to that in which it occurs in other biological systems such as evolutionary dynamics or the immune system [15,16].

Given the random structure of reinforcement in variable interval schedules of reinforcement (VI), the animal lever press in VI schedule of reinforcement would be particularly suited for modeling and analyzing random processes. In VI, there is a variable time for a response to be reinforced, and the response reinforced can occur at any time in the interval, following a scheduled random pattern inside the interval. In a VI with a 60 s period, the reinforced response occurs at any time in this period, as selected by random number generators in the Skinner box controller program. In this study [10], it was checked if stochasticity appears in the response behavior of rats in VI schedules by computing the autocorrelation function and the probability density function of the IRTs. In fact, the IRT series showed no autocorrelation, as tested through autocorrelation values inside the so-called Bartlett bands (Figure 1).

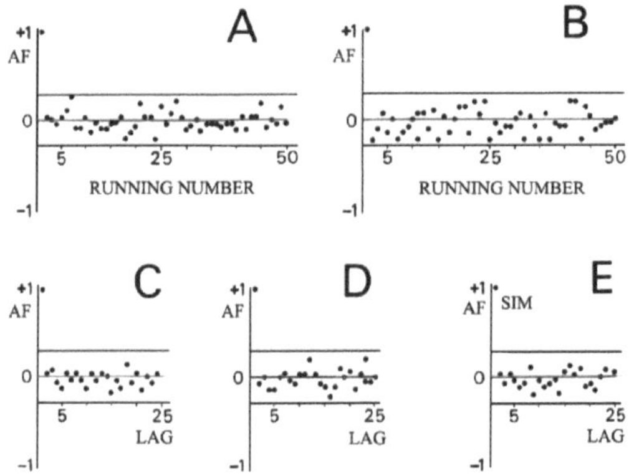

Figure 1. Autocorrelation function (AF) of the individual IRT values (**A**,**B**) and cumulated IRT values in time bins (**C**,**D**). The same for randomly generated IRT values (**E**). Notice that autocorrelation values are always inside the Bartlett bands, indicating no time dependency in the IRT series. Adapted with permission from [10]. 1992, Springer.

These statistical results can be interpreted as if the IRT series could be modeled via a random process with no time serial structure [17]. The latter analysis suggests that an internal non-periodic process underlies the decision process of lever response during VI. In this case, it is possible that, as suggested by Staddon and Simmelhag [15], the inner behavioral dynamics would profit from using a random internal generator to adaptively adjust its behavior to the unpredictable VI schedule of reinforcement. Interestingly, the

IRT frequency distribution was modeled as a gamma and/or a Poisson distribution, both distributions considered as distributions modeling random events:

The gamma distribution following Johnson and Kotz [18] is modeled as follows:

$$Px(X) = \frac{(X-y)^{\alpha-1} \cdot e^{-(X-y)/\beta}}{\beta^{\alpha} \cdot \Gamma(\alpha)} \tag{1}$$

Alpha = arithmetic mean/2 * (arithmetic mean − geometric mean)
Beta = arithmetic mean/Alpha.
y = when n is large, it is reasonable to estimate y as a value smaller than the smallest observed value [18].

And

$$\Gamma(\alpha) = \int_{0}^{\infty} t^{\alpha-1} \cdot e^{-t} dt \tag{2}$$

The mean/variance ratio close to 1 (Fano factor) suggested that the IRT time series could be distributed as a Poisson process. The Poisson probability density function is defined [19] as follows:

$$P(X = x) = e^{-h} \cdot h^{x}/x! \tag{3}$$

The quantity h is the only parameter of the distribution and can be estimated as the variance or the mean of the data. And x is the number of events in a certain time or space. Both distributions fit the IRT values, although the Poisson distribution did so in a significant manner in a higher percentage of cases than the gamma distribution (90% and 60% of cases, respectively).

Given that a Poisson probability density function represents the probability that a random variable appears n times in a fixed period of time or space [19], as well as that the gamma probability density function of IRT in this context would be generated as produced by the counting of n events from a random variable, both models of IRTs are based on random variables. The latter observation, joined to the previous comment about the lack of a time serial structure, points to the idea that IRTs in VI behavioral reinforcement schedules could be modeled as random variables. From this perspective, learning would permit us to reduce non-adaptive random responses and amplify adaptive random responses [15,20].

3. Ocular Tremor

The ocular tremor is a micromovement of the eye (several min of arc). This tiny movement has been proved to be functional to prevent vision fading, which can be produced by the adaptation of photoreceptors if the light entering the eye is static during eye fixations [21]. The tiny tremor ensures that a given photoreceptor would receive constant light in terms of intensity and quality [22], avoiding adaption and vision fading. The purpose of this study [7] was to demonstrate that the statistical properties of the neural activity of the extraocular muscles were the cause of this movement, after filtering neural activity through the ocular mechanics. Taking into account the possible stochastic nature of the behavior of abducens motoneurons [6], as well as the fact that the statistical properties of the spike count of the abducens nerve [7] would also be a random process due to the central limit theorem, then a random behavior for the ocular tremor movement is implied. Notice that ocular tremor movement is caused by the activity of the extraocular muscles as a by-product of the oculomotor nerves.

This endeavor was possible given the large amount of data about the activity of individual motoneurons in the abducens nucleus and nerve [6,23]. The abducens nucleus and nerve control the lateral rectus muscle, which controls the eye's outward direction (abduction), while the motoneurons of the oculomotor nucleus control the inward direction (adduction). The here-reviewed simulation only took into consideration neural spiking at the abducens nerve, assuming that the activity of the oculomotor nerve would be symmetric to the pattern of activity of the abducens nucleus, and then both would achieve a similar result. Therefore, the objective of the simulation of the abducens nerve was recreate possible global activity during eye fixations to define if there was any serial time dependency (rhythmic activity), as well as to compute the possible effects of the abducens neural activity in the eye position, and then determine if a pattern similar to this pattern of ocular tremor was replicated.

Figure 2 shows the procedure followed to compute the spiking rate of the abducens nerve from the sum of the spikes generated by individual neurons (Figure 2A). The parameters used for computing the spike train for each neuron are described in Table 1 of Gómez et al. [7] and depicted in Figure 2A. The abducens nerve global activity was computed as the sum of the spike trains of individual motoneurons (Figure 2B). The frequency distribution of the global activity was fitted by a Poisson distribution, despite resembling a Gaussian distribution when a high number of neurons were active, as occurs for the more eccentric positions. Then, the Power Spectral Density (PSD) of the spiking global activity carried by the abducens nerve was computed for different eye positions (Figure 3A,C), producing a rather flat spectrum (Figure 3A), being flat for high frequencies and peaking at the mean frequency of motoneuron activity for a given eye position (Figure 3B). A final analysis was performed to observe the spectrum of the eye position. The PSD function of eye position falls monotonically with frequency due to the filtering of the ocular mechanics model, but for the eccentric position of the eye, a peak remained at the same mean frequency as the motoneurons (Figure 3D).

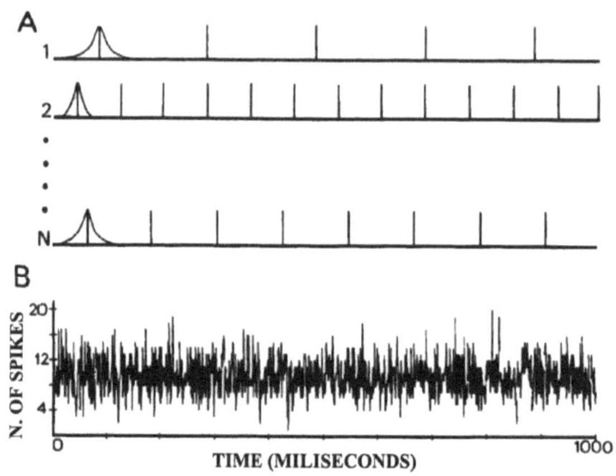

Figure 2. Computation of the total activity of the abducens nerve. (**A**) The activity of three motoneurons of the left abducens nucleus during eye fixation. The clock-like firing activity is modulated by random noise following a normal distribution. (**B**) The simulation of the neural discharge carried by the abducens nerve for a central position of the eye in the orbit computed as the sum of the spike trains of single neurons. Adapted with permission from [7]. 1989, Elsevier.

To compute the PSD of eye tremor (Figure 3B,D), the PSD of the simulated abducens nerve spike count was filtered using a model of the ocular mechanics (following Robinson [24]) by multiplying the latter neural signal by the squared transfer function (TFz) of the second-order differential equation modeling ocular mechanics.

$$TF^2 = 1/((1 + 2\pi f \cdot T1)^2) \cdot (1 + 2\pi f \cdot T2)^2) \quad (4)$$

T1 and T2 were calculated from empirical data providing the position, velocity, and acceleration coefficients (k, r, and m, respectively) of the second-order differential equation modeling the ocular mechanics:

$$R(t) = R0 + k \cdot Pos + r \cdot dPos/dt + m \cdot dPos^2/dt^2 \quad (5)$$

R(t): Force exerted by the muscle (function of abducens nerve discharge);

R0: Constant;
Pos: Eye angular position;
dPos/dt: Eye angular velocity;
$dPos^2/dt^2$: Eye angular acceleration.

Then, T1 was calculated as r/k and T2 as m/k.

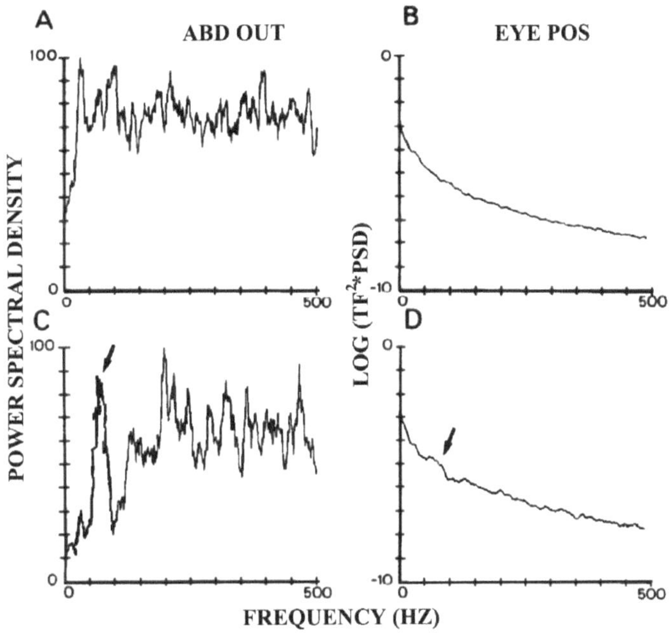

Figure 3. Power Spectral density. (**A**,**C**) Averaged Power Spectral Density (PSD) of 50 sequences of the simulated neural discharge of the abducens nerve (ABD OUT) for 0° (**A**) and 10° eye positions (**C**). (**B**,**D**) PSD functions of the 0° (**B**) and 10° eye positions (EYE POS) (**D**). Adapted with permission from [7]. 1989, Elsevier.

Apart from the demonstration in the present report that the ocular tremor would be produced by the statistical properties of the oculomotor nerves, given the objective of the present report, it is important to consider that at the organismic level, there are traces of random neural noise activity, so it can be proposed that the deviations from the strict pacemaker activity of abducens neurons (following a normal distribution), producing a relatively flat frequency spectrum of the neural nerve discharge, except for the mean of the neural discharge of neurons in a certain eye position, would be an indication of a random process in the global activity of the macroscopic abducens cranial nerve. This suggestion can also be sustained by the Poisson distribution of the nerve spike discharge, approaching a Gaussian distribution as the number of active motoneurons increases.

4. The Perceptual Transitions during the Perception of Ambiguous Figures and Eye Rivalry

During perception of ambiguous figures such as the Necker cube, as well as the perception of incompatible images presented in each eye (eye rivalry phenomenon), the conscious perceptions of the two incompatible images alternate. It is possible, given the broad variability in the perceptual duration of the presented images, that transitions between percepts would be animated via random processes. In fact, at least two models have proposed such random influence at the organismic behavioral level. Logothetis [25] proposed that perceptual duration histograms are fitted by gamma distributions and Lehky [26] by a log-normal probability. Lehky [26] also showed that the time series presented no autocorrelations and that this time series was not explained by a chaotic system. The demonstration of not being a chaotic system was based on the correlation dimensions and nonlinear forecasting of the time series, giving ground for a random process to govern the transitions of perceptual states. Gómez et al. [12] proposed that alternation between percepts occurs as a form of competition between the neural networks representing the two percepts. Such networks should have a mutual inhibition, as depicted in Figure 4A, and each network representing an image should have a stochastic dynamic as represented in Figure 4B.

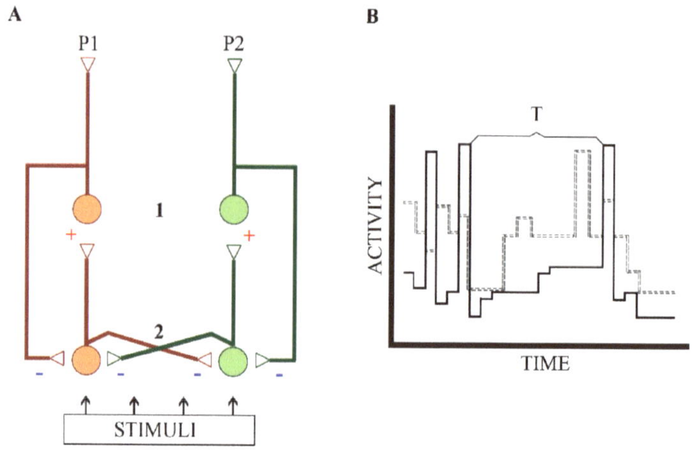

Figure 4. Competition model for the perception of ambiguous figures and eye rivalry. (**A**) Two independent networks that are competing (mutually inhibiting each other) to emerge in the perceptual field (P1 and P2). 1 represents the conscious perceptual level and 2 the sensory processing level. (**B**). Only the network presenting the higher activity would reach the perceptual threshold, with T representing the conscious perception of the network represented by the dotted line.

The competitive model assumes that the expression of a given percept occurs when the activity of its underlying neural network obtains a value higher than that of the alternative network [27]. The probability p is estimated such that the network that allows the representation of a certain percept wins the competition, and (1 − p) is the probability of the opposite percept winning the competition.

Then, the frequency of cases in which a certain percept lasts a certain duration (f(t)): Duration Time) obtains a particular value between time 0 and time t, which is as follows:

$$f(t) = (1 - p)^{(t - 1)} \cdot p \cdot N \tag{6}$$

This equation implies that the frequency of cases in which a perception time obtains a particular value (f(t)) depends on how many times it wins the competition on a time scale between t = 1 and t. Equation (6) corresponds to the geometric probability density function (Figure 5A) multiplied by the number of individual perceptions (N).

f(t) = frequency: The number of times a perception lasts a period of t (N = total number of individual perceptions).
$(1 - p)^{(t - 1)}$ = The probability that the current percept network wins the competition. It must be multiplied by N to obtain the frequency histogram of the duration of a percept.
p: the alternative percept);
t − 1 = The number of times that the network of one of the percepts wins in a row and current perception is maintained

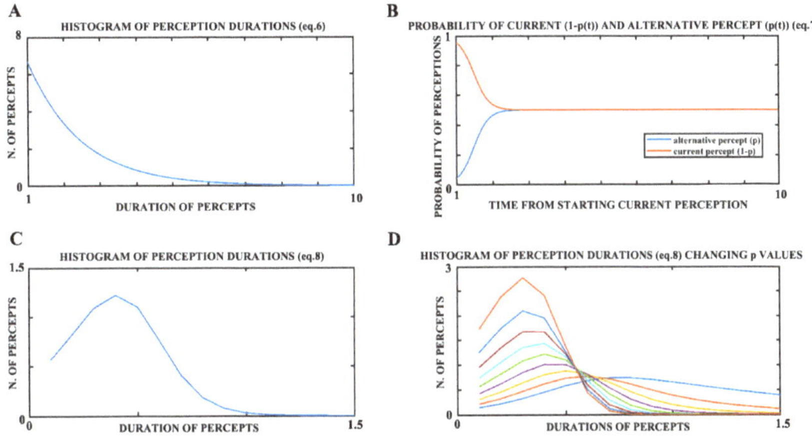

Figure 5. Durations of perceptions with p = 0.5. (**A**) Frequency histogram of perception durations following a geometric distribution (Equation (6)). (**B**) Change in p values with time elapsed from the previous perceptual change (Equation (7)). (**C**) Frequency histogram of perception durations following a geometric distribution, modulating p as a function of the time elapsed from the previous perceptual change (p(t)) (Equation (8)). (**D**) Same as (**C**) but changing p values from 0.1 to 0.9.

The values of (1 − p) and p can be made to be dependent on the time in which the transition occurs. This modification of the geometric distribution permitted us to increase the fitting of experimental data [12] and supposes, that after a transition, there is a higher probability for the network representing the current perceived object to be defeated in the competition. Equation (7) and Figure 5B imply that the average probability that the win probability of the neural network representing the alternative percept (p) relative to the one that won the competition (1 − p) increases over time, while the probability that the current perception would be defeated in the competition increases with time. The most probable physiological mechanism would be the habituation of the neural network representing the current perception. Therefore, the probability that a percept is consciously perceptible during a certain time will be a function of the time since its perception was established (Equation (8)). Therefore, the probability of perception is now a function of time since the last change in perception in this bistable perceptual system (p(t)) (Figure 5B).

$$p(t) = (1/(1 + (e\hat{}((-t \cdot A) + e^2))) \cdot p \tag{7}$$

A = A parameter used to modulate the curvature of the sigmoid;
e^2 is introduced into the sigmoid equation to have the origin at zero.

Finally, the frequency distribution of perception times (Figure 5C) should be

$$f(t) = (1 - p(t))^{(t-1)} \cdot p(t) \cdot N; \tag{8}$$

The parameter A must be estimated from the empirical distribution of f from time 0 to the mode of the empirical distribution of f. To estimate the parameter p, only the right tail of the f distribution will be considered, because in this part of Equation (7), p(t) and (1 − p(t)) have reached the asymptotic level, and p(t) has a value close to the asymptotic value p.

The fitting of this model was successful for fitting the frequency histograms of the perceptual times of both the Necker cube and eye rivalry [12]. Interestingly, the mean time of perception was modulated by attention. This experimental result can be accommodated in the model by modulating the value of parameter p (Figure 5D), which changes the shape of the frequency histogram of perception durations. The latter finding suggests that attention, a higher cognitive function, is able to modulate the neural activity of a given representational network.

The model has also been successfully applied to another situation: the behavior of the lever pressing in the VI schedule of reinforcement [11]. In this approach, the same model is applied to motor networks representing the lever pressing (p), as well as any other possible alternative behavior (1 − p). The use of this modified geometric distribution (with the parameter p being dependent of time) would be an alternative to the gamma and Poisson distributions for fitting the IRTs, as previously described [10]. This fitting approach would not only highlight the random process for the IRTs, but it would also be based on the competition between the lever pressing behavior and any other possible behavior, in a similar manner to the competition between alternative percepts in ambiguous figures and eye rivalry.

5. The Psychometric Function (Also Called Psychophysics Function)

According to the America Psychological Association, this function refers to "the relationship between a stimulus and judgments about the stimulus, as expressed in a mathematical formula". The relationship between the intensity of the stimulation and the probability of reaching the perceptually conscious threshold is described by an asymptotic sigmoid [28]. This relationship indicated a continuous increase in the probability of perception with stimulus intensity increase.

One possibility of generating such probabilistic perceptual threshold would be related to the neural noise of sensory channels, in which, on certain occasions, the external energy is sufficient for reaching the threshold, or not in other cases. The probabilistic threshold would be more critical for low-intensity sensory stimulation, in which the intensity of neural noise can be critical for reaching the perceptual threshold. This possibility is exemplified by the phenomenon of stochastic resonance, which refers to the fact that the addition of a certain amount of noise leads to better information transmission [29]. In the case of low-intensity stimuli, stochastic resonance would facilitate reaching the perceptual threshold.

This study tried to explain the probabilistic nature of the psychometric function as a consequence of the random behavior of voltage-gated ionic channels. The approach followed tried to capture the stochastic nature of the perceptual threshold based on neural noise, without entering into an almost impossibly detailed description of the process in neurons and neural assemblies but keeping the essentials of the already-demonstrated threshold process approach, which holds for ionic channels, action potential generation, and the perceptual threshold. This approach permits the exclusion of many of the mechanistic details, without losing the dynamic essentials. In fact, the influence of neural noise on the discharge of neurons [8], as well as the influence of noisy spike trains on perception [30] has already been addressed. The simulation, best defined as numerical exploration, brought from the random opening of ionic channels to the psychometric function is described in Figures 6–8 (details in Gómez, [13]).

Figure 6A shows a series of simulated closings and openings of ionic channels selected from a Bernoulli random process. The frequency histogram of the opening durations of an ionic channel is displayed in Figure 6B, showing an exponential decay, as has been observed empirically [1]. The time series of the number of open channels are represented in Figure 6C for three different stimuli intensities. This time series presents a Gaussian distribution (Figure 6D). The increase in the number of open channels for higher stimuli intensity induced an increase in variability (Figure 6E,F).

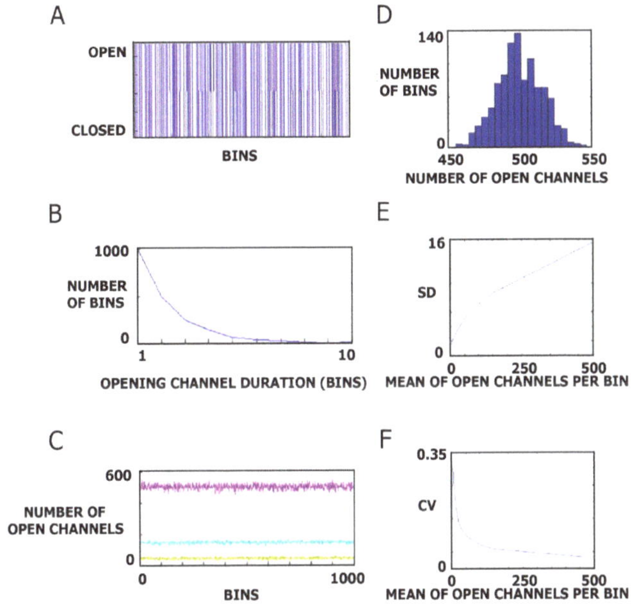

Figure 6. Simulation of ionic channel behavior. Series of openings and closings of an individual ionic channel modeled by a Bernoulli random process (**A**). (**B**) Exponential-like decay of the ionic channel opening durations. (**C**) Time series of the number of ionic channels open at different stimulus intensities. (**D**) Bell-shaped distribution of the number of open channels per time bin. (**E**) Increase in absolute variability (Standard Deviation: SD) and decrease in relative variability (coefficient of variation: CV) (**F**) with the number of open channels. Adapted with permission from [13]. 2008, Springer.

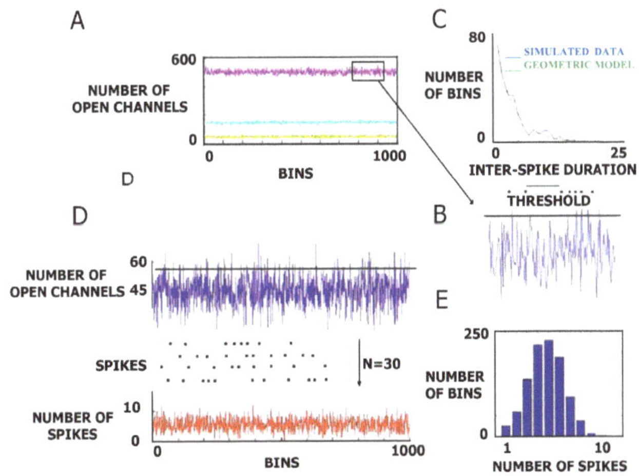

Figure 7. Simulation of the number of spikes in a sensory network representing a perceptual object. (**A**) Same as in Figure 6C. (**B**) The number of open channels in each time bin with a line indicating the hypothetical action potential threshold. (**C**) The geometric model of the frequency distribution of the inter-spike intervals. (**D**) On the upper side, the open channel time series, and on the middle and lower sides, the number of spikes in the sensory network that would represent the perceived object. (**E**) Gaussian-like distribution of the number of spike series of the lower (**D**). Adapted with permission from [13]. 2008, Springer.

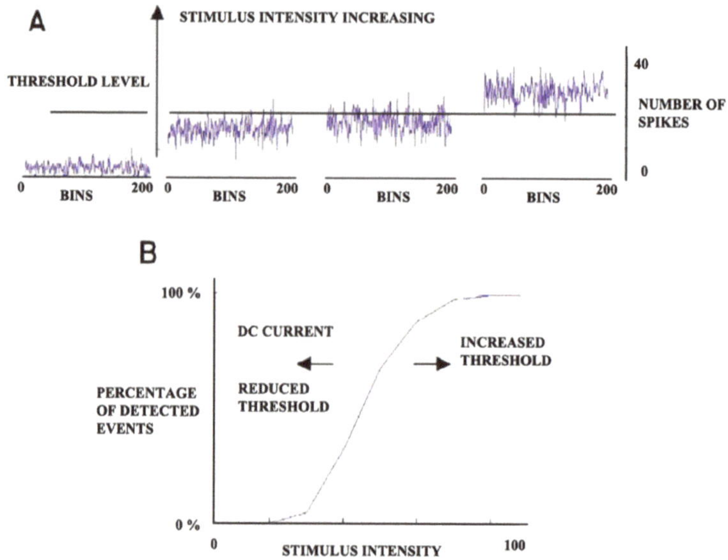

Figure 8. Psychometric or psychophysics function. (**A**) Number of spikes for the increasing intensity of the stimulation. The line shows the arbitrary threshold for stimulus detection. With stimulation intensity, the number of spikes per time bin increases and then the event detection probability also increases. (**B**) Psychometric or psychophysics function. The predicted displacement of the psychometric function caused by reducing or increasing the threshold for event detection or injecting a certain level of DC current is indicated by arrows. Adapted with permission from [13]. 2008, Springer.

Although the relative variability in the number of open channels per time unit decreases with the number of open channels (measured as a coefficient of variation) (Figure 6F), the absolute variability measured as the standard deviation increases monotonically with the number of ionic channels and, consequently, with the stimulus intensity (Figure 6E). The important point in the simulation of the opening and closing of the channels, modeled via a random Bernoulli process, is that the addition of the channel does not ride out the neural noise intrinsic to the neuron, and then the neuron discharge to a constant current input is also influenced by neural noise. This is shown in Figure 7. Figure 7A shows the time series of the number of ionic channels openings induced by a stimulus near the threshold. The number of open channels per time bin appears in Figure 7B. The line indicates the neuron fixed threshold for inducing an action potential. As a consequence of the imposed threshold and the number of open channels, spikes are generated. Figure 7C shows the frequency distribution of the inter-spike interval obtained. The results for the frequency distribution of the inter-spike intervals conform to the expected geometrical distribution, similar to Equation (6), which, in this case, is as follows:

$$\Pr(X = t) = (1 - p)^{t-1} \cdot p \quad (9)$$

Therefore, the probability of the duration of a given inter-spike interval (X) having a value (t) is equal to the probability of the number of open channels being equal to or above the threshold in a certain time bin (p). Then, p must be multiplied by the probability that in the (t − 1) previous time bins, the number of open channels had a value lower than the threshold value (1 − p). Figure 7D shows the number of spikes in a neural network representing the stimulus (below), and Figure 7E shows the bell-shaped frequency distribution of the number of spikes by time unit. The last part of the simulation tries to obtain the psychometric function using a fixed perceptual threshold of the number of spikes in the neural net representing the presented stimulus. Figure 8A shows the time series of the total number of spikes relative to the increasing intensity of the stimulus, and in Figure 8B, the psychometric function is obtained by plotting the frequency of cases in which the number of spikes is higher than the perceptual threshold.

The whole model can be validated at the different levels of description from ionic channels to the psychometric function; I would concentrate here on the psychometric function. The validation of more basic levels can be found in the original publication [13]. The obtained shape in the simulation is similar to the sigmoid psychometric function obtained experimentally [31,32]. The simulation is able to predict the direct current injection of cortical columns representing the direction of moving targets' bias psychometric function to lower stimulus intensities (Figure 8B) [33]. The latter experiment would be, in some sense, an electrophysiological substitute for attention to certain stimuli, which also reduce the perceptual threshold [34]. By increasing the amount of noise, it was possible to bias the sigmoidal psychometric function with a perceptual improvement similar to "stochastic resonance" [29]. Interestingly the stochastic resonance results open the possibility that reducing the threshold for a certain stimulus through attention could possibly be implemented not only via stochastic resonance but also by the increase in the neural activity arriving at the neural net representing the observed percept [34].

From the perspective of the present review, the main point of this numerical exploration is that, at least at a theoretical level, the neural noise activity cannot be ridden out, and it can be observed at a macroscopic perceptual level.

6. Discussion

The aim of this report was to demonstrate that the probabilistic natures of the basic electrophysiological processes of neurons can still be observed at the level of the organism. It must be always considered that given the huge complexity of biological processes, we can only suggest that random modeling is a good approximation of the studied phenomenon, but it is very difficult, if not impossible, to overlook the fact that some hidden or unknown variables are deterministically defining the process at hand. Also, the distinction between chaotic and random behavior is very difficult [26]. The approach followed in the different studies presented here is (i) to model the data via probability density functions, which, theoretically, are the by-products of random processes, and (ii) show no time dependence in the data time series.

The latter two characteristics are the main properties of random processes [9] The neural global discharge of the abducens nerve [7] and the IRTs of lever pressing in a VI schedule of reinforcement fulfilled both conditions. The perception times of ambiguous figures and eye rivalry [12] were fitted by the modification of the geometric distribution, and. qualitatively, there were no time dependencies in the successive events. The obtained psychometric function was based on a simulation in which the organismic response (the psychometric or psychophysics function) was constructed from its more basic underpinnings [13], in the interaction between the stimulus intensity and the random closing and opening of ionic channels. We will comment first on the psychometric function study given its more detailed description extending from the molecular to the organismic level.

The results of the simulation of the psychometric function could lie in the stochastic nature of voltage-gated ionic channels. The computation of a probabilistic psychometric function suggests that variability at the microscopic level is conserved at the perceptual organismic level. An interesting point derived from the simulations is the alternation between geometric, exponential, and Gaussian distribution across levels: exponential for the opening time of channel distributions (please notice the great similarity between geometric and exponential distributions), Gaussian for the number of channels open at a given time, geometric for the inter-spike time intervals of single neurons, and Gaussian for the number of spikes in the neural net representing the perceptual level. Finally, and as a result of the whole process, the sigmoidal shape of the psychometric function, as a consequence of the collective behavior of sensory modules was obtained from the last level of representation: the number of spikes in the neural net representing the percept that reached the perceptual threshold (i.e., [30]).

The exponential and geometric distributions would be obtained from a threshold process on a random variable. Here, the random variable, which can be proposed to underlie the consecution of a voltage sufficient for overriding the energy barrier for the transitions between closed and open ionic channels, would be the thermally induced local movement of ions in the vicinity of the channels. In fact, the voltage and ion concentrations in ionic solutions possibly follow an Ornstein–Uhlenbeck process, with a mean voltage value due to the ion concentration of intra- and extracellular medium and random processes due to the thermal noise and electromagnetic interactions between ions and other ions or water. It is expected that the voltage at a given position should be random due to the Brownian motion of ions in the extra- and intracellular fluid.

Therefore, it can be speculated that a certain concentration of ions due to largely random electrochemical processes could create an electric field across the ionic channel to determine its opening. In fact, it is not strictly necessary for the fluctuation of voltage across ionic channels to have any particular distribution, just that it reaches a certain threshold to induce the channel opening. The geometric distributions of the inter-spike intervals would be obtained from the Gaussian distribution of intracellular voltages created from the normal distribution of open channels per time unit, from which, once a certain threshold is obtained, spikes are generated. The latter argument is supported by more detailed simulations in which subthreshold voltage fluctuations increased with the mean voltage and the neuronal noise was normally distributed [35]. The Gaussian distributions obtained for the number of open ionic channels and the number of spikes in the neural network representing the percept would be a direct consequence of the central limit theorem.

Therefore, a succession of distributions across levels of integrations would represent a Gaussian distribution of voltage around channels, an exponential (may be geometric given the shape similarity with the exponential distribution) distribution for the opening times of channels, a Gaussian distribution for the intracellular voltages of neurons, a geometric distribution for inter-spike intervals, and a normal distribution for the global number of spikes in the network representing the percept. An interesting prediction for the model would be that for a stimulus of low intensity, the time intervals between two perceived stimuli should follow a geometrical distribution. With respect to the probability density functions of the different variables presented in this review, the perceptions of ambiguous figures [12], IRTs [10,11], and abducens nerve spike count [7] were all tested via chi-squared and Kolmogorov–Smirnov tests. However, for the simulation of the psychometric function [13], only an approximate graphical method was applied.

The Gaussian distribution for the global activity of a neural structure in response to a continuous input was validated by the computation of the number of spikes in the abducens nerve at the neuronal level, and the abducens motoneurons show a Gaussian distribution [6]. However, for constant inputs, Poisson distributions [36,37] and unimodal left-skewed distributions [38] have been obtained. This variety of probability distributions for inter-spike intervals can be explained by changes in ionic conductances after spikes, not only producing phenomena such as after-hyperpolarization potential [39,40] but also producing short- and long-range adaptation and the activation of recurrent neural activity through reverberant circuits. Therefore, it is not surprising that during neuronal steady inputs, different empirical inter-spike frequency distributions are obtained in neuron discharge.

With regard to the duration of perception in eye rivalry and ambiguous figures [12] and the IRTs of lever pressing in VI [11], both were explained by the geometric distribution and refractoriness for very short durations due to the change in the probabilities as a function of time from the preceding event (perception or response). Perception durations did not show qualitatively long-range time series dependency, and the IRTS proved by means of autocorrelation the absence of time dependency.

The possible random behavior for ambiguous figures' perceptual changes over chaotic deterministic dynamics has previously been explored [26], and for ocular tremor and the psychometric function, the compelling argument that the constituents of the model are random processes (inter-spike modeling and the opening and closing of ionic channels, respectively) suggest a random basis for these organismic phenomena. But, of course, it does not discard other alternative models based on deterministic chaotic dynamics. In the case of IRT, probability density functions have been fitted to the data and no time dependence in the data time series has been found. Although these results suggest random behavior, more compelling evidence, as this suggested by Lehky [26], is needed.

By integrating all the previous data, it can be proposed that at the organismic level, there are clear traces of random processes, probably due to the random basic electrophysiological processes of neurons. It has been proposed that the presence of this unavoidable random activity would be used by organisms to generate behavioral variability that could be used to reinforce the most adaptive behaviors, as in other selective systems such as species evolution [41], clonal selection in the immune system [42], or neural network selection in the so-called "selection of neuronal groups theory" [43]. In the present report, the notion of alternation across levels of Gaussian distributions (ion-mediated microscopic voltage distributions, intracellular voltages mediated by ionic channels, global activity in macroscopic structures such as the abducens nerve), as well as geometric distributions (and exponential distributions) mediated by a threshold to be overcome (ionic channels opening, spike firing, perception and responses), would be a functional characteristic in the transition from the microscopic to the macroscopic in the nervous system. It must be highlighted that the stance taken in the present report is to consider that the presence of thermal noise in the ionic channels and ionic solution across the membrane does not constitute just hidden variables but also intrinsic noise in which the neurons, networks, and behavior have to co-live and somehow thrive for generating behavioral flexibility [44,45]. In this sense, neural and behavioral variability would be caused not only by complex neural dynamics [45] but also by intrinsic neural random processes. The phenotypical variability in life is huge for the organization levels presented in this review, as well as in any other order of complexity level of life. We have tried only to highlight the influence of neural noise based on biophysical concepts of behavioral variability. Future studies should delimitate the relative importance of each of the factors influencing behavioral freedom of degrees in similar sensory contexts.

Author Contributions: Conceptualization, C.M.G.; resources, C.M.G., E.I.R.-M. and M.A.A.-M.; writing—original draft preparation, C.M.G., E.I.R.-M. and M.A.A.-M.; writing—review and editing, C.M.G., E.I.R.-M. and M.A.A.-M.; funding acquisition, C.M.G. and E.I.R.-M. All authors have read and agreed to the published version of the manuscript.

Funding: This work was supported by the Spanish Agencia Estatal de Investigación (AEI) (PID2022-139151OB-I00) (FEDER funds), and the Consejería de Innovación, Ciencia y Empresa of the Junta de Andalucía (P20_00537).

Institutional Review Board Statement: Not applicable.

Informed Consent Statement: Not applicable.

Data Availability Statement: The original contributions presented in the study are included in the article, further inquiries can be directed to the corresponding author.

Conflicts of Interest: The authors declare no conflict of interest.

References

1. Hille, B. *Ion Channels of Excitable Membranes*, 3rd ed.; Sinauer Associates: Sunderland, UK, 2001; ISBN 978-0878933211.
2. Del Castillo, J.; Katz, B. Quantal components of the endplate potential. *J. Physiol.* **1954**, *124*, 560–573. [CrossRef] [PubMed]
3. Destexhe, A.; Rudolph, M.; Fellows, J.M.; Sejnowski, T.J. Fluctuating synaptic conductance recreate in vivo-like activity in neocortical neurons. *Neuroscience* **2001**, *107*, 13–24. [CrossRef] [PubMed]

4. Strassberg, A.F.; De Felice, L.J. Limitations of the Hodgkin–Huxley formalism: Effect of single channel kinetics on transmembrane voltage dynamics. *Neural Comput.* **1993**, *5*, 843–855. [CrossRef]
5. Chow, C.; White, J. Spontaneous action potentials due to channel fluctuations. *Biophys. J.* **1996**, *71*, 3013–3021. [CrossRef] [PubMed]
6. Gómez, C.; Canals, J.; Torres, B.; Delgado-García, J.M. Analysis of the fluctuations in the inter-spike of abducens nucleus neurons during ocular fixations in the alert cat. *Brain Res.* **1986**, *381*, 401–404. [CrossRef]
7. Gómez, C.; Quero, J.M.; Escudero, M. Computer simulation of the neural discharge carried by the abducens nerve during eye fixation in the cat. *Int. J. Biomed. Comput.* **1989**, *24*, 207–215. [CrossRef] [PubMed]
8. Schneidman, E.; Freedman, B.; Segev, I. Ion channel stochasticity may be critical in determining the reliability and precision of spike timing. *Neural Comput.* **1998**, *10*, 1679–1703. [CrossRef]
9. Pishro-Nik, H. *Introduction to Probability, Statistics and Random Processes*; Kappa Investigación, LLC: Miami, FL, USA, 2014; ISBN 978-0990637202.
10. Gómez, C.; Ruiz-Adán, A.; Llosa, M.; Ruiz, G. Quantitative Analysis of IRT Variability during the First Training Stages of a Variable-Interval Schedule in Rats. *Psychol. Rec.* **1992**, *42*, 273–284. [CrossRef]
11. Gómez, C. A Competition Model of IRT Distributions during the First Training Stages of Variable-Interval Schedule. *Psychol. Rec* **1992**, *42*, 285–293. [CrossRef]
12. Gómez, C.; Argandoña, E.D.; Solier, R.G.; Angulo, J.C.; Vázquez, M. Timing and competition in networks representing ambiguous figures. *Brain Cogn.* **1995**, *2*, 103–114. [CrossRef]
13. Gómez, C.M. Numerical exploration of the influence of neural noise on the psychometric function at low stimulation intensity levels. *J. Biosci.* **2008**, *33*, 743–753. [CrossRef]
14. Waschke, L.; Kloosterman, N.A.; Obleser, J.; Garrett, D.D. Behavior needs neural variability. *Neuron* **2021**, *109*, 751–766. [CrossRef]
15. Staddon, J.E.R.; Simmelhag, B. The superstition experiment: A reexamination of its implications for the principles of adaptative behavior. *Psychol. Rev.* **1971**, *78*, 3–43. [CrossRef]
16. *Laszlo, L The System View of the World*; Braziller: New York, NY, USA, 1972; ISBN 9780807606360.
17. Gottman, J.M. *Time-Series Analysis. A Comprehensive Introduction for Social Scientists*; Cambridge University Press: London, UK, 1981; ISBN 9780521235976.
18. Johnson, N.I.; Kotz, S. *Continuous Univariate Distributions-1*; Houghton Mifflin Company: Boston, MA, USA, 1970.
19. Sokal, R.R.; Rohlf, F.S. *Biometry*; W. H. Freeman: New York, NY, USA, 1981.
20. Deneubourg, J.L.; Aron, S.; Goss, S.; Pasteels, J.M.; Duerinck, G. Random behavior, amplification processes and number of participants: How they contribute to the foraging properties of ants. *Physica* **1986**, *22D*, 176–186.
21. Carpenter, R.H.S. *Movements of the Eyes*; Pion: London, UK, 1977.
22. Steinman, R.M.; Haddad, G.M.; Skavensky, A.A.; Wyman, D. Minature eye movement. *Science* **1973**, *181*, 810–819. [CrossRef]
23. Delgado-Garcia, J.M.; del Pozo, F.; Baker, R. Behavior of neurons in the abducens nucleus of the alert cat. I. Mononeurons *Neuroscience* **1986**, *17*, 929–952. [CrossRef]
24. Robinson, D.A. The use of control systems analysis in the neurophysiology of eye movements. *Annu. Rev. Neurosci.* **1981**, *4*, 463–503. [CrossRef]
25. Logothetis, N.K. Vision: A window on consciousness. *Sci. Am.* **1999**, *281*, 69–75. [CrossRef] [PubMed]
26. Lehky, S.R. Binocular rivalry is not chaotic. *Proc. Biol. Sci.* **1995**, *259*, 71–76. [CrossRef] [PubMed]
27. Feldman, J.A.; Ballard, D.H. Modelos conexionistas y sus propiedades. *Cienc. Cogn.* **1982**, *6*, 205–254. [CrossRef]
28. Parker, A.J.; Newsome, W.T. Sense and the single neuron: Probing the physiology of perception. *Ann. Rev. Neurosci.* **1998**, *21*, 227–277. [CrossRef]
29. Moss, F.; Ward, L.M.; Sannita, W.G. Stochastic resonance and sensory information processing: A tutorial and review application *Clin. Neurophysiol.* **2004**, *115*, 267–281. [CrossRef]
30. Bruce, I.C.; White, M.W.; Irlicht, L.S.; O'Leary, S.J.; Dynes, S.; Javel, E.; Clark, G.M. A stochastic model of the electrically stimulated auditory nerve: Single-pulse response. *IEEE Trans. Biomed. Eng.* **1999**, *46*, 617–629. [CrossRef]
31. Hecht, S.; Shlaer, S.; Pirenne, M.H. Energy, quanta and vision. *J. Gen. Physiol.* **1942**, *25*, 819–840. [CrossRef]
32. Johanson, R.S.; Vallbo, A.B. Skin mechanoreceptorsin the human hand: An inference of some population properties. In *Sensory Functions of the Skin in Primates*; Pergamon: Oxford, UK, 1976; pp. 171–184. [CrossRef]
33. Salzman, C.D.; Murasugui, C.M.; Britten, K.H.; Newsome, W.T. Microstimulation in visual area M.T.: Effects on direction discrimination performance. *J. Neurosci.* **1992**, *2*, 2331–2355. [CrossRef]
34. Hawkins, H.L.; Hillyard, S.A.; Luck, S.J.; Mouloua, M.; Downing, C.J.; Woodward, D.P. Visual attention modulates signal detectability. *J. Exp. Psychol. Hum. Percept. Perform.* **1990**, *16*, 802–811. [CrossRef]
35. Steinmetz, P.N.; Manwani, A.; Koch, C.; London, M.; Segev, I. Subthreshold voltage noise due to channel fluctuations in active neuronal membranes. *J. Comp. Neurosci.* **2000**, *9*, 133–148. [CrossRef]
36. Gummer, A.W. Postsynaptic inhibition can explain the concentration of short inter-spike-intervals in avian auditory nerve fibers. *Hear. Res.* **1991**, *55*, 231–243. [CrossRef]
37. Richter, C.P.; Sauer, G.; Hoidis, S.; Klinke, R. Development of activity patterns in auditory nerve fibres of pigeons. *Hear. Res.* **1996**, *95*, 77–86. [CrossRef] [PubMed]

38. Barlow, H.B.; Levick, W.R.; Yoon, M. Responses to single quanta of light in retinal ganglion cells of the cat. *Vis. Res.* **1971**, *11* (Suppl. S3), 87–101. [CrossRef] [PubMed]
39. Lowen, S.B.; Liebovitch, L.S.; White, J.A. Fractal ion channels behavior generates fractal firing patterns in neuronal models. *Phys. Rev. Lett. E* **1999**, *59*, 5970–5980. [CrossRef] [PubMed]
40. Durand, J. Electrophysiological and morphological properties of rat abducens motoneurones. *Exp Brain Res.* **1989**, *76*, 141–152. [CrossRef] [PubMed]
41. Darwin, C.; Kebler, L. *On the Origin of Species by Means of Natural Selection, or, the Preservation of Favoured Races in the Struggle for Life*; J. Murray: London, UK, 1859.
42. Burnet, F.M. A Modification of Jerne's Theory of Antibody Production using the Concept of Clonal Selection. *CA Cancer J. Clin.* **1976**, *26*, 119–121. [CrossRef]
43. Edelman, G.M. Neural Darwinism: Selection and reentrant signaling in higher brain function. *Neuron* **1993**, *10*, 115–125.
44. Garrett, D.D.; Samanez-Larkin, G.R.; MacDonald, S.W.; Lindenberger, U.; McIntosh, A.R.; Grady, C.L. Moment-to-moment brain signal variability: A next frontier in human brain mapping? *Neurosci. Biobehav. Rev.* **2013**, *37*, 610–624. [CrossRef]
45. Renart, A.; Machens, C.K. Variability in neural activity and behavior. *Curr. Opin. Neurobiol.* **2014**, *25*, 211–220. [CrossRef]

applied sciences

MDPI

Article

A LeViT–EfficientNet-Based Feature Fusion Technique for Alzheimer's Disease Diagnosis

Abdul Rahaman Wahab Sait

Department of Archives and Communication, Center of Documentation and Administrative Communication, King Faisal University, P.O. Box 400, Hofuf 31982, Al-Ahsa, Saudi Arabia; asait@kfu.edu.sa

Abstract: Alzheimer's disease (AD) is a progressive neurodegenerative condition. It causes cognitive impairment and memory loss in individuals. Healthcare professionals face challenges in detecting AD in its initial stages. In this study, the author proposed a novel integrated approach, combining LeViT, EfficientNet B7, and Dartbooster XGBoost (DXB) models to detect AD using magnetic resonance imaging (MRI). The proposed model leverages the strength of improved LeViT and EfficientNet B7 models in extracting high-level features capturing complex patterns associated with AD. A feature fusion technique was employed to select crucial features. The author fine-tuned the DXB using the Bayesian optimization hyperband (BOHB) algorithm to predict AD using the extracted features. Two public datasets were used in this study. The proposed model was trained using the Open Access Series of Imaging Studies (OASIS) Alzheimer's dataset containing 86,390 MRI images. The Alzheimer's dataset was used to evaluate the generalization capability of the proposed model. The proposed model obtained an average generalization accuracy of 99.8% with limited computational power. The findings highlighted the exceptional performance of the proposed model in predicting the multiple types of AD. The recommended integrated feature extraction approach has supported the proposed model to outperform the state-of-the-art AD detection models. The proposed model can assist healthcare professionals in offering customized treatment for individuals with AD. The effectiveness of the proposed model can be improved by generalizing it to diverse datasets.

Keywords: feature extraction; deep learning; transformer; LeViT; hyperparameter tuning; model optimization; neuroimaging; neurodegenerative diseases

Citation: Sait, A.R.W. A LeViT–EfficientNet-Based Feature Fusion Technique for Alzheimer's Disease Diagnosis. *Appl. Sci.* **2024**, *14*, 3879. https://doi.org/10.3390/app14093879

Academic Editor: Alexander N. Pisarchik

Received: 25 March 2024
Revised: 25 April 2024
Accepted: 29 April 2024
Published: 30 April 2024

1. Introduction

According to the World Health Organization, the total number of individuals aged 60 and older is expected to double by 2050, reaching approximately 2.1 billion people, 22% of the global population [1]. Alzheimer's disease (AD) is a neurodegenerative condition that primarily affects the elderly population [2]. However, it may manifest in younger individuals. It is the primary cause of dementia. Mild cognitive impairment may occur in the initial stages of AD [2]; this is a transitional stage from normal functioning to AD in which an individual has moderate cognitive abnormalities [3]. The individuals may experience difficulties in performing their routine tasks [4]. They may face challenges in remembering recent events, names, and conversations. In addition, they may exhibit agitation and aggression. With an anticipated increase in AD cases, the disease has become one of the significant global concerns of the modern era. Despite massive efforts to find a cure, AD is still a non-preventable and irreversible form of dementia that impairs an individual's daily life [5]. It is complicated and progressive, necessitating early discovery, diagnosis, therapy, and family support [6]. As the condition progresses, AD patients increasingly rely on their caretakers and require assistance with routine activities.

The primary etiology of AD remains unclear. However, genetics, environment, and lifestyle may contribute to AD [6]. Medical treatment and assistance can place a financial burden on individuals with AD and their families. Globally, governments, healthcare organizations, and research institutes are focusing on the development of practical approaches to address the challenges associated with AD [6]. Researchers investigate AD's backgrounds, risk factors, prevention, and therapy to identify successful strategies to reduce its progression. Most cases of dementia are based on neurodegeneration caused by AD. Increasing evidence from neuropathological and neuroimaging studies shows that mixed etiologies cause many dementia cases, especially in people over 80 years [7]. There has been more variation in the findings regarding the prevalence of dementia and AD in populations older than 90 years compared to younger populations [7]. Healthy aging, dementia care, and caregiver assistance are being applied to improve AD patients' quality of life [8].

Cognitive tests and assessments investigate memory, attention, language, reasoning, and problem-solving [8]. The Mini–Mental State Examination, Montreal Cognitive Assessment, and AD Assessment Scale–Cognitive Subscale are frequently used for AD detection [8]. Cognitive function may be assessed in greater detail with neuropsychological tests [9]. These tests demonstrate cognitive strengths and limitations and may distinguish AD from distinct dementias [10]. Lumbar puncture may collect cerebrospinal fluid from the lower back [11–14]. Elevated beta-amyloid and tau proteins in cerebrospinal fluid may indicate AD pathology. In a few instances, genetic testing may be utilized to diagnose AD, particularly among individuals with a family history of AD [15]. An in-depth neuropsychological evaluation is a crucial diagnostic component in the diagnosis of dementia [16]. It analyzes magnetic resonance imaging (MRI) scans for signs of regional brain atrophy and determines the AD biomarkers using the cerebrospinal fluid biomarker profile. It evaluates individuals' memory, attention, language, and emotional performance. Healthcare practitioners' subjective interpretation of cognitive and neuropsychological testing can result in diagnostic discrepancies [16–18].

Several imaging modalities may reveal the brain's structure and function, highlighting abnormalities associated with AD [18]. The diagnosis of AD relies on a wide variety of biomarkers, including genetic and biological data and neuroimaging techniques, MRI, amyloid positron emission tomography (PET), and diffusion tensor imaging [19–21]. The brain structural changes, including hippocampal shrinkage and other AD-related changes in addition to malignancies and strokes, can be identified using MRI [22]. These changes can be used to determine brain abnormalities associated with mild cognitive impairment, which may indicate AD. PET imaging can identify AD's beta-amyloid plaques and tau protein tangles in the brain. PET scans utilizing florbetapir, flutemetamol, or florbetaben may confirm AD [23]. Chin-Yun Kuo et al. (2023) [24] discussed the significance of integrating neuropsychological assessment with neuroimaging in order to identify AD in its initial stages. Researchers can obtain valuable information on brain anatomical components from high-resolution MRI images. In addition, MRI images have been made available through public open access databases. These datasets are frequently updated, and researchers can utilize them to develop automated AD detection.

DL models can improve early detection, understand disease pathology, integrate image features, leverage large-scale datasets, and advance personalized medicine for individuals with AD [25]. These models can capture complex and high-dimensional patterns in medical imaging data, including MRI and PET, assisting in diagnosing and understanding AD [26]. By identifying biomarkers and subtypes of AD, DL-based models may enable individualized treatments [26]. DL algorithms can learn complex representations from massive data, providing improved precision and generalizable AD detection models [27–31]. Islam and Zhang (2017) [32] employed a multi-class classification model to detect AD. Hussain et al. (2020) [33] introduced a binary classification to distinguish individuals with and without AD using MRI data. Murugan et al. (2021) [34] proposed a DL model to predict AD and dementia. Raees and Thomas (2021) [35] used a Support Vector Machine and Deep Neural Network to detect AD using MRI. Mamun et al. (2022) [36] used a DL model for AD detection. Helaly et al. (2022) [37] proposed a DL model to predict AD in the early stages. Liu et al. (2022) [38] employed a three-dimensional deep convolutional neural network (CNN) to differentiate individuals with mild AD from those without AD. El-Latif et al. (2023) [39] pre-processed the MRI scans and improved the CNN model's capability in identifying AD. However, the existing AD detection models demand high-performance graphical or tensor processing units and large-scale computing infrastructure for training and inference. An effective fine-tuning algorithm is required to find the optimal hyperparameters for optimal outcomes. Hyperparameter selection requires substantial testing and manual adjustment, which is time-consuming and computationally expensive. Overfitting or poor generalization may result from inadequate data for deep learning models.

Furthermore, researchers and practitioners with restricted computing resources may encounter challenges in model implementation. Existing AD detection approaches using MRI require human interpretation or essential feature extraction, limiting diagnosis accuracy and reliability. There is a demand for advanced and automated techniques to detect subtle disease-specific AD patterns. Transformer-based architectures and CNNs have produced promising results in medical image processing. The integration of transformers and CNNs can extract local and global spatial information from complex images. Combining these architectures may strengthen AD detection feature extraction frameworks. These features have motivated the author to build a hybrid transformer and CNN-based AD detection model. The contributions of the study are as follows:

A feature fusion-driven LeViT–EfficientNet B7-based feature extraction model to extract the crucial features of AD.

An enhanced Dartbooster XGBoost (DXB)-based AD detection model using a Bayesian optimization hyperband (BOHB) optimization algorithm.

The structure of the proposed study is organized as follows: The proposed methodology for detecting AD using MRI images is described in Section 2. Section 3 outlines the findings of the performance validation. The study's contribution is discussed in Section 4 Lastly, Section 5 concludes the study by outlining the limitations and future direction.

2. Materials and Methods

The author introduced an integrated approach that combines a vision transformer (ViT), CNN, and gradient-boosting model. A ViT can capture global spatial relationships and long-range interdependence in images [40]. To identify AD anomalies in MRI scans, determining the spatial context of brain regions is crucial. Based on task relevance, a ViT utilizes self-attention mechanisms to rank image patches. The model's interpretability enables researchers and clinicians to observe its regions of interest, allowing them to comprehend AD detection characteristics. A pre-trained ViT model can be fine-tuned on smaller MRI datasets for AD detection [40]. LeVit [40] is a ViT based on a hybrid neural network [37]. Using a transfer learning approach, a feature extraction can be developed to extract crucial AD patterns in order to improve AD detection generalization. LeViT can be seamlessly integrated with CNN to a diverse set of features. CNN can recognize edges, textures, shapes, and structures in MRI images using multiple layers of convolutional and pooling processes. It can identify AD-related regional anomalies in MRI images using attention mechanisms and spatial pooling. EfficientNet B7 is a state-of-the-art CNN model with a compound scaling technique [41]. It is widely used for extracting features from medical images. The capability of LeViT and EfficientNet B7 in extracting the intricate patterns has motivated the author to employ a hybrid feature extraction approach. In addition, the author employed a DXB, which is a gradient-boosting model, to identify the type of AD using the extracted features. Figure 1 reveals the proposed methodology for identifying AD using MRI images.

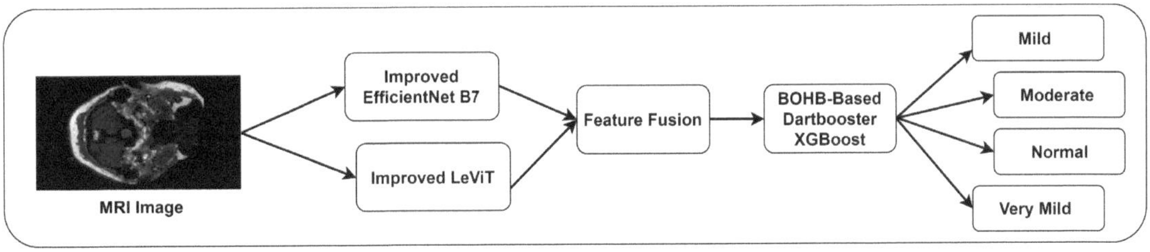

Figure 1. The Proposed AD Detection Methodology.

2.1. Dataset Acquisition

Open Access Series of Imaging Studies (OASIS) Alzheimer's dataset contains a cross-sectional collection of T1-weighted MRI scans of 416 subjects aged 18 to 96. The subjects include males and females. The dataset provides cognitive scores and the diagnosis status of individuals. OASIS Alzheimer's dataset is freely accessible through the repository [42]. Alzheimer's dataset consists of 5000 T1-weighted MRI images [43]. The images were categorized based on the disease severity. The characteristics of the datasets are presented in Table 1.

Table 1. Dataset Characteristics.

Classes	OASIS Alzheimer's Dataset	Alzheimer's Dataset
Mild	5002	896
Moderate	488	64
Normal	67,200	3200
Very mild	13,700	2240

The datasets were highly imbalanced. EfficientNet B7 and LeViT models may require considerable data augmentation to boost robustness and minimize overfitting. Qi et al. [44] proposed a data augmentation technique for brain MRI images. They applied generative adversarial networks to generate the synthetic images. Thus, the author employed the data augmentation technique [44] to overcome the limitation. In addition, traditional data augmentation techniques, including rotation, translation, scaling, flipping, gamma correction, shearing, and histogram equalization, were used in this study.

2.2. EfficientNet B7-Based Feature Extraction

EfficientNet B7 excels in image categorization [41]. It captures complex MRI characteristics and patterns for AD diagnosis using the depth, width, and resolution scaling features. It can handle massive amounts of MRI data with less computation cost. By revealing MRI image representations, EfficientNet B7's hierarchical structure can facilitate model interpretation. Clinicians and researchers may use these representations to understand AD's unique characteristics and provide personalized treatment. EfficientNet B7 may struggle to gain long-range relationships and contextual information in MRI images. This shortcoming may impair the model's detection of AD symptoms. In order to improve the efficiency of the EfficientNet B7 model, the author employed an attention mechanism and mixed-precision training. Figure 2 highlights the recommended feature extraction model.

Using the EfficientNet B7 backbone, a feature extraction model was constructed. An attention mechanism was introduced to capture the long-range dependencies and contextual information. Residual connections were incorporated to overcome the vanishing gradients during the training phase.

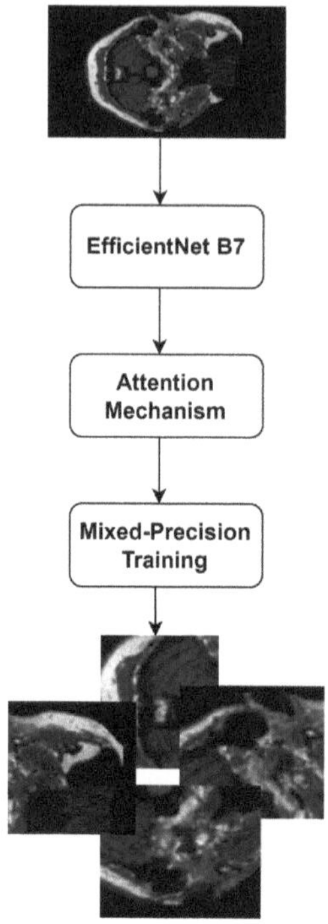

Figure 2. The recommended EfficientNet B7-Based Feature Extraction.

Furthermore, the author employed mixed-precision training to accelerate the training and reduce memory consumption. Activation functions, gradients, and accumulation were performed in a single precision format to prevent numerical underflow or overflow challenges. In addition, a loss scaling factor was dynamically integrated into the loss function to address vanishing gradients.

2.3. *LeViT-Based Feature Extraction*

LeViT offers a powerful platform to handle a wide range of medical image processing tasks, including classification, object detection, and segmentation [40]. It demands fewer parameters compared to traditional CNN models. The self-attention mechanism can learn interpretable representations of the MRI images. The global context modeling technique captures holistic information associated with the MRI images. The patch extractor transforms the image shape from 224 × 224 × 3 into 250 × 14 × 14. A shrinking attention block is used to reduce the size of the activation maps. These features have motivated the author to employ LeViT to extract AD patterns from the MRI images. However, LeViT faces challenges in capturing fine-grained local details, affecting the ability to locate the smaller objects. To overcome this limitation and improve the performance of LeViT-based feature extraction, the author integrated spatial transformer networks (STNs) [45] with LeViT architecture. Initially, an STN is built to perform spatial transformation on the MRI images and extract features based on the region of interest. A feature extraction model is constructed using the LeViT backbone. The extracted features are passed through the LeViT in order to capture high-level representations of the spatially transformed features. Figure 3 highlights the enhanced LeViT model for the feature extraction.

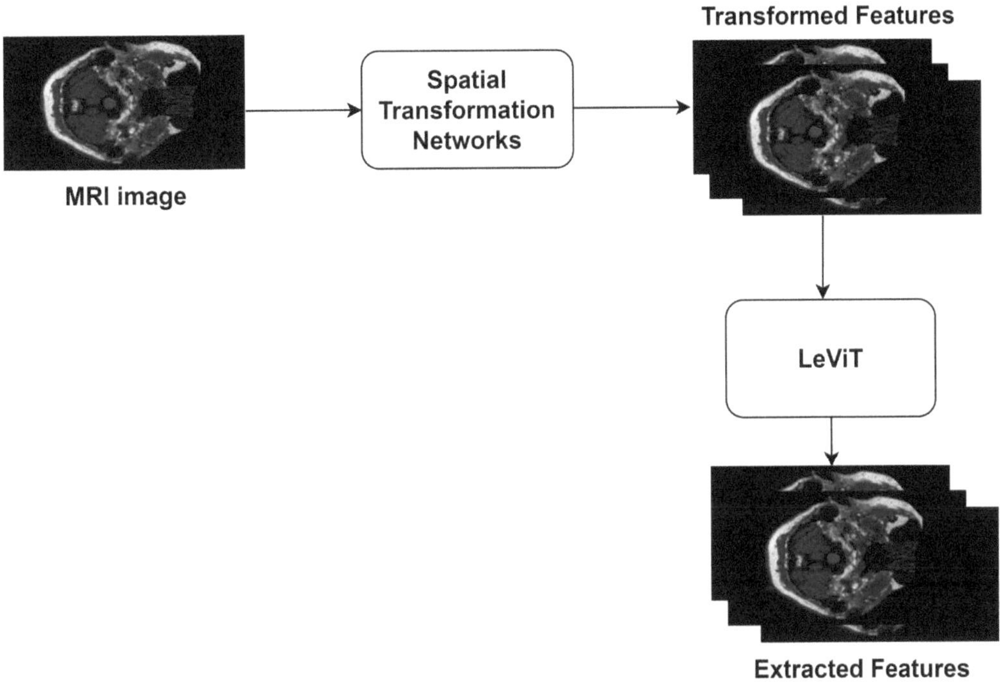

Figure 3. The Enhanced LeViT Model.

A fully connected layer with the Softmax function is used to classify the features based on the severity. Equations (1) and (2) show the computational forms of *STN* and *LeViT* models.

$$F = STN(C, I) \tag{1}$$

where F is the image feature, *STN*() is the spatial transformer network function, C is the input channel, and I is the image.

$$F = LeViT(C, Cl, F) \tag{2}$$

where F is the image feature, C is the input channel, Cl is the AD classes, and *LeViT*() is the function for implementing the *LeViT* model.

After fusing the features, the author normalized the features using feature-wise normalization to prevent numerical instability. Finally, a fully connected layer with the Softmax function was used to generate the outcome. The outcomes were stored as a vector.

2.4. Feature Fusion Layer

The author combined a fusion layer with *LeViT* to fuse the features using an element-wise addition approach. A dimension-matching process was used to identify the features with different dimensions. A reshape function was applied to reshape the feature maps into unique dimensions. Subsequently, element-wise addition combines the elements of *EfficientNet B7* and *LeViT*. Equation (3) shows the mathematical form of feature fusion.

$$\sum_{i=1}^{n} f_{fused} = \sum_{i=1}^{n} f_{EfficientNet\ B7} + \sum_{i=1}^{n} f_{LeViT} \tag{3}$$

where n is the number of features, f_{fused} is the fused features, $f_{EfficientNet\ B7}$ is the *EfficientNet B7* features, and f_{LeViT} is the *LeViT* features.

2.5. Dartbooster XGBoost-Based AD Detection

DXB is an enhanced version of the traditional XGBoost algorithm [46]. It focuses on dropout regularization to prevent overfitting by randomly dropping units during training. Compared to the existing gradient-boosting algorithms, DXB achieves a considerable outcome with limited computational power. In this study, the author employed a DXB model to predict the AD type using the extracted features. However, DXB may face challenges maintaining exploration–exploitation trade-offs in high-dimensional search space. In addition, it may struggle to scale to complex models due to the increased computational requirements. To overcome these limitations, the author employed the BOHB algorithm to fine-tune the model. The hyperband algorithm follows a strategy to allocate computational resources to unique hyperparameter optimization. Bayesian optimization uses a probabilistic surrogate function to control the performance of the DXB hyperparameters. During the training phase, a resource budget (hyperparameters) was initialized. A Gaussian process was updated with the observed performance data. Multiple rounds of optimization were performed until computational resources were exhausted. Equations (4) and (5) show the mathematical forms of the BOHB and DXB hyperparameter tuning processes.

$$BOHB = \arg_{a \in A}^{max} \propto (a) \tag{4}$$

$$O = BOHB(DXB(f), A) \tag{5}$$

where A is the number of hyperparameters, $\propto (a)$ is the acquisition function that controls the selection of hyperparameters, f is the feature, BOHB() is the Bayesian optimization and hyperband function, DXB() is the Dartbooster XGBoost function, and O is the outcomes.

Furthermore, the author included SHapley Additive exPlanations (SHAP) values in the DXB model to improve the model's interpretability. The integration of SHAP values can assist healthcare professionals in gaining deeper insights into the model's prediction.

2.6. Performance Validation

The author validates the proposed model's performance using widely applied evaluation metrics. Accuracy represents the overall correctness of the proposed model's predictions. Specificity indicates the model's ability to detect negative instances. Sensitivity measures the model's capability of detecting positive classes. Precision indicates the proposed model's capability to prevent false positives, whereas recall represents the model's ability to identify positive instances. Cohen's Kappa is used to assess the reliability and consistency of the model's findings. In addition, the area under the receiver–operating characteristics curve (AUROC) and the area under the precision–recall curve (AUPRC) are used to evaluate the effectiveness of the proposed AD detection model.

3. Results

The performance evaluation of the proposed model was conducted using Windows 11 Pro, Intel i9-12900k, 16 GB RAM, NVIDIA RTX 4090, and Python 3.8.0. The libraries, including Pytorch 1.9, TensorFlow 2.11.0, Theano 1.0.5, and Keras 2.12.0, were used for model development. The OASIS Alzheimer's dataset was divided into a train set (70%), a validation set (15%), and a test set (15%). Alzheimer's dataset (20%) was used to generalize the proposed AD detection model. Table 2 reveals the experimental settings for implementing the proposed AD detection model.

The performance of the proposed AD detection during the training and validation phase is highlighted in Figure 4a,b. Compared to the training phase, there was a significant improvement in the validation phase. The recommended early-stopping strategies and regularization techniques have improved the model performance by monitoring the validation loss. The model has attained an optimal performance at the 77th epoch.

Table 2. Experimental Settings.

Model	Parameters	Values
LeViT	Image Size	224 × 224 × 3
	Decay Factor	0.1 every 10 epochs
	Initial Learning Rate	0.001
	Batches	43
	Epochs	75
	Loss Function	Cross-Entropy
	Optimizer	Adam
	Fusion Layer	Element-wise addition
EfficientNet B7	Image	224 × 224 × 3
	Optimizer	Adam
	Loss Function	Cross-Entropy
	Validation Loss Monitor	Early Stopping
	Regularization	Dropout, L1, and L2
	Convolutional Layers	5
	Activation Function	Softmax
DXB	Learning Rate	(η, [0, 1])
	Minimum Split Loss	(γ, [0, ∞])
	Maximum Tree Depth	([0, ∞])
	Optimizer	BOHB

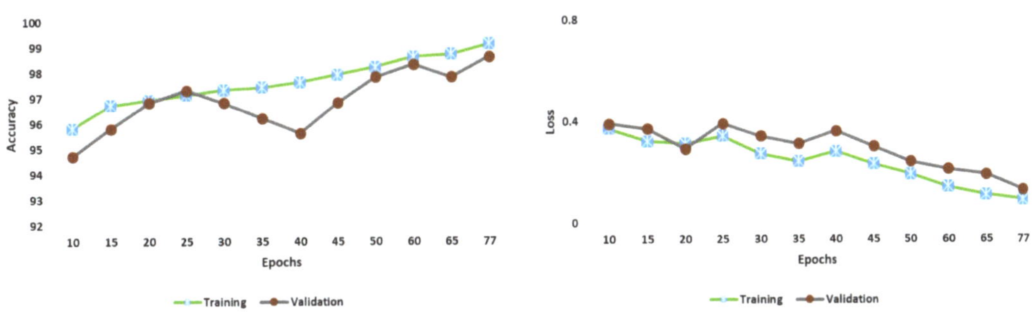

(**a**) (**b**)

Figure 4. (**a**) Prediction Accuracy and (**b**) Loss.

The findings of the performance validation using dataset 1 are outlined in Table 3. The recommended LeViT–EfficientNet B7 feature extraction has improved the prediction accuracy of the proposed model. In addition, the data augmentation has supported the model in identifying the critical patterns associated with AD.

Figure 5 presents the findings of a comparative analysis of the existing transformer and CNN backbones. The proposed model has outperformed the existing models by obtaining an optimal generalization accuracy of 99.8%. The recommended fine-tuning processes assisted the proposed model in addressing the overfitting, vanishing gradient, and amplification effects. Figure 6 highlights the computational loss of the AD detection models. The proposed model produced a minimal loss compared to the existing models.

Table 3. Outcomes of Performance Validation.

Classes	Accuracy	Specificity	Kappa	Precision	Recall	F1-Score
Mild	99.8	99.9	97.5	99.3	99.5	99.4
Moderate	99.9	99.8	96.8	98.6	99.4	99.0
Normal	99.6	100	97.3	99.3	99.5	99.4
Very mild	99.8	99.8	97.9	99.5	99.6	99.5

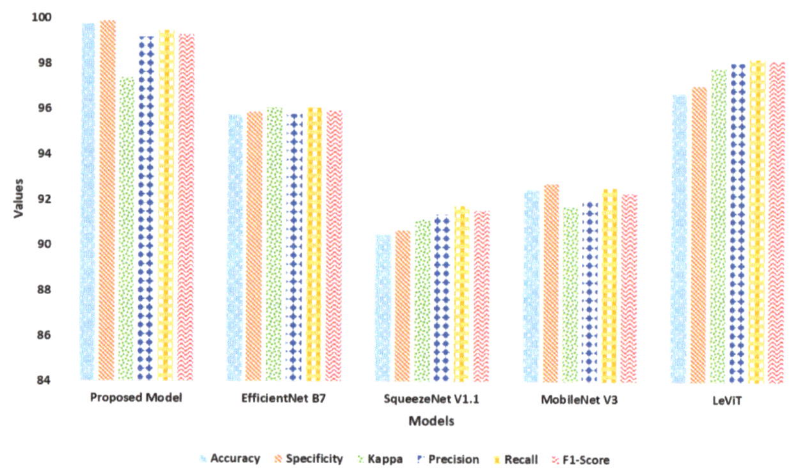

Figure 5. The Comparative Analysis Outcomes.

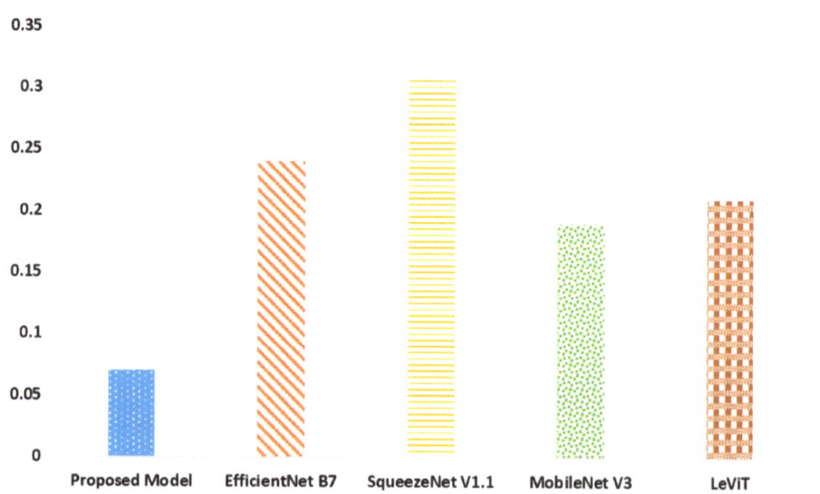

Figure 6. Computational Loss.

Table 4 presented that the proposed model required a few parameters and FLOPs to deliver a remarkable outcome compared to the existing backbones. The findings indicated that the model can be implemented in a resource-constrained healthcare environment. The BOHB algorithm has supported the proposed model in maintaining a trade-off between high generalization accuracy and limited computational resources.

Table 4. Computational Configurations.

Model	Parameters (in Millions (m))	FLOPs (in Millions (m))	Testing Time (Seconds)
Proposed Model	27	42	1.02
EfficientNet B7	39	53	2.15
SqueezeNet V1.1	46	59	1.23
MobileNet V3	47	61	2.08
SWIN Transformer	52	59	1.36
LeViT	37	45	1.56

Table 5 highlights the findings of the reliability and consistency analysis. The proposed model has achieved excellent AUROC and AUPRC, indicating high discrimination in distinguishing the multiple classes of AD. High AUROC and AUPRC highlight the reliability of the proposed AD detection model. The proposed model achieved an exceptional SD and CI, indicating a reliable and consistent outcome. In addition, a smaller SD shows that the model's performance is consistent across diverse data points. Clinicians can benefit from the model and reduce unnecessary medical interventions. The recommended feature extraction approach has produced highly discriminative features by capturing subtle patterns associated with AD. The suggested BOHB-based hyperparameter tuning has selected appropriate DXB parameters to prevent overfitting and enhance the model's robustness.

Table 5. Reliability and Consistency Analysis.

Model	AUROC	AUPRC	SD	CI
Proposed Model	0.99	0.97	0.0004	[95.8–96.8]
EfficientNet B7	0.91	0.93	0.0005	[95.1–97.5]
SqueezeNet V1.1	0.89	0.91	0.0007	[94.8–95.9]
MobileNet V3	0.85	0.86	0.0011	[96.1–97.7]
SWIN Transformer	0.91	0.90	0.0006	[95.7–96.9]
LeViT	0.92	0.91	0.0007	[96.1–96.9]

Table 6 presents the performance of the AD detection models. The utilization of improved LeViT enhances the proposed model's ability to detect long-range dependencies and spatial relationships associated with AD. The scaling coefficient of the EfficientNet B7 model enables the model to handle inherent complexities and variations in the MRI image resolutions.

Table 6. Findings of Comparative Analysis.

Model	Accuracy	Specificity	Sensitivity	AUROC	AUPRC
Proposed Model	99.8	99.8	99.4	0.99	0.97
Raees & Thomas (2021) [35]	90.1	88.7	87.6	0.84	0.81
Mamun et al. (2022) [36]	97.8	95.8	96.1	0.91	0.90
Helaly et al. (2022) [37]	97.1	92.4	91.5	0.90	0.91
El-Latif et al. (2023) [39]	95.9	91.5	92.3	0.91	0.88
Liu et al. (2022) [38]	86.1	78.1	80.2	0.85	0.83

4. Discussions

In this study, the author introduced an EfficientNet B7 and LeViT-based feature fusion technique for extracting key features from MRI images. The EfficientNet B7 model was improved by integrating the attention mechanism. In addition, the author trained the EfficientNet B7 model using mixed-precision training to reduce the computational cost. A fine-tuned DXB model was used to detect AD using the extracted features. The model was trained and tested using the OASIS Alzheimer's dataset. A data augmentation technique was employed in order to provide adequate training to the model to learn intricate patterns of AD. The author generalized the model using the Alzheimer's dataset.

Table 3 highlights the performance of the proposed AD detection model. The model produced an outstanding performance by achieving an accuracy of 98.9% and specificity of 98.7%. Tables 4 and 5 reveal the findings of the comparative analysis using the existing backbones. Table 6 outlines the findings of the existing AD detection models. The proposed model has outperformed the existing AD detection models. It required less computational power to identify AD. The recommended feature fusion technique has supported the proposed model in delivering an optimal outcome. In addition, the suggested BOHB optimization has fine-tuned the parameters of the DXB model to make an effective decision with limited resources. The proposed model demonstrated remarkable performance with limited computational costs. Models with exceptional AUROC and AUPRC can assist healthcare professionals in diagnostic interpretation and treatment options.

The proposed AD detection model can empower clinicians to make effective decisions and offer personalized care to individuals. It holds promise for improving patient outcomes and advancing the understanding of AD symptoms in the earlier stages. By integrating computational approaches with clinical practice, this study enhanced AD detection using MRI images. The proposed model's accuracy and efficiency have significant clinical implications. Effective AD detection enables physicians to diagnose, schedule, and track disease development. Reliable diagnostic techniques and timely intervention can enhance patient outcomes and quality of life. Moreover, scientific communities may benefit from the study findings to extend the research in medical imaging analysis and DL methods.

The author trained the proposed model using the OASIS dataset that covers the MRI images with biomarkers, including an individual's age, sex, cognitive score, and diagnosis status. Researchers can gain insights into the underlying AD pathology and build effective diagnostic and therapeutic strategies. The proposed model allows researchers to identify critical biomarkers, including brain atrophy, cortical thickness changes, hippocampus alterations, white matter integrity alterations, and abnormalities in specific brain regions. Integrating SHAP values facilitates healthcare professionals to identify the significance of MRI biomarkers (features) associated with AD. The proposed model assigns a positive and negative SHAP value to each feature. Healthcare professionals can use SHAP values to understand the importance of features in AD prediction. For instance, a SHAP value of 0.7 related to brain atrophy feature indicates that higher activation in the brain atrophy region is associated with AD prediction. In contrast, a negative SHAP value is associated with a decreased likelihood of AD.

Raees and Thomas (2021) [35] employed AlexNet, Visual Geometry Group (VGG)-16, and ResNet-50 to extract features from MRI images. They used a Support Vector Machine to predict AD. The pre-trained CNN models may produce biased predictions, leading to false positives. The limited generalization ability has reduced the model's performance in the context of AD prediction. The class imbalances have reduced the Support Vector Machine model's capability of detecting AD. In addition, the lack of interpretability may cause challenges to healthcare professionals in understanding the results. The proposed AD detection model integrated the SHAP values in order to provide the results with interpretability. With the recommended feature extraction, it generated an exceptional outcome.

Mamun et al. (2022) [36] employed ResNet-101, DenseNet-121, and VGG-16 models to detect AD. These models achieved an average accuracy of 97.8%. VGG-16 required parameters of 138 M to generate the outcome, leading to high computational cost. It is less expensive compared to the proposed AD model. ResNet-101 architecture was complex, resulting in high training time. It required additional computational power due to the residual connections. DenseNet-121 model required a substantial memory during the training phase. The dense connectivity pattern has reduced the ability to find AD patterns compared to the proposed model.

Helaly et al. (2022) [37] used VGG-10 to classify the AD classes using MRI images. They fine-tuned VGG-19's performance to improve the prediction accuracy. The fixed architecture of the VGG-19 has reduced the model's performance. The vanishing gradient problem has affected the model's learning ability. The depth and complexity enabled the model to produce results with high computational cost. In addition, VGG-19 demanded substantial memory to store the intermediate results. In contrast, the proposed AD detection model has employed mixed-precision training to reduce the computational power. Moreover, the self-attention mechanism has supported the proposed model's remarkable outcome.

Liu et al. (2022) [38] used free surfer segmentation to locate AD patterns. They constructed a gradient-boosting classifier for detecting AD statuses. The processing time of free surfer segmentation may vary depending on the hardware specification. The limited spatial resolutions of MRI have reduced the performance of the model. In addition, augmented samples of 3D MRI were complex, limiting the effectiveness of the AD detection model. In contrast, the proposed AD model combined LeViT and EfficientNet B7 to improve prediction accuracy by producing complex AD patterns.

EL-Latif et al. (2023) [39] constructed a shallow CNN model to classify the AD types. They employed 2D CNN for multi-class classification. The model comprised seven convolutional layers trained using the weights of the pre-trained model. It required extensive image pre-processing in order to maintain a considerable performance. The lack of generalization has reduced the model's prediction accuracy. The model's performance was low compared to the proposed model.

The author encountered challenges in managing and optimizing the feature extraction processes. The high-dimensional and heterogeneous MRI images caused challenges in extracting intricate AD patterns. However, the EfficientNet B7 and LeViT backbones were fine-tuned to overcome the image complexities. The high risk of overfitting due to integrating LeViT and EfficientNet B7 models was reduced using regularization and effective data augmentation techniques. The authors applied the mixed-precision training strategy to minimize the computational costs for the feature extraction.

The proposed AD detection model was generalized on two datasets. A rigorous validation and generalization test is essential in order to ensure the proposed model's robustness and reliability across diverse populations. It can improve the model's trustworthiness in a real-time environment. The integration of the proposed model into the clinical workflow may demand substantial validation, standardization, and flexible user interfaces. The variations in MRI images may influence the model's robustness and generalization. Continuous monitoring and updating are essential in order to adapt to technical advancements and clinical guidelines. AD detection is challenging and requires coordination between computer scientists, neuroscientists, radiologists, and medical professionals. To enhance the model's diagnostic accuracy, multiple data modalities, including PET, genetic information, and cerebrospinal fluid biomarkers can be explored. Investigating advanced data augmentation techniques can enhance the model's robustness to variations in the image quality. The proposed AD prediction models can be improved through unique differences in risk factors, disease progression, and symptom presentation by incorporating language abilities, societal impact, and cognitive abilities as predictor variables. Researchers and clinicians can improve AD prediction, diagnosis, and treatment by combining these factors.

5. Conclusions

The study presented a novel approach, integrating the strengths of LeViT, EfficientNet B7, and the DXB model with the BOHB algorithm to identify different types of AD using MRI images. The proposed model achieved a remarkable accuracy of 99.8% and specificity of 99.8% with limited computational resources. The improved LeViT and EfficientNet B7 with attention mechanisms have produced critical features of AD. The BOHB algorithm has strengthened the DXB model to deliver a superior generalization capability compared to the existing models. The findings indicate that the proposed model can be deployed in healthcare and rehabilitation centers to diagnose AD. The lightweight nature of the proposed model can reduce the complexities in the model implementation. However, the author encountered challenges integrating STN with LeViT and fine-tuning the DXB model using the BOHB algorithm. Integrating multimodal data sources, including PET and genetic data, can unveil novel biomarkers of AD. In addition, enhancing the model's interpretability can foster trust and understanding among clinicians and individuals with AD. Advanced data augmentation techniques can improve the proposed model's generalization capability.

Funding: This work was supported by the Deanship of Scientific Research, Vice Presidency for Graduate Studies and Scientific Research, King Faisal University, Saudi Arabia (Grant No. GrantA035).

Institutional Review Board Statement: Not applicable.

Informed Consent Statement: Not applicable.

Data Availability Statement: OASIS Alzheimer's dataset. https://www.kaggle.com/datasets/ninadaithal/imagesoasis, accessed on 21 March 2023. Alzheimer's dataset. https://www.kaggle.com/datasets/tourist55/alzheimers-dataset-4-class-of-images, accessed on 25 March 2023.

Conflicts of Interest: The author declares no conflict of interest.

References

1. Mehanna, A. Healthy Ageing: Reviewing the Challenges, Opportunities, and Efforts to Promote Health Among Old People. *J. High Inst. Public Health* **2022**, *52*, 45–52. [CrossRef]
2. Ebrahimighahnavieh, M.A.; Luo, S.; Chiong, R. Deep learning to detect Alzheimer's disease from neuroimaging: A systematic literature review. *Comput. Methods Programs Biomed.* **2020**, *187*, 105242. [CrossRef] [PubMed]
3. Altinkaya, E.; Polat, K.; Barakli, B. Detection of Alzheimer's disease and dementia states based on deep learning from MRI images: A comprehensive review. *J. Inst. Electron. Comput.* **2020**, *1*, 39–53.
4. Al-Shoukry, S.; Rassem, T.H.; Makbol, N.M. Alzheimer's diseases detection by using deep learning algorithms: A mini-review. *IEEE Access* **2020**, *8*, 77131–77141. [CrossRef]
5. Kivipelto, M.; Mangialasche, F.; Ngandu, T. Lifestyle interventions to prevent cognitive impairment, dementia and Alzheimer disease. *Nat. Rev. Neurol.* **2018**, *14*, 653–666. [CrossRef] [PubMed]
6. Arafa, D.A.; Moustafa, H.E.D.; Ali-Eldin, A.M.; Ali, H.A. Early detection of Alzheimer's disease based on the state-of-the-art deep learning approach: A comprehensive survey. *Multimed. Tools Appl.* **2022**, *81*, 23735–23776. [CrossRef]
7. Kuo, C.-Y.; Stachiv, I.; Nikolai, T. Association of late life depression, (non-)modifiable risk and protective factors with dementia and Alzheimer's disease: Literature review on current evidences, preventive interventions and possible future trends in prevention and treatment of dementia. *Int. J. Environ. Res. Public Health* **2020**, *17*, 7475. [CrossRef] [PubMed]
8. Khojaste-Sarakhsi, M.; Haghighi, S.S.; Ghomi, S.F.; Marchiori, E. Deep learning for Alzheimer's disease diagnosis: A survey. *Artif. Intell. Med.* **2022**, *130*, 102332. [CrossRef]
9. Mggdadi, E.; Al-Aiad, A.; Al-Ayyad, M.S.; Darabseh, A. Prediction Alzheimer's Disease from MRI Images using Deep Learning. In Proceedings of the 12th International Conference on Information and Communication Systems (ICICS), Valencia, Spain, 24–26 May 2021; IEEE: Piscataway, NJ, USA, 2021; pp. 120–125.
10. Hamdi, M.; Bourouis, S.; Rastislav, K. Evaluation of neuro images for the diagnosis of Alzheimer's disease using deep learning neural network. *Front. Public Health* **2022**, *10*, 834032.
11. Balaji, P.; Chaurasia, M.A.; Bilfaqih, S.M.; Muniasamy, A.; Alsid, L.E.G. Hybridized deep learning approach for detecting Alzheimer's disease. *Biomedicines* **2023**, *11*, 149. [CrossRef]
12. Mehmood, A.; Yang, S.; Feng, Z.; Wang, M.; Ahmad, A.S.; Khan, R.; Maqsood, M.; Yaqub, M. A transfer learning approach for early diagnosis of Alzheimer's disease on MRI images. *Neuroscience* **2021**, *460*, 43–52. [CrossRef] [PubMed]
13. Saratxaga, C.L.; Moya, I.; Picón, A.; Acosta, M.; Moreno-Fernandez-de-Leceta, A.; Garrote, E.; Bereciartua-Perez, A. MRI deep learning-based solution for Alzheimer's disease prediction. *J. Pers. Med.* **2021**, *11*, 902. [CrossRef]
14. Salehi, A.W.; Baglat, P.; Gupta, G. Alzheimer's disease diagnosis using deep learning techniques. *Int. J. Eng. Adv. Technol.* **2020**, *9*, 874–880. [CrossRef]
15. Yamanakkanavar, N.; Choi, J.Y.; Lee, B. MRI segmentation and classification of human brain using deep learning for diagnosis of Alzheimer's disease: A survey. *Sensors* **2020**, *20*, 3243. [CrossRef] [PubMed]

16. Reul, S.; Lohmann, H.; Wiendl, H.; Duning, T.; Johnen, A. Can cognitive assessment really discriminate early stages of Alzheimer's and behavioural variant frontotemporal dementia at initial clinical presentation? *Alzheimer's Res. Ther.* **2017**, *9*, 61. [CrossRef] [PubMed]
17. Acharya, U.R.; Fernandes, S.L.; WeiKoh, J.E.; Ciaccio, E.J.; Fabell, M.K.M.; Tanik, U.J.; Rajinikanth, V.; Yeong, C.H. Automated detection of Alzheimer's disease using brain MRI images—A study with various feature extraction techniques. *J. Med. Syst.* **2019**, *43*, 302. [CrossRef] [PubMed]
18. Bi, X.; Li, S.; Xiao, B.; Li, Y.; Wang, G.; Ma, X. Computer aided Alzheimer's disease diagnosis by an unsupervised deep learning technology. *Neurocomputing* **2020**, *392*, 296–304. [CrossRef]
19. Battineni, G.; Hossain, M.A.; Chintalapudi, N.; Traini, E.; Dhulipalla, V.R.; Ramasamy, M.; Amenta, F. Improved Alzheimer's disease detection by MRI using multimodal machine learning algorithms. *Diagnostics* **2021**, *11*, 2103. [CrossRef]
20. Li, H.; Habes, M.; Wolk, D.A.; Fan, Y. Alzheimer's Disease Neuroimaging Initiative. A deep learning model for early prediction of Alzheimer's disease dementia based on hippocampal magnetic resonance imaging data. *Alzheimer's Dement.* **2019**, *15*, 1059–1070. [CrossRef]
21. Balne, S.; Elumalai, A. Machine learning and deep learning algorithms used to diagnosis of Alzheimer's. *Mater. Today Proc.* **2021**, *47*, 5151–5156. [CrossRef]
22. Sathiyamoorthi, V.; Ilavarasi, A.K.; Murugeswari, K.; Ahmed, S.T.; Devi, B.A.; Kalipindi, M. A deep convolutional neural network based computer aided diagnosis system for the prediction of Alzheimer's disease in MRI images. *Measurement* **2021**, *171*, 108838. [CrossRef]
23. Ullah, Z.; Jamjoom, M. A Deep Learning for Alzheimer's Stages Detection Using Brain Images. *Comput. Mater. Contin.* **2023**, *74*, 1457–1473. [CrossRef]
24. Kuo, C.-Y.; Tseng, H.-Y.; Stachiv, I.; Tsai, C.-H.; Lai, Y.-C.; Nikolai, T. Combining Neuropsychological Assessment with Neuroimaging to Distinguish Early-Stage Alzheimer's Disease from Frontotemporal Lobar Degeneration in Non-Western Tonal Native Language-Speaking Individuals Living in Taiwan: A Case Series. *J. Clin. Med.* **2023**, *12*, 1322. [CrossRef]
25. Ghazal, T.M.; Abbas, S.; Munir, S.; Ahmad, M.; Issa, G.F.; Zahra, S.B.; Khan, M.A.; Hasan, M.K. Alzheimer Disease Detection Empowered with Transfer Learning. *Comput. Mater. Contin.* **2022**, *70*, 5005–5019. [CrossRef]
26. Yagis, E.; De Herrera AG, S.; Citi, L. Convolutional autoencoder based deep learning approach for Alzheimer's disease diagnosis using brain mri. In Proceedings of the 2021 IEEE 34th International Symposium on Computer-Based Medical Systems (CBMS), Aveiro, Portugal, 7–9 June 2021; IEEE: Piscataway, NJ, USA, 2021; pp. 486–491.
27. Liu, J.; Li, M.; Luo, Y.; Yang, S.; Li, W.; Bi, Y. Alzheimer's disease detection using depthwise separable convolutional neural networks. *Comput. Methods Programs Biomed.* **2021**, *203*, 106032. [CrossRef]
28. Han, R.; Chen, C.P.; Liu, Z. A novel convolutional variation of broad learning system for Alzheimer's disease diagnosis by using MRI images. *IEEE Access* **2020**, *8*, 214646–214657. [CrossRef]
29. AlSaeed, D.; Omar, S.F. Brain MRI analysis for Alzheimer's disease diagnosis using CNN-based feature extraction and machine learning. *Sensors* **2022**, *22*, 2911. [CrossRef]
30. Tuan, T.A.; Pham, T.B.; Kim, J.Y.; Tavares, J.M.R.S. Alzheimer's diagnosis using deep learning in segmenting and classifying 3D brain MR images. *Int. J. Neurosci.* **2022**, *132*, 689–698. [CrossRef]
31. Shamrat, F.M.J.M.; Akter, S.; Azam, S.; Karim, A.; Ghosh, P.; Tasnim, Z.; Hasib, K.M.; De Boer, F.; Ahmed, K. AlzheimerNet: An effective deep learning based proposition for Alzheimer's disease stages classification from functional brain changes in magnetic resonance images. *IEEE Access* **2023**, *11*, 16376–16395. [CrossRef]
32. Islam, J.; Zhang, Y. A novel deep learning based multi-class classification method for Alzheimer's disease detection using brain MRI data. In Proceedings of the Brain Informatics: International Conference 2017, BI 2017, Beijing, China, 16–18 November 2017; Springer International Publishing: Berlin/Heidelberg, Germany, 2017; pp. 213–222.
33. Hussain, E.; Hasan, M.; Hassan, S.Z.; Azmi, T.H.; Rahman, M.A.; Parvez, M.Z. Deep learning based binary classification for Alzheimer's disease detection using brain mri images. In Proceedings of the 2020 15th IEEE Conference on Industrial Electronics and Applications (ICIEA), Kristiansand, Norway, 9–13 November 2020; IEEE: Piscataway, NJ, USA, 2020; pp. 1115–1120.
34. Murugan, S.; Venkatesan, C.; Sumithra, M.G.; Gao, X.Z.; Elakkiya, B.; Akila, M.; Manoharan, S. DEMNET: A deep learning model for early diagnosis of Alzheimer diseases and dementia from MR images. *IEEE Access* **2021**, *9*, 90319–90329. [CrossRef]
35. Raees, P.M.; Thomas, V. Automated detection of Alzheimer's Disease using Deep Learning in MRI. *J. Phys. Conf. Ser.* **2021**, *1921*, 012024. [CrossRef]
36. Mamun, M.; Shawkat, S.B.; Ahammed, M.S.; Uddin, M.M.; Mahmud, M.I.; Islam, A.M. Deep Learning Based Model for Alzheimer's Disease Detection Using Brain MRI Images. In Proceedings of the 2022 IEEE 13th Annual Ubiquitous Computing, Electronics Mobile Communication Conference (UEMCON), New York, NY, USA, 26–29 October 2022; IEEE: Piscataway, NJ, USA, 2022; pp. 0510–0516.
37. Helaly, H.A.; Badawy, M.; Haikal, A.Y. Deep learning approach for early detection of Alzheimer's disease. *Cogn. Comput.* **2022**, *14*, 1711–1727. [CrossRef] [PubMed]
38. Liu, S.; Masurkar, A.V.; Rusinek, H.; Chen, J.; Zhang, B.; Zhu, W.; Fernandez-Granda, C.; Razavian, N. Generalizable deep learning model for early Alzheimer's disease detection from structural MRIs. *Sci. Rep.* **2022**, *12*, 17106. [CrossRef] [PubMed]
39. El-Latif, A.A.A.; Chelloug, S.A.; Alabdulhafith, M.; Hammad, M. Accurate detection of Alzheimer's disease using lightweight deep learning model on MRI data. *Diagnostics* **2023**, *13*, 1216. [CrossRef]

40. LeViT Model. Available online: https://github.com/facebookresearch/LeViT (accessed on 2 May 2023).
41. PyTorch. Available online: https://github.com/lukemelas/EfficientNet-PyTorch (accessed on 12 May 2023).
42. OASIS Alzheimer's Dataset. Available online: https://www.kaggle.com/datasets/ninadaithal/imagesoasis (accessed on 21 March 2023).
43. Alzheimer's Dataset. Available online: https://www.kaggle.com/datasets/tourist55/alzheimers-dataset-4-class-of-images (accessed on 25 March 2023).
44. Qi, C.; Chen, J.; Xu, G.; Xu, Z.; Lukasiewicz, T.; Liu, Y. SAG-GAN: Semi-supervised attention-guided GANs for data augmentation on medical images. *arXiv* **2020**, arXiv:2011.07534.
45. Spatial Transformer Network Model. Available online: https://github.com/topics/spatial-transformer-network (accessed on 15 May 2023).
46. XGBoost Model. Available online: https://github.com/dmlc/xgboost (accessed on 15 May 2023).

applied sciences

MDPI

Article

PadGAN: An End-to-End dMRI Data Augmentation Method for Macaque Brain

Yifei Chen [1], Limei Zhang [2], Xiaohong Xue [2], Xia Lu [1], Haifang Li [1,2,*] and Qianshan Wang [1,*]

1 College of Computer Science and Technology (College of Data Science), Taiyuan University of Technology, Taiyuan 030024, China; chenyifei0465@link.tyut.edu.cn (Y.C.); luxia0482@link.tyut.edu.cn (X.L.);

2 College of Computer and Information Engineering, Shanxi Technology and Business University, Taiyuan 030031, China; 13633456237@163.com (L.Z.); 18234073562@163.com (X.X.)

* Correspondence: lihaifang@tyut.edu.cn (H.L.); wangqianshan0203@gmail.com (Q.W.); Tel.: +86-153-4068-2795 (H.L.)

Abstract: Currently, an increasing number of macaque brain MRI datasets are being made publicly accessible. Unlike human, publicly accessible macaque brain datasets suffer from data quality in diffusion magnetic resonance imaging (dMRI) data. Typically, dMRI data require a minimum ratio of 1:10 between low b-value (b < 10) volumes and high b-value (b > 300) volumes. However, the currently accessible macaque datasets do not meet this ratio. Due to site differences in macaque brain images, traditional human brain image-to-image translation models struggle to perform well on macaque brain images. Our work introduces a novel end-to-end primary-auxiliary dual generative adversarial network (PadGAN) for generating low b-value images. The auxiliary generator in the PadGAN is responsible for extracting the latent space features from peak information maps and transmitting them to the primary generator, enabling the primary generator to generate images with rich details. Experimental results demonstrate that PadGAN outperforms existing methods both qualitatively and quantitatively (mean SSIM increased by 0.1139). Diffusion probabilistic tractography using dMRI data augmented by our method yields superior results.

Keywords: medical image-to-image translation; generative adversarial networks; dMRI data augmentation; macaque brain image

Citation: Chen, Y.; Zhang, L.; Xue, X.; Lu, X.; Li, H.; Wang, Q. PadGAN: An End-to-End dMRI Data Augmentation Method for Macaque Brain. *Appl. Sci.* **2024**, *14*, 3229. https://doi.org/10.3390/app14083229

Academic Editor: Alexander N. Pisarchik

Received: 4 March 2024
Revised: 6 April 2024
Accepted: 8 April 2024
Published: 11 April 2024

1. Introduction

Studying the macaque brain provides a crucial avenue for understanding human brain mechanisms in neuroscience research [1]. Currently, the macaque monkey serves as a prominent primate model and has become a vital subject for investigating the human brain using various medical imaging techniques [2,3].

Diffusion magnetic resonance imaging (dMRI) technology detects the movement direction of water molecules in the brain, utilizing the anisotropic diffusion characteristics of water molecules in the white matter to reconstruct the white matter in the brain. The b-value represents the intensity of the diffusion-sensitive gradient field, which, along with its corresponding three b-vectors, reflects the influence of microstructural tissue on water diffusion within living tissue in dMRI. Researchers commonly refer to the images corresponding to different b-value intensities in the dMRI volume as b-value images. Diffusion tensor imaging (DTI) estimation and probabilistic tractography techniques are established methods for reconstructing major white matter fiber bundles in brain imaging [4]. Typically, dMRI images consist of multiple b-value images, with low b-value (b < 10, recommend b = 0) volumes serving as the basis for DTI, which is crucial for data analysis in neuroscience research. Nowadays, to mitigate interference such as head motion during acquisition, one low b-value image often corresponds to 5–10 high b-value (b > 300,commonly b = 1000) volumes [5]. However, in some publicly accessible macaque brain dMRI datasets, the ratio of low b-value to high b-value volumes may be below 1:5 or even 1:10, which could be due to the use of early acquisition protocol configurations [6]. The reliability of computed results, such as DTI estimation, from analyses using data that do not meet the required ratio needs further confirmation. Therefore, it is necessary to generate and optimize low b-value volumes in macaque brain dMRI data.

Medical image-to-image translation refers to the method of translation images from an input image modality to an output modality through a mapping relationship. This approach can be used to acquire additional data or complete missing data [7], and it can be applied to downstream tasks in medical image processing, such as image registration and segmentation [8,9], as well as image classification [10].

Generative adversarial networks (GANs) are network models based on game theory, consisting of a generator and a discriminator [11]. The generator attempts to generate high-quality images to deceive the discriminator, while the discriminator distinguishes between real and generated images. Both sides become stronger in the adversarial process, resulting in the generator producing increasingly realistic images. With the emergence of GANs, the performance of medical image-to-image translation has been greatly improved. Initially, GANs were only used to generate images from random noise. With researchers attempting to use Transformer as the generator of GANs [12], Transformer is being applied in the field of medical image-to-image translation. The advent of pix2pix and CycleGAN propelled the performance of GANs in image-to-image translation tasks [13,14]. New methods harness the powerful generative capabilities of GANs to produce visually and objectively superior images.

Some researchers have explored the application scenarios of CycleGAN in medical image-to-image translation [15–17], but more efforts have been devoted to improving CycleGAN for better application in unsupervised learning settings [18–23]. Methods based on CycleGAN are unsupervised approaches, with the advantage of being able to perform mutual translation between two domains without requiring paired images. However, because CycleGAN serves two image translation tasks, its performance on the generation task in a single target domain is generally inferior compared with supervised methods.

Compared with CycleGAN, methods based on Transformer are a supervised learning approach. Some researchers have employed Transformer for medical segmentation [24], MRI reconstruction [25], and medical image-to-image translation [8,26,27]. However, Transformer requires a large amount of data, but publicly accessible samples of macaque dMRI data are limited, making it challenging to fully leverage the advantages of Transformer [28].

The pix2pix-based method is also a widely used supervised learning approach for medical image-to-image translation. The Synb0-DisCo method applies the pix2pix technique to correct distorted b0 images [29]. pGAN and Ea-GAN, respectively, enhance the image detail capability by improving the loss function and considering edge information [30,31]. MedGAN [32] employs a cascaded U-Net as its generator for various medical image translation tasks. As pix2pix-based methods are designed for the generation task of a single target domain with paired image data, they often exhibit higher generation accuracy in medical image-to-image translation tasks. However, since such methods typically rely on a single generative adversarial network, they lack in detail learning.

Furthermore, all these methods share a common issue. Currently, most studies on modality translation of brain MRI images are based on human brain GRAY color space, with the aim of providing visually interpretable images [21,26,31,33]. Medical imaging signal intensity values have absolute significance [8], and are required for probabilistic tractography calculations, rather than the typical GRAY color range of bitmap images. Therefore, the images generated by the aforementioned methods cannot meet the requirements of computational neuroscience research.

In this work, we introduce the concept of peak information maps and propose a novel end-to-end primary-auxiliary dual GAN network (PadGAN), which can extract latent space features from peak information maps to translate high-quality low b-value images. The generated low b-value images can be used for augmenting dMRI image data, improving the quality of dMRI images. The results show that PadGAN outperforms existing methods in qualitative observations and quantitative metrics, and the effectiveness of each module is validated through ablation experiments. Finally, we use the Xtract toolbox [34] in FSL6.0 (FMRIB Software Library) tools [35] to perform probabilistic tractography and use FSL tools to conduct DTI estimation on dMRI data augmented. The Xtract calculation results of dMRI data augmented using our method are more satisfactory. In summary, the specific contributions of this paper are as follows:

1. We introduce the concept of peak information maps and design a corresponding method for calculating peak information maps.
2. We propose a novel end-to-end primary-auxiliary dual GAN network to translate high b-value images to low b-value images. In this network, the auxiliary generator extracts latent space features from peak information maps and transfers these features to the primary generator. The primary network integrates the latent space features and multi-scale features to generate low b-value images.
3. Through DTI estimation and Xtract probabilistic tractography experiments, we validate the effectiveness of generating low b-value images for augmenting original dMRI data, providing new validation approaches for quality assessment in brain science research and offering optimized dMRI data for brain science studies.

2. Materials and Methods

2.1. Datasets

We obtained human brain dMRI images from the WU-Minn public dataset released by the Human Connectome Project (HCP) in 2016 [36]. We selected 96 dMRI data with the following specific parameters: echo-planar imaging (EPI) sequence, TR/TE = 5520/89.5 ms, flip angle (FA) = 78°, and voxel resolution of 1.25 × 1.25 × 1.25 mm.

We used the publicly accessible macaque brain imaging dataset from The PRIMatE Data Exchange (PRIME-DE) [6]. This dataset contains data from different sites, and we collected 8 data samples from Aix-Marseille Université (AMU), 12 data samples from Mount Sinai School of Medicine-Philips (MountSinai-P), 5 data samples from Mount Sinai School of Medicine-Siemens (MountSinai-S), 38 data samples from University of California, Davis (UCDavis), and 582 data samples from University of Wisconsin–Madison (UWM). The parameters of macaque datasets from different sites are shown in Table 1. These datasets all suffer from varying degrees of imbalance between the number of low b-value and high b-value images. The quantities and ratios of low b-value images to high b-value images in the dMRI images from different data sites are shown in Table 2.

Table 1. Specific parameters of macaque datasets.

Datasets	Scanner (3T)	Voxel Resolution (mm)	TE (ms)	TR (ms)	b-Values (s/mm^2)
AMU	Siemens Prisma	1 × 1 × 1	87.6	7520	5, 500
MountSinai-P	Philips Achieva	1.5 × 1.5 × 1.5	19	2600	0, 1000
MountSinai-S	Siemens Skyra	1.0 × 1.0 × 1.0	95	5000	10, 1005
UCDavis	Siemens Skyra	1.4 × 1.4 × 1.4	115	6400	5, 1600
UWM	GE DISCOVERY_MR750	2.1875 × 3.1 × 2.1875	94.3	6100	0, 1000

Table 2. The number of low b-value and high b-value images in the macaque dataset.

Datasets	Number of Low b-Value Images	Number of High b-Value Images	Ratio
AMU	4	67	1:17
MountSinai-P	2	120	1:60
MountSinai-S	10	80	1:8
UCDavis	6	60	1:10
UWM	1	12	1:12

2.2. Preprocessing

The series of preprocessing steps applied to all the datasets are as follows:

1. Head motion correction and eddy current correction were performed using the FSL tool.
2. Non-brain tissues were removed from human brain images using the FSL tool, while non-brain tissues were removed from macaque brain images using a deep learning method developed by our research group [37].
3. Paired high b-value and low b-value images were extracted from the dMRI images where the high b-value images served as inputs to the model, and the low b-value images served as reference images. The task of extracting b-value images was accomplished using the FSL tool.
4. All high b-value images were scaled to the range of 0 to 1 using the min–max normalization method, and their dimensions were resampled to 256 × 256 × 256.
5. The data were divided into pre-training, training, and testing sets: the pre-training set included 96 pairs of human brain images and 467 pairs of UWM images. The remaining data from UWM, AMU, MountSinai-P, MountSinai-S, and UCDavis sites were divided into training and testing sets, with a ratio of 8:2.

2.3. PadGAN

We propose a primary-auxiliary dual generative adversarial network called PadGAN, consisting of two generative adversarial networks: the primary network and the auxiliary network, both targeting the domain of low b-value images. Figure 1 illustrates the training data flow of PadGAN. During training, the peak information map is input into the auxiliary generator, which learns towards the domain of low b-value images through adversarial learning while simultaneously passing latent space features to the primary generator. The high b-value images are input into the primary generator, which maps them to the domain of low b-value images through feature fusion modules by merging latent space features. The low b-value images generated by the auxiliary generator and the primary generator are passed to the auxiliary discriminator and the primary discriminator, respectively, to discriminate between real and generated images, thereby enhancing the generation capabilities of both generators through adversarial learning.

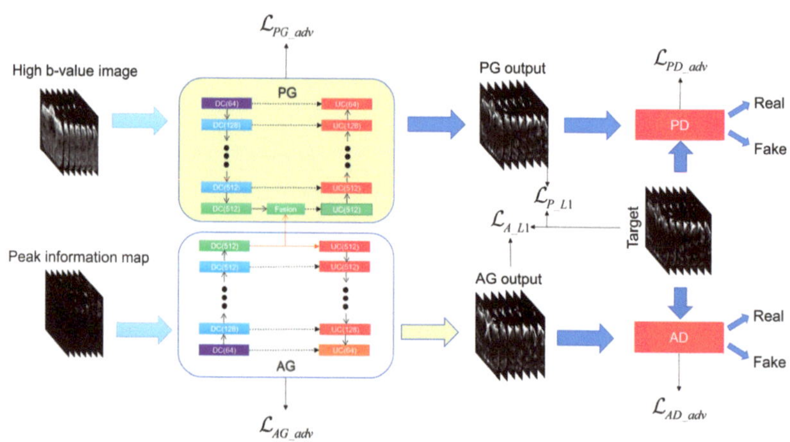

Figure 1. The training data flow diagram of PadGAN. The auxiliary discriminator (AD) discriminates between the images generated by the auxiliary generator (AG) and the real images, while the primary discriminator (PD) discriminates between the images generated by the primary generator (PG) and the real images. The auxiliary network uses three losses, $\mathcal{L}_{AG\_adv}$, $\mathcal{L}_{AD\_adv}$, and $\mathcal{L}_{A\_L1}$, for backpropagation, while the primary network uses three losses, $\mathcal{L}_{PG\_adv}$, $\mathcal{L}_{PD\_adv}$, and $\mathcal{L}_{A\_L1}$, for backpropagation.

2.3.1. Peak Information Maps

Recently, latent space has flourished in the field of image generation [38]. Latent space can generate diverse high-resolution images [39] and can also be used for re-editing images by extracting latent space features [40,41]. In order to introduce diversity into generated images, random Gaussian noise is commonly used as the input for extracting latent space features. However, unlike works focused on enhancing image diversity, this paper places high demands on the accuracy of generated image details. Therefore, random Gaussian noise as the input for latent space feature extraction may not be suitable.

Max-pooling layers are widely used in image classification, segmentation, and other fields [42]. They can preserve texture features and edge information of images while reducing information redundancy. However, there is also a risk of losing important information. Due to the high demand for image details in end-to-end image-to-image translation tasks, max-pooling layers are rarely used to prevent information loss during training [43]. To reduce the risk of losing other information while preserving texture and other detailed information, we introduced the concept of peak information maps.

Given the assumption that brain images from the same data site and the same species exhibit a certain degree of similarity, we perform a per-pixel maximum extraction operation on the low b-value brain images of macaques within the same site. All extracted maximum values are concatenated into a 3D image, which represents the peak information map of that site, as illustrated in Figure 2. Additionally, Equations (1) and (2) demonstrate this process. The peak information maps from different sites serve as inputs to the auxiliary network for the respective site's data, facilitating the extraction of latent space features.

$$vox_{ij} = MAX(img_{i1}(vox_j), img_{i2}(vox_j), ...img_{in}(vox_j)) \tag{1}$$

$$ref_i = PConcat(vox_{ij}), i = 1, 2, ..., k, j = 1, 2, ..., m. \tag{2}$$

where vox_{ij} represents the j-th voxel selected at the i-th site, $MAX(\cdot)$ represents the peak extraction operation, $img_{in}(vox_j)$ represents the j-th pixel of the n-th image at the i-th site, and ref_i represents the peak information map of the i-th site, of which there are k such peak information maps. $PConcat(\cdot)$ represents the pixel concatenation operation, which concatenates individual pixels into the entire image. Iterate over all i and j values to obtain the peak information map for each site.

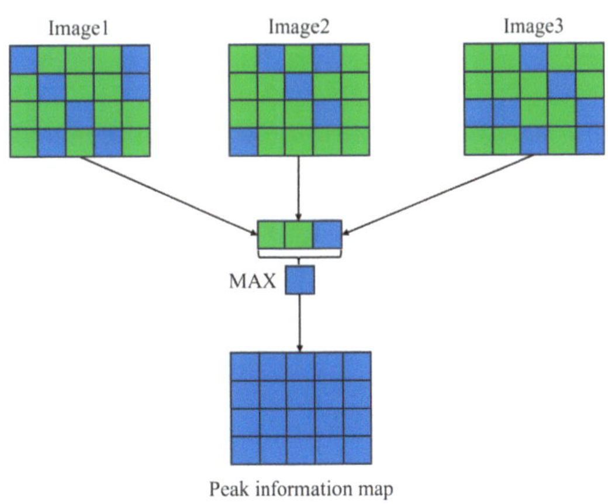

Figure 2. Schematic diagram of peak information map. Image1, Image2, and Image3 represent three images within the same site. The green and blue rectangles represent the pixels of the image, where the blue rectangles represent the maximum pixel values at the same position in the three images. Concatenating the maximum value pixels at each position yields the peak information map.

2.3.2. Auxiliary GAN

In the field of image generation, there is typically no end-to-end training data accessible. The extraction of latent space features is often achieved through several fully connected layers to decouple Gaussian noise and generate more diverse images [39]. The peak information map proposed in this paper provides end-to-end training data for extracting latent space features. We adopted an adversarial learning approach to extract latent space features to enhance the details of the generated images.

The role of the auxiliary generative adversarial network is to provide high-quality latent space information to the primary generative adversarial network. To achieve this, the auxiliary GAN continuously maps from the peak information map to the low b-value images through adversarial learning. The main architecture of the auxiliary generator network adopts a U-Net convolutional neural network, which is divided into an encoder and a decoder. The encoder consists of 8 down-sampling convolutional blocks, while the decoder consists of 8 up-sampling convolutional blocks. After encoding through the 8 down-sampling convolutional blocks, the input data obtain a $512 \times 1 \times 1$ latent space feature, as shown in Equation (3).

$$Latent = 8 * DC(x) \tag{3}$$

where $Latent$ represents latent space features, $8 * DC(\cdot)$ represents the 8 down-sampling convolution operations, and x represents the input image, where the output of each down-sampling convolution operation serves as the input to the next down-sampling convolution operation. The specific down-sampling convolution operation is shown in Equation (4).

$$fea = LReLU(BN(Conv(input))) \tag{4}$$

where fea represents the feature map obtained from a down-sampling convolution operation $DC(\cdot)$, $LReLU(\cdot)$ represents the LeakyReLU activation function, $BN(\cdot)$ represents the batch normalization operation, and $Conv(\cdot)$ represents the convolution operation with a kernel size of 4×4, stride of 2, and padding of 1. $input$ denotes the input image or feature map. It should be noted that there is no activation function operation in the first down-sampling convolutional layer, and the ReLU activation function is used instead of LeakyReLU in the last down-sampling convolutional layer.

The latent space features have two destinations: The first one is sent to the primary network to enhance its generation capability, and the second one is sent to the auxiliary network to strengthen the inherent properties of the latent space features. Within the auxiliary network, 8 up-sampling convolution modules decode the latent space features and map them to the low b-value space, as shown in Equation (5).

$$\hat{y} = 8 * UC(Latent) \tag{5}$$

where $\hat{y}$ represents the output image of the auxiliary generator, and $8 * UC(\cdot)$ denotes 8 up-sampling transpose convolution operations, where the output of each up-sampling transpose convolution operation serves as the input to the next up-sampling transpose convolution operation. The specific details of the up-sampling transpose convolution operation are outlined in Equations (6) and (7).

$$fea_{Ci} = \begin{cases} fea_{DC(9-i)}, i = 1 \\ Concat(fea_{DC(9-i)}, fea_{UC(i-1)}), i = 2, 3, ..., 8 \end{cases} \tag{6}$$

$$fea_{UCi} = \begin{cases} ReLU(BN(ConvT(fea_{Ci}))), i = 1, 2, ..., 7 \\ Tanh(ConvT(fea_{Ci})), i = 8 \end{cases} \tag{7}$$

where $fea_{DC(9-i)}$ represents the features of the (9-i)-th down-sampling convolutional module, and fea_{UCi} represents the features of the i-th up-sampling transpose convolutional module. $Concat(\cdot)$ represents the operation of concatenating feature dimensions. If this is the first up-sampling transpose convolutional module, the $Concat(\cdot)$ operation is ignored. $ReLU(\cdot)$ represents the ReLU activation function, and $ConvT(\cdot)$ denotes the transpose convolution operation, with a kernel size of 4×4, a stride of 2, and padding of 1. $Tanh(\cdot)$ represents the Tanh activation function. It is worth noting that different equations are executed for different values of i, and finally, when $i = 8$, the final generated image is output.

The discriminator of the auxiliary generator adopts the PatchGAN architecture [13], which consists of 5 convolutional layers. Each convolutional layer performs down-sampling on the feature map. Eventually, it obtains a feature map size that is $\frac{1}{2^5} \times \frac{1}{2^5}$ times larger than the original image, where each intensity value in this feature map corresponds to the discriminative result of a certain region in the input image. PatchGAN divides the image into small patches for discrimination, which allows for accurate reflection of local information and enhances accuracy.

2.3.3. Primary GAN

The generator of the primary network consists of down-sampling convolutional blocks, feature fusion modules, and up-sampling convolutional blocks. The down-sampling convolutional blocks and up-sampling convolutional blocks have the same architecture as those in the auxiliary generator, with the only difference being that the input to the primary generator is the high b-value image. The details and connections between the auxiliary generator and the primary generator are illustrated in Figure 3. The feature fusion module combines the encoded features from the auxiliary network's latent space and the primary generator, as specified in Equation (8).

$$fea_{out} = Fusion(Latent, fea_{MDC}) \tag{8}$$

where $Fusion(\cdot)$ represents the feature fusion operation, $Latent$ denotes the latent space feature map from the auxiliary generator, fea_{MDC} represents the encoded features from the primary generator, and fea_{out} represents the output feature map after the fusion operation. The specific feature fusion operation is illustrated in Equations (9) and (10).

$$fea_{Cout} = Concat(Latent, fea_{MDC}) \tag{9}$$

$$fea_{out} = ReLU(Linear(fea_{Cout})) \tag{10}$$

where $Linear(\cdot)$ denotes the linear fusion operation. The linear layer not only reduces the dimensionality of the features but also effectively integrates the useful features according to weights. fea_{Cout} represents the output features after the concatenation operation.

The latent space features and the features from the primary generator are combined through the feature fusion module to obtain richer texture details. After passing through the feature fusion module, the features are processed by 8 up-sampling convolutional modules to output the generated images. During the training process, the generated images and the real images are evaluated by the primary discriminator, promoting the model's generation capability through adversarial learning. Similar to the architecture of the auxiliary generator's discriminator, the primary discriminator also adopts the PatchGAN network.

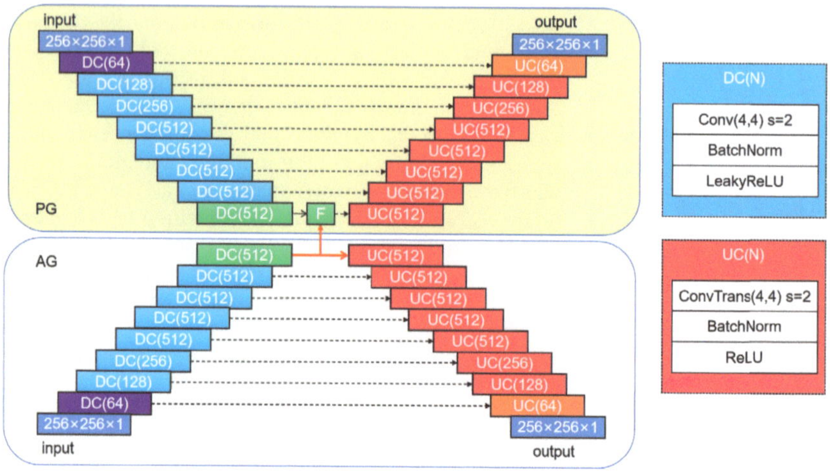

Figure 3. The structure of the two generators. The upper part represents the primary generator (PG), while the lower part represents the auxiliary generator (AG). In the primary generator, the letter "F" represents the feature fusion layer, UC(N) represents the up-sampling transpose convolution operation, and DC(N) represents the down-sampling convolution operation. N represents the number of convolutional channels.

2.3.4. Loss

The loss function consists of both the primary network loss and the auxiliary network loss. Both are trained together but independently backpropagated. The loss function equations of the primary network and the auxiliary network are the same and include both generator adversarial loss, discriminator adversarial loss, and pixel reconstruction loss. Equation (11) represents the generator adversarial loss:

$$\mathcal{L}_{G\_adv} = E[D(x, G(x)) - 1]^2 \tag{11}$$

where $\mathcal{L}_{G\_adv}$ represents the generator adversarial loss, $E(\cdot)$ denotes the expectation, $D(\cdot)$ represents the discriminator's output result, x denotes the input image, and $G(x)$ represents the generator's output result. Theoretically, the generator's adversarial loss is minimized when the discriminator identifies the generated result as 1. Equation (12) shows the discriminator adversarial loss:

$$\mathcal{L}_{D\_adv} = E[D(x, y) - 1]^2 + E[D(x, G(x))]^2 \tag{12}$$

where $\mathcal{L}_{D\_adv}$ represents the discriminator adversarial loss and y represents the real image. The discriminator adversarial loss consists of two parts: the first part minimizes when the concatenated real image with the source image dimension, after being passed through the discriminator, approaches 1; the second part minimizes when the concatenated generated image with the source image dimension, after being passed through the discriminator, approaches 0.

The generator loss and discriminator loss have opposite objective functions, and, during training, one should be fixed while the other is trained in an alternating manner to achieve the adversarial goal. Furthermore, to enhance the authenticity of the generated images, pixel-wise reconstruction loss should be introduced, as shown in Equation (13):

$$\mathcal{L}_1 = E[\|y - G(x)\|_1] \tag{13}$$

where $\mathcal{L}_1$ represents the pixel-wise reconstruction loss and $\|\cdot\|_1$ represents the $L1$ norm. Therefore, the overall loss for both the main network and the auxiliary network is represented as Equation (14):

$$\mathcal{L} = \lambda_{L1}\mathcal{L}_1 + \lambda_{adv}(\mathcal{L}_{G\_adv} + \mathcal{L}_{D\_adv}) \quad (14)$$

where $\mathcal{L}$ represents the overall loss, λ_{L1} represents the pixel-wise reconstruction loss coefficient, and λ_{adv} represents the adversarial loss coefficient.

2.4. Process of dMRI Images Augmentation

After training, the entire dMRI image augmentation process using the final PadGAN model is as follows:

1. Preprocess the dMRI images.
2. Segment the data into 2-dimensional images along the second dimension and input them into PadGAN for processing to generate low b-value images.
3. Multiply the generated images' signal intensity values by the maximum value of the images before normalization to restore the original signal intensity range.
4. Merge the generated two-dimensional images into three-dimensional images and resample all data to the original size.
5. The synthesized three-dimensional images are incorporated into the 4-D dMRI images using FSL tools, effectively improving the quality of the dMRI data. The entire process is illustrated in Figure 4.

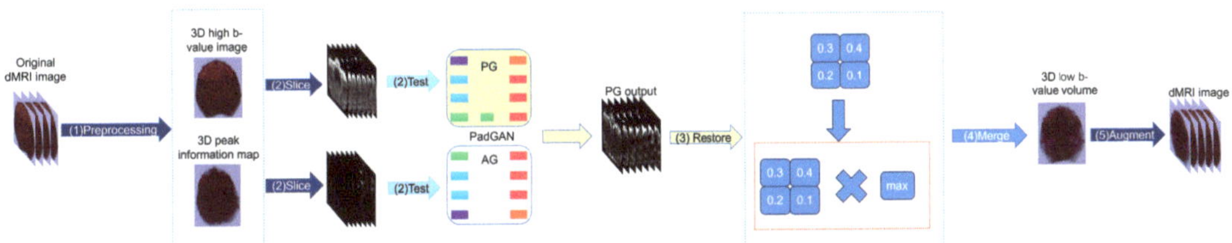

Figure 4. The overall processing flow of dMRI images augmentation. The figure provides a detailed description of the steps outlined in Section 2.4. During testing, the auxiliary generator no longer outputs results, as there is no need to further optimize the latent space through backpropagation.

3. Results

3.1. Comparison Experiments and Results

The method proposed in this paper is compared with five existing methods that have shown good performance in the field of medical image-to-image translation research. Specifically:

1. Pix2pix [13] network adopts the U-Net architecture as the main framework of the generator.
2. CycleGAN [14] network shares the same generator architecture as pix2pix, but it involves two generators and two discriminators for cyclic generation tasks.
3. SwinUnet [24] utilizes the Swin Transformer as the main framework for medical image segmentation tasks, adapted for application in this paper.
4. ResViT [26] builds upon the Vision Transformer architecture as the main generator framework.
5. pGAN [30] adopts ResNet as the main framework.

For the comparative experiments, the original models' architectures and training parameters are used during the training process. All models are pre-trained for 20 epochs and trained for an additional 80 epochs on an NVIDIA GeForce RTX 3090. Structural similarity (SSIM), peak signal-to-noise ratio (PSNR), and mutual information (MI) are selected as quantitative evaluation metrics in this paper.

Table 3 lists the comprehensive results of the AMU, Mount Sinai-P, Mount Sinai-S, UCDavis, and UWM sites, each containing non-brain tissue. To compare the results with only brain tissue, the non-brain tissue is removed from all results, as shown in Table 4, which displays the results after excluding non-brain tissue for the five sites. The overall results are consistent with Table 1, but there is a slight decrease. Subsequent experiments show results after excluding non-brain tissue. The specific results for the five datasets are shown in Table 5, and Figures 5 and 6. The CycleGAN method produces results closer to the source images on most datasets. Although this method employs a dual-generator and dual-discriminator structure, with each generative adversarial network serving separate tasks for generating target and source images, it is suitable for scenarios where paired images are not required in both domains. In contrast, both generative adversarial networks in our method are dedicated to generating low b-value images, resulting in better visual observations and evaluation metrics. The pGAN method fails to generate detail-rich images, as it uses ResNet as the basic generator architecture with a deeper network structure, but lacks the capability to retain encoder feature map information like U-Net. Our method utilizes the advantages of the U-Net architecture to capture features from different layers, thereby preserving detailed image information. Transformer-based ResViT and SwinUnet methods exhibit relatively generic performance due to the differences in global information from different sites in the macaque brain image dataset and the limited data samples. In contrast, our method, a fully convolutional neural network, maximizes the local generation capabilities of convolutional neural networks. The Pix2pix method, a single generator adversarial network based on the U-Net generator architecture, performs well in generating global structural features but lacks detailed features. Our method addresses this limitation by using the auxiliary generative adversarial network to provide hidden space containing more detailed features, thus compensating for the shortcomings of the single generator adversarial network in capturing detailed features.

Table 3. Quantitative comparison results including non-brain tissue.

Methods	PSNR	SSIM	MI
pix2pix	33.7100	0.9285	1.4313
CycleGAN	28.7177	0.8681	1.3716
pGAN	25.9224	0.8534	1.3467
SwinUnet	28.7114	0.8799	1.3786
ResViT	24.6464	0.8428	1.3614
Ours	**38.8700**	**0.9556**	**1.5005**

The bold font indicates the best result.

Table 4. Quantitative comparison of non-brain tissue removal.

Methods	PSNR	SSIM	MI
pix2pix	27.6511	0.7683	1.3144
CycleGAN	22.7904	0.5211	1.2528
pGAN	20.0104	0.4600	1.2275
SwinUnet	23.0855	0.5583	1.2623
ResViT	18.7379	0.4161	1.2376
Ours	**32.2587**	**0.8822**	**1.3828**

The bold font indicates the best result.

Table 5. Quantitative comparison between PadGAN and other translation frameworks across five independent sites.

Model	UCDavis			MountSinai-P			MountSinai-S			AMU			UWM		
	PSNR	SSIM	MI	PSNR	SSIM	MI	PSNR	SSIM	MI	PSNR	SSIM	MI	PSNR	SSIM	MI
pix2pix	29.0037	0.7994	1.3353	22.4200	0.6227	1.2560	25.9427	0.8027	1.3347	**29.6144**	0.7919	1.2938	29.0558	0.8367	1.3045
CycleGAN	23.1076	0.5038	1.2514	19.4039	0.3975	1.2299	25.2008	0.6765	1.2872	26.2187	0.7005	1.2778	19.1597	0.4966	1.2367
pGAN	19.2235	0.4242	1.2284	19.9858	0.4199	1.2128	23.8789	0.6279	1.2511	22.0872	0.5719	1.2402	18.0913	0.4958	1.2018
SwinUnet	23.3034	0.5619	1.2584	17.4287	0.3400	1.2495	25.9383	0.7381	1.2993	26.8461	0.6911	1.2759	28.1703	0.7316	1.2662
ResViT	17.4558	0.3666	1.2398	20.1088	0.4487	1.2268	23.9303	0.5899	1.2568	20.2219	0.4879	1.2489	16.8820	0.4320	1.2023
Ours	**35.7479**	**0.9188**	**1.4185**	**24.7701**	**0.8027**	**1.3379**	**30.6730**	**0.9068**	**1.3826**	28.9085	**0.8150**	**1.3033**	**29.5930**	**0.8753**	**1.3229**

The bold font indicates the best result.

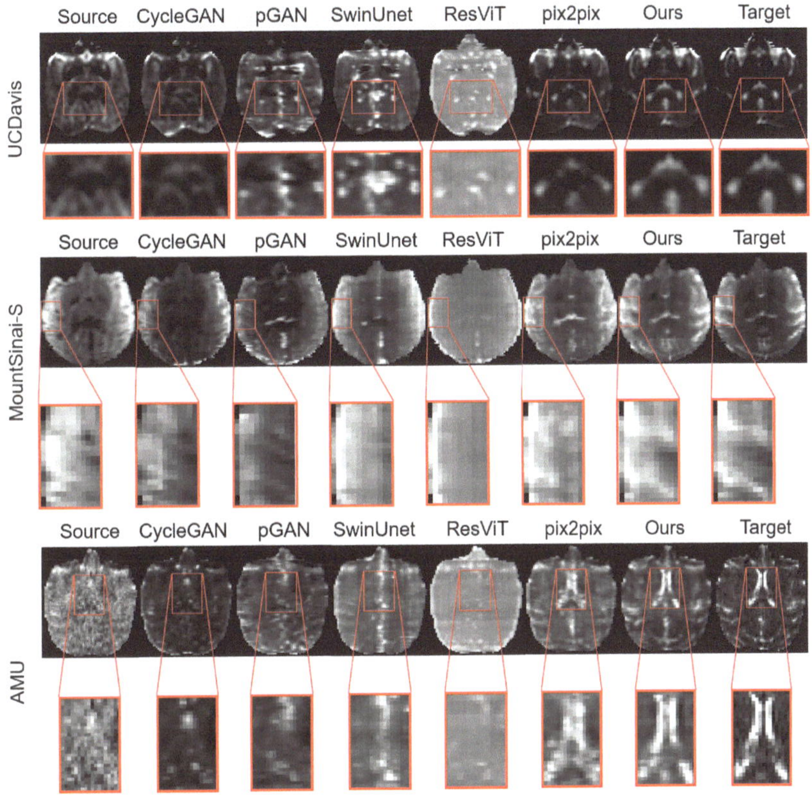

Figure 5. Visualization of 3 site datasets. These are 3 randomly selected data samples from the 3 datasets. The first column represents the source image, the last column represents the target image, and the middle column represents the comparative result. The red box highlights some details.

3.2. Ablation Experiments and Results

We conducted three ablation experiments to further investigate the role and effectiveness of the auxiliary generator in our proposed method. The details of the experiments are as follows: (1) removing the auxiliary network and retaining only the encoder part of the auxiliary network to encode the peak information map, to verify the role of the auxiliary network; (2) replacing the latent space features extracted by the auxiliary generator with random Gaussian noise to explore the role of latent space features; and (3) directly reusing the weights of the main generator in the auxiliary network to verify whether the auxiliary network needs to be trained separately.

The results are shown in Table 6. (1) After removing the auxiliary network, PSNR decreased by 5.1256, SSIM decreased by 0.1225, and MI decreased by 0.0736. This indicates that the auxiliary generator plays an important role in improving the network performance. (2) When replacing the auxiliary generator with noise, PSNR decreased by 4.2291, SSIM decreased by 0.0649, and MI decreased by 0.0445. This suggests that the auxiliary generator can effectively extract latent space features from the peak information map. (3) When reusing the main network's network weights in the auxiliary network, PSNR decreased by 1.8627, SSIM decreased by 0.0385, and MI decreased by 0.0371, fully demonstrating that the latent space learned by the auxiliary generator is different from that of the main generator, and the auxiliary generator has a necessary existence.

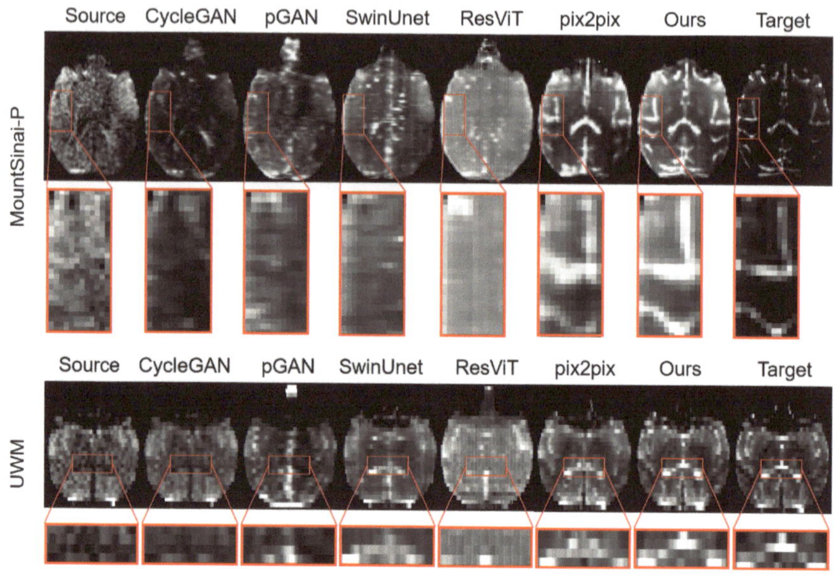

Figure 6. Visualization of 2 site datasets. These are 2 randomly selected data samples from the 2 datasets. The first column represents the source image, the last column represents the target image, and the middle column represents the comparative result. The red box highlights some details.

Table 6. Quantitative comparison of ablation experiments.

Methods	PSNR	SSIM	MI
PadGAN	**32.3856**	**0.8825**	**1.3857**
Setting (1)	27.2600	0.7600	1.3121
Setting (2)	28.1565	0.8176	1.3412
Setting (3)	30.5229	0.8440	1.3486

The bold font indicates the best result.

3.3. *Xtract and DTI Estimation Results*

Xtract is a robust probabilistic tractography method integrated into the FSL6.0 software package. It utilizes dMRI data to estimate the trajectories and connectivity patterns of white matter tracts. To assess the effectiveness of the augmented macaque dMRI brain images through our proposed method, we employed Xtract to compute the structural connectivity of dMRI brain images. Eight subjects were selected from the UCDavis dataset, and the images generated by pix2pix and PadGAN were respectively added to the corresponding dMRI data. Subsequently, we conducted Xtract tractography experiments on the dMRI images augmented by the pix2pix and PadGAN methods, as well as the original reference dMRI images, resulting in a total of 42 fiber tracts.

As shown in Figure 7, the fiber bundle visualization results demonstrate that, compared with pix2pix, our method captures more fiber bundles visually, and the shapes are similar to the reference results. It is worth noting that our results display more and clearer fiber bundles within the white rectangular area.

DTI is a magnetic resonance imaging technique used to study the diffusion properties of water molecules within tissues. DTI offers various diffusion parameters, with the most commonly used being fractional anisotropy (FA) and mean diffusivity (MD). FA represents the degree of directional diffusion of water molecules within the tissue, while MD represents the average strength of water molecule diffusion. To better evaluate the quality of the generated images, this study conducted DTI estimation on dMRI images augmented by the PadGAN and pix2pix methods. Figure 8 displays the DTI estimation results using FA and MD as examples. In the low b-value replacement experiment, our method demonstrates higher similarity to the original reference dMRI images compared with the pix2pix method. In the experiment of augmenting the original reference dMRI, our method shows smoother results. The last column in the figure demonstrates that the absence of low b-value volumes in dMRI images significantly affects the DTI estimation results. Therefore, low b-value images are crucial for DTI computation.

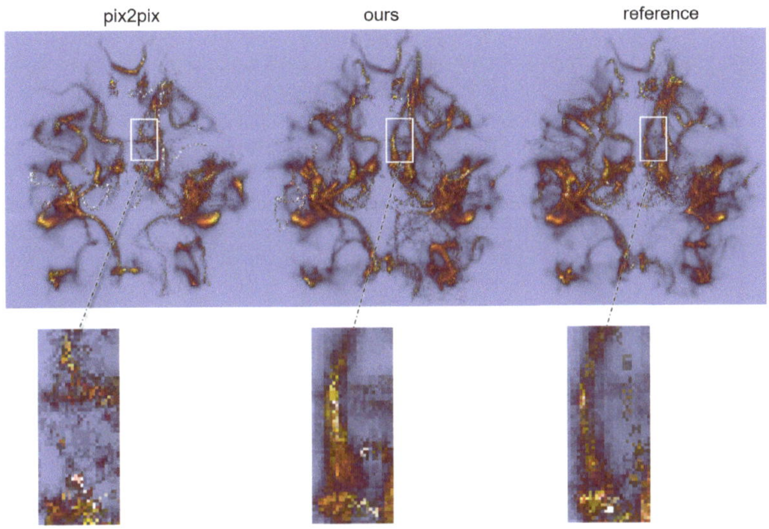

Figure 7. Fiber bundle visualization results. The left and middle columns respectively show the results after data enhancement with the pix2pix and PadGAN methods, while the right column shows the results of the reference original dMRI image. The part inside the white rectangle is zoomed in for comparison.

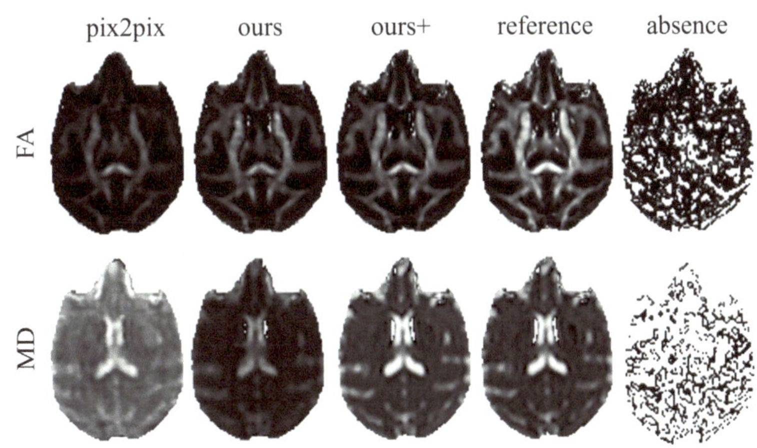

Figure 8. DTI estimation results. The first row displays FA, and the second row shows MD. The first and second columns respectively show the DTI estimation results after replacing the original low b-value volume with volumes generated using pix2pix and our method. The third column shows the DTI estimation results after augmenting the original reference dMRI images using our method for data augmentation. The fourth and fifth columns respectively display the DTI estimation results for the reference images and dMRI without the low b-value volume.

The experiments above indicate that Xtract and DTI estimation results can reflect the quality of different macaque image generation methods. Therefore, Xtract and DTI estimation are expected to become further validation methods for assessing the quality of generated macaque or medical images.

4. Discussion

In this work, we propose a method for dMRI brain image data augmentation using PadGAN to generate low b-value images. The introduction of peak information maps creates end-to-end conditions for extracting latent space features, allowing the auxiliary network to obtain latent space features through adversarial learning. On the basis of the U-Net network, a feature fusion module is added to the primary generator to merge latent space features and multi-scale information, thus generating images with rich details. Additionally, various generative adversarial network models are explored, and the strengths and weaknesses of each model are analyzed. PadGAN is creatively proposed and compared with comparative models in qualitative, quantitative, Xtract probabilistic tractography and DTI estimation to demonstrate its overall performance. Finally, ablation experiments are conducted on each module of PadGAN to demonstrate the importance of each part.

Both generators in PadGAN adopt the encoder–decoder architecture based on U-Net, preserving multi-scale information through skip connections, and the introduction of latent space features enables PadGAN to learn fine-grained image features. As shown in Figures 5 and 6, unlike previous studies on human brain datasets where Transformer-based network models yield poor results, typically due to the large volume of data in human brain datasets resulting in different model parameters for each dataset, our approach uses a unified training strategy for the limited datasets of macaque brain images from each site. For datasets collected from each site, there are significant differences in acquisition parameters. Therefore, attention mechanisms are difficult to perform effectively for multi-site datasets. While ResNet can maintain model learning capability, even with deep network layers, it does not preserve multi-scale features like U-Net, resulting in deficiencies in detail generation. The pix2pix method based on U-Net demonstrates good performance, but, as a single generator and discriminator method, it still lacks in generating image details. Although CycleGAN has two generative adversarial networks, these networks are tasked with mutual conversion between two modal data samples and do not leverage both networks to generate images in one target domain. The auxiliary network in PadGAN provides latent space information to the primary network to enhance the detail generation of the generated images, utilizing U-Net's skip connections to preserve multi-scale information, resulting in superior performance in image details.

Unlike the typical computer vision image-to-image translation domain, the signal intensity values of MRI images have absolute significance and can be used for DTI estimations or neuroimaging studies. Common images in daily life are usually RGB images with a maximum pixel intensity value of 255, while the signal intensity value range of macaque brain images is typically in the range of thousands to tens of thousands. Therefore, when evaluating the quality of MRI image generation, we can go beyond quantitative metrics and qualitative observations. For medical MRI, some researchers conduct Turing tests with expert radiologists to assess the authenticity of generated images [32]. For macaque and human brain images used in research, we can further evaluate the quality of generated images by calculating neural tracing or DTI estimation results, which presents a novel validation approach.

In future work, we can explore the generation of realistic images using multi-modal data. Although the macaque brain imaging dataset is limited, with few data within each site, many sites have at least two modalities of data. Utilizing network models that can effectively leverage multi-modal information may lead to the generation of higher quality images. Additionally, our method also has the potential for application in human brain imaging. Firstly, our method can be used for data augmentation of human brain dMRI images. Although human brain images typically have a higher spatial resolution and signal-to-noise ratio, and there are more publicly available datasets with better data quality, there may still be issues with insufficient collection of low b-value images due to operator and configuration issues. In such challenges, applying our method directly to human brain images is a good choice. Secondly, our method has the potential for application in classification studies of normal and diseased brain images. By using the PadGAN method to generate more images of a certain modality, the image sample size can be expanded, thereby improving classification accuracy. However, diseased images typically require higher precision in a certain region, and it may be a good choice to introduce attention mechanisms to enhance contextual information.

5. Conclusions

PadGAN is employed to translate high b-value images of macaque brains to low b-value images and augment dMRI image data. Visually, the low b-value images generated by PadGAN exhibit richer detail information. In terms of evaluation metrics, both image quality and structural similarity show significant improvement. Results from Xtract probabilistic tractography and DTI estimation indicate that the dMRI images obtained through our data augmentation method yield better outcomes. This work can provide data augmentation and optimization services for neuroscience, and also offers insights into quality assessment methods for macaque dMRI brain imaging data.

Author Contributions: conceptualization, Y.C. and L.Z.; methodology, Y.C. and X.L.; software, Y.C.; validation, X.X. and L.Z.; formal analysis, Y.C. and L.Z.; investigation, X.X. and L.Z.; resources, X.X. and L.Z.; data curation, X.L.; writing—original draft preparation, Y.C.; writing—review and editing, Q.W. and H.L.; visualization, X.L. and Q.W.; supervision, Q.W. and H.L.; project administration, H.L.; funding acquisition, H.L. All authors have read and agreed to the published version of the manuscript.

Funding: This project is supported by the ShanXi Science and Technology Department (YDZJSX2021C005), the Natural Science Foundation of Shanxi (2022ZDYF128) (2023L490), and the National Natural Science Foundation of China (61976150).

Institutional Review Board Statement: Our data were collected by the Human Connectome Project and International Neuroimaging Data-Sharing Initiative (INDI) PRIMatE Data Exchange Project. We follow the data usage polices Attribution-NonCommercial Share Alike (CC-BY-NC-SA).

Informed Consent Statement: Not applicable.

Data Availability Statement: In this research, all datasets were collected from publicly accessible datasets. The human data are accessed at https://www.humanconnectome.org/study/hcp-young-adult, accessed on 1 March 2024. The Macaque data are accessed at http://fcon_1000.projects.nitrc.org/indi/PRIMEdownloads.html, accessed on 1 March 2024. The FSL6.0 tools are accessed at https://fsl.fmrib.ox.ac.uk/fsl/fslwiki/, accessed on 7 April 2024.

Conflicts of Interest: The authors declare that the research was conducted in the absence of any commercial or financial relationships that could be construed as a potential conflict of interest. And the funders had no role in the design of the study; in the collection, analyses, or interpretation of data; in the writing of the manuscript; or in the decision to publish the results.

References

1. Passingham, R. How good is the macaque monkey model of the human brain? *Curr. Opin. Neurobiol.* **2009**, *19*, 6–11. [CrossRef]
2. Neubert, F.X.; Mars, R.B.; Sallet, J.; Rushworth, M.F. Connectivity reveals relationship of brain areas for reward-guided learning and decision making in human and monkey frontal cortex. *Proc. Natl. Acad. Sci. USA* **2015**, *112*, E2695–E2704. [CrossRef]
3. Wang, Q.; Wang, Y.; Chai, J.; Li, B.; Li, H. A review of homologous brain regions between humans and macaques. *J. Taiyuan Univ. Technol.* **2021**, *52*, 274–281.
4. Bauer, M.H.; Kuhnt, D.; Barbieri, S.; Klein, J.; Becker, A.; Freisleben, B.; Hahn, H.K.; Nimsky, C. Reconstruction of white matter tracts via repeated deterministic streamline tracking–initial experience. *PLoS ONE* **2013**, *8*, e63082. [CrossRef]
5. Soares, J.M.; Marques, P.; Alves, V.; Sousa, N. A hitchhiker's guide to diffusion tensor imaging. *Front. Neurosci.* **2013**, *7*, 31. [CrossRef]

6. Milham, M.P.; Ai, L.; Koo, B.; Xu, T.; Amiez, C.; Balezeau, F.; Baxter, M.G.; Blezer, E.L.A.; Brochier, T.; Chen, A.H. An Open Resource for Non-human Primate Imaging. *Neuron* **2018**, *100*, 61–74.e2. [CrossRef]
7. Yurt, M.; Dar, S.U.; Erdem, A.; Erdem, E.; Oguz, K.K.; Cukur, T. mustgan: Multi-stream generative adversarial networks for mr image synthesis. *Med. Image Anal.* **2021**, *70*, 101944. [CrossRef]
8. Shin, H.C.; Ihsani, A.; Mandava, S.; Sreenivas, S.T.; Forster, C.; Cha, J. Ganbert: Generative adversarial networks with bidirectional encoder representations from transformers for mri to pet synthesis. *arXiv* **2020**, arXiv:2008.04393.
9. Huang, J.H. Swin transformer for fast mri.Neurocomputing. *Neurocomputing* **2022**, *493*, 281–304. [CrossRef]
10. Sikka, A.; Virk, J.S.; and Bathula, D.R. Mri to pet cross-modality translation using globally and locally aware gan (gla-gan) for multi-modal diagnosis of alzheimer's disease. *arXiv* **2021**, arXiv:2108.02160.
11. Goodfellow, I.J.; Pouget-Abadie,J.; Mirza, M.; Xu, B.; Warde-Farley, D.; Ozair, S.; Courville, A.;Bengio, Y. Generative Adversarial Networks. *arXiv* **2014**, arXiv:1406.2661.
12. Jiang, Y.F.; Chang, S.Y.; Wang, Z.Y. TransGAN: Two Pure Transformers Can Make One Strong GAN, and That Can Scale Up. *arXiv* **2021**, arXiv:2102.07074.
13. Isola, P.; Zhu, J.Y.; Zhou, T.; Efros, A.A. Image-to-image translation with conditional adversarial networks. In Proceedings of the IEEE Conference on Computer Vision and Pattern Recognition, Honolulu, HI, USA, 21–26 July 2017; pp. 1125–1134.
14. Zhu, J.Y.; Park, T.; Isola, P.; Efros, A.A. Unpaired image-to-image translation using cycle-consistent adversarial networks. In Proceedings of the IEEE International Conference on Computer Vision, Venice, Italy, 22–29 October 2017; pp. 2242–2251.
15. Welander, P.; Karlsson, S.; Eklund, A. Generative adversarial networks for image-to-image translation on multi-contrast mr images-a comparison of cyclegan and unit. *arXiv* **2018**, arXiv:1806.07777.
16. Gu, X.; Knutsson, H.; Nilsson, M.; Eklund, A. Generating diffusion mri scalar maps from t1 weighted images using generative adversarial networks. In *Image Analysis*; Lecture Notes in Computer Science (Including Subseries Lecture Notes in Artificial Intelligence and Lecture Notes in Bioinformatics); Springer: Cham, Switzerland, 2019; pp. 489–498.
17. Abramian, D.; Eklund, A. Generating fmri volumes from t1-weighted volumes using 3d cyclegan. *arXiv* **2019**, arXiv:1907.08533
18. Zhao, P.; Pan, H.; Xia, S. Mri-trans-gan: 3d mri cross-modality translation. In Proceedings of the 2021 40th Chinese Control Conference (CCC), Shanghai, China, 26–28 July 2021; pp. 7229–7234.
19. Armanious, K.; Jiang, C.M.; Abdulatif, S.; Kustner, T.; Gatidis, S.; Yang, B. Unsupervised Medical Image Translation Using Cycle-MedGAN. In Proceedings of the 2019 27th European Signal Processing Conference (EUSIPCO), A Coruña, Spain, 2–6 September 2019; pp. 1–5.
20. Benoit, A.R. Manifold-Aware CycleGAN for High-Resolution Structural-to-DTI Synthesis. In *Computational Diffusion MRI: International MICCAI Workshop*; Springer: Cham, Switzerland, 2021; pp. 213–224.
21. Kearney, V.; Ziemer, B.P.; Perry, A.; Wang, T; Chan, J.W.; Ma, L.; Morin, O.; Yom, S.S.; Solberg, T.D. Attention-Aware Discrimination for MR-to-CT Image Translation Using Cycle-Consistent Generative Adversarial Networks. *Radiol. Artif. Intell.* **2020**, *2*, e190027 [CrossRef]
22. Bui, T.D.; Nguyen, M.; Le,N.; Luu, K. Flow-Based Deformation Guidance for Unpaired Multi-contrast MRI Image-to-Image Translation. In Proceedings of the Medical Image Computing and Computer Assisted Intervention—MICCAI 2020, Lima, Peru, 4–8 October 2020; pp. 728–737.
23. Zhang, H.; Li, H.; Parikh, N.A.; He, L. Multi-contrast mri image synthesis using switchable cycle-consistent generative adversarial networks. *Diagnostics* **2022**, *12*, 816. [CrossRef]
24. Cao, H.; Wang, Y.Y.; Chen, J.; Jiang, D.S.; Zhang, X.P.; Tian, Q.; Wang M.N. Swin-unet: Unet-like pure transformer for medical image segmentation. In Proceedings of the European Conference on Computer Vision, Tel Aviv, Israel, 23–27 October 2022; pp. 205–218.
25. Huang, J.; Xing, X.; Gao, Z.; Yang, G. Swin Deformable Attention U-Net Transformer (SDAUT) for Explainable Fast MRI for explainable fast mri. In Proceedings of the International Conference on Medical Image Computing and Computer-Assisted Intervention, Singapore, 18–22 September 2022; pp. 538–548.
26. Dalmaz, O.; Yurt, M.; Cukur, T. ResViT: Residual vision transformers for multi-modal medical image synthesis. *IEEE Trans. Med. Imaging* **2022**, *41*, 2598–2614. [CrossRef]
27. Yan, S.; Wang, C.; Chen, W.; Lyu, J. Swin transformer-based GAN for multi-modal medical image translation. *Front. Oncol.* **2022**, *12*, 942511. [CrossRef]
28. Dosovitskiy, A.; Beyer, L.; Kolesnikov, A; Weissenborn, D.; Zhai, X; Unterthiner, T.; Dehghani, M.; Minderer, M.; Heigold, G.; Gelly, S.; et al. An Image is Worth 16x16 Words: Transformers for Image Recognition at Scale. *arXiv* **2021**, arXiv:2010.11929.
29. Schilling, K.G.; Blaber, J.; Hansen, C.; Cai, L.; Rogers, B.; Anderson, A.W.; Smith, S.; Kanakaraj, P.; Rex, T; Resnick S.M.; et al. Distortion correction of diffusion weighted MRI without reverse phase-encoding scans or field-maps Distortion correction of diffusion weighted mri without reverse phase-encoding scans or field-maps. *PLoS ONE* **2020**, *15*, e0236418. [CrossRef]
30. Pgan Dar, S.U.; Yurt, M.; Karacan, L.; Erdem, A.; Erdem, E.; Cukur, T. Image synthesis in multi-contrast mri with conditional generative adversarial networks. *IEEE Trans. Med. Imaging* **2019**, *38*, 2375–2388.
31. Yu, B.T.; Zhou, L.P.; Wang, L.; Shi, Y.H.; Fripp, J.; Bourgeat, P. Ea-GANs: Edge-Aware Generative Adversarial Networks for Cross-Modality MR Image Synthesis. *IEEE Trans. Med. Imaging* **2019**, *38*, 1750–1762. [CrossRef] [PubMed]
32. Armanious, K.; Jiang, C.M.; Fischer, M.; Kustner, T.; Hepp, T.; Nikolaou, K.; Gatidis, S.; Yang, B. MedGAN: Medical image translation using GANs. *Comput. Med. Imaging Graph.* **2020**, *79*, 101684. [CrossRef] [PubMed]

33. Yang, Q.; Li, N.; Zhao, Z.; Fan, X.; Chang, E.; Xu, Y. Mri cross-modality image-to-image translation. *Sci. Rep.* **2020**, *10*, 3753. [CrossRef]
34. Warrington, S.; Bryant, K.L.; Khrapitchev, A.A.; Sallet, J.; Charquero-Ballester, M.; Douaud, G.; Jbabdi, S.; Mars, R.B.; Sotiropoulos, S.N. Xtract-standardised protocols for automated tractography in the human and macaque brain. *NeuroImage* **2020**, *217*, 116923. [CrossRef] [PubMed]
35. Jenkinson, M.; Beckmann, C.F.; Behrens, T.E.; Woolrich, M.W.; Smith, S.M. FSL. *NeuroImage* **2012**, *62*, 782–790. [CrossRef]
36. Van Essen, D.C.; Smith, S.M.; Barch, D.M.; Behrens, T.E.; Yacoub, E.; Ugurbil, K. The wu-minn human connectome project: An overview. *NeuroImage* **2013**, *80*, 62–79. [CrossRef] [PubMed]
37. Wang, Q; Fei, H; Abdu, N.S.; Xia, X.; Li, H. A Macaque Brain Extraction Model Based on U-Net Combined with Residual Structure. *Brain Sci.* **2022**, *12*, 260. [CrossRef] [PubMed]
38. Abdal, R.; Qin, Y.; Wonka, P. Image2stylegan: How to embed images into the stylegan latent space? In Proceedings of the IEEE/CVF International Conference on Computer Vision, Seoul, Republic of Korea, 27 October–2 November 2019; pp. 4432–4441.
39. Karras, T.; Laine, S.; Aila, T. A style-based generator architecture for generative adversarial networks. *IEEE Trans. Pattern Anal. Mach. Intell.* **2019**, *43*, 4217–4228. [CrossRef]
40. Wang, T.; Zhang, Y.; Fan, Y.; Wang, J.; Chen, Q. High-Fidelity GAN Inversion for Image Attribute Editing. In Proceedings of the IEEE/CVF Conference on Computer Vision and Pattern Recognition, New Orleans, LA, USA, 18–24 June 2022; pp. 11379–11388.
41. Richardson, E.; Alaluf, Y.; Patashnik, O.; Nitzan, Y.; Azar, Y.; Shapiro, S.; Cohen-Or, D. Encoding in Style: A StyleGAN Encoder for Image-to-Image Translation. In Proceedings of the IEEE/CVF Conference on Computer Vision and Pattern Recognition, Nashville, TN, USA, 20–25 June 2021; pp. 2287–2296.
42. Gholamalinezhad, H.; Khosravi, H. Pooling Methods in Deep Neural Networks, a Review. *arXiv* **2020**, arXiv:2009.07485.
43. Radford, A.; Metz, L.; Chintala, S. Unsupervised Representation Learning with Deep Convolutional Generative Adversarial Networks. *arXiv* **2015**, arXiv:1511.06434.

applied sciences

MDPI

Article

Exploring EEG Emotion Recognition through Complex Networks: Insights from the Visibility Graph of Ordinal Patterns

Longxin Yao [1,2], Yun Lu [2,3,*], Mingjiang Wang [1,2,*], Yukun Qian [1,2] and Heng Li [1,2]

1 School of Electronic and Information Engineering, Harbin Institute of Technology, Shenzhen 518055, China; 19b952005@stu.hit.edu.cn (L.Y.); qianyukun@stu.hit.edu.cn (Y.Q.); 19s152117@stu.hit.edu.cn (H.L.)
2 Shenzhen Key Laboratory of IoT Key Technology, Harbin Institute of Technology, Shenzhen 518055, China
3 School of Computer Science and Engineering, Huizhou University, Huizhou 516007, China
* Correspondence: luyun_hit@163.com (Y.L.); mjwang@hit.edu.cn (M.W.)

Abstract: The construction of complex networks from electroencephalography (EEG) proves to be an effective method for representing emotion patterns in affection computing as it offers rich spatiotemporal EEG features associated with brain emotions. In this paper, we propose a novel method for constructing complex networks from EEG signals for emotion recognition, which begins with phase space reconstruction to obtain ordinal patterns and subsequently forms a graph network representation from the sequence of ordinal patterns based on the visibility graph method, named ComNet-PSR-VG. For the proposed ComNet-PSR-VG, the initial step involves mapping EEG signals into a series of ordinal partitions using phase space reconstruction, generating a sequence of ordinal patterns. These ordinal patterns are then quantified to form a symbolized new sequence. Subsequently, the resulting symbolized sequence of ordinal patterns is transformed into a graph network using the visibility graph method. Two types of network node measures, average node degree (AND) and node degree entropy (NDE), are extracted from the graph networks as the inputs of machine learning for EEG emotion recognition. To evaluate the effectiveness of the proposed construction method of complex networks based on the visibility graph of ordinal patterns, comparative experiments are conducted using two types of simulated signals (random and Lorenz signals). Subsequently, EEG emotion recognition is performed on the SEED EEG emotion dataset. The experimental results show that, with AND as the feature, our proposed method is 4.88% higher than the existing visibility graph method and 12.23% higher than the phase space reconstruction method. These findings indicate that our proposed novel method for constructing complex networks from EEG signals not only achieves effective emotional EEG pattern recognition but also exhibits the potential for extension to other EEG pattern learning tasks, suggesting broad adaptability and application potential for our method.

Keywords: emotion recognition; complex network; ordinal patterns

Citation: Yao, L.; Lu, Y.; Wang, M.; Qian, Y.; Li, H. Exploring EEG Emotion Recognition through Complex Networks: Insights from the Visibility Graph of Ordinal Patterns. *Appl. Sci.* **2024**, *14*, 2636. https://doi.org/10.3390/app14062636

Academic Editor: Alexander N. Pisarchik

Received: 29 January 2024
Revised: 13 March 2024
Accepted: 18 March 2024
Published: 21 March 2024

1. Introduction

The emotional dimensions of electroencephalography (EEG) have garnered increasing recognition, owing to its extensive applications in diagnosing mental illnesses and facilitating human–computer interaction [1,2]. By delving into the study of emotional patterns within EEG, we can enrich our comprehension of human behavior, refine psychological health treatment methodologies, and cultivate more intelligent and responsive systems within the realm of human–computer interaction. In recent years, the efficacy of complex networks in unraveling the spatiotemporal characteristics and dynamic shifts in emotional EEG has become evident. EEG signals, serving as physiological indicators of brain activity, contribute significantly to this exploration. Given the intricate structure and interconnections within the brain network, the adoption of complex networks for analyzing both brain networks and emotional EEG has garnered increasing attention. Research grounded in complex networks offers a more holistic insight into the intricate topology and information transmission among distinct brain regions [3]. In [4], Lu et al. constructed a new complex network pattern based on the arrangement characteristics of time series and achieved excellent results in brain state recognition based on EEG signals in physiology and pathology. In [2], Yao et al. constructed a complex network of EEG signals using a viewable approach and extracted spatial network features, achieving high resolution in EEG emotion recognition. The transformation of EEG data into complex networks proves to be a valuable approach, providing a more effective representation of the complexity and dynamics inherent in brain activity. This transformation enhances our capacity to accurately capture the neural mechanisms associated with emotions. Consequently, this avenue of research holds the promise of advancing our understanding of emotional EEG, paving the way for innovative developments in neuroscience and human–computer interaction.

Complex network methods have the capability to unveil intricate interactions and connectivity patterns among various brain regions, a collaboration crucial in emotional processing. By scrutinizing connection patterns within complex networks, a deeper comprehension of the functions and interactions among different brain regions during emotional processing is attained. Unlike time-domain or frequency-domain methods applied in EEG signal analysis, complex networks can encapsulate both global and local features within the brain network, thus surpassing the constraints of localized time- or frequency-domain features [3]. The dynamic fluctuations within the brain network across different time points are observable through the construction of complex networks, providing a more profound insight into the spatiotemporal characteristics of brain activity during emotional processes.

To comprehensively analyze the connection density of nodes in complex networks from both local and global perspectives, effective measures, such as average node degree (AND) and node degree entropy (NDE), come into play. The AND serves as a valuable metric to offer overall insights into the connection density of nodes in a network, providing a descriptive overview of the network's general properties. Meanwhile, NDE plays a pivotal role in the analysis of complex networks, aiming to articulate the uncertainty and diversity inherent in the degree distribution among nodes. The degree of a node denotes the number of edges connected to it, and NDE takes into consideration the distribution of these degrees, shedding light on the quantity and relative frequency of nodes with varying degrees in the network. This metric offers crucial information about the degree distribution across nodes, allowing for a deeper understanding of how nodes interconnect and the prevalence of nodes with similar or distinct degrees. By capturing the uncertainty inherent in degree distribution, NDE becomes a powerful tool for unraveling the intricacies of network structure. It operates as a metric for gauging the complexity of the network with highly structured networks exhibiting higher node degree entropy. Additionally, NDE can be harnessed to scrutinize the correlation between node degrees, uncovering connections between nodes with specific degrees. This aspect proves instrumental in capturing features of degree correlation, providing valuable insights into the network's organization. In essence, NDE, by encapsulating the diversity and uncertainty present in degree distribution, contributes supplementary information for a more profound and nuanced analysis of complex networks.

The phase space reconstruction method involves deriving a set of multidimensional vectors from the original time series using embedding dimensions and delay time estimation techniques [5]. These vectors serve as nodes in the complex network, and the edges connecting these nodes are determined based on the similarity between vectors. However, this method faces instability issues during the embedding dimensions and delay time estimation process. Additionally, establishing the optimal threshold for edge relationship judgment proves challenging, resulting in diminished robustness in practical applications [4,6]. On the other hand, the visibility graph construction method regards data points in the original time series as nodes in the network with the visual relationships between these data points serving as edges [7,8]. In contrast to the phase space reconstruction method, the visibility graph construction method boasts fewer parameters and enhanced algorithmic robustness [9–11]. However, it is important to note that the size of the network in this approach is directly proportional to the length of the time series. Consequently, when analyzing longer time series, the complexity of the network increases correspondingly, leading to heightened computational complexity in extracting subsequent features from the complex network [12–15].

In our research, we propose a pioneering method for constructing complex networks, which diverges from traditional approaches. The novelty of our method lies in the fusion of phase space reconstruction techniques and visibility graph methods, enabling the simultaneous depiction and analysis of complex network structures and dynamic behaviors from both temporal and spatial viewpoints. Phase space reconstruction delves into the internal relationships and dynamic behaviors of network nodes, while visibility graph construction highlights the overarching structure and connectivity patterns of nodes [16,17].

By amalgamating the phase space reconstruction and visibility graph methods, we harness the advantages of both approaches, thereby enhancing the accuracy and robustness of complex network construction. The specific implementation can be tailored and fine-tuned according to practical needs [18]. Through the integration of these two methods, we attain a more comprehensive comprehension of network properties and patterns. By concurrently leveraging the benefits of phase space reconstruction and visibility graph construction, we augment the efficiency and precision of our analyses. In summary, the main contributions of our work include the following:

(1) A novel method for constructing complex networks from EEG signals, named ComNet-PSR-VG, is introduced by exploiting both the phase space reconstruction method and the visibility graph method;
(2) Employing the proposed ComNet-PSR-VG method to effectively identify EEG emotion states, obtaining outstanding classification outcomes of emotion recognition.

The remainder of the paper is as follows: The second part presents the proposed new method for constructing complex networks and the extracted network structure features, the third part presents the results of the data analysis and EEG emotion classification experiments; the fourth part compares our method with existing related research through experiments and results; and the last part is the conclusion of the article.

2. Materials and Methods

Our proposed method includes several key steps, as shown in Figure 1:

- Recording the corresponding emotional EEG signals generated by different emotional stimuli;
- Constructing complex networks for each channel of EEG signals using the proposed method and proposing network structure entropy features;
- Extracting entropy features of network structure;
- Inputting these features as feature sequences into the machine-learning model to obtain the corresponding classification results.

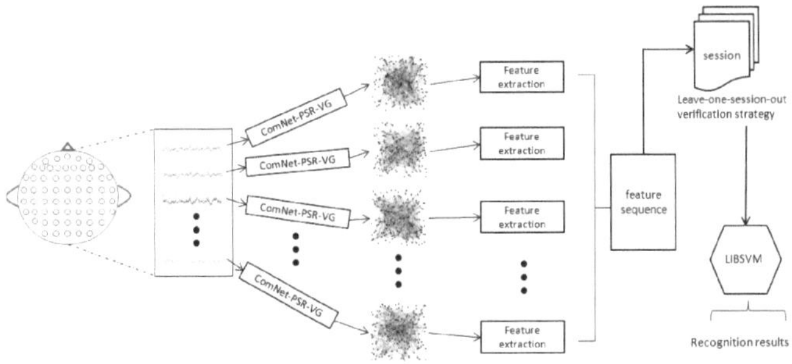

Figure 1. Framework of the proposed EEG emotion recognition.

2.1. Experimental Dataset

In our study, we utilized the SEED (SJTU Emotion EEG Dataset), an openly available dataset for thorough analysis. This dataset encompasses data from 15 Chinese subjects with a gender distribution of 7 males and 8 females and an average age of 23.27 years (standard deviation: 2.37). The emotional stimuli for the participants were derived from 15 Chinese film clips, each designed to elicit positive, neutral, or negative emotions, and each film lasted approximately 4 min. To execute our experiments, each participant engaged in 15 trials, resulting in a total of 45 trials (15 trials for each of the three emotional categories: positive, neutral, and negative). The experimental design comprised three distinct groups of experiments [19]. In each trial, the subjects were exposed to emotional stimuli through designated film clips, inducing the specified emotion (positive, neutral, or negative). This rigorous experimental setup aimed to comprehensively capture the varied responses to emotional stimuli across the different emotional categories.

2.2. Construction of Complex Networks from EEG Signals Based on Visibility Graph of Ordinal Patterns

The signal from each channel in EEG can be treated as a time series $\{x_i\}$, where i $=$ $1,2,\ldots N$. Initially, the phase space reconstruction method is used to reconstruct this time series into a sequence [4] using embedding dimensions d and time delay τ. The resulting sequence can be written as follows:

$$v_j = \left(x_j, x_{j+\tau}, x_{j+2\tau}\right)\ j = 1,2,\ 3,\ldots,L \tag{1}$$

where $L = N - (d-1) * \tau$ and denotes the number of partitions v_j in the resulting sequence.

Subsequently, each partition v_j is mapped into an ordinal pattern $O^{(i)} = (\pi_0,\ \pi_1,\ \pi_2,\ \ldots,\ \pi_{d-1})$ where $\pi_i \in \{0,\ 1,\ 2,\ \ldots,\ d-1\}$ ($\pi_i \neq \pi_j$ if $i \neq j$). Specifically, the indices of each element in the partition $v_i = (x_i,\ x_{i+\tau},\ x_{i+2\tau},\ \ldots,\ x_{i+(d-1)\tau})$ are rewritten to $v_i = (x_{i+\pi_0},\ x_{i+\pi_1},\ x_{i+\pi_2},\ \ldots,\ x_{i+\pi_{d-1}})$, according to the ascending order of the values of elements in the partition v_i:

$$x_{i+\pi_0} \le x_{i+\pi_1} \le x_{i+\pi_2} \cdots \le x_{i+\pi_{d-1}}, \forall x_{i+\pi_k} \in v_i \ and\ \pi_k = \{0,1,2,\cdots,d-1\} \tag{2}$$

For example, taking the $\{18,9,5,11\}$ as a partition, it can be mapped to an ordinal pattern $\{2,1,3,0\}$.

Finally, we introduce a metric for quantifying the ordinal patterns, denoted as the ordinal pattern number (OPN) [4]. Its formulation is articulated as follows:

$$OPN\left(O^{(i)}\right) = Inv\ (\pi_0) \times (d-1)\,! + Inv\ (\pi_1) \times (d-2)\,! + \cdots + Inv\ (\pi_{d-2}) \times (1)\,! + 1 \tag{3}$$

where $(\cdot)!$ denotes the factorial function, and $Inv\ (\pi_i)$ represents the inverse number of each element π_i in the ordinal pattern $O^{(i)} = (\pi_0,\ \pi_1,\ \pi_2,\ \ldots,\ \pi_{d-1})$. In accordance with Equation (3), the minimum value of the OPN is 1, which corresponds to the permutation

$\pi = (0, 1, 2, \ldots, d-1)$ in ascending order, the maximum value of the *OPN* is $d!$, and descending order is $\pi = (d-1, \ldots, 2, 1, 0)$.

Following the aforementioned time series transformation and employing the phase space reconstruction method and ordinal pattern quantization, the time series with a data length of N is transformed into a symbol sequence with a length of L. Subsequently, utilizing the visibility graph method [4], the resulting symbol sequence is mapped into a graph network. To clearly illustrate the proposed method, which constructs a network, the basic process of constructing a complex network from a time series is shown in Figure 2a. Figure 2b presents the proposed method for time-series mapping to the OPN of network nodes.

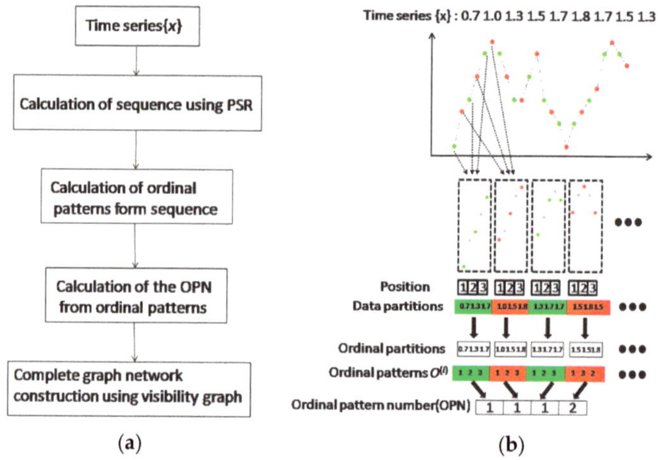

Figure 2. (**a**) Construction of complex networks from time series based on the visibility graph of ordinal patterns; (**b**) The proposed mapping algorithm for time-series mapping to the OPN of network nodes.

2.3. Extracting Network Entropy Measures from Complex Networks

Network measures are commonly expressed through diverse network structural parameters, such as nodes and links, which typically represent network-related features and characterize the patterns of the network. As one of the classical network measures, the average node degree (AND) serves as a valuable tool for offering comprehensive insights into the connection density of nodes within the network. This network node measure serves as an effective descriptor of the overall properties of the network. It captures the average connection strength among neighboring nodes, facilitating an understanding of the distribution of node degrees and the characteristics of connections in the network. The calculation expression for AND is as follows:

$$k_{nn} = \frac{1}{N}\sum_{i=1}^{N} k_{nn}^{i} \tag{4}$$

where k_{nn}^{i} indicates the degree of neighboring nodes for a node.

Network entropy, derived from information theory, is a measure of disorder used to quantify the information content encoded within a graph network. It provides a quantitative metric to assess network complexity. As one of the crucial network structure entropies, the strength of node degree entropy (NDE) lies in its comprehensive and unified depiction of the degree distribution within the network structure, determined through the consideration of neighbor degrees of nodes. The NDE proves highly effective in assessing node heterogeneity concerning neighbor degrees with its calculation expressed as follows:

$$H = -\sum_{i} p_i \log p_i \tag{5}$$

p_i is the probability description of the node degree, which can be expressed in the following form:

$$p_i = \frac{d_i}{\sum\limits_{j=1}^{N} d_j} \tag{6}$$

d_i is the number of neighbors in a node network.

2.4. Machine-Learning Model

The support vector machine (SVM) stands as a pivotal classification model in the realm of machine learning with the primary goal of delineating samples by identifying an optimal hyperplane. Its fundamental objective centers around maximizing intervals for effective segmentation. In our research, we leveraged individual channels of EEG signals as distinctive structural attributes within a network. The SVM served as our classifier, adept at distinguishing between positive and negative emotions. Harnessing kernel-based capabilities, the SVM exhibited prowess in achieving both linear and nonlinear classifications, thanks to diverse kernel functions with varying performance characteristics. Our study meticulously scrutinized multiple prevalent SVM kernels, ultimately identifying the radial basis function (RBF) as the most efficient performer. For our SVM classifier, we utilized the LIBSVM software package (https://www.csie.ntu.edu.tw/~cjlin/libsvm/), specifically implementing the RBF kernel. The configuration of SVM parameters involved values such as S, T, and C alongside default settings. S is the model setting type for SVM, T is the kernel function type, and C is the cost. Notably, T was set at 2, while S stood at 0. Determining the optimal C value entailed a meticulous one-step search within the parameter space ($10^{-3:2}$). Our methodological framework, which integrates complex network feature measures for emotive recognition via the SVM classifier, is comprehensively illustrated in Figure 3.

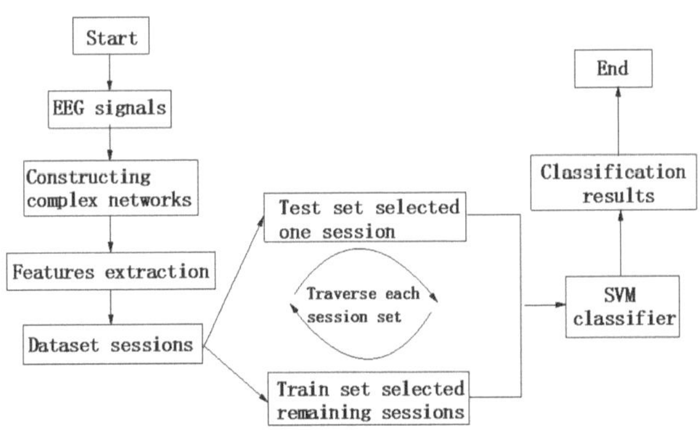

Figure 3. Flowchart of machine-learning classification using complex network features.

2.5. Performance Evaluation

In our study, accuracy, sensitivity, and specificity serve as the performance metrics for evaluating the EEG emotion recognition task. Positive emotion is designated as positive instances, while negative emotion is designated as negative instances. The mathematical definitions of these evaluation metrics are expressed as follows:

$$Accuracy = \frac{TP + TN}{TP + FP + TN + FN} \quad (7)$$

$$Sensitivity = \frac{TP}{TP + FN} \quad (8)$$

$$Specificity = \frac{TN}{TN + FP} \quad (9)$$

where TP represents the number of the true positive test samples correctly classified as positive, FN represents the number of the true positive test samples incorrectly classified as negative, TN represents the number of the true negative test samples correctly classified as negative, and FP represents the number of the true negative test samples incorrectly classified as positive.

3. Results

In our experiment, we first evaluate the performance of the proposed complex network construction method using simulated signals. We employ numerically generated time series with well-defined properties to initiate our empirical exploration. Within our investigation we delve into the analysis of numerically simulated chaotic signals, widely acknowledged as robust approximations of numerous real-world datasets. Furthermore, we evaluate the performance of the proposed method, which constructs a complex network using EEG emotion signals. We broaden the scope of our proposed approach for network construction to analyze EEG signals, thus shedding light on its prospective applications.

3.1. Performance Evaluation of the Proposed Complex Network Method Using the Simulated Signals

The purpose of the experiment is to use Lorenz signals and random signals as examples to verify the ability of our method to convert time series into network representations. Random time series are comprised of sequences of sequentially uncorrelated random variables. In our study, the random signals utilized consist of uniformly distributed pseudo-random numbers within the interval (0, 1). Figure 4a illustrates an example of the random time series used in our study, comprising 2000 samples (data points). To further underscore the robust applicability of the proposed method, which constructs a complex network for time series analysis, we extend our investigation to constructing networks for chaotic signals. In our experimentation, simulated chaotic signals are generated using a Lorenz system with the system function defined by Equation (10). This equation yields components x, y, and z, corresponding to the convection velocity, temperature difference, and temperature gradient components, respectively. Figure 4b portrays an example of the x component of the Lorenz system employed in our experiment, comprising 2000 samples.

$$\begin{cases} \frac{dx}{dt} = -10 \times (x - y) \\ \frac{dy}{dt} = 30 \times x - y - x \times z \\ \frac{dz}{dt} = x \times y - \frac{8}{3} \times z \end{cases} \quad (10)$$

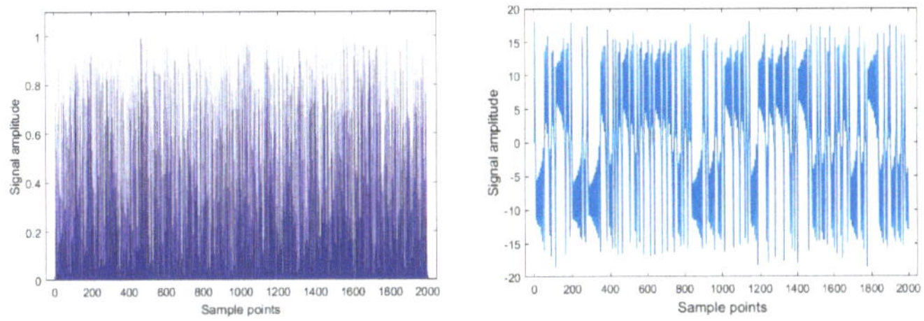

(a) (b)

Figure 4. Experimental results from random signals and Lorenz signals by the proposed network construction method. (**a**) An example of random time series; (**b**) An example of x component of Lorenz system by the proposed network construction method.

In the context of our proposed complex network methodology for constructing networks from random signals, the scalar time series undergo an initial reconstruction process into a sequence of ordinal partitions. This reconstruction is based on the phase space reconstruction method, utilizing different embedded dimensions (d = 6) with a fixed time lag (τ = 2). In accordance with the definition of the proposed method, which constructs a complex network, each ordinal partition is considered a network node, characterized by a specific set of ordinal patterns.

As shown in Figure 5a,b, the experimental results of the adjacency matrix of the unweighted network structure for the random signal and Lorenz signal x components of two thousand samples are presented, based on the proposed new method with embedded dimension d = 6 with time lag τ = 2.

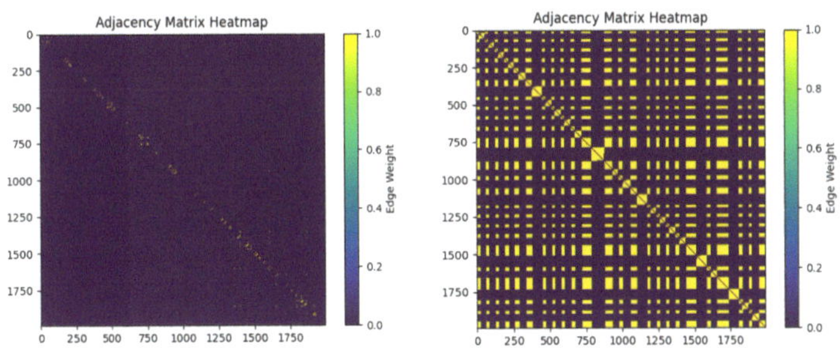

(a) (b)

Figure 5. Experimental results for the adjacency matrix of the network construction from random signals and Lorenz signals by the proposed network construction method using embedded dimension d = 6 with time lag τ = 2. (**a**) The result of the adjacency matrix for the random signal; (**b**) The result of the adjacency matrix for the Lorenz signal x components.

We established 10 sets of Lorenz signals and 10 sets of random signals, employing the proposed method to extract the NDE and AND network features from these respective signal sets. A comparative analysis of the feature results was conducted. Figure 6a shows the AND results for the 10 sets of Lorenz signals; the range of the AND values is from 330 to 390. Figure 6b shows the AND results for the 10 sets of random signals; the range of the AND values is from 5.35 to 5.55. From Figure 6a,b, it can be concluded that the AND value of the Lorenz signal is significantly higher than that of the random signal. In Figure 6c, a box plot is presented for the NDE results, illustrating a comparison between the Lorenz signals and random signals. The median NDE value of the Lorenz signal is 4.34, while the median NDE value of the random signal is 3.42. The NDE value of the Lorenz signal is larger than that of the random signal. The time series with different characteristics exhibit significant differences in their network parameters, which is the significance demonstrated by Figure 6.

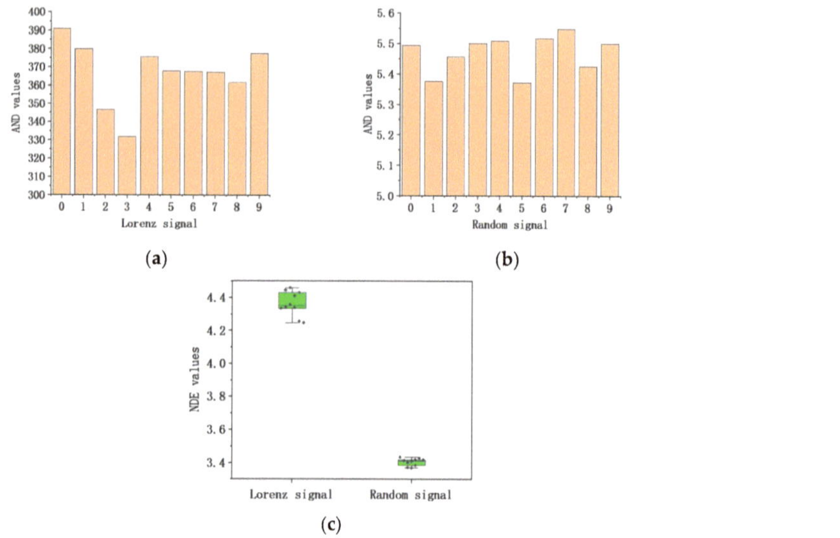

Figure 6. Experimental results of extracting NDE and AND from Lorenz signals and random signals using the proposed method. (**a**) AND results based on Lorenz signals, (**b**) AND results based on random signals, (**c**) NDE results based on Lorenz signals and random signals.

3.2. *Performance Evaluation of EEG Emotion Recognition Based on the Proposed Complex Network Construction Method*

In Figure 7, the EEG data utilized in this study span a duration of 2 min, carefully selected from the midpoint of the 62-channel EEG signals (specifically, from 60 s to 180 s). The SEED dataset encompasses EEG signals from 15 subjects, each with 62 channels. For each channel, we embarked on constructing a complex network using three distinct methods. Subsequently, we extracted the network node degree entropy, employed it as the input for the machine-learning models, and garnered the ensuing classification results. Figure 7 and Table 1 elucidate the comparative outcomes of the three methods for classifying positive and negative emotions within the SEED dataset. Figure 7a contrasts the outcomes for positive and negative emotions based on the AND features, while Figure 7b compares the results based on the NDE features. Upon scrutinizing the classification results in Figure 7, it becomes evident that the proposed method's performance in classifying positive and negative emotions outshines significantly when compared to the outcomes of the other two conventional complex network construction methods.

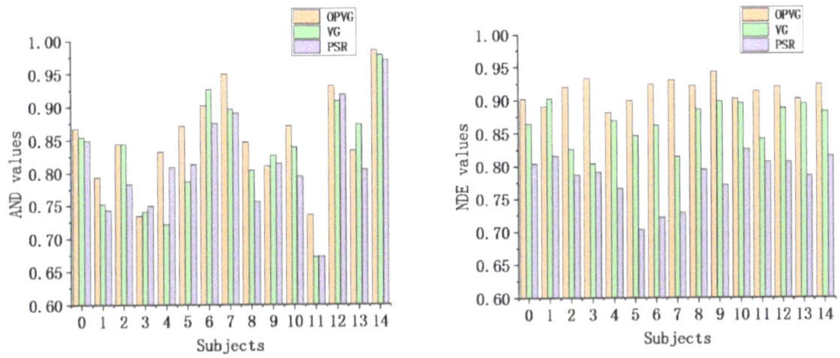

(**a**) (**b**)

Figure 7. The intra-individual comparison of classification results (both binary and triple) among the different methods is based on the SEED dataset. (**a**) Comparison of outcomes for positive and negative emotions based on the AND features; (**b**) Comparison of outcomes for positive and negative emotions based on the NDE features.

Table 1. The performance of the ComNet-PSR-VG method in constructing both AND and NDE features to identify positive and negative emotions from EEG signals in the SEED dataset is detailed. The values are presented as means ± standard deviations with positive emotions specified as positive instances and negative emotions as negative instances.

Feature	Method	Sensitivity (%)	Specificity (%)	Accuracy (%)
AND	OPVG	90.96 ± 4.06	91.24 ± 6.46	91.39 ± 4.69
	VG	84.69 ± 8.63	86.54 ± 9.04	86.51 ± 3.19
	PSR	79.3 ± 8.49	80.08 ± 8.40	79.16 ± 3.69
NDE	OPVG	84.84 ± 6.97	85.99 ± 7.29	85.39 ± 7.09
	VG	82.40 ± 7.98	83.36 ± 8.79	82.84 ± 8.35
	PSR	81.01 ± 7.23	82.40 ± 8.01	81.66 ± 7.57

In this study, we conducted a comparative analysis of the impact of various data lengths on classification outcomes. Figure 8 presents our exploration using data spans of 30 s (from 60 to 90 s), 45 s (from 60 to 105 s), 60 s (from 60 to 120 s), 75 s (from 60 to 135 s), 90 s (from 60 to 150 s), 105 s (from 60 to 175 s), and 120 s (from 60 to 180 s) extracted from the SEED dataset. We employed our proposed methodology to construct complex networks for each of these seven data lengths. Subsequently, we derived the NDE feature from the constructed complex networks and inputted them into machine-learning models for classification. The outcomes depicted in Figure 8 reveal that, concerning the NDE feature, the classification performance for the 2 min data surpasses that of other durations. Our experimental findings indicate that selecting longer-duration data yields improved classification outcomes compared to shorter durations. Notably, with a data duration of 45 s, the proportion of redundant information increases, resulting in a slight decline in classification accuracy. Nevertheless, the overarching trend illustrates that, as the duration of the data increases, classification accuracy tends to enhance, reaching its pinnacle and experiencing minimal variance with a 120 s data duration.

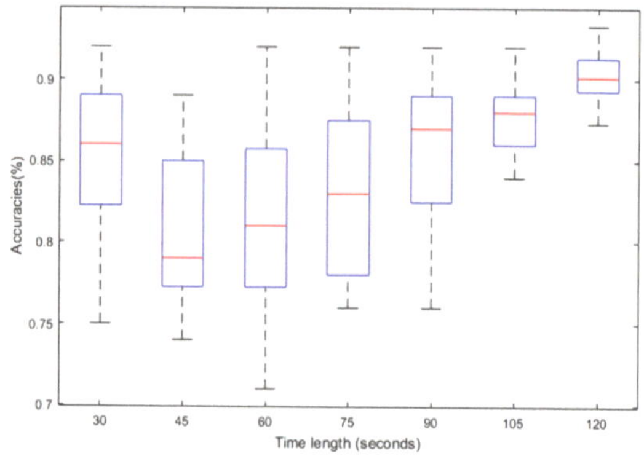

Figure 8. Comparison of classification results using the proposed method on different data durations.

4. Discussion

In our research, we propose an innovative approach to constructing complex networks for EEG analysis, specifically targeting emotion recognition. This method synergizes the features of phase space and visibility, demonstrating remarkable performance in emotion recognition based on EEG signals across two categories. Our proposed method differs from the existing approaches in several key aspects.

First, the selection of nodes for constructing complex networks diverges from the conventional methods. While existing approaches typically select ordinal numbers of time series as nodes, our method employs two parameters—dimension and delay—to map time series to phase space. Nodes in this phase space then serve as the foundation for constructing complex networks. Subsequently, these nodes are mapped into complex networks using visibility methods. The rationale behind the success of our method lies in the belief that the amalgamation of temporal and spatial features captures more physiological information than relying solely on temporal features. To elucidate the distinctions in representations of EEG signals in separate time-domain features and in combination with spatiotemporal features, we conducted a comprehensive analysis. Specifically, we performed time-domain feature analysis and spatiotemporal feature analysis on EEG emotional signals and EEG epilepsy signals separately. Subsequently, we compared the results obtained from these analyses. This meticulous approach provides insights into the efficacy of our proposed method, shedding light on its potential advantages in understanding and categorizing EEG signals related to emotions.

To demonstrate the superior performance of our proposed method in EEG emotion classification, we conducted a comprehensive comparison with recent studies that utilized the same SEED dataset. Our evaluation involved benchmarking against studies conducted by Zheng, Li, and Song.

In Zheng's research, the group sparse canonical correlation analysis (GSCCA) method was introduced to perform simultaneous electroencephalogram (EEG) channel selection and emotion recognition. Li's study utilized the graph regularized sparse linear regression (GRSLR) approach to address EEG emotion recognition problems, while Song's study employed dynamical graph convolutional neural networks (DGCNN) for EEG emotion recognition.

Upon analyzing the results, as depicted in Table 2, the individual EEG emotion classification accuracies for Zheng's study, Li's study, and Song's study were 82.96%, 87.39%, and 90.40%, respectively. Notably, our proposed ComNet-PSR-VG method achieved an outstanding individual EEG emotion classification accuracy of 91.39%, signifying a significant enhancement in classification performance. These outcomes suggest that our method outperforms the benchmark studies in EEG emotion classification. The proposed ComNet-PSR-VG method effectively preserves crucial spatial structural information within the EEG, enabling more accurate and efficient classification of emotions. The experimental results underscore the method's robustness and its ability to achieve superior performance in the realm of EEG emotion recognition.

Table 2. The results of classification accuracy from Zheng's study, Li's study, and Song's study.

Title 1	Dataset	Methodology	Mean Accuracy	StdACC
Zheng's study [20]	SEED	GSCCA	82.96%	9.95%
Li's study [21]	SEED	GRSLR	87.39%	8.64%
Song's study [22]	SEED	DBN-CRF	90.40%	8.49
Our work	SEED	NEM	91.39%	4.69%

In the realm of EEG emotion recognition, the temporal and spatial characteristics of features harbor abundant information, enabling a more comprehensive depiction of brain activity patterns and subsequently enhancing the precision of emotion recognition [23,24]. Tao's investigation [25] employs attention-based convolutional recurrent neural networks (ACRNN) to dynamically assign weights to different channels, integrating extended self-attention into the RNN. This methodology yields features that retain rich information across channels and time, demonstrating significant superiority over traditional emotion recognition methods. In Wang's study [26], a hybrid spatial–temporal feature fusion neural network (STFFNN) is introduced, amalgamating extracted features through convolutional neural networks (CNN) for spatial learning and utilizing Bi LSTM for network storage by merging temporal and spatial features. In our study, we also extract features preserving rich spatial and temporal information. However, our approach involves constructing a new spatial network for EEG signals within the framework of complex networks to enhance the extraction of EEG information.

Emotion recognition based on EEG signals holds promising applications, including auditory attention research and clinical psychiatric investigations. Despite these prospects, there are inherent limitations in the current research. This article presents a novel complex network achieved through the fusion of phase space reconstruction and visibility graph, thereby retaining the intricate temporal and spatial features of EEG signals. The absence of a standardized criterion for selecting spatial dimensions and time-delay parameters in phase space construction necessitates a discussion tailored to different signals and research contexts. Moreover, emotional stimulation introduces a certain impact on the selection of EEG patterns and features. In Chen's study [27], a discernible relationship between emotion and cognition was identified in specific regions during emotional interference, encompassing the bilateral dorsal anterior cingulate cortex, anterior insula, left inferior frontal gyrus, and superior parietal lobule, which exhibit sustained effects in these areas. Research affirms the nervous system's involvement in various interference processing types with the regulation of emotional and cognitive interference relying on interactions within extensive distributed brain networks. In Di Plinio's investigation [28], the pivotal role of the default mode network (DMN) region and executive region in emotional interference processes was demonstrated. Negative emotional interference prompts activity regulation in diverse regions, such as the frontal and parietal lobes, correlating with the regulation of functional connections between these task-activation regions and DMN regions. Both studies highlight that emotional interference triggers engagement in emotional processing activities in specific brain regions, influencing characteristic responses within the brain network. Consequently, subsequent EEG emotion classification research should factor in the impact of emotional interference and opt for suitable classification modes and features.

5. Conclusions

In this paper, we present a novel approach to construct complex networks for EEG emotion recognition by synergizing the phase space reconstruction and visibility graph methods. The main innovation in our proposed method lies in the seamless integration of the phase space reconstruction and visibility graph methods. From the perspective of the visibility graph of ordinal patterns, we proposed a new construction method of complex networks from EEG signals, ComNet-PSR-VG. With the help of the phase space reconstruction method, EEG signals are mapped to a series of ordered partitions and symbolized to obtain a sequence of ordinal patterns. Subsequently, the generated symbolic sequence of ordinal patterns is transformed into a graph network using the visibility graph method. To validate the effectiveness and versatility, we constructed the experiment on random signals, Lorenz signals, and the SEED emotion dataset by the proposed method. Two types of network node measures, AND and NDE, are extracted from the resulting graph networks. These extracted network features are then utilized as the input features for emotion classification, employing SVM as the pattern classifier to discern positive and negative emotions. The experimental results demonstrated outstanding classification performance, reinforcing the effectiveness and universality of our method. Furthermore, we compared our experimental results with existing research methods, showcasing the superior performance of our proposed entropy measure in EEG emotion recognition. The outstanding generalization observed in our proposed method suggests its significant practical potential in the field of EEG emotion recognition. Overall, our method stands out as a promising and effective approach for EEG emotion recognition, paving the way for advancements in the broader domain of EEG pattern-learning research.

Author Contributions: Conceptualization, L.Y.; methodology, L.Y. and Y.L.; software, L.Y. and Y.L.; validation, L.Y.; formal analysis, L.Y., H.L. and Y.Q.; writing—original draft preparation, L.Y. and Y.L.; supervision, M.W. All authors have read and agreed to the published version of the manuscript.

Funding: This research was funded in part by National Natural Science Foundation of China, under grants No. 62276076 and No. 62176102, as well as by the Guangdong Basic and Applied Basic Research Foundation (grants No. 2020A1515110498 and No. 2023A1515140109), the Professorial and Doctoral Scientific Research Foundation of Huizhou University under Grant 2020JB058.

Data Availability Statement: Publicly available datasets were analyzed in this study. This data can be found here: https://bcmi.sjtu.edu.cn/~seed/seed.html.

Acknowledgments: The authors gratefully acknowledge the use of the SJTU Emotion EEG Dataset (SEED) provided by Wei-Long Zheng, Shanghai Jiao Tong University, 800 Dong Chuan Road, Shanghai 200240, China.

Conflicts of Interest: The authors declare no conflicts of interest.

References

1. Yong, Z.; Donner, R.V.; Marwan, N.; Donges, J.F.; Kurths, J. Complex network approaches to nonlinear time series analysis. *Phys. Rep.* **2018**, *787*, 1–97. [CrossRef]
2. Yao, L.; Wang, M.; Lu, Y.; Li, H.; Zhang, X. EEG-Based Emotion Recognition by Exploiting Fused Network Entropy Measures of Complex Networks across Subjects. *Entropy* **2021**, *23*, 984. [CrossRef] [PubMed]
3. McCullough, M.; Small, M.; Iu, H.H.C.; Stemler, T. Multiscale ordinal network analysis of human cardiac dynamics. *Philos. Trans. R. Soc. A Math. Phys. Eng. Sci.* **2017**, *375*, 20160292. [CrossRef]
4. Lu, Y.; Yao, L.; Li, H.; Tasleem, K.; Zhang, Z.; Gao, P.; Wang, M. A new network representation for time series analysis from the perspective of combinatorial property of ordinal patterns. *Heliyon* **2023**, *9*, e22455. [CrossRef] [PubMed]
5. Lacasa, L.; Luque, B.; Ballesteros, F.; Luque, J.; Nuño, J.C. From time series to complex networks: The visibility graph. *Proc. Natl. Acad. Sci. USA* **2008**, *105*, 4972–4975. [CrossRef]
6. Pessa, A.A.B.; Ribeiro, H.V. Characterizing stochastic time series with ordinal networks. *Phys. Rev. E* **2019**, *100*, 042304. [CrossRef]
7. Zheng, W.-L.; Zhu, J.-Y.; Lu, B.-L. Identifying Stable Patterns over Time for Emotion Recognition from EEG. *IEEE Trans. Affect. Comput.* **2019**, *10*, 417–429. [CrossRef]
8. Pham, T.D. Quantification analysis of fuzzy recurrence plots. *Europhys. Lett.* **2022**, *137*, 62002. [CrossRef]
9. Gao, Z.; Jin, N. Complex network from time series based on phase space reconstruction. *Chaos* **2009**, *19*, 033137. [CrossRef]
10. Liu, X.; Fu, Z. A Novel Recognition Strategy for Epilepsy EEG Signals Based on Conditional Entropy of Ordinal Patterns. *Entropy* **2020**, *22*, 1092. [CrossRef]
11. Himmel, A.-S.; Hoffmann, C.; Kunz, P.; Froese, V.; Sorge, M. Computational complexity aspects of point visibility graphs. *Discret. Appl. Math.* **2018**, *254*, 283–290. [CrossRef]
12. McCullough, M.; Small, M.; Stemler, T.; Iu, H.H.-C. Time lagged ordinal partition networks for capturing dynamics of continuous dynamical systems. *Chaos* **2015**, *25*, 053101. [CrossRef]

13. Kulp, C.W.; Chobot, J.M.; Freitas, H.R.; Sprechini, G.D. Using ordinal partition transition networks to analyze ECG data. *Chaos* **2016**, *26*, 073114. [CrossRef] [PubMed]
14. Donner, R.V.; Small, M.; Donges, J.F.; Marwan, N.; Zou, Y.; Xiang, R.; Kurths, J. Recurrence-Based Time Series Analysis by Means of Complex Network Methods. *Int. J. Bifurc. Chaos* **2011**, *21*, 1019–1046. [CrossRef]
15. Zhang, J.; Sun, J.; Luo, X.; Zhang, K.; Nakamura, T.; Small, M. Characterizing pseudoperiodic time series through the complex network approach. *Phys. D Nonlinear Phenom.* **2008**, *237*, 2856–2865. [CrossRef]
16. Marques, J.A.L.; Cortez, P.C.; Madeiro, J.P.V.; De Albuquerque, V.H.C.; Fong, S.J.; Schlindwein, F.S. Nonlinear characterization and complexity analysis of cardiotocographic examinations using entropy measures. *J. Supercomput.* **2018**, *76*, 1305–1320. [CrossRef]
17. Yang, Y.; Yang, H. Complex network-based time series analysis. *Phys. A Stat. Mech. Its Appl.* **2008**, *387*, 1381–1386. [CrossRef]
18. Knuth, D.E. *The Art of Computer Programming*; Pearson Education: London, UK, 1981.
19. Zheng, W.-L.; Lu, B.-L. Investigating Critical Frequency Bands and Channels for EEG-Based Emotion Recognition with Deep Neural Networks. *IEEE Trans. Auton. Ment. Dev.* **2015**, *7*, 162–175. [CrossRef]
20. Zheng, W. Multichannel EEG-based emotion recognition via group sparse canonical correlation analysis. *IEEE Trans. Cogn. Dev. Syst.* **2017**, *9*, 281–290. [CrossRef]
21. Li, Y.; Zheng, W.; Cui, Z.; Zong, Y.; Ge, S. EEG emotion recognition based on graph regularized sparse linear regression. *Neural Process. Lett.* **2019**, *49*, 555–571. [CrossRef]
22. Song, T.; Zheng, W.; Song, P.; Cui, Z. EEG emotion recognition using dynamical graph convolutional neural networks. *IEEE Trans. Affect. Comput.* **2018**, *11*, 532–541. [CrossRef]
23. Gao, Z.; Li, R.; Ma, C.; Rui, L.; Sun, X. Core-Brain-Network-Based Multilayer Convolutional Neural Network for Emotion Recognition. *IEEE Trans. Instrum. Meas.* **2021**, *70*, 2510209. [CrossRef]
24. Jafari, M.; Shoeibi, A.; Khodatars, M.; Bagherzadeh, S.; Shalbaf, A.; García, D.L.; Gorriz, J.M.; Acharya, U.R. Emotion recognition in EEG signals using deep learning methods: A review. *Comput. Biol. Med.* **2023**, *165*, 107450. [CrossRef]
25. Tao, W.; Li, C.; Song, R.; Cheng, J.; Liu, Y.; Wan, F.; Chen, X. EEG-Based Emotion Recognition via Channel-Wise Attention and Self Attention. *IEEE Trans. Affect. Comput.* **2023**, *14*, 382–393. [CrossRef]
26. Wang, Z.; Wang, Y.; Zhang, J.; Hu, C.; Yin, Z.; Song, Y. Spatial-Temporal Feature Fusion Neural Network for EEG-Based Emotion Recognition. *IEEE Trans. Instrum. Meas.* **2022**, *71*, 2507212. [CrossRef]
27. Chen, T.; Becker, B.; Camilleri, J.; Wang, L.; Yu, S.; Eickhoff, S.B.; Feng, C. A domain-general brain network underlying emotional and cognitive interference processing: Evidence from coordinate-based and functional connectivity meta-analyses. *Brain Struct. Funct.* **2018**, *223*, 3813–3840. [CrossRef] [PubMed]
28. Plinio, S.D.; Ferri, F.; Marzetti, L.; Romani, G.L.; Northoff, G.; Pizzella, V. Functional connections between activated and deactivated brain regions mediate emotional interference during externally directed cognition. *Hum. Brain Mapp.* **2018**, *39*, 3597–3610. [CrossRef] [PubMed]

applied sciences

MDPI

Article

Analysis of Head Micromovements and Body Posture for Vigilance Decrement Assessment

Dario Rossi [1], Pietro Aricò [2,3], Gianluca Di Flumeri [1,2], Vincenzo Ronca [2,3], Andrea Giorgi [4], Alessia Vozzi [4], Rossella Capotorto [4], Bianca M. S. Inguscio [1,2], Giulia Cartocci [1,2], Fabio Babiloni [2,5,6,*] and Gianluca Borghini [1,2]

1 Department of Molecular Medicine, Sapienza University of Rome, Viale Regina Elena, 291, 00161 Rome, Italy; dario.rossi@uniroma1.it (D.R.); gianluca.diflumeri@uniroma1.it (G.D.F.); biancams.inguscio@uniroma1.it (B.M.S.I.); giulia.cartocci@uniroma1.it (G.C.); gianluca.borghini@uniroma1.it (G.B.)

2 BrainSigns srl, Via Tirso, 14, 00198 Rome, Italy; pietro.arico@uniroma1.it (P.A.); vincenzo.ronca@uniroma1.it (V.R.)

3 Department of Computer, Control, and Management Engineering "Antonio Ruberti", Sapienza University of Rome, Piazzale Aldo Moro 5, 00185 Rome, Italy

4 Department of Anatomical, Histological, Forensic & Orthopedic Sciences, Sapienza University of Rome, Piazzale Aldo Moro 5, 00185 Rome, Italy; andrea.giorgi@uniroma1.it (A.G.); alessia.vozzi@uniroma1.it (A.V.); rossella.capotorto@uniroma1.it (R.C.)

5 Department of Physiology and Pharmacology "Vittorio Erspamer", Sapienza University of Rome, Piazzale Aldo Moro 5, 00185 Rome, Italy

6 Department of Computer Science, Hangzhou Dianzi University, Hangzhou 310018, China

* Correspondence: fabio.babiloni@uniroma1.it; Tel.: +39-3287697914

Abstract: Vigilance refers to the capability of humans to respond accordingly to relevant and unpredictable tasks and surrounding environment changes over prolonged periods of time. Identifying vigilance decrements can, therefore, have huge and vital impacts on several operational environments in which a simple slip of mind or a deficit in attention can bear life-threatening and disastrous consequences. Several methodologies have been proposed to assess and characterize vigilance, and the results have indicated that the sole measure of performance and self-reports are not enough to obtain reliable and real-time vigilance measure. Nowadays, monitoring head and body movements to obtain information about performance in daily activities, health conditions, and mental states has become very simple and cheap due to the miniaturization of inertial measurement units and their widespread integration into common electronic devices (e.g., smart glasses, smartwatches). The present study aimed to understand the relationship between head micromovements and body posture changes to vigilance decrease while performing the psychomotor vigilance task. The results highlighted that head micromovements can be employed to track vigilance decrement during prolonged periods of time and discriminate between conditions of high or low vigilance.

Keywords: vigilance; inertial measurement units; psychomotor vigilance task; head micromovements; body posture

Citation: Rossi, D.; Aricò, P.; Di Flumeri, G.; Ronca, V.; Giorgi, A.; Vozzi, A.; Capotorto, R.; Inguscio, B.M.S.; Cartocci, G.; Babiloni, F.; et al. Analysis of Head Micromovements and Body Posture for Vigilance Decrement Assessment. *Appl. Sci.* **2024**, *14*, 1810. https://doi.org/10.3390/app14051810

Academic Editor: Alexander N. Pisarchik

Received: 29 December 2023
Revised: 16 February 2024
Accepted: 20 February 2024
Published: 22 February 2024

1. Introduction

A vast corpus of studies has highlighted that cognitive processing (e.g., visuospatial ability, memory, attention and executive functions) appears to be influenced by the contribution of the vestibular system (for a review see [1]). Typically, the role of this system is to maintain gaze stability and body position and stabilize head movements, but attention-demanding tasks have consistently shown a decrease in performance (e.g., response latency, accuracy) when the vestibular system is challenged [2–4]. In particular, the results showed that body posture did not worsen when cognitive tasks were added, indicating that the brain prioritizes balance and posture stability. This indicates that cognitive tasks are not simply reflexive but compete with attention for cognitive resources [1]. This evidence suggests that vigilance and the vestibular response (in term of head movements and body posture) could require processing functions by similar cognitive networks.

The definition of vigilance in the scientific literature is ambiguous. In fact, it varies according to the field that is being studied. For example, in psychology and cognitive neuroscience, vigilance is described as the ability of the observer to sustain attention over a prolonged period of time under monotonous stimulus [5]. The vigilance level is a concept to which clinical neurophysiologists usually refer when describing the activity of the corticothalamic networks implied in the sleep–wake dimension [6]. Vigilance has also been defined through the degradation of performance results over time while involved in cognitively simple tasks [7], indicating the involvement of time as a factor to define and/or measure it.

Common definitions of vigilance usually contain terms closely related to arousal, alertness, and sustained attention. Arousal refers to a nonspecific activation of the cerebral cortex and the neurobiological mechanism behind vigilance itself, with low levels related to sleep and high levels related to the vigilant state [8]. In a hyper-aroused condition, it is also associated with models describing insomnia [9]. Alertness has been described as the state of maintained high sensitivity to incoming stimuli [10] or the quantitative measure of the state of the mind to being sensitive to internal or external stimuli [8]. Attention is one of the basic human cognitive abilities, allowing for the discrimination of relevant parts of information and the ability to discard the others. Attention is related to a focused activation of cerebral cortex that enhances information processing [11]. By extension, the concept of sustained attention is the ability to maintain a mental state of focused attention and alertness over time [12], a definition that is very close to that of vigilance, which is typically used as a synonym [13].

Most vigilance definitions refer to the capability of humans to respond accordingly to relevant and unpredictable changes over prolonged periods of time while dealing with tasks. We have, therefore, considered this definition for the proposed study [14,15].

Vigilance can also be influenced by cognitive processing, motivation, and stress [16]. An important factor to motivation is the dopamine system related to the reward [17], indicating that performance and vigilance can decrease in non-stimulating environments. Stress also has impacts on vigilance and relative associated performance [18], and several neurophysiological markers associated to stress (e.g., EEG, skin temperature, electrodermal activity, heart rate variability, blood pressure and breathing) revealed that stress and cognitive functions have a U-shaped curve relationship. This suggests that opportune levels of stress can, in fact, improve vigilance-related performance [16].

Although, over the last 50 years, automation technology has profoundly changed human–machine interactions (HMI), high levels of automation can have negative consequences due to, for example, excessive trust in autonomous systems' abilities [19] or the loss of situational awareness [20], which causes the well-known out-of-the-loop phenomenon (OOTL) [21,22]. This leads to a decrease in operator vigilance and contributes to the failure to detect and understand the problem and make the right decision. A conspicuous number of accidents caused by vigilance decrement, in particular in the aviation field [23,24] or during vehicle driving [25,26], has been widely recognized. For example, a recent study by Greenlee et al. [27] highlighted the importance of monitoring vigilance in drivers engaged with automated vehicles. The results showed that the drivers experienced a decrease in sensitivity to hazards and an increase in false alarms in the automated control condition in respect to the manual control condition. Because the presence on the streets of automated or partially automated vehicles for consumers' use is projected to increase, the importance of identifying and tracking states of low vigilance appears to be crucial. Thus, the capability of identifying vigilance degradations can have many benefits in all contexts in which a simple slip of mind or a deficit in attention can bear life-threatening and disastrous consequences.

Several methodologies and markers have been proposed to characterize and assess vigilance changes. The psychomotor vigilance task (PVT) [28] is a reliable and widely used method to monitor users' vigilance over a prolonged period of time (at least 10 min). PVT measures vigilance degradation by recording reaction times to visual or auditory stimuli that occur at random intervals (typically ranging from 1 to 10 s). In the PVT, cognitive impairments due to vigilance degradation is reflected not only by the identification and response to the target stimuli (i.e., reaction time) but also by the number of missed targets and the false response in the case that no stimulus is presented. Moreover, subjective measures, such as the visual analogue scale (VAS) [29], have also been used to rate perceived vigilance level. Significant differences in vigilance ratings were observed between participants, and the results were not consistent for different tasks [30,31]. These results indicate that the sole measure of performance and self-reported questionnaires are not enough to have generalizable results on vigilance [32]. Most importantly, questionnaires cannot be acquired during the execution of a task, with the drawback of compromising the accuracy and reliability of the measure they intend to evaluate [33,34]. However, performance data, although available during the execution of the task, are strongly related to the task at hand, so it is difficult to compare results obtained from different settings [35].

Data collected using inertial measurement units (IMUs), such as head micromovements and body posture changes, and neurophysiological measures, such as electroencephalography signal (EEG), can allow us to overcome the drawbacks of subjective measures by objectively assessing the user's cognitive states in real time during the execution of a task. The application of EEG and its reliability has been already well explored by the authors in a variety of laboratory and realistic settings by evaluating human–robot interaction in training assessments, driving, and air traffic control [36–40]. Although commercial and cheap EEG devices are available on the market, their correct usage requires specialized personnel to be able to check the correct position of the sensors (electrodes located exactly over the corresponding brain area) and quality of the EEG signals to achieve the results [41,42]. Therefore, the possibility of monitoring vigilance degradations through the analysis of head and body movements could be a valid alternative, especially due to the simple and cheap integration of IMUs (accelerometers, gyroscopes) into electronic devices such as smartwatches, virtual reality, and biosignal recording systems (e.g., EEG). This aspect also allows for tracking movements in environments in which global positioning system (GPS) tracking is not possible or difficult [43,44]. IMUs have been extensively used in clinical applications to monitor patients' rehabilitation both in conditions in which robotic exoskeletons are involved and during free movement rehabilitation [45], stroke rehabilitation [46], and posture evaluation and rehabilitation progression in children with cerebral palsy [47,48]. Also, in operational environments, the recognition of human activity (HA) by the means of wearable sensors has gained high importance to safely assess the position in time and space of the operators and improve their performance, especially where human–robot interaction is involved [49]. For example, Ramirez et al. [50] used inertial sensors to spot the visual focus of attention of a driver, while Lee et al. [51] embedded an inertial sensor in a custom-made glove to assess driver stress based on driving behavior.

Previous studies have highlighted how it is possible to identify different human activities [52], discriminate stress conditions [51] or variations in vigilance and drowsiness [53] based on drivers' steering behavior, or identify different positions of the head with the possibility of linking them to different attentional states [54] with the use of inertial sensors. However, to the best of our knowledge, the micromovements of the head and variations in body posture with a decrease in vigilance have not yet been addressed. The present study, therefore, aims to understand the relationship between the micromovements of the head and changes in body posture with vigilance decrease by analyzing data from inertial sensors. In particular, given the potential of IMU devices, the present study aims to develop and validate a vigilance index based on the user's head micromovements and body posture. In fact, data collected through IMUs are easily available and do not require professional personnel for setting up the sensors on the user's body, as the neurophysiological measures do.

2. Materials and Methods

2.1. Sample Population

Thirteen healthy participants (27 ± 3 years old, 7 males and 6 females) were enrolled on a voluntary basis in this study. The selection of the participants has been performed accurately to ensure the same mental and physical status (homogeneity of the experimental sample). They have been asked about past neurological and physical disorders and instructed to maintain a specific kind of lifestyle. For example, they have been asked to avoid alcohol, heavy meals, and caffeine right before the experiments (homogeneity of the "internal conditions" of the subjects during the experiments). The lab environment has been kept under control (lights intensity, room temperature, seat position) across the different days of the experiments (homogeneity of the "external conditions" during the experiment). Written informed consent was obtained from each participant after the explanation of this study and before the start of the data acquisition. The experiment was conducted following the principles outlined in the Declaration of Helsinki of 1975, as revised in 2008. It received a favorable opinion from the Ethical Committee of Fondazione Santa Lucia (Prot. CE/PROG.604 dated 5 April 2017). Moreover, the participants were informed on how to complete the tasks proposed later during the experimentation, and all of them took part in a practice session before starting with the experiment to avoid compromised results due to learning and familiarization effects. Then, a resting phase was considered before the start of the actual experiment session. Due to missing data, one participant has been excluded from the analysis. Thus, the final sample population is composed of 12 participants completely balanced between males and females.

2.2. Psychomotor Vigilance Task (PVT)

All the participants performed two separated PVT with a conjunction visual search task in between (not considered in this study, as it is specific for selective attention functions). After each task, a resting period was considered according to the participants' disposition to avoid causing visual strain that could confound and impair the correct evaluation of vigilance decrease. Therefore, the entire protocol consisted of three phases, with a total duration of about 45 min and each phase lasting a maximum of 15 min. Moreover, screen distance, luminance, and contrast were adjusted according to participant's demands [55,56]. The PVT, which is a specific test to induce vigilance degradation, in this study consisted of 10 min of continuous stimuli presentation on a monitor with random *interstimulus intervals* (ISI) ranging from 1 to 10 s. The duration of the PVT was set based on the results obtained from Loh et al. [57], in which a significant vigilance degradation was observed after 10 min. The participants had to press the space bar on a keyboard in front of them as fast as possible in response to the stimuli presentation (a red circle in the center of the screen) after the appearance of a fixation cross (Figure 1). Thus, the PVT was composed of multiple subsequent repetitions of trials, which included (1) ISI; (2) a fixation cross; and (3) target stimulus. During the entire protocol, the participants were seated on a comfortable chair in front of a computer screen. Moreover, high-resolution electroencephalography (HR-EEG) signal was acquired using a 61-channel system (see Sebastiani et al. [58] for more details on data acquisition and analysis), and micro and macro movements were recorded using 2 IMU devices composed of a 3-axis accelerometer and a 3-axis gyroscope placed on the chest and forehead of the participants. Moreover, the participants' reaction time (RT) in response to the correct target stimuli was collected to measure participants' performance. The RT was obtained by timing the time between the target onset (red dot) and the participant hit on the keyboard. Participants' EEG signals, IMU data and on-screen stimuli were synchronized for the entire duration of the protocol. Before the beginning of the experiment, one minute of resting state with open eyes in front of the blank monitor (OA) was recorded to obtain head micromovements and body posture measures from a *movement-free* baseline condition.

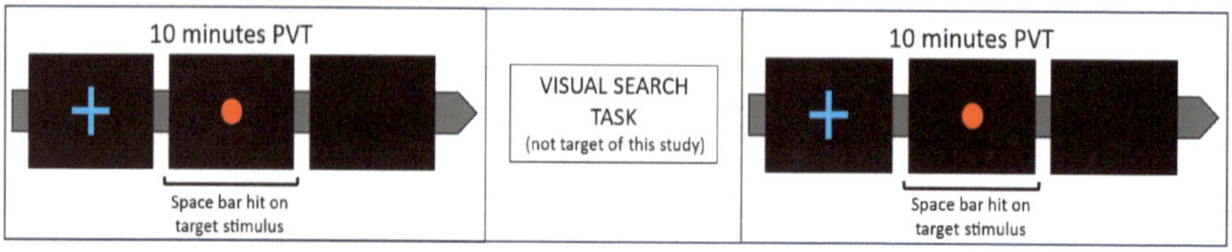

Figure 1. Graphical representation of stimuli presented to participants and the flow of the protocol proposed, consisting of two separate psychomotor vigilance tasks (PVT) and a visual search task in between. Resting periods have been considered at the end of each task to avoid visual strain. The PVT required the participants to press the space bar as fast as possible after the onset of a red dot which appears on the screen after a blue fixation cross.

2.3. Acceleration Data Recording and Processing

Head and chest acceleration were recorded by using two Shimmer GSR3+ systems (Shimmer Sensing, Dublin, Ireland) with integrated inertial sensing via accelerometer and gyroscope at a sampling rate of 100 Hz. Before the beginning of each experimental session, the accelerometer and gyroscope of the IMU devices were calibrated on a flat and stable surface according to manufacturer guidelines to obtain their relative correct offsets, sensitivity, and alignment matrices. Once the calibration was completed, we positioned one device on the forehead and one on the chest of each participant. Then, from each device, the linear acceleration was obtained through data fusion of the acceleration and angular velocity with the use of the Madgwick filter [59] implemented in the *imufusion* Python package. Finally, we used the linear acceleration from the three axes to calculate the modulus of the acceleration of the head and the chest. To describe the data distribution related to possible micromovements associated to vigilance, for each minute of the PVT, the median value and the median absolute deviation (MAD) of the modulus of the accelerations were estimated as a measure of the intensity and variability of participants movements respectively. In other words, the median value of the acceleration describes the magnitude (i.e., how much) over time of the participants' head micromovements and body postures changes, while the MAD describes the variation over time in the magnitude exhibited by the participants' head micromovements and body posture variations due to vigilance variations. Additionally, the use of statistical parameters like the median and MAD allowed us to obtain distributions that were not heavily influenced by possible outliers due to sporadic large movement of the head and the body. In particular, values derived from the sensor on the head were used as estimations of head micromovements because they are related to attentional states [54], while the sensors on the chest were used as estimations of body posture changes (for a review see [60]). Acceleration median and MAD values were then normalized with respect to the corresponding values obtained during the movement-free condition (open eyed phase) by using median and median absolute deviation (MMAD) normalization due to its robustness to outliers [61].

2.4. Behavioral Data: Reaction Time

During the entire PVT, participants' reaction times (RT) in response to each trial were recorded. RTs were defined as the time elapsed from the onset of target stimulus to the spacebar press. For each minute of the PVT, only the RTs related to correct answers and trials within the corresponding minute were averaged.

2.5. Statistical Analyses

Participants' RTs, acceleration median, and MAD of head micromovements and body posture of both PVTs were averaged. Then, trends over time of these parameters were analyzed using Page's trend test [62] to assess the significance of the trends over time to understand if there was a decrease or increase in those parameters from the beginning to the end of PVT.

The Wilcoxon signed rank test was used [63] to confirm any vigilance decrement between the first and last minute of the PVT of the parameters with statistically significant trends. Statistical analyses were performed using Python *scipy* [64] and *pingouin* [65] packages.

3. Results

3.1. High- and Low-Vigilance Conditions

First, we performed the Page's trend test to verify a possible decline of vigilance over time in terms of behavioral data (RT). The results confirmed a statistically significant increase in RT from the first to the last minute of the PVT (L = 4059, $p < 0.001$). In other words, this result indicates that the participants experienced a vigilance decrement while performing the PVT (Figure 2A). Moreover, we wanted to find out whether such a vigilance decrement between the beginning and end of the PVT was significant. In this regard, the Wilcoxon signed rank test reported a statistically significant increase (W = 63, $p = 0.03$) of the participants' RT between the first and last minute of the PVT (Figure 2B). Vigilance decreases over time (Page's trend test results: L = 4271, $p < 0.01$) and, between the first and last minute of the task, was also confirmed using the EEG–based vigilance index (Wilcoxon signed rank test results: W = 77, $p < 0.01$), which was calculated as proposed by the authors in a previous study [58]. The results obtained from behavioral data and the vigilance index confirmed the presence of vigilance degradation between the first (high-vigilance condition) and the last minute of PVT (low-vigilance condition).

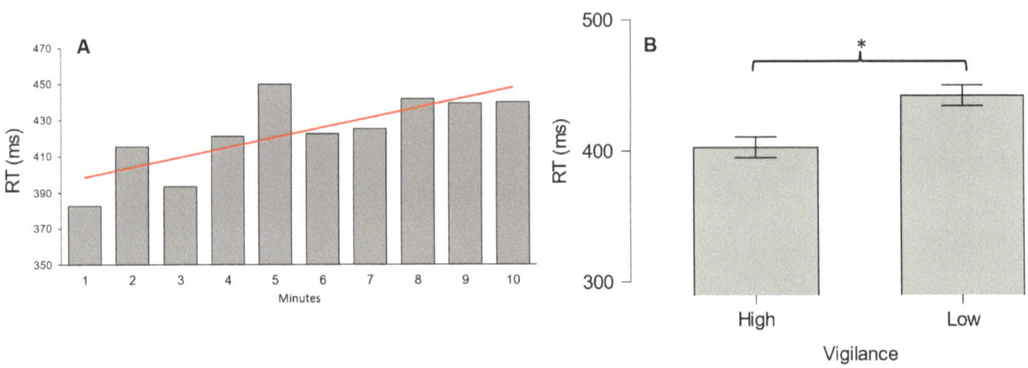

Figure 2. (**A**) Vigilance decrement over time during PVT indicated by RT increase; (**B**) RT increase in low-vigilance condition. * denotes $p < 0.05$. Red line indicates significant data trend.

3.2. Acceleration Results: Head Micromovements

The results for the head micromovements showed a statistically significant positive trend over time both for the median (intensity) and the MAD (variability) of the acceleration (L = 4041, $p < 0.01$ and L = 3866, $p = 0.01$, respectively) during the execution of the PVT (Figures 3A and 4A). Figures 3B and 4B show that there was a significant increase in median and MAD acceleration of head micromovements (W = 6, $p < 0.01$ and W = 13, $p = 0.02$, respectively) when the vigilance decreased significantly.

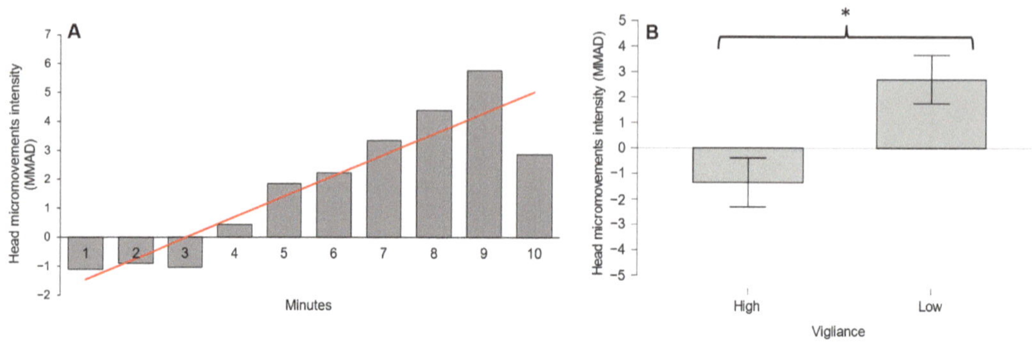

Figure 3. (**A**) Head micromovement intensity increase over time during PVT; (**B**) Head micromovement intensity increase in low-vigilance condition. * denotes $p < 0.05$. Red lines indicate significant data trends.

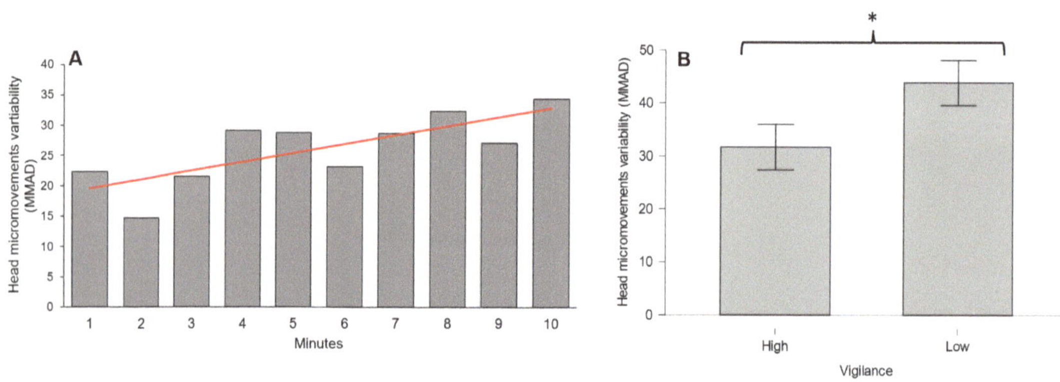

Figure 4. (**A**) Head micromovement variability increase over time during PVT; (**B**) Head micromovement variability increase in low-vigilance condition. * denotes $p < 0.05$. Red lines indicate significant data trends.

3.3. Acceleration Results: Body Posture

The results for the body posture showed a statistically significant positive trend over time (L = 3881, $p < 0.01$) for the median (intensity), but no significant trend ($p = 0.32$) was found for the MAD (variability). In addition, the statistical analysis between the beginning and the end of the PVT did not return any significant changes for the median or MAD acceleration of the body posture ($p = 0.08$ and $p = 0.28$, respectively). For this reason, body posture was not considered as a possible marker of vigilance decrease in the subsequent analysis.

3.4. Repeated Measures Correlations

Based on the results derived by head micromovements data, we performed a repeated measures correlation analysis to better understand the relationships between RT and head micromovements. We calculated a moving root mean square (rms) over time for each of the two variables analyzed. The results indicated a weak positive statistically significant correlation between RT and acceleration measures with an r = 0.27 ($p < 0.01$) for head micromovements intensity (Figure 5A), while showed a strong positive statistically significant correlation with an r = 0.72 ($p < 0.01$) for variability (Figure 5B).

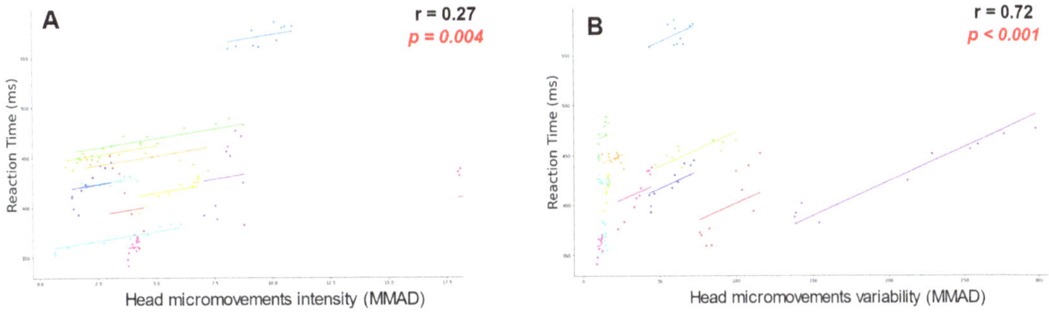

Figure 5. Repeated measures scatterplots (subject data in different colors). (**A**) Correlation between Reaction Times and head micromovements intensity; (**B**) Correlation between Reaction Times and head micromovements variability. All correlations are significant with $p < 0.05$, highlighted in red in the figure.

4. Discussion

The results obtained in this study confirmed the possibility of assessing vigilance degradations through the analysis of participants' head micromovements. A significant vigilance reduction at the end of the PVTs [57] was highlighted by participants' reaction times (RTs) and also confirmed by a previous study [58] (Figure 2). In this regard, participants' head micromovements were able to discriminate low- and high-vigilance conditions. In fact, when the vigilance decreased significantly, we found increased intensity and variability of micromovement acceleration over time (Figures 3 and 4). Meanwhile, the results derived from body postures showed that, although there was a positive trend for acceleration intensity, we did not find any significant differences between the high- and low-vigilance conditions, both for acceleration intensity and variability.

Moreover, the analysis of head micromovement variability showed a positive and significant correlation with participants' RT (Figure 5), indicating that this parameter could be used to assess the current state of operators' vigilance while dealing with tasks.

Taken together, the results hint at the possible implication of the vestibular system with the vigilance assessment. In particular, it seems that in a state of high vigilance, the cognitive resources are sufficiently balanced between PVT execution and the request of the head and the body to maintain their position, which is most likely in an apparent nonconscious way. Meanwhile, the low-vigilance condition leads to higher cognitive demand to hold the head steady, leading to an increase in conscious control of head micromovements, while body position is not affected. Lower resources are instead dedicated to PVT execution, leading to a decrease in performance, as highlighted by the increase in RT. As a matter of fact, previous studies have pointed out that, when the vestibular system is challenged, maintaining the same balance and posture starts to be cognitively demanding [1,2,66]. However, on the other hand, the incorporation of vestibular stimulation, like swinging on a swing during leisure time, as suggested in [67], over time could instead improve cognitive function. Although plausible, the link between vigilance and the vestibular system is far from being resolved by the present study. More focused studies should address the actual causal relationship between the two.

The results of the present study also demonstrate that IMUs endow a cheap, direct, and reliable measure of participants' vigilance levels. Due to the integration of IMUs with neurophysiological devices, these kinds of data could also be combined for assessing vigilance changes in terms of participants' behavior and cognitive response. This integration could help to identify and, if necessary, alert operators of conditions of vigilance deterioration. There are plenty of operation environments in which it could be possible to integrate inertial measurements on already existing equipment, such as in aviation on traffic controller operator or pilots' headphones or in environments in which security helmets are used, for example, on construction sites for heavy machinery operators or quality control. In this way, no additional device that could modify or impede a normal day-to-day work routine would be required, and the ergonomics of the existing ones would not be compromised, allowing researchers to overcome a possible barrier to the adoption of such devices. The outcome can be used to create a closed loop between the user and the machine and make them continuously interact to mitigate OOTL phenomena and improve both users' performance and task safety.

Moreover, the measure that we proposed here in this study does not require extensive computational time (14.45 ± 1.79 ms is the average time for the elaboration of 60 s of data with an Apple M1 CPU), allowing for online vigilance monitoring. This means that, with proper calibration on a movement-free phase, the index could be also implemented as a direct measure that can be used to adapt user interfaces when vigilance is lowered below a predefined threshold [68].

Despite the innovative and interesting results, some limitations must be discussed. First, the sample size consisted of 12 participants within a narrow age range, and those factors could be a limit to a broader generalization of the findings reported in this study. Therefore, in future studies, we will enlarge the population and include different age ranges to provide more reliable evidence and substantiate the findings reported in this study. Secondly, we want to estimate the time resolution by which the IMUs can assess vigilance decrement. In this regard, we need a task that is able to provide performance data with high temporal resolution (e.g., every second) so that we will be able to identify vigilance degradation with a resolution of seconds and perform correlations between participants' performance and other kinds of data (i.e., IMUs). Finally, future studies on head micromovements related to vigilance decrease will have to take into account the role of motivation and stress and their effect on the possible degradation of performance and how they affect vigilance [17,18].

5. Conclusions

Given the miniaturization of inertial sensors, the use of IMUs can be easily implemented in all operating environments in which is crucial to adequately evaluate vigilance degradation and prevent accidents related to it [24,69], such as in aviation [36], while driving a vehicle [70], or in a surgery room [71]. The capability of using head micromovements as described in this study, paves the way to extending this strategy to track other kinds of mental states (e.g., mental workload, stress) by exploiting technological progress and the integration of IMUs in commercial and personal devices.

Author Contributions: Conceptualization, D.R. and G.B.; Data curation, D.R., B.M.S.I. and G.B.; Formal analysis, D.R., P.A., G.D.F., B.M.S.I. and G.B.; Funding acquisition, F.B. and G.B.; Investigation, D.R. and G.B.; Methodology, D.R., P.A., G.D.F., V.R., A.G., A.V., R.C., G.C. and G.B.; Project administration, F.B. and G.B.; Resources, D.R. and G.B.; Supervision, F.B. and G.B.; Visualization, D.R., B.M.S.I., G.C. and G.B.; Writing—original draft, D.R.; Writing—review and editing, D.R., P.A., G.D.F., V.R., A.G., A.V., R.C., B.M.S.I., G.C. and G.B. All authors have read and agreed to the published version of the manuscript.

Funding: This work was financed by the European Commission by Horizon 2020 projects Sesar-01-2015 Automation in ATM, "STRESS" (GA No. 699381).

Institutional Review Board Statement: The experiments were conducted following the principles outlined in the Declaration of Helsinki of 1975, as revised in 2013. The experiments have been approved by the Ethical Committee of Fondazione Santa Lucia (Prot. CE/PROG.604 dated 5 April 2017).

Informed Consent Statement: Informed consent was obtained from all subjects involved in this study.

Data Availability Statement: The data presented in this study are available on request from the corresponding author. The data are not publicly available due to privacy reasons.

Acknowledgments: This work was co-funded by the individual grants "CHALLENGES: CompreHensive frAmework to aLLEge and analyze surGEons' Stress" (Bando Ateneo Medio 2021), "BRAINORCHESTRA: Multimodal teamwork assessment through hyperscanning technique" (Bando Ateneo Medio 2022) provided by Sapienza University of Rome to Gianluca Borghini, "AI-DRIVE: AI-based multimodal evaluation of car drivers' performance for onboard assistive systems" (Avvio alla ricerca 2021), provided by Sapienza University of Rome to Gianluca Di Flumeri, "The Smelling Brain: discovering the unconscious effect of the odours in industrial contexts" (Avvio alla ricerca 2022), provided by Sapienza University of Rome to Alessia Vozzi, "REMES–Remote tool for emotional states evaluation" provided to Vincenzo Ronca, and "HF AUX-Aviation: Advanced tool for Human Factors evaluation for the AUXiliary systems assessment in Aviation" provided by Sapienza University of Rome to Vincenzo Ronca.

Conflicts of Interest: All authors declare that the research was conducted in the absence of any commercial or financial relationships that could be construed as a potential conflict of interest.

References

1. Bigelow, R.T.; Agrawal, Y. Vestibular Involvement in Cognition: Visuospatial Ability, Attention, Executive Function, and Memory. *J. Vestib. Res.* **2015**, *25*, 73–89. [CrossRef]
2. Redfern, M.S.; Müller, M.L.; Jennings, J.R.; Furman, J.M. Attentional Dynamics in Postural Control during Perturbations in Young and Older Adults. *J. Gerontol. Ser. A Biol. Sci. Med. Sci.* **2002**, *57*, B298–B303. [CrossRef]
3. Furman, J.M.; Müller, M.L.; Redfern, M.S.; Jennings, J.R. Visual–Vestibular Stimulation Interferes with Information Processing in Young and Older Humans. *Exp. Brain Res.* **2003**, *152*, 383–392. [CrossRef] [PubMed]
4. Yardley, L.; Gardner, M.; Bronstein, A.; Davies, R.; Buckwell, D.; Luxon, L. Interference between Postural Control and Mental Task Performance in Patients with Vestibular Disorder and Healthy Controls. *J. Neurol. Neurosurg. Psychiatry* **2001**, *71*, 48. [CrossRef] [PubMed]
5. Davies, D.R.; Parasuraman, R. *The Psychology of Vigilance*; Academic Press: London, UK, 1982.
6. Steriade, M.; Steriade, M. Coherent Oscillations and Short-Term Plasticity in Corticothalamic Networks. *Trends Neurosci.* **1999**, *22*, 337–345. [CrossRef]
7. Sanders, A.F.; Sanders, A. *Elements of Human Performance: Reaction Processes and Attention in Human Skill*; Psychology Press: London, UK, 2013.
8. van Schie, M.K.; Lammers, G.J.; Fronczek, R.; Middelkoop, H.A.; van Dijk, J.G. Vigilance: Discussion of Related Concepts and Proposal for a Definition. *Sleep Med.* **2021**, *83*, 175–181. [CrossRef] [PubMed]
9. Riemann, D.; Spiegelhalder, K.; Feige, B.; Voderholzer, U.; Berger, M.; Perlis, M.; Nissen, C. The Hyperarousal Model of Insomnia: A Review of the Concept and Its Evidence. *Sleep Med. Rev.* **2010**, *14*, 19–31. [CrossRef]
10. Posner, M.I. Measuring Alertness. *Ann. N. Y. Acad. Sci.* **2008**, *1129*, 193–199. [CrossRef]
11. Petersen, S.E.; Posner, M.I. The Attention System of the Human Brain: 20 Years after. *Annu. Rev. Neurosci.* **2012**, *35*, 73–89. [CrossRef]
12. Helge Johnsen, B.; Christian Laberg, J.; Eid, J.; Hugdahl, K. Dichotic Listening and Sleep Deprivation: Vigilance Effects. *Scand. J. Psychol.* **2002**, *43*, 413–417. [CrossRef]
13. Parasuraman, R. *The Attentive Brain*; MIT Press: Cambridge, MA, USA, 2000.
14. Warm, J.S.; Parasuraman, R.; Matthews, G. Vigilance Requires Hard Mental Work and Is Stressful. *Hum. Factors* **2008**, *50*, 433–441. [CrossRef]
15. Al-Shargie, F.; Tariq, U.; Mir, H.; Alawar, H.; Babiloni, F.; Al-Nashash, H. Vigilance Decrement and Enhancement Techniques: A Review. *Brain Sci.* **2019**, *9*, 178. [CrossRef]
16. Oken, B.S.; Salinsky, M.C.; Elsas, S.M. Vigilance, Alertness, or Sustained Attention: Physiological Basis and Measurement. *Clin. Neurophysiol.* **2006**, *117*, 1885–1901. [CrossRef] [PubMed]
17. Schultz, W. Getting Formal with Dopamine and Reward. *Neuron* **2002**, *36*, 241–263. [CrossRef] [PubMed]
18. Hancock, P.A. A Dynamic Model of Stress and Sustained Attention. *Hum. Factors* **1989**, *31*, 519–537. [CrossRef]
19. Parasuraman, R.; Molloy, R.; Singh, I.L. Performance Consequences of Automation-Induced 'complacency'. *Int. J. Aviat. Psychol.* **1993**, *3*, 1–23. [CrossRef]
20. Endsley, M.R.; Kiris, E.O. The Out-of-the-Loop Performance Problem and Level of Control in Automation. *Hum. Factors* **1995**, *37*, 381–394. [CrossRef]
21. Kaber, D.B.; Endsley, M.R. Out-of-the-Loop Performance Problems and the Use of Intermediate Levels of Automation for Improved Control System Functioning and Safety. *Process Saf. Prog.* **1997**, *16*, 126–131. [CrossRef]
22. Jones, E.E.; Carter-Sowell, A.R.; Kelly, J.R.; Williams, K.D. I'm out of the Loop': Ostracism through Information Exclusion. *Group Process. Intergroup Relat.* **2009**, *12*, 157–174. [CrossRef]
23. Molloy, R.; Parasuraman, R. Monitoring an Automated System for a Single Failure: Vigilance and Task Complexity Effects. *Hum. Factors* **1996**, *38*, 311–322. [CrossRef]

24. Gerbert, K.; Kemmler, R. The Causes of Causes: Determinants and Background Variables of Human Factor Incidents and Accidents. *Ergonomics* **1986**, *29*, 1439–1453. [CrossRef] [PubMed]
25. Stutts, J.; Feaganes, J.; Reinfurt, D.; Rodgman, E.; Hamlett, C.; Gish, K.; Staplin, L. Driver's Exposure to Distractions in Their Natural Driving Environment. *Accid. Anal. Prev.* **2005**, *37*, 1093–1101. [CrossRef] [PubMed]
26. Sajan, S.; Ray, G.G. Human Factors in Safe Driving-A Review of Literature on Systems Perspective, Distractions and Errors. In Proceedings of the 2012 IEEE Global Humanitarian Technology Conference, Seattle, WA, USA, 21–24 October 2012; pp. 83–88.
27. Greenlee, E.T.; DeLucia, P.R.; Newton, D.C. Driver Vigilance Decrement Is More Severe during Automated Driving than Manual Driving. *Hum. Factors* **2024**, *66*, 574–588. [CrossRef]
28. Dinges, D.F.; Powell, J.W. Microcomputer Analyses of Performance on a Portable, Simple Visual RT Task during Sustained Operations. *Behav. Res. Methods Instrum. Comput.* **1985**, *17*, 652–655. [CrossRef]
29. Wewers, M.E.; Lowe, N.K. A Critical Review of Visual Analogue Scales in the Measurement of Clinical Phenomena. *Res. Nurs. Health* **1990**, *13*, 227–236. [CrossRef] [PubMed]
30. Davies, D.R.; Tune, G.S. *Human Vigilance Performance*; American Elsevier Pub. Co.: New York, NY, USA, 1969.
31. Baker, C. *Consistency of Performance in Two Human Vigilance Task*; McGraw-Hill: New York, NY, USA, 1963.
32. Parasuraman, R. Consistency of Individual Differences in Human Vigilance Performance: An Abilities Classification Analysis. *J. Appl. Psychol.* **1976**, *61*, 486. [CrossRef] [PubMed]
33. Arico, P.; Borghini, G.; Di Flumeri, G.; Sciaraffa, N.; Colosimo, A.; Babiloni, F. Passive BCI in Operational Environments: Insights, Recent Advances, and Future Trends. *IEEE Trans. Biomed. Eng.* **2017**, *64*, 1431–1436. [CrossRef] [PubMed]
34. Moustafa, K.; Luz, S.; Longo, L. Assessment of Mental Workload: A Comparison of Machine Learning Methods and Subjective Assessment Techniques. In Proceedings of the Human Mental Workload: Models and Applications: First International Symposium, H-WORKLOAD 2017, Dublin, Ireland, 28–30 June 2017; Springer: Berlin/Heidelberg, Germany, 2017; pp. 30–50.
35. Pidun, T.; Felden, C. Limitations of Performance Measurement Systems Based on Key Performance Indicators; AMCIS 2011 Proceedings-Al. 2011. Available online: https://www.google.com/url?sa=t&rct=j&q=&esrc=s&source=web&cd=&ved=2ahUKEwjChpbAqMCEAxUuma8BHVq0BxgQFnoECBEQAQ&url=https://aisel.aisnet.org/cgi/viewcontent.cgi?article=1013&context=amcis2011_submissions&usg=AOvVaw2OznW_R95PnRPhKPJKTRO2&opi=89978449 (accessed on 19 February 2024).
36. Aricò, P.; Borghini, G.; Di Flumeri, G.; Bonelli, S.; Golfetti, A.; Graziani, I.; Pozzi, S.; Imbert, J.-P.; Granger, G.; Benhacene, R.; et al. Human Factors and Neurophysiological Metrics in Air Traffic Control: A Critical Review. *IEEE Rev. Biomed. Eng.* **2017**, *10*, 250–263. [CrossRef]
37. Brookings, J.B.; Wilson, G.F.; Swain, C.R. Psychophysiological Responses to Changes in Workload during Simulated Air Traffic Control. *Biol. Psychol.* **1996**, *42*, 361–377. [CrossRef]
38. Borghini, G.; Aricò, P.; Graziani, I.; Salinari, S.; Sun, Y.; Taya, F.; Bezerianos, A.; Thakor, N.V.; Babiloni, F. Quantitative Assessment of the Training Improvement in a Motor-Cognitive Task by Using EEG, ECG and EOG Signals. *Brain Topogr.* **2016**, *29*, 149–161. [CrossRef]
39. Di Flumeri, G.; Borghini, G.; Aricò, P.; Sciaraffa, N.; Lanzi, P.; Pozzi, S.; Vignali, V.; Lantieri, C.; Bichicchi, A.; Simone, A.; et al. EEG-Based Mental Workload Neurometric to Evaluate the Impact of Different Traffic and Road Conditions in Real Driving Settings. *Front. Hum. Neurosci.* **2018**, *12*, 509. [CrossRef]
40. Borghini, G.; Aricò, P.; Di Flumeri, G.; Sciaraffa, N.; Colosimo, A.; Herrero, M.-T.; Bezerianos, A.; Thakor, N.V.; Babiloni, F. A New Perspective for the Training Assessment: Machine Learning-Based Neurometric for Augmented User's Evaluation. *Front. Neurosci.* **2017**, *11*, 325. [CrossRef] [PubMed]
41. Sciaraffa, N.; Di Flumeri, G.; Germano, D.; Giorgi, A.; Di Florio, A.; Borghini, G.; Vozzi, A.; Ronca, V.; Babiloni, F.; Aricò, P. Evaluation of a New Lightweight EEG Technology for Translational Applications of Passive Brain-Computer Interfaces. *Front. Hum. Neurosci.* **2022**, *16*, 901387. [CrossRef]
42. Giorgi, A.; Ronca, V.; Vozzi, A.; Aricò, P.; Borghini, G.; Capotorto, R.; Tamborra, L.; Simonetti, I.; Sportiello, S.; Petrelli, M.; et al. Neurophysiological Mental Fatigue Assessment for Developing User-Centered Artificial Intelligence as a Solution for Autonomous Driving. *Front. Neurorobot.* **2023**, *17*, 1240933. [CrossRef] [PubMed]
43. Tanenhaus, M.; Carhoun, D.; Geis, T.; Wan, E.; Holland, A. Miniature IMU/INS with Optimally Fused Low Drift MEMS Gyro and Accelerometers for Applications in GPS-Denied Environments. In Proceedings of the 2012 IEEE/ION Position, Location and Navigation Symposium, Myrtle Beach, SC, USA, 23–26 April 2012; pp. 259–264.
44. Ahmad, N.; Ghazilla, R.A.R.; Khairi, N.M.; Kasi, V. Reviews on Various Inertial Measurement Unit (IMU) Sensor Applications. *Int. J. Signal Process. Syst.* **2013**, *1*, 256–262. [CrossRef]
45. Giggins, O.M.; Sweeney, K.T.; Caulfield, B. Rehabilitation Exercise Assessment Using Inertial Sensors: A Cross-Sectional Analytical Study. *J. Neuroeng. Rehabil.* **2014**, *11*, 158. [CrossRef] [PubMed]
46. Eriksson, J.; Mataric, M.J.; Winstein, C.J. Hands-off Assistive Robotics for Post-Stroke Arm Rehabilitation. In Proceedings of the 9th International Conference on Rehabilitation Robotics, ICORR 2005, Chicago, IL, USA, 28 June–1 July 2005; pp. 21–24.
47. Velasco, M.A.; Raya, R.; Muzzioli, L.; Morelli, D.; Otero, A.; Iosa, M.; Cincotti, F.; Rocon, E. Evaluation of Cervical Posture Improvement of Children with Cerebral Palsy after Physical Therapy Based on Head Movements and Serious Games. *Biomed. Eng. Online* **2017**, *16*, 74. [CrossRef]

48. Rossi, D.; Billeci, L.; Bonfiglio, L.; Aliboni, S.; Posteraro, F.; Bortone, I. Combining Biosignals to Assess and Monitor VR-Assisted Rehabilitation of Children with Cerebral Palsy: A Machine Learning Approach. In Proceedings of the 2023 IEEE EMBS Special Topic Conference on Data Science and Engineering in Healthcare, Medicine and Biology, Malta, 7–9 December 2023; pp. 139–140.
49. Ramasubramanian, A.K.; Aiman, S.M.; Papakostas, N. On Using Human Activity Recognition Sensors to Improve the Performance of Collaborative Mobile Manipulators: Review and Outlook. *Procedia CIRP* **2021**, *97*, 211–216. [CrossRef]
50. Ramirez, J.M.; Rodriguez, M.D.; Andrade, A.G.; Castro, L.A.; Beltran, J.; Armenta, J.S. Inferring Drivers' Visual Focus Attention through Head-Mounted Inertial Sensors. *IEEE Access* **2019**, *7*, 185422–185432. [CrossRef]
51. Lee, D.S.; Chong, T.W.; Lee, B.G. Stress Events Detection of Driver by Wearable Glove System. *IEEE Sens. J.* **2016**, *17*, 194–204. [CrossRef]
52. Seenath, S.; Dharmaraj, M. Conformer-Based Human Activity Recognition Using Inertial Measurement Units. *Sensors* **2023**, *23*, 7357. [CrossRef]
53. Lee, B.-G.; Lee, B.-L.; Chung, W.-Y. Wristband-Type Driver Vigilance Monitoring System Using Smartwatch. *IEEE Sens. J.* **2015**, *15*, 5624–5633. [CrossRef]
54. Peng, Y.; He, C.; Xu, H. Attachable Inertial Device with Machine Learning toward Head Posture Monitoring in Attention Assessment. *Micromachines* **2022**, *13*, 2212. [CrossRef]
55. Chi, C.-F.; Lin, F.-T. A Comparison of Seven Visual Fatigue Assessment Techniques in Three Data-Acquisition VDT Tasks. *Hum. Factors* **1998**, *40*, 577–590. [CrossRef]
56. Lin, Y.-H.; Chen, C.-Y.; Lu, S.-Y.; Lin, Y.-C. Visual Fatigue during VDT Work: Effects of Time-Based and Environment-Based Conditions. *Displays* **2008**, *29*, 487–492. [CrossRef]
57. Loh, S.; Lamond, N.; Dorrian, J.; Roach, G.; Dawson, D. The Validity of Psychomotor Vigilance Tasks of Less than 10-Minute Duration. *Behav. Res. Methods Instrum. Comput.* **2004**, *36*, 339–346. [CrossRef] [PubMed]
58. Sebastiani, M.; Di Flumeri, G.; Aricò, P.; Sciaraffa, N.; Babiloni, F.; Borghini, G. Neurophysiological Vigilance Characterisation and Assessment: Laboratory and Realistic Validations Involving Professional Air Traffic Controllers. *Brain Sci.* **2020**, *10*, 48. [CrossRef] [PubMed]
59. Madgwick, S.O.; Harrison, A.J.; Vaidyanathan, R. Estimation of IMU and MARG Orientation Using a Gradient Descent Algorithm. In Proceedings of the 2011 IEEE International Conference on Rehabilitation Robotics, Zurich, Switzerland, 29 June–1 July 2011; pp. 1–7.
60. Rahmani, M.H.; Berkvens, R.; Weyn, M. Chest-Worn Inertial Sensors: A Survey of Applications and Methods. *Sensors* **2021**, *21*, 2875. [CrossRef] [PubMed]
61. Kappal, S. Data Normalization Using Median Median Absolute Deviation MMAD Based Z-Score for Robust Predictions vs. Min–Max Normalization. *Lond. J. Res. Sci. Nat. Form.* **2019**, *19*, 10–13140.
62. Page, E.B. Ordered Hypotheses for Multiple Treatments: A Significance Test for Linear Ranks. *J. Am. Stat. Assoc.* **1963**, *58*, 216–230. [CrossRef]
63. Conover, W.J. *Practical Nonparametric Statistics*; John Wiley & Sons: Hoboken, NJ, USA, 1999; Volume 350.
64. Virtanen, P.; Gommers, R.; Oliphant, T.E.; Haberland, M.; Reddy, T.; Cournapeau, D.; Burovski, E.; Peterson, P.; Weckesser, W.; Bright, J.; et al. SciPy 1.0: Fundamental Algorithms for Scientific Computing in Python. *Nat. Methods* **2020**, *17*, 261–272. [CrossRef] [PubMed]
65. Vallat, R. Pingouin: Statistics in Python. *J. Open Source Softw.* **2018**, *3*, 1026. [CrossRef]
66. Belluscio, V.; Cartocci, G.; Terbojevich, T. Facilitating or Disturbing? An Investigation about the Effects of Auditory Frequencies on Prefrontal Cortex Activation and Postural Sway. *Front. Neurosci.* **2023**, *17*, 1197733. [CrossRef] [PubMed]
67. Rajagopalan, A.; Kumar, S.S.; Mukkadan, J.K. Effect of Vestibular Stimulation on Auditory and Visual Reaction Time in Relation to Stress. *J. Adv. Pharm. Technol. Res.* **2017**, *8*, 34.
68. Dennerlein, J.T.; Yang, M.C. Haptic Force-Feedback Devices for the Office Computer: Performance and Musculoskeletal Loading Issues. *Hum. Factors* **2001**, *43*, 278–286. [CrossRef] [PubMed]
69. Reinerman-Jones, L.; Matthews, G.; Mercado, J.E. Detection Tasks in Nuclear Power Plant Operation: Vigilance Decrement and Physiological Workload Monitoring. *Saf. Sci.* **2016**, *88*, 97–107. [CrossRef]
70. Körber, M.; Cingel, A.; Zimmermann, M.; Bengler, K. Vigilance Decrement and Passive Fatigue Caused by Monotony in Automated Driving. *Procedia Manuf.* **2015**, *3*, 2403–2409. [CrossRef]
71. Glavin, R.; Maran, N. Integrating Human Factors into the Medical Curriculum. *Med. Educ.* **2003**, *37*, 59–64. [CrossRef]

applied sciences

MDPI

Article

Somatosensory Mismatch Response in Patients with Cerebral Palsy

Susmita Roy [1,†], Stefan K. Ehrlich [2,†] and Renée Lampe [1,3,*]

1 Research Unit of the Buhl-Strohmaier Foundation for Cerebral Palsy and Pediatric Neuroorthopaedics, Department of Orthopaedics and Sports Orthopaedics, TUM School of Medicine and Health, University Hospital Rechts der Isar, Technical University of Munich, Ismaninger Strasse 22, 81675 Munich, Germany; susmita.roy@tum.de

2 SETLabs Research GmbH, Healthcare Technologies, Elsenheimerstrasse 55, 80687 Munich, Germany; stefan.kh.ehrlich@gmail.com

3 Markus Würth Professorship, Technical University of Munich, Ismaninger Strasse 22, 81675 Munich, Germany

* Correspondence: renee.lampe@tum.de

† These authors contributed equally to this work.

Abstract: Background: Mismatch negativity (MMN), an event-related potential (ERP) component occurring at specific recording sites and latency, is associated with an automatic change detection response, generally elicited using oddball paradigms wherein infrequent stimuli are embedded in repeated, frequent stimuli. To verify the presence of mismatch-related ERP responses to somatosensory stimulation in individuals with cerebral palsy (CP), we conducted a preliminary study involving healthy participants and patients with CP. Methods: Both groups underwent 'frequent' and 'infrequent' stimulation applied to the ring finger and thumb of their left hand, respectively. ERPs were recorded at frontal, central, and parietal scalp locations using electroencephalography. A healthy cohort tested the experimental protocol and showed evidence that mismatch-related ERP responses were observable. Subsequent analysis focused on the patient group. Results: Statistically significant differences between the two types of stimuli were observed on the frontocentral and parietal channels between 150 and 250 ms after the stimulus onset in the patient group. Furthermore, a late discriminative response was observed in the frontal and parietal channels. Conclusion: The results demonstrate the presence of mismatch-related ERP responses in individuals with CP.

Keywords: mismatch negativity; EEG; somatosensory stimuli; cerebral palsy; cognitive enhancement

Citation: Roy, S.; Ehrlich, S.K.; Lampe, R. Somatosensory Mismatch Response in Patients with Cerebral Palsy. *Appl. Sci.* **2024**, *14*, 1030. https://doi.org/10.3390/app14031030

Academic Editor: Alexander N. Pisarchik

Received: 6 November 2023
Revised: 19 January 2024
Accepted: 21 January 2024
Published: 25 January 2024

1. Introduction

Event-related potential (ERP) is a measured brain response to a specific sensory, cognitive, or motor event. ERPs are measured by means of electroencephalography (EEG), which is a noninvasive electrophysiological method of monitoring the brain's electrical activities. Mismatch negativity (MMN) is a component of an ERP observable at specific recording sites (e.g., frontal) and latency (∼120–200 ms) relative to the moment of stimulus presentation, as detected in the EEG signal. MMN occurs when a sequence of repetitive standard stimuli is interrupted by an odd or deviant stimulus, i.e., when the brain detects a change in a background of homogeneous events [1]. For example, if one hears a series of identical tones (i.e., standard stimuli), and then a slightly different tone (i.e., odd stimulus) is introduced, MMN reflects the brain's automatic response to this deviation. The odd stimulus must differ from the standard stimulus in at least one stimulus attribute, in frequency, duration, or intensity. MMN is associated with automatic, subconscious memory processes [1], through its role in detecting and registering deviations from expected stimuli. A sequence of frequent stimuli causes a track or a regular pattern of stimuli in the sensory memory, and any new incoming stimulus is then compared with that created memory track. When a stimulus breaks that regular pattern, MMN is elicited, indicating that the brain has detected a mismatch. This process is thought to be related to preattentive, automatic memory mechanisms. MMN reflects the brain's ability to compare incoming sensory information with stored memory representations, contributing to the early stages of memory formation and updating [2].

MMN has been originally discovered and studied intensively by auditory stimuli [3], but there is evidence of MMN in the visual [4,5] and somatosensory [6–8] modalities also. In the visual domain, it involves changes in visual stimuli, such as color, shape, orientation, and emotional expression [9,10]. In the somatosensory domain, it has been assessed most commonly using vibrotactile stimuli and is known as somatosensory MMN (sMMN). For example, different durations, different frequencies of stimulation, or stimulation of different body parts is used for standard and odd stimuli [11,12].

The better a subject can distinguish the deviant from the standard stimuli, the larger the MMN [13]. Especially in the auditory domain, MMN has been widely employed to study speech and language development. This research includes both typical cases, such as infants exposed to one or two languages with normal development [14], and atypical cases, like children having problems with a specific language impairment [15]. Visual MMN in the context of cognitive impairment and aging can provide information about age-related changes in sensory processing and cognitive function [16,17]. Furthermore, visual MMN seems to be sensitive enough to disclose gender differences [18,19]. In the tactile domain, few studies reported the existence of somatosensory MMN in healthy individuals [6,11,20,21]. However, to our knowledge, contrary to MMN studies in the auditory and visual modalities, much less is known about the effects of mismatch in the sMMN modalities, and related developmental studies are very sparse. The reported somatosensory ERPs showed different results depending on the stimulus properties, such as duration, spatial location, and vibrotactile frequencies. For example, Kekoni et al. [11] used a vibratory mismatch paradigm and observed sMMN as a negative deflection at 100–200 ms, while Shinozaki et al. [22] found sMMN as a positive deflection at 100–200 ms, using a topographical mismatch paradigm. Tamura et al. adopted a two-point discrimination paradigm and obtained a negative potential at ~140 ms (N140) and two positive components at around 300 and 500 ms [23]. On the other hand, using a temporal discrimination task, Akatsuka et al. found a negative component peaking at approximately 60 ms (N60) and a large positive peak at around 100–200 ms (P150) [7]. In most cases, sMMN appeared over the frontocentral regions [6,11].

In addition to MMN, a later negative mismatch-related ERP component is often observed in both auditory [24–26] and somatosensory [21] oddball paradigms and is referred to as late discriminative negativity (LDN). Generally, LDN is observed at around 400 ms, following MMN [27]. Although less is known about LDN, its amplitude is typically higher in infants and children, but it has also been observed in adults [24,28].

Irrespective of the modalities of mismatch-related ERP components, they are widely used for monitoring treatment adequacy in cognitive-impairment-related diseases like schizophrenia, Alzheimer's, and vascular dementia [23,29,30]. While the potential uses of mismatch responses are noteworthy, studies involving patients with cerebral palsy (CP) are generally lacking to date, and to our knowledge. CP leads to a movement disorder caused by nonprogressive damage in the developing brain during early childhood. The movement disorder associated with CP is classified into three types: spasticity, dyskinesia, and ataxic CP. Depending on the extent of the brain damage, patients with CP may also exhibit additional symptoms such as cognitive, communicative, and/or behavioral deficits. CP is mainly characterized by motor abnormalities. Any correctly initiated movement requires an intact sensory motor system. Therefore, it is evident and supported by research also that somatosensory dysfunction plays a crucial role in movement control in the case of CP. Furthermore, children with CP often have difficulties processing somatosensory information, which can also lead to difficulties in learning and movement execution. For example, in the case of hemiplegia, the hand on the less affected side may process touch differently than the one on the more affected side. This difference in sensory information processing is frequently manifested in more pronounced disparities in movement execution and strength between the two sides of the body [31–33].

Mismatch-related ERP responses have the potential to serve as a valuable tool for probing the neural mechanisms underlying somatosensory processing in patients with CP. By studying those, researchers and clinicians can gain insights into the nature of sensory abnormalities. Knowledge of the sensory process can help to develop different therapies tailored to the problem. For instance, Restuccia et al. [34] demonstrated that the cerebellum plays a role in the automatic detection of changes in somatosensory input. Their study not only validated the reliability of somatosensory mismatch negativity (sMMN) recordings but also suggested that individuals with cerebellar damage might experience difficulties in processing incoming somatosensory information in the cortex.

Conducting ERP experiments would be particularly suitable for patients with CP, including young adults, considering potential challenges in concentration and attention span during tasks [35,36]. Mismatch-related ERP responses, in particular, provide an opportunity to look at cognitive processing even when a patient faces concentration deficit, as the patient does not need to focus on the task. The appearance of mismatch-related responses in the EEG allows one to detect changes related to sensory or cognitive processes in the brain. In this study, we chose an adult population of patients with CP who were able to read and had sufficient concentration. Our goal was to assess the tolerability of the EEG cap preparation. Additionally, we aimed to know whether it was feasible to derive any mismatch-related ERP responses at all in the case of CP with our experimental protocol (i.e., can the participants perform the tasks?). Furthermore, the study is based on the assumption that sMMN is elicited between 150 and 250 ms, and LDN is elicited at around 400 ms after the stimulus onset, as either a negative or positive component [6,11,12,24,27,28]. These assumptions are based on the literature, where most of the studies involve healthy children and adults. The further goal of the present study was to verify the assumption in the case of CP. The study first recorded the EEG responses in four healthy adults and then in seven patients diagnosed with CP. All participants experienced mechanical vibrations on their middle finger ('standard' stimulus) interrupted by frequent vibrations on the thumb ('deviant' stimulus) while reading a text. With the healthy cohort, the experimental protocol and the EEG cap were initially tested; further, mismatch-related ERP responses were confirmed. Subsequent analysis focused on the patient group.

2. Methods

2.1. Participants

EEG was collected from 4 healthy volunteers and 7 patients with CP. Healthy volunteers were aged between 29 and 55 years and recruited among personnel working at the hospital. Patients were aged between 23 and 53 years and recruited from a special center for people diagnosed with CP. The diagnosis of CP was confirmed by a senior orthopedic specialist before the start of the study. Additionally, the inclusion criteria of the selected patients included the ability to read and maintain adequate concentration during reading. Table 1 shows the patients' information in addition to their diagnosis. The degree of the patient's mobility was expressed according to the Gross Motor Function Classification System (GMFCS) [37]. This system defines five different levels of mobility from a GMFCS of I, when the person can walk freely without the need of a walking aid, to a GMFCS level of V, when a person has substantial motor limitations and requires a wheelchair permanently, not being able to move by himself or herself. The GMFCS levels of the participants of this study varied from a GMFCS of I to IV (Table 1). The Manual Ability Classification System (MACS) describes how patients use their hands to handle objects in daily activities. MACS ranges from 0 to 5, with a higher MACS level indicating a higher level of spasticity [38].

The ability to read is independent of the severity of the disability. This is also the reason why the selected group was very inhomogeneous concerning GMFCS. In principle, the participants did not need to focus on the vibratory stimulus while reading because mismatch-related ERP signals occur when participants are not focused on the task. One should not be able to detect mismatch-related responses if the participants are focused on the task, or the observed EEG signals occurring in an atypical time course, unrelated to the mismatch-related signal.

Table 1. Demographic characteristics of enrolled participants and diagnosis classifications of their mobility according to GMFCS and according to their hand's performance in daily life, i.e., MACS level.

Participant	Gender	Age Year	GMFCS	MACS L	MACS R	Diagnosis
P01	Female	23	I	4	1	Unilateral CP
P02	Male	45	III	1	1	Bilateral CP
P03	Male	39	III	3	3	Bilateral CP
P04	Male	36	IV	4	4	Bilateral CP
P05	Female	53	IV	2	2	Ataxic CP
P06	Female	47	II	2	4	Unilateral spastic CP
P07	Male	25	III	2	2	Bilateral spastic CP

2.2. *Experimental Procedure and Data Analysis*

2.2.1. Stimuli and Procedure

Mechanical vibrations were delivered via vibration motors placed on the fingers. 'Frequent' (or 'standard') and 'infrequent' (or 'deviant') stimulations were delivered to the ring finger and thumb of the left hand, respectively. Frequent and infrequent stimulation occurred at a ratio of 90% and 10%, respectively, with pseudo-randomized occurrence. Figure 1a shows a schematic diagram of the protocol for stimulus delivery. Three successive runs of 500 stimuli were delivered with 1 s of interstimulus interval. During the experiment, the participants sat comfortably in a chair in a quiet room and were asked to read a text displayed on a screen while stimuli were delivered to their fingers. The vibration motors were attached to the nail side of the finger (Figure 1a) because of the more direct transmission of vibration. A microcontroller provided an interface between the vibration motors and a computer. A Viewablewritten software controlled the delivery of vibrations, i.e., the sequence of stimulation. All subjects received the same vibration amplitude and frequency (1 G at 200 Hz).

All participants could easily perceive the intensity of the vibration and did not report any pain or discomfort resulting from the stimulation. The participants were advised not to pay attention to their hands during the session but to relax with the text reading on the screen.

2.2.2. EEG Acquisition

Electroencephalogram was recorded with an Enobio wireless EEG system [39] at 8 scalp locations (Figure 1b). Electrodes drained with saline solution were placed on the electrode cap (Enobio 8 EEG cap) at the F3, Fz, F4, C3, Cz, C4, P3, and P4 positions according to the international 10–20 system, referenced to an electrode placed on the left mastoid. The ground electrode was placed in the middle of the forehead. The signal was stored on a hard disk at a sampling rate of 500 Hz. At first, the experimental protocol was tested on healthy participants, and then the same procedure was followed for the patients.

2.2.3. Data Processing

Data preprocessing was performed using the EEGlab v2023.0 toolbox running on Matlab R2020a. Data were first band-pass-filtered between 1 and 30 Hz and then re-referenced to the common average. After re-referencing, the three sets of 500 trials were concatenated together. The data were examined for possible bad channels using Kurtosis statistics with a threshold value of 2. Since no channels were found bad, data from all the channels were used for the analysis. The continuous merged data were then decomposed by independent component analysis (ICA) using the 'runica' function of the EEGlab toolbox. The decomposed data were manually inspected individually, and the non-neuronal originated artifacts such as components related to muscle activity and eye blinks were identified on the basis of their scalp topography and component activity power spectrum, and removed from the data set. Artifact-corrected data were then used to study event-related EEG responses. The data epochs time-locked to standard and deviant events were extracted from −200 to 800 ms relative to stimulus onset from the resulting continuous data signals. A baseline correction of 200 ms was applied. Event segments with amplitudes larger than $\pm 120\ \mu Volt$ were removed for further analysis. On average, 1315 epochs per subject were accepted for further analyses.

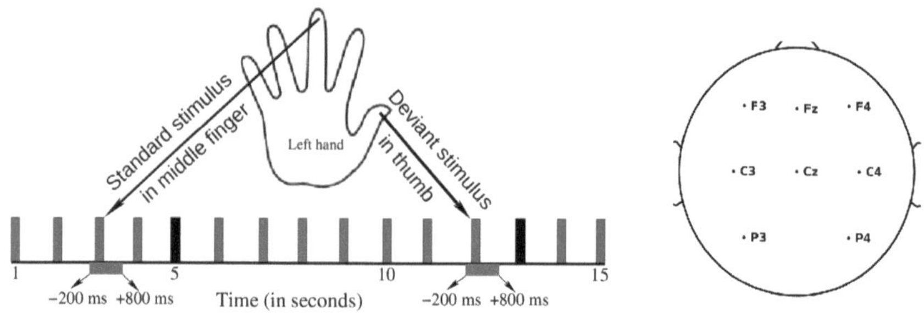

(a) (b)

Figure 1. (**a**) Schematic illustration of the stimulus sequence of the standard and deviant stimulus used for the ERP experiment. A sequence of vibrations was delivered with 1 s intervals mostly to the ring finger denoted as standard stimulus (gray filled block); 10% of them were delivered to the thumb finger, denoted as deviant stimulus (black filled block). The difference in waveforms between the ERP responses to deviant stimuli and standard stimuli is mismatch-related ERP response. The processed signals were separated from −200 to +800 ms by the deviant and standard stimulus. (**b**) An EEG montage with 8 electrodes (frontal, central, and parietal) was used in the experiment.

In line with our primary objectives, the key analysis strategy involved determining our ability to extract somatosensory mismatch-related ERP responses. Additionally, we aimed to verify the following assumption in the case of adult patients with CP: the presence of sMMN, elicited at about 150–250 ms after the stimulus onset over the frontocentral regions as either a negative or positive component [6,11,12], followed by LDN, a second component in the difference signal, at an approximate latency of 350 ms [25,40]. This assumption finds support in studies mostly involving healthy young and elderly adults. Therefore a further statistical test was performed exclusively on the data obtained from patients. Wilcoxon signed-rank tests were performed on standard and deviant responses for each channel within two predefined time ranges, averaged across patients. The selected time ranges were between 150–250 ms for sMMN [12,41] and 350–450 ms for LDN [24] based on literature where mismatch-related activities are expected. The nonparametric test statistics determined the sum of the ranks of positive differences between the observations in the samples, in this case, the differences between the two traces obtained for frequent and infrequent stimuli.

3. Results

Figure 2 presents the stimulus onset-locked segments, separated from −200 to +800 ms taking the median, across the healthy participants. Each block in the top and middle panels represents the results from individual channels, showcasing ERPs for the standard ('STD') and deviant ('DEV') events along with their differences ('DIFF'). The lower panel displays the scalp topographies of differences (deviant minus standard ERPs) in different time windows. The figure illustrates noticeable differences between the two ERP traces, confirming our experimental protocol qualitatively in the healthy cohort. This supports our further tests for the patients.

Equivalent to Figure 2, Figure 3 shows the median across patient stimulus onset-locked segments. The figure illustrates that there exist obvious differences between the two ERP traces. Although not shown in the figure, analysis of subject-specific ERPs reveals that irrespective of the age and gender of the participants in this study, the amplitude due to deviant stimulus response was higher than the amplitude due to standard stimulus response.

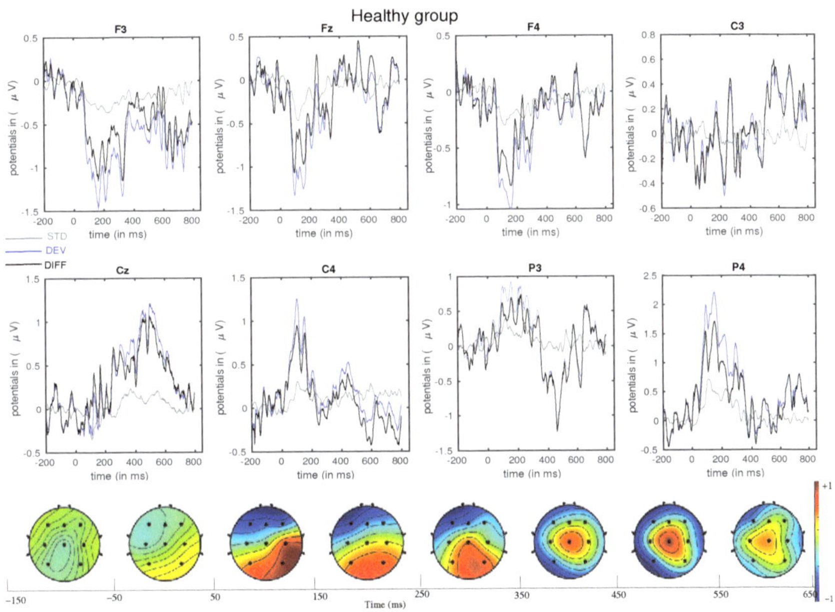

Figure 2. Top and middle panels: median traces of event-related potentials and their differences across the healthy participants for different channels. Traces for standard stimuli are represented with a gray line, for deviant stimuli with a blue line, and for differences with a black line. STD—standard; DEV—deviant; DIFF—difference. Lower panel: spatial topography of activation patterns (difference: deviant minus standard) for different time frames.

The distributions of the detected significant channels over the resulting time windows are visualized with the box plots presented in Figure 4 along with the scalp topographic maps. The left panel of Figure 4a shows the median trace for 'STD' and 'DEV' events and their differences for the channel *Fz* ($p = 0.046$) and the box plot for them over a time window of 150–250 ms. The right panel shows the same but for the channel *P*4 ($p = 0.031$). The distributions of the boxes and the separation of the medians of the boxes specified for 'STD' and 'DEV 'events indicate the differences between the events. The middle panel illustrates the scalp topographic maps for the standard and deviant stimuli along with their differences over a time range of 150–250 ms. The upper rows of the middle panel show the median of the topography map for standard and deviant stimuli, while the lower row shows the difference between them. A difference in distributions for the standard and the deviant stimulus, especially larger variation due to deviant stimulus, supports the existence of sMMN within this time window. Additionally, the topographic map for the difference trace also supports that finding. Figure 4b shows the same but for the time window 350–450 ms for the statistically significant channels, i.e., *F*4 ($p = 0.031$) and *P*4 ($p = 0.046$).

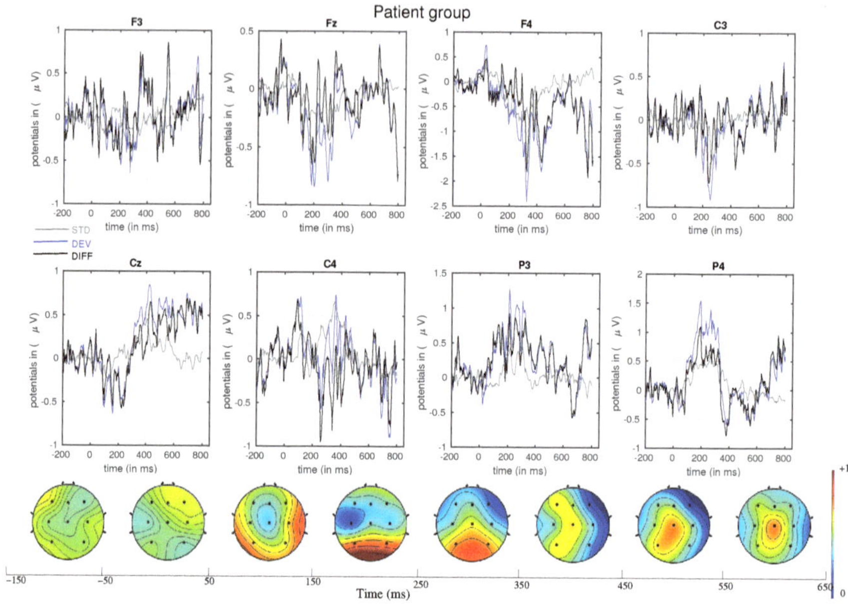

Figure 3. Top and middle panels: median traces of event-related potentials and their differences across the patients for different channels. Traces for standard stimuli are represented with a gray line, for deviant stimuli with a blue line, and for differences with a black line. STD—standard; DEV—deviant; DIFF—difference. Lower panel: spatial topography of activation patterns (difference: deviant minus standard) on different time windows.

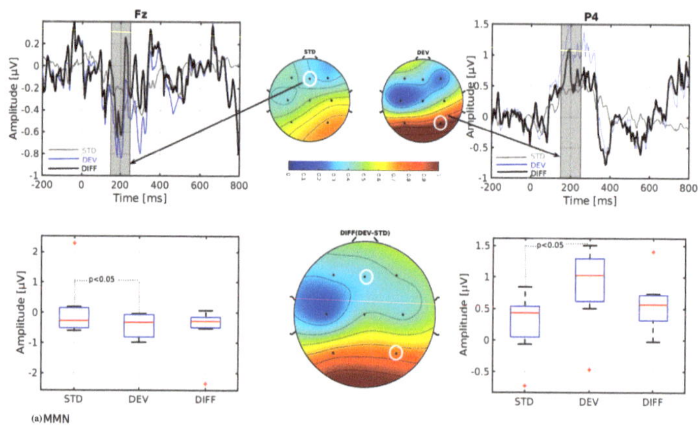

Figure 4. *Cont.*

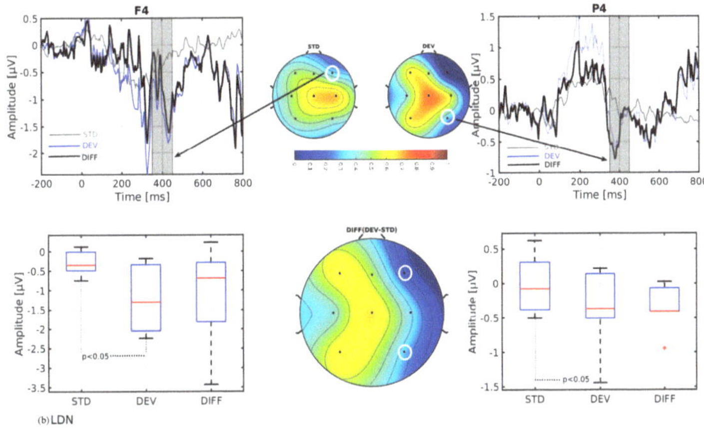

Figure 4. (**a**) The left and right panels show the distributions for the statistically significant channels, e.g., *Fz* and *P4*, along with their box plots for the time window 150–250 ms. The middle panel shows the scalp topographic maps for this time window. STD—standard; DEV—deviant; DIFF—deviant minus standard. (**b**) The same but for the statistically significant channels, e.g., *Fz* and *P4*, for the time window 350–450 ms.

4. Discussion

Somatosensory evoked event-related potentials were recorded in healthy individuals and patients with cerebral palsy to verify the presence of mismatch-related ERP components especially in patients with CP. The central findings of the study are as follows:

- The mismatch-related somatosensory responses can be observed in patients with cerebral palsy.
- In line with our assumption, the observed mismatch-related ERP components from frontal and parietal channels were statistically significant at two predefined latency ranges: *Fz* and *P4* channels at 150–250 ms and *F4* and *P4* channels at 350–450 ms after the stimulus onset. The observed response in the time range of 150–250 ms is considered as sMMN, and the response in the time range of 350–450 ms is considered as LDN. In terms of time window and channel location, these findings are qualitatively in good agreement with the studies involving healthy adults [21,25].

In an early study on auditory MMN, Giar et al. [42] proposed that MMN rises at around 100–150 ms after stimulus onset and peaks at around 200–250 ms over the frontocentral areas of the scalp. Later on, further studies confirmed that auditory MMN is generated in the temporal and frontal areas [34]. In most somatosensory studies, MMN has been confirmed over the frontocentral regions as either a negative or positive component at about 100–250 ms of latency [7,11,12]. Some other studies found mismatch-related ERP responses at two separate latencies. For example, Strömmer et al. [6] found sMMN centroparietally at 180–220 ms and frontocentrally at 250–290 ms after the stimulus onset in adults (22–36 years). Spackman et al. [12] reported a frontocentral negative peak at 100–200 ms, followed by a centroparietal positive shift at 150–250 ms to vibrotactile presented changes in duration and frequency. On the other hand, Akatsuka et al. [7,43] found a significantly enhanced sMMN in early negativity (30–70 ms) and later a positive peak at 100–200 ms after stimulus. Similarly, Butler et al. [41] reported an sMMN response over the frontal midline scalp with two phases of MMN waveform: an earlier negative peak at ~145 ms, followed by a positive peak at ~235 ms. Our present sMMN peak between 150 and 250 ms agrees with the findings of Strömmer et al. [6], and the appeared peak between 350 and 450 ms is likely representing LDN [25,44].

The results indicate the presence and observability of mismatch-related components in the case of patients with CP. Nevertheless, this study has certain limitations that should be explored and addressed in subsequent research.

A central limitation is the relatively small and inhomogeneous patient cohort. While this limitation does not affect the main finding of this study, the data provide insufficient power for more in-depth investigations. A larger patient cohort would, for instance, enable the examination of relationships between clinical parameters, such as somatosensory impairments, and the expression of mismatch-related ERP responses, providing an important basis for the establishment of mismatch-related ERP responses as diagnostic or monitoring biomarkers in CP. Another limitation relates to the fixed interstimulus interval (ISI) rather than a randomized ISI, which is typically favored in the design of ERP studies. While a definite confirmation of whether this has an effect on the observed mismatch-related ERP responses requires further research, we expect, if at all, a randomized ISI to lead to a rather larger effect size relative to a fixed ISI due to the habituation effect. Therefore, we expect the findings to be equivalent, if not more pronounced in case of a randomized ISI.

Despite the limitations, our study provides evidence for the reliable measurement of sMMN in patients with CP. With this simplistic experimental setup, our results indicate the feasibility of successfully measuring sMMN in patients with CP, who typically have limited attention span. The paradigm may also be suitable for children with CP, which, however, requires further thorough investigation. Due to the ongoing development of a child's brain, it is reasonable to assume that brain waves are distributed differently than in adults, and therefore, the response to stimulation is likely to be different. Therefore experiments should be conducted separately across different age groups, and in general, a larger number of patients are needed. While the current study does not establish a basis for predicting the extent of somatosensory impairments through sMMN, our results still offer a valuable starting point for the advancement of diagnostic or therapeutic tools. Mismatch responses may be used to probe for somatosensory impairments in CP patients, or monitor changes in somatosensory perception as a result of sensorimotor rehabilitation, which has been performed in healthy subjects [45,46].

Author Contributions: Conceptualization, S.R., S.K.E. and R.L.; methodology, S.R. and S.K.E.; formal analysis, S.R.; resources, R.L.; data curation, R.L.; writing—original draft preparation, S.R. writing—review and editing, S.K.E. and R.L.; supervision, R.L.; project administration, R.L.; funding acquisition, R.L. All authors have read and agreed to the final version of the manuscript.

Funding: This research was supported by funds provided by the Buhl-Strohmaier Foundation and the Würth Foundation. The experimental work was supported by the Leonhard Lorenz Foundation

Institutional Review Board Statement: The study was conducted in accordance with the Declaration of Helsinki, and the experiment and all its procedures related to patients' data privacy and personal interests were approved by the Ethics Committee of the Faculty of Medicine of the Technical University of Munich (Number: 582/16 S).

Informed Consent Statement: Participation in this study was voluntary. Informed written consent was obtained from all the participants involved in the study.

Data Availability Statement: The data set generated and/or analyzed during the current study is available from the corresponding author on reasonable request.

Acknowledgments: The authors would like to thank the participants for their time and support for this study. The authors sincerely acknowledge the support received from Ana Alves Pinto, for the EEG data collection, part of the experimental fund acquisition, and her valuable suggestions.

Conflicts of Interest: Stefan K. Ehrlich was employed by the company SETLabs Research GmbH Healthcare Technologies. The remaining authors declare that the research was conducted in the absence of any commercial or financial relationships that could be construed as a potential conflict of interest. The funders had no role in the design of the study; in the collection, analyses, or interpretation of data; in the writing of the manuscript; or in the decision to publish the results.

References

1. Näätänen, R.; Jacobsen, T.; Winkler, I. Memory-based or afferent processes in mismatch negativity (MMN): A review of the evidence. *Psychophysiology* **2005**, *42*, 25–32. [CrossRef]
2. Schröger, E. Measurement and interpretation of the mismatch negativity. *Behav. Res. Methods Instrum. Comput.* **1998**, *30*, 131–145. [CrossRef]
3. Näätänen, R.; Gaillard, A.; Mäntysalo, S. Early selective-attention effect on evoked potential reinterpreted. *Acta Psychol.* **1978**, *42*, 313–329. [CrossRef]
4. Tales, A.; Troscianko, T.; Wilcock, G.K.; Newton, P.; Butler, S.R. Agerelated changes in the preattentional detection of visual change. *Neuroreport* **2002**, *13*, 969–972. [CrossRef] [PubMed]
5. Lorenzo-Lopez, L.; Amenedo, E.; Pazo-Alvarez, P.; Cadaveira, F. Preattentive detection of motion direction changes in normal aging. *Neuro Rep.* **2004**, *15*, 2633–2636. [CrossRef]
6. Strömmer, J.M.; Tarkka, I.M.; Astikainen, P. Somatosensory mismatch response in young and elderly adults. *Front. Aging Neurosci.* **2014**, *6*, 293. [CrossRef]
7. Akatsuka, K.; Wasaka, T.; Nakata, H.; Inui, K.; Hoshiyama, M.; Kakigi, R. Mismatch responses related to temporal discrimination of somatosensory stimulation. *Clin. Neurophysiol.* **2005**, *116*, 1930–1937. [CrossRef]
8. Astikainen, P.; Ruusuvirta, T.; Korhonen, T. Somatosensory event-related potentials in the rabbit cerebral and cerebellar cortices: a correspondence with mismatch responses in humans. *Neurosci. Lett.* **2001**, *298*, 222–224. [CrossRef] [PubMed]
9. Czigler, I.; Balázs, L.; Winkler, I. Memory-based detection of task-irrelevant visual changes. *Psychophysiology* **2002**, *39*, 869–873. [CrossRef] [PubMed]
10. Li, X.; Lu, Y.; Sun, G.; Gao, L.; Zhao, L. Visual mismatch negativity elicited by facial expressions: New evidence from the equiprobable paradigm. *Behav. Brain Funct.* **2012**, *8*, 7. [CrossRef]
11. Kekoni, J.; Hämäläinen, H.; Saarinen, M.; Gröhn, J.; Reinikainen, K.; Lehtokoski, A.; Näätänen, R. Rate effect and mismatch responses in the somatosensory system: ERP-recordings in humans. *Biol. Psychol.* **1997**, *46*, 125–142. [CrossRef]
12. Spackman, L.; Boyd, S.; Towell, A. Effects of stimulus frequency and duration on the somatosensory mismatch negativity. *Clin. Neurophysiol.* **2007**, *118*, e175. [CrossRef]
13. Novitski, N.; Tervaniemi, M.; Huotilainen, M.; Näätänen, R. Frequency discrimination at different frequency levels as indexed by electrophysiological and behavioral measures. *Cogn. Brain Res.* **2004**, *20*, 26–36. [CrossRef]
14. Shafer, V.; Yu, Y.; Garrido-Nag, K. Neural mismatch indices of vowel discrimination in monolingually and bilingually exposed infants: does attention matter? *Neurosci. Lett.* **2012**, *526*, 10–14. [CrossRef]
15. Rinker, T.; Kohls, G.; Richter, C.; Maas, V.; Schulz, E.; Schecker, M. Abnormal frequency discrimination in children with SLI as indexed by mismatch negativity (MMN). *Neurosci. Lett.* **2007**, *413*, 99–104. [CrossRef]
16. Stothart, G.; Tales, A.; Kazanina, N. Evoked potentials reveal age-related compensatory mechanisms in early visual processing. *Neurobiol. Aging* **2013**, *34*, 1302–1308. [CrossRef] [PubMed]
17. Tales, A.; Haworth, J.; Wilcock, G.; Newton, P.; Butler, S. Visual mismatch negativity highlights abnormal pre-attentive visual processing in mild cognitive impairment and Alzheimer's disease. *Neuropsychologia* **2008**, *46*, 1224–1232. [CrossRef] [PubMed]
18. Whittle, S.; Yücel, M.; Yap, M.B.; Allen, N.B. Sex differences in the neural correlates of emotion: Evidence from neuroimaging. *Biol. Psychol.* **2011**, *87*, 319–333. [CrossRef] [PubMed]
19. Yang, X.; Yu, Y.; Chen, L.; Sun, H.; Qiao, Z.; Qiu, X.; Zhang, C.; Wang, L.; Zhu, X.; He, J.; et al. Gender differences in pre-attentive change detection for visual but not auditory stimuli. *Clin. Neurophysiol.* **2016**, *127*, 431–441. [CrossRef]
20. Näätänen, R. Somatosensory mismatch negativity: A new clinical tool for developmental neurological research? *Dev. Med. Child Neurol.* **2009**, *51*, 930–931. [CrossRef]
21. Shen, G.; Weiss, S.M.; Meltzoff, A.N.; Marshall, P.J. The somatosensory mismatch negativity as a window into body representations in infancy. *Int. J. Psychophysiol.* **2018**, *134*, 144–150. [CrossRef] [PubMed]
22. Shinozaki, N.; Yabe, H.; Sutoh, T.; Hiruma, T.; Kaneko, S. Somatosensory automatic responses to deviant stimuli. *Brain Res. Cogn. Brain Res.* **1998**, *7*, 165–171. [CrossRef]
23. Tamura, Y.; Hoshiyama, M.; Inui, K.; Nakata, H.; Wasaka, T.; Ojima, S.; Inoue, K.; Kakigi, R. Cognitive processes in two-point discrimination: An ERP study. *Clin. Neurophysiol.* **2004**, *115*, 1875–1884. [CrossRef] [PubMed]
24. Cheour, M.; Korpilahti, P.; Martynova, O.; Lang, A.H. Mismatch Negativity and Late Discriminative Negativity in Investigating Speech Perception and Learning in Children and Infants. *Audiol. Neurotol.* **2001**, *6*, 2–11. Available online: https://karger.com/aud/article-pdf/6/1/2/2249646/000046804.pdf (accessed on 26 October 2023). [CrossRef] [PubMed]
25. Petermann, M.; Kummer, P.; Burger, M.; Lohscheller, J.; Eysholdt, U.; Döllinger, M. Statistical detection and analysis of mismatch negativity derived by a multi-deviant design from normal hearing children. *Hear. Res.* **2009**, *247*, 128–136. [CrossRef] [PubMed]
26. Bishop, D.V.; Hardiman, M.J.; Barry, J.G. Is auditory discrimination mature by middle childhood? A study using time-frequency analysis of mismatch responses from 7 years to adulthood. *Dev. Sci.* **2011**, *14*, 402–416. Available online: https://onlinelibrary.wiley.com/doi/pdf/10.1111/j.1467-7687.2010.00990.x (accessed on 27 October 2023). [CrossRef] [PubMed]
27. Čeponienė, R.; Lepistö, T.; Soininen, M.; Aronen, E.; Alku, P.; Näätänen, R. Event-related potentials associated with sound discrimination versus novelty detection in children. *Psychophysiology* **2004**, *41*, 130–141. Available online: https://onlinelibrary.wiley.com/doi/pdf/10.1111/j.1469-8986.2003.00138.x (accessed on 26 October 2023). [CrossRef]

28. Schulte-Körne, G.; Deimel, W.; Bartling, J.; Remschmidt, H. Speech perception deficit in dyslexic adults as measured by mismatch negativity (MMN). *Int. J. Psychophysiol.* **2001**, *40*, 77–87. [CrossRef]
29. Kropotov, J.D. *Chapter 18—Schizophrenia*; Academic Press: San Diego, CA, USA, 2009; pp. 420–431. [CrossRef]
30. Rüsseler, J.; Münte, T. *Kognitive Potenziale (ereigniskorrelierte Potenziale, EKP). Evozierte Potenziale, Neurovegetative Diagnostik, Okulographie*; Buchner, H., Noth, J., Eds.; Georg Thieme Verlag KG, Stuttgart, Germany, 2005; Chapter 7, pp. 80–94. [CrossRef]
31. Kurz, M.J.; Heinrichs-Graham, E.; Arpin, D.J.; Becker, K.M.; Wilson, T.W. Aberrant synchrony in the somatosensory cortices predicts motor performance errors in children with cerebral palsy. *J. Neurophysiol.* **2014**, *111*, 573–579. [CrossRef]
32. Hoon, A.H., Jr.; Stashinko, E.E.; Nagae, L.M.; Lin, D.D.M.; Keller, J.; Bastian, A.; Campbell, M.L.; Levey, E.; Mori, S.; Johnston, M.V. Sensory and motor deficits in children with cerebral palsy born preterm correlate with diffusion tensor imaging abnormalities in thalamocortical pathways. *Dev. Med. Child Neurol.* **2009**, *51*, 697–704. [CrossRef]
33. Matusz, P.J.; Key, A.P.; Gogliotti, S.; Pearson, J.; Auld, M.L.; Murray, M.M.; Maitre, N.L. Somatosensory Plasticity in Pediatric Cerebral Palsy following Constraint-Induced Movement Therapy. *Neural Plast.* **2018**, *2018*, 1891978. [CrossRef]
34. Restuccia, D.; Della Marca, G.; Valeriani, M.; Leggio, M.G.; Molinari, M. Cerebellar damage impairs detection of somatosensory input changes. A somatosensory mismatch-negativity study. *Brain* **2007**, *130 (Pt 1)*, 276–287. [CrossRef] [PubMed]
35. Hakkarainen, E.; Pirilä, S.; Kaartinen, J.; Meere, J. Stimulus evaluation, event preparation, and motor action planning in young patients with mild spastic cerebral palsy: An event-related brain potential study. *J. Child Neurol.* **2012**, *27*, 465–470. [CrossRef]
36. Lackner, C.; Gorter, J.; Segalowitz, S. Cognitive Event-Related Potentials in Young Adults With Cerebral Palsy: A Proof-of-Concept Study. *Clin. EEG Neurosci.* **2024**, *55*, 64–75. [CrossRef]
37. Palisano, R.; Rosenbaum, P.; Walter, S.; Russell, D.; Wood, E.; Galuppi, B. Development and reliability of a system to classify gross motor function in children with cerebral palsy. *Dev. Med. Child Neurol.* **1997**, *39*, 214–223. [CrossRef] [PubMed]
38. Eliasson, A.; Krumlinde-Sundholm, L.; Rösblad, B.; Beckung, E.; Arner, M.; Ohrvall, A.; Rosenbaum, P. The Manual Ability Classification System (MACS) for children with cerebral palsy: Scale development and evidence of validity and reliability. *Dev. Med. Child Neurol.* **2006**, *48*, 549–554. [CrossRef] [PubMed]
39. Ruffini, G.; Dunne, S.; Farres, E.; Marco-Pallares, J.; Ray, C.; Mendoza, E.; Silva, R.; Grau, C. A dry electrophysiology electrode using CNT arrays. *Sens. Actuators A Phys.* **2006**, *132*, 34–41. [CrossRef]
40. Garrido, M.; Kilner, J.; Stephan, K.; Friston, K. The mismatch negativity: A review of underlying mechanisms. *Clin. Neurophysiol.* **2009**, *120*, 453 –463. [CrossRef]
41. Butler, J.; Molholm, S.; Fiebelkorn, I.; Mercier, M.; Schwartz, T.; Foxe, J. Common or redundant neural circuits for duration processing across audition and touch. *J. Neurosci.* **2011**, *31*, 3400–3406. [CrossRef]
42. Giard, M.; Perrin, F.; Pernier, J.; Bouchet, P. Brain generators implicated in the processing of auditory stimulus deviance: A topographic event-related potential study. *Psychophysiology* **1990**, *27*, 627–640. [CrossRef]
43. Akatsuka, K.; Wasaka, T.; Nakata, H.; Kida, T.; Kakigi, R. The effect of stimulus probability on the somatosensory mismatch field. *Exp. Brain Res.* **2007**, *181*, 607–614. [CrossRef] [PubMed]
44. Makeig, S.; Jung, T.; Bell, A.; Gharemani, D.; Sejnowski, T. Blind separation of auditory event-related brain responses into independent components. *Proc. Natl. Acad. Sci. USA* **1997**, *94*, 10979–10984. [CrossRef] [PubMed]
45. Xide, Y.; Tao, L.; Dingguo, G. The Mismatch Negativity: An Indicator of Perception of Regularities in Music. *Behav. Neurol.* **2015**, *2015*, 469508. [CrossRef]
46. Näätänen, R.; Paavilainen, P.; Rinne, T.; Alho, K. The mismatch negativity (MMN) in basic research of central auditory processing: A review. *Clin. Neurophysiol.* **2007**, *118*, 2544–2590. [CrossRef] [PubMed]

applied sciences

Article

Partial Parallelism Plots

Axel Petzold [1,2]

1 Queen Square Institute of Neurology, UCL, London WC1N 3BG, UK; a.petzold@ucl.ac.uk
2 Amsterdam UMC, De Boelelaan 1117, 11081 HX Amsterdam, The Netherlands

Featured Application: This article proposes a novel graphical approach for the the assessment of parallelism of biomarker tests that takes into consideration situations where parallelism is partially lacking. The new approach expands on earlier observations and criticism of the limitations of statistical methods included in the guidelines of regulatory authorities. Researchers in the field concur on emphasising the importance of ensuring the accuracy and reliability; a pertinent point which still remains to be addressed. To this purpose, two primary computational approaches are discussed: (a) statistical assessment and (b) visual assessment. Statistical methods, such as regression analysis and parallelism/non-parallelism indexes, offer precision and objectivity, making them suitable for large datasets and high accuracy requirements. They can detect subtle differences in parallelism that may be missed by visual assessment. However, they assume a linear relationship between analyte concentration and assay response, which may not always hold true. Visual assessment relies on interpreting graphs or charts depicting the biomarker–concentration–response relationship. It is intuitive and can quickly identify gross deviations from partial parallelism, making it useful for screening biomarker assays. Visual assessment may detect non-parallelism due to confounding factors that statistical methods might miss. The graphical method proposed here suggests using partial parallelism plots, which visually depict the relationship between biomarker concentration and assay response for each sample. These plots enable the identification of non-parallelism caused by analytical issues or confounding factors. They assist in determining the optimal range of dilutions for each sample and provide a language that is easily understood by researchers, regulatory authorities, and technicians. For regulatory authorities, this document provides valuable insights into the assessment of partial parallelism for biomarker tests. It highlights the need for both statistical and visual assessment methods to evaluate parallelism accurately. The proposed use of partial parallelism plots can aid in visualising and understanding the relationship between biomarker concentration and assay response. By considering these plots during the evaluation of biomarker assays, regulatory authorities can ensure the accuracy, reliability, and suitability of these tests as trial outcome measures and for clinical use.

Abstract: Demonstrating parallelism in quantitative laboratory tests is crucial to ensure accurate reporting of data and minimise risks to patients. Regulatory authorities make the demonstration of parallelism before clinical use approval mandate. However, achieving statistical parallelism can be arduous, especially when parallelism is limited to a subrange of the data. To address potential biases and confounds, I propose a simple graphical method, the Partial Parallelism Plot, to demonstrate partial parallelism. The proposed method offers ease of understanding, intuitiveness, and graphical simplicity. It enables the graphical assessment of quantitative data risk when parallelism is lacking within a defined range. As parallelism may not be consistent across the entire analytical range, the plots focus on partial parallelism. The method can readily be programmed into graphical applications for enhanced interactivity. By providing a clear graphical representation, the method allows researchers to ascertain the presence of parallelism in laboratory tests, thus aiding in the validation process for trials and clinical applications.

Keywords: parallelism; biomarker; laboratory; test; graphical statistics

Citation: Petzold, A. Partial Parallelism Plots. *Appl. Sci.* **2024**, *14*, 602. https://doi.org/10.3390/app14020602

Academic Editor: Alexander N. Pisarchik

Received: 25 September 2023
Revised: 3 January 2024
Accepted: 4 January 2024
Published: 10 January 2024

1. Introduction

In clinical medicine, precise determination of the concentration of a given compound is essential. For instance, in the case of suspected heart attack, the concentration of specific biomarkers must be determined accurately in a blood sample of the patient to support the diagnosis. Inaccurate mathematical calculations based on laboratory measurements may lead to erroneous biomarker concentrations and misdiagnosis. Therefore, mathematical calculations are heavily relied upon in daily clinical and laboratory practices, but it is crucial to ensure that the underlying assumptions of these calculations are satisfied. This report addresses one such assumption, namely parallelism.

Demonstration of parallelism is crucial for the accuracy of any test based on calculating sample concentration from a standard curve [1]. However, it has been noted that there is no widely adopted universal strategy for assessing parallelism in bioassays. Without assurance of parallelism, investigators are unable to calculate reliable estimates for serum antibody concentrations [1]. To address this issue, it has been suggested to visually compare the slope of logistic-log curves, for which a series of excellent examples have been provided. The authors cautioned against purely statistical assessments of parallelism, as the methods of computation are complex, not readily available in software packages, prone to error unless interpreted correctly, and overly sensitive to negligible departures from parallelism when model precision is high. Furthermore, no guidance was provided on how to interpret the data in cases where there is partial non-parallelism, which may make it challenging for users to determine the appropriate course of action. Notwithstanding this constraint, the parallelism plots initially proposed [1] continue to serve as a valuable graphical tool for evaluating parallelism in laboratory tests, and their significance has been acknowledged in subsequent research. According to this authoritative perspective [2], the experimental validation of parallelism remains a challenging and pivotal aspect in the validation of bioanalytical methods to this day, an assertion that was reiterated in a highly influential white paper [3].

Regulatory authorities impose strict requirements for the approval of an assay, including the demonstration of parallelism. As per the latest guidelines by the Food and Drug Administration (FDA) and European Medical Agency (EMA), parallelism is defined as "Parallelism demonstrates that the serially diluted incurred sample response curve is parallel to the calibration curve" [4]. The guideline provides explicit laboratory instructions for conducting the study, involving the dilution of a high-concentration study sample to at least three concentrations with a blank matrix. However, the interpretation of results becomes more ambiguous. The guideline states that the consistency of back-calculated concentrations between samples in a dilution series should not exceed a 30% coefficient of variation (CV). Nevertheless, it is essential to carefully monitor the data, as results meeting this criterion may still indicate trends of non-parallelism. In cases where the sample does not dilute linearly, a predefined procedure for reporting results should be established. In this report, I propose a simple graphical approach for such a procedure, as it may offer greater intuitiveness and be less susceptible to the limitations previously recognised in purely numerical methods [1,2].

The concept of parallelism may appear simple at first glance, but it can be difficult to understand upon further examination. Additionally, numerically driven, statistical representations of parallelism may not be intuitive for individuals without a statistical background. This lack of understanding can be problematic for regulatory authorities and mixed expertise panels tasked with making decisions in laboratory-based research.

2. The Range of Accuracy and Effect Size in the Assessment of Laboratory Tests

Experimental evidence indicates a significant impact of the lack of parallelism on the quantification of neurofilaments, a well-established biomarker for neurodegeneration [5,6]. The FDA and the EMA approved the use of two novel drugs based on laboratory results quantifying neurofilaments. A state-of-the-art randomised controlled trial (RCT) demonstrated a reduction in neurofilament blood levels as proof of efficacy for a novel disease-modifying treatment in multiple sclerosis [7], and another RCT [8] lead to rapid FDA approval for the antisense oligonucleotide tofersen to treat amyotrophic lateral sclerosis. However, neither study considered the possibility of partial non-parallelism of neurofilaments. Although not currently relevant in studies with large effect sizes, such as [7,8], non-parallelism becomes more pertinent in studies with smaller effect sizes, such as those encountered in the large number of trials on Alzheimer's disease which employ biomarkers as an outcome measure.

Accepting that parallelism is a vital factor in the evaluation of laboratory tests for biomarkers, it needs to be acknowledged that parallelism is just one among several other factors influencing test reliability [9]. Pum emphasised that analytical and clinical specificity and sensitivity are additional critical factors [2]. Various biological and technical factors, such as matrix effects, variations in biomarker metabolism, or variations in laboratory test procedures, can also influence the accuracy of laboratory tests for biomarkers. A large international consortium underscored the importance of using high-quality samples [10]. Furthermore, prospective experimental evidence highlighted that the inter-laboratory reproducibility and technician skills are other key factors affecting test outcomes [11].

Assessing the range of accuracy of laboratory tests for biomarkers is a complex task that depends on multiple factors in addition to parallelism. It is crucial to be aware of these factors and to critically evaluate laboratory tests to determine their suitability as diagnostic tools and trial outcome measures in medicine.

3. The Definition of Parallelism and Partial Parallelism

The term parallelism, in its simplest definition, describes the relationship between the concentration of an analyte (such as a biomarker) in a sample and the signal produced by the reference standard of the laboratory test used to measure that analyte as earlier introduced [1,2]. When the relationship between concentration and signal is linear, parallelism is said to be present. Hence, the other term used in the literature for parallelism is linearity. This is important because it means that the laboratory test accurately reflects the concentration of the analyte in the sample and, therefore, provides a reliable measurement of the biomarker.

However, if there is non-parallelism (i.e., a non-linear relationship between concentration and signal), the accuracy of the laboratory test may be compromised. This can occur if there is interference from other substances in the sample, or if the laboratory test is not able to accurately detect the analyte, for example, at higher concentrations. This is a frequent problem with biomarker assays requiring use of a non-linear standard curve, as reviewed theoretically in [2] and demonstrated experimentally in [5].

In order to test for linearity, a regression analysis is performed to determine the slope of the line of best fit. The formula for the slope of the line is:

$$slope = \frac{\sum_{i=1}^{n} (x_i - \bar{x}) \times (y_i - \bar{y})}{\sum_{i=1}^{n} (x_i - \bar{x})^2} \tag{1}$$

where

x_i is the concentration of the biomarker in the sample;
y_i is the signal produced by the test used to measure the biomarker;
n is the number of data points;
$\bar{x}$ is the mean concentration of the biomarker;
$\bar{y}$ is the mean signal produced by the laboratory test.

If the slope is not significantly different from 1 (i.e., if $|slope - 1| \leq SE$ where SE is the standard error), then parallelism is present. The values of x_i are the given concentrations (i.e., ng/mL, pg/mL, g/L), and formula (1) uses those values to calculate the slope of the line of best fit, which represents the relationship between the concentration of the analyte and the signal produced by the laboratory test, as detailed in an entire book chapter [9]. It follows that for biomarker assays with proof of linearity, a parallelism coefficient close to 1 indicates that the patient sample and standard curve have similar slopes:

$$\text{Parallelism coefficient} = \frac{\text{Slope of Patient Dilution}}{\text{Slope of Standard Curve}} \tag{2}$$

In absence of linearity, determination of parallelism was defined for bioassay dilution curves in absence of a standard curve by a logistic-log model in which the signal for the test are optic densities (OD) as:

$$OD = d + \frac{a-d}{1+(\frac{dilution}{c})^b} \tag{3}$$

where

a is the upper asymptote of the curve of the OD at a theoretical infinite concentrations;
d is the lower asymptote of the curve of the OD at a theoretical zero concentrations;
b is a curvature parameter;
c is the symmetry point of the sigmoid.

Linear transformation of the curve is achieved through a logistic function where OD_{min} and OD_{max} correspond to bespoke upper and lower asymptotes as

$$Logit(OD)_{fs} = \log\left(\frac{OD - OD_{min}}{OD_{max} - OD}\right) \tag{4}$$

The formula can be reduced to express a partially specified logit model introduced in [1]:

$$Logit(OD)_{ps} = \log\left(\frac{OD}{OD_{max} - OD}\right) \tag{5}$$

The visualisation of this approach is illustrated in Figure 1. Clearly, none of these curves in Figure 1A meet the criteria for linearity as defined at the onset of this section. Only bespoke logistic-log transformation permits to demonstrate parallelism (Curves 1–4 in Figure 1B) and lack of parallelism (Curve 5 in Figure 1B). The logistic-log transformation is the basis for the statistical analysis of a dilution series intended to facilitate visualisation as intended [1].

$$\text{Relative dilution}_i = 100 \times \left(\frac{\text{actual sample dilution}_i}{\text{maximum dilution in series}}\right) \tag{6}$$

where i indicates the dilution step. Finally, Plikaytis et al. used Generalised Linear Models (one-way analysis of covariance) for determination of parallelism.

In summary, parallelism is described as a critical aspect of the analytical test validation in the context of analytical linearity (i.e., regression analysis (Formula (1)), parallelism coefficient (Formula (2))) and non-linearity (i.e., a logistic-log model (Formulas (3)–(5))). The logistic-log transformation ensures that the test can accurately measure the concentration of the biomarker over a range of concentrations [1]. Overall, the choice of method for assessing parallelism depends on the data distribution (i.e., linear or nonlinear).

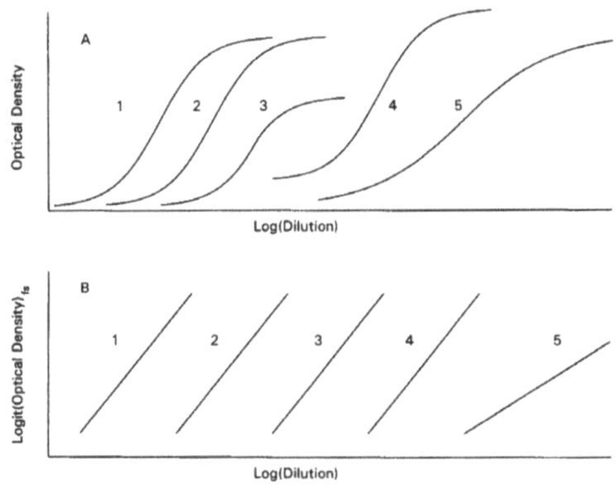

Figure 1. Comparison of logistic-log curves and their fully specified logit-log transformed counterparts. (**A**) Lines 1 and 2, logistic-log curves with identical slopes and asymptotes; Lines 3 and 4, logistic-log curves with identical slopes and different asymptotes; Line 5, logistic-log curve with different slope and asymptotes. (**B**) Corresponding straight lines formed by using the fully specified logit-log transformation. [Reproduced with permission from [1]].

4. Graphical Presentations of Parallelism

Graphical presentations of parallelism can provide a visual representation of the accuracy and reliability of laboratory tests for biomarkers. Graphical presentations can help healthcare providers and researchers to rapidly (i.e., at a glance) identify potential issues with biomarker tests, such as interference from other substances in the sample or limitations with the analytical sensitivity of the test. They can also be useful for comparing the accuracy of different laboratory tests for the same biomarker.

One common graphical presentation of parallelism is the parallelism plot, which involves plotting the signal produced by the laboratory test on the y-axis and the concentration of the biomarker on the x-axis. A seminal example from the literature was presented in Figure 1. If the lines are parallel, it suggests that the laboratory test accurately reflects the concentration of the biomarker in the sample and that parallelism is present.

One development in the biomarker field, since the introduction of the logistic-log transformation [1], has been the use of calibrated and quality controlled protein standard curves. Consequently, reported biomarker concentrations are derived from the curve between the symmetry point (c as defined for Formula (3)), but never from the asymptotes. The lower asymptotes (d) indicate non-measurable data. This is either because the detection limit of the assay is insufficient or because there is nothing there to be measured. For the upper asymptote (a), the concentration of the biomarker is too high to be estimated reliably. Extrapolation is not permitted. Sample dilution is required. Taken together parallelism of a biomarker is therefore only determined for

$$(x_i^{std}, y_i^{std}), \in \{1, \ldots, n^{std}\}, \text{where } y \neq a \vee d \quad (7)$$

Table 1 shows the data used for calculation of the graphical presentation of the curves in Figures 2 and 3. The first step of the data transformation used for the graphical presentation of the partial parallelism plots is to adjust the calculated concentration at each dilution step (i) as follows:

$$z_i = x_i^{std} \times i \quad (8)$$

This is followed by normalisation of each value of the transformed series to the value for the lowest dilution step (i.e., dilution 1:1, Table 2) as

$$\bar{z}_i = \frac{z_i}{z_1} \quad (9)$$

Table 1. Raw data for the doubling dilution curves used for Figures 2 and 3.

Dilution	Standard	Sample-A	Sample-B
1:1	10	8	4
1:2	5	4	6
1:4	2.5	2	5
1:8	1.25	1	4
1:16	0.625	0.5	3
1:32	0.3125	0.25	1.5
1:64	0.15625	0.125	0.75
1:128	0.078125	0.0625	0.375
1:256	0.0390625	0.03125	0.1875
1:512	0.01953125	0.015625	0.09375

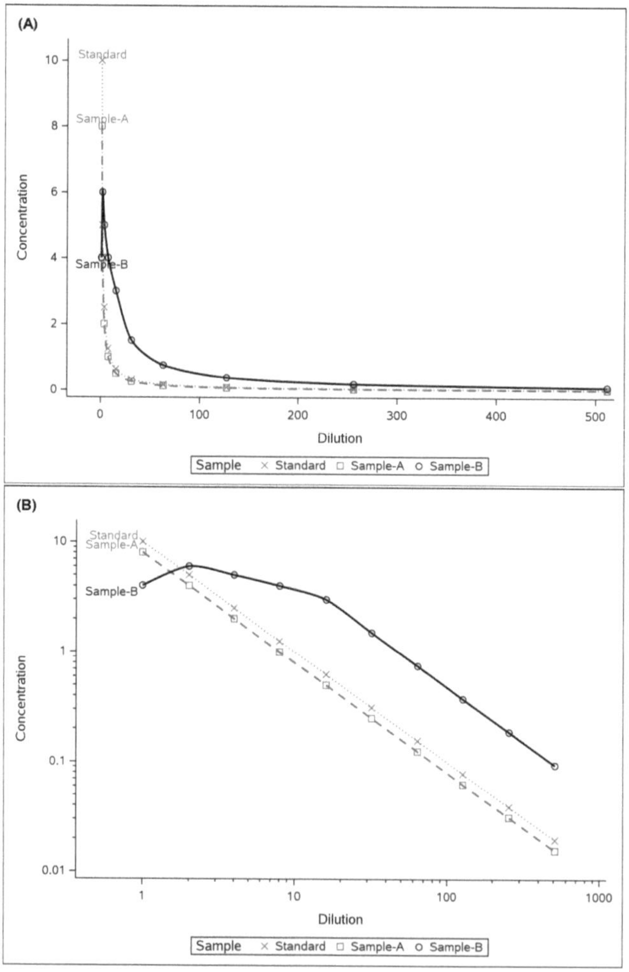

Figure 2. Conventional presentations of a doubling dilution curve for demonstration of parallelism between a standard and a sample. This graph illustrates how the concentration of a compound (y-axis) decreases with subsequent dilution steps (x-axis) either presented as a continuous variable on (**A**), a linear scale as used in [12], and (**B**) on the logarithmic scale derived from Formula (5) [1]. The standard curve (cross, dotted grey line) and dilution curves (Sample-A, open square, dashed grey line; Sample-B, open circle, black line).

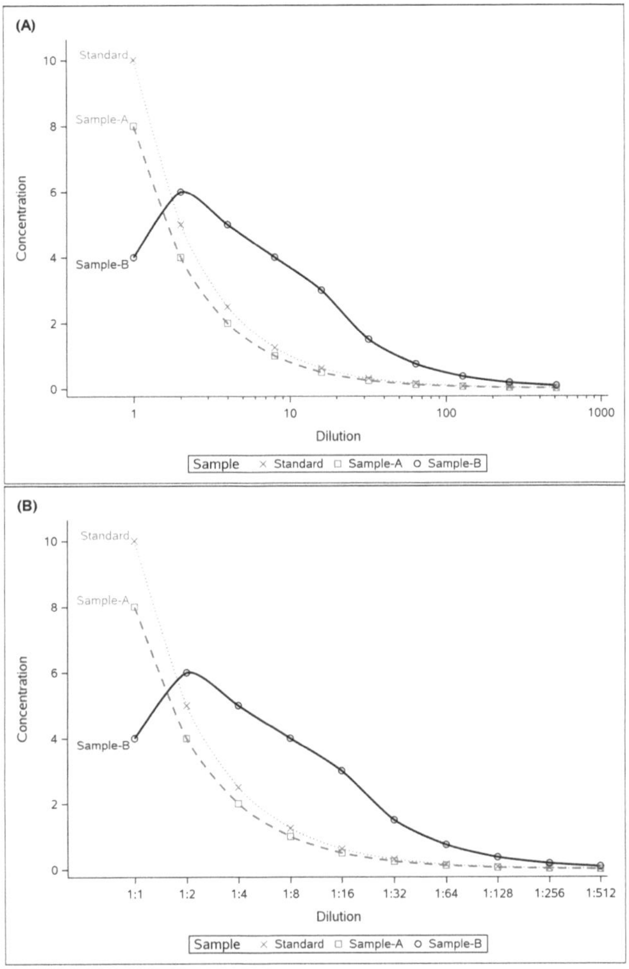

Figure 3. Graphical comparison of the partial logistic presentation on the x-axis only [1] in (**A**) switched for a categorical variable in (**B**). For any test, the standard (cross, dotted grey line) is used as the main comparator. In this example, the dilution curve for Sample-A (open square, dashed grey line) is parallel to the standard curve. There is a small offset on the y-axis between the standard and Sample-A because the starting concentration for Sample-B was less than for the standard. In contrast, the dilution curve for Sample-B (open circle, black line) is not parallel to the standard. For Sample-B, there is an increase in the concentration with the first dilution step. For dilution steps 1:4 to 1:16, the concentration in Sample-B reduces to a lesser degree than for the standard. After Dilution Step 1:32, there is parallelism between Sample-B and the Standard, but this is not clearly visible with this format of graphical presentation.

Overall, graphical presentations of parallelism are an important tool for evaluating the accuracy and reliability of laboratory tests for biomarkers, but there are important practical limitations to their interpretability. The next section will illustrate how this can be overcome in a standardised way which will simplify the interpretation of the graphical presentation.

Table 2. Transformed data from Table 1 as needed to develop the partial parallelism plots shown in Figure 4. Abbreviations: Standard = Std, Sample-A = a, Sample-b = b. The horizontal bar above the abbreviation (e.g., $\overline{\text{Std}}$) indicates the normalised values.

Dil.	Std [1]	A [1]	B [1]	$\overline{\text{Std}}$ [2]	$\overline{\text{A}}$ [2]	$\overline{\text{B}}$ [2]
1:1	10	8	4	1	1	1
1:2	10	8	12	1	1	3
1:4	10	8	20	1	1	5
1:8	10	8	32	1	1	8
1:16	10	8	48	1	1	12
1:32	10	8	48	1	1	12
1:64	10	8	48	1	1	12
1:128	10	8	48	1	1	12
1:256	10	8	48	1	1	12
1:512	10	8	48	1	1	12

[1] Data of concentrations from Table 1 multiplied by dilution step from series (i.e., $5 \times 2 = 10, 2.5 \times 4 = 10$, etc.) as summarised in Formula (8). [2] Data normalised to concentration at lowest dilution step of the series (i.e., $\frac{10}{10} = 1$, etc.) as summarised in Formula (9).

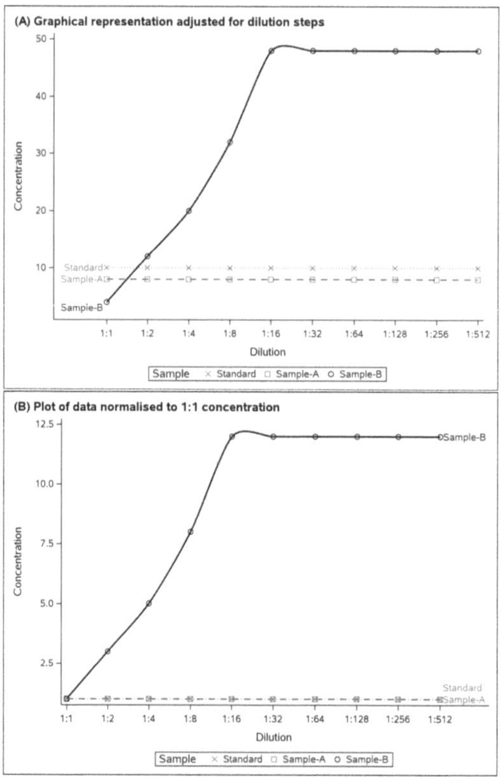

Figure 4. (**A**) illustrates that the parallelism between the Standard and Sample-A is visually more intuitive compared to Figure 2. For this presentation, the value of the concentration was corrected by multiplication with the dilution. The offset on the y-axis between the Standard and Sample-A is explained by the difference in concentration. This can be a problem for the graphical representation if this difference is very large. Therefore, in example (**B**), all values were normalised to the baseline concentration. Now parallelism between the Standard and Sample-A is illustrated by the overlay of the horizontal lines at the y-axis value of one. The consequence of the absence of parallelism for Sample-B in the initial dilution steps leads to an overestimation of a factor of ≈12.5 once parallelism is achieved after a dilution of 1:32.

5. Partial Parallelism Plots

In a partial parallelism plot, the laboratory test results are plotted against each other on the x- and y-axes. A line of unity is then added to the plot to represent perfect parallel agreement between the measurements. The slope of the line, normalised to the first dilution step of the standard curve, is zero with an intercept of one. Therefore, a horizontal reference line at once permits the comparison of the slope of the samples to visually analyse the degree of parallelism for the biomarker in question. If the line of unity (horizontal reference line) is significantly different from the slope of the line of best fit for the sample, it suggests that partial parallelism is absent.

One advantage of partial parallelism plots is that they can be used to assess parallelism between samples over a limited, thought to represent the clinically useful, range of concentrations. This can provide a more practical evaluation of parallelism as relevant for routine healthcare practice.

As a first step towards this goal, Figures 3 and 4 introduce the graphical representation of the line of unity. The result of normalisation for subsequent dilution steps (Formula (7)) is shown graphically for the data from Table 1. In this presentation, there is a similar graphical pattern for the plots in Figure 3A,B. The difference between the two plots can be seen on the x-axis. Note that the x-axis in Figure 3A is log based. Whilst mathematically correct, this presentation does not make for an easy laboratory, clinical, or health authority-tuned assessment of the biomarker concentration. A much more common notation is the dilution step as used on the categorical scaled x-axis in Figure 3B. A limitation of both graphical presentations is that it cannot readily be seen that parallelism between Sample-B and the line of unity is only achieved after a dilution step of 1:32.

The graphical presentation can be improved to better visualise when parallelism is achieved. Figure 4A gives a graphical representation of the same two plots as in Figure 3, adjusted for the dilution steps. For generalisation, the intercept is normalised to one at the baseline in Figure 4B. This graphical presentation is the basis for the development of partial parallelism plots.

The term partial parallelism plot shall be defined as a defined range of biomarker concentrations for which parallelism between sample and standard can be demonstrated. In laboratory practice, parallelism may only be achieved after a certain dilution step because of, for example, a matrix effect (Table 3). Figure 5A illustrates a theoretical situation with a small matrix effect which persists up to a dilution of 1:4 (see vertical reference line). The graphical presentation for a mildly stronger matrix effect persisting up to a dilution of 1:8 is shown in Figure 5B.

Importantly, lack of parallelism can also be present at later stages of the dilution curve (Table 4), for example, because the concentration of the biomarker is below the detection limit of the assay (i.e., d in Formula (3)). Figure 6A shows the graphical presentation for lack of parallelism after a dilution step of 1:128. It would be physically impossible to see the developing lack of parallelism with the curves presented in Figure 3. Finally, Figure 6B illustrates the presence of partial parallelism between a dilution of 1:8 to 1:128. At lower or higher dilution steps there is non-parallelism. Again, this pattern cannot be visually extracted from the graphical presentation in Figure 3.

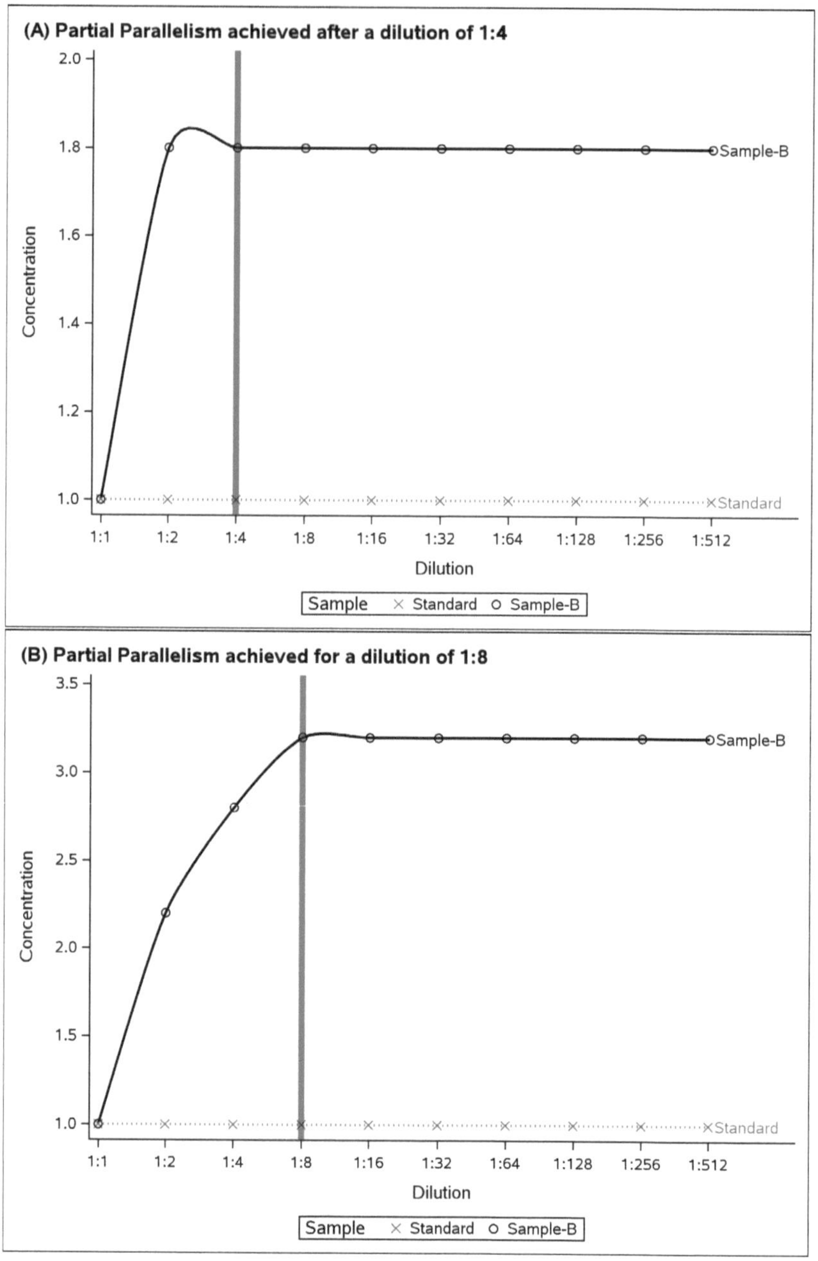

Figure 5. These two examples show that parallelism is achieved after (**A**) a dilution of 1:4 and (**B**) a dilution of 1:8. The data are from Table 3.

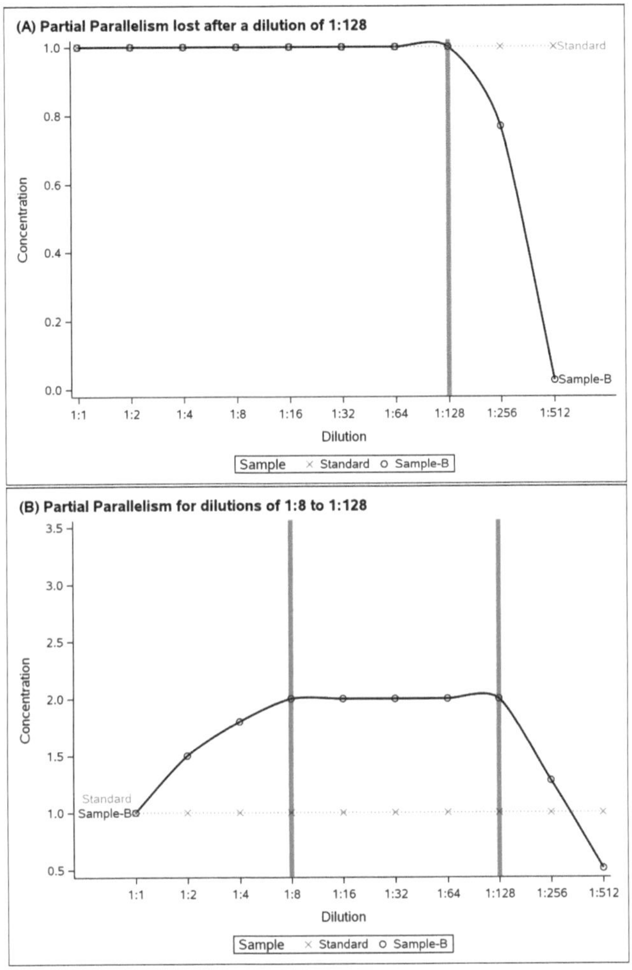

Figure 6. These two examples shows that parallelism is lost after (**A**) a dilution of 1:128. Finally, (**B**) illustrates that parallelism was only achieved between a dilution of 1:8 to 1:128. Data are from Table 4.

Table 3. Raw and transformed data for the partial parallelism plots shown in Figure 5A,B. Subsequent steps of data transformation are indicated by the superscript in the Table (e.g., Std^1, Std^2, etc.). The numbers for $\overline{Std^2}$ and $\overline{B^2}$ were used to draw Figure 5.

Dil.	Std	B	Std^1	B^1	$\overline{Std^2}$	$\overline{B^2}$
(A)						
1:1	10	9	10	9	1	1
1:2	5	8.1	10	16.2	1	1.8
1:4	2.5	4.05	10	16.2	1	1.8
1:8	1.25	2.025	10	16.2	1	1.8
1:16	0.625	1.0125	10	16.2	1	1.8
1:32	0.3125	0.50625	10	16.2	1	1.8
1:64	0.15625	0.253125	10	16.2	1	1.8
1:128	0.078125	0.1265625	10	16.2	1	1.8
1:256	0.0390625	0.06328125	10	16.2	1	1.8
1:512	0.01953125	0.031640625	10	16.2	1	1.8

Table 3. *Cont.*

Dil.	Std	B	Std^1	B^1	$\overline{Std^2}$	$\overline{B^2}$
(B)						
1:1	10	1	10	1	1	1
1:2	5	1.1	10	2.2	1	2.2
1:4	2.5	0.7	10	2.8	1	2.8
1:8	1.25	0.4	10	3.2	1	3.2
1:16	0.625	0.2	10	3.2	1	3.2
1:32	0.3125	0.1	10	3.2	1	3.2
1:64	0.15625	0.05	10	3.2	1	3.2
1:128	0.078125	0.025	10	3.2	1	3.2
1:256	0.0390625	0.0125	10	3.2	1	3.2
1:512	0.01953125	0.00625	10	3.2	1	3.2

Table 4. Raw and transformed data for the partial parallelism plots shown in Figure 6A,B. The numbers for $\overline{Std^2}$ and $\overline{B^2}$ were used to draw Figure 6.

Dil.	Std	B	Std^1	B^1	$\overline{Std^2}$	$\overline{B^2}$
(A)						
1:1	10	2	10	2	1	1
1:2	5	1	10	2	1	1
1:4	2.5	0.5	10	2	1	1
1:8	1.25	0.25	10	2	1	1
1:16	0.625	0.125	10	2	1	1
1:32	0.3125	0.0625	10	2	1	1
1:64	0.15625	0.03125	10	2	1	1
1:128	0.078125	0.015625	10	2	1	1
1:256	0.0390625	0.006	10	1.536	1	0.768
1:512	0.01953125	0.0001	10	0.0512	1	0.0256
(B)						
1:1	10	2	10	2	1	1
1:2	5	1.5	10	3	1	1.5
1:4	2.5	0.9	10	3.6	1	1.8
1:8	1.25	0.5	10	4	1	2
1:16	0.625	0.25	10	4	1	2
1:32	0.3125	0.125	10	4	1	2
1:64	0.15625	0.0625	10	4	1	2
1:128	0.078125	0.03125	10	4	1	2
1:256	0.0390625	0.01	10	2.56	1	1.28
1:512	0.01953125	0.002	10	1.024	1	0.512

6. Examples from the Literature

In a test comparison study [12], parallelism was investigated for allopregnanolone in saliva samples from pregnant women. The sample dilution curves were plotted as in Figure 2A (Figures 1 and 2 in [12]). The conclusion was that the first kit (pg/mL) requires a minimal dilution of 1:5 for an acceptable mean percentage parallelism of 104.3%. The authors accepted that the second kit's test performance met the criteria for parallelism. Using the partial parallelism plot approach, Figure 7A illustrates a lack of parallelism for the first kit. The interpretation of Figure 7A is different to the proposed 1:5 dilution to achieve parallelism [12]. For the second kit, (ng/mL) partial parallelism can be achieved for a dilution range from 1:1 to 1:16, as illustrated by the two vertical reference lines in Figure 7B. At higher dilutions, there is a floor effect of the data suggesting that the test has reached its lower detection limit; hence, the incorrect overestimation of higher concentrations with ever more dilution steps.

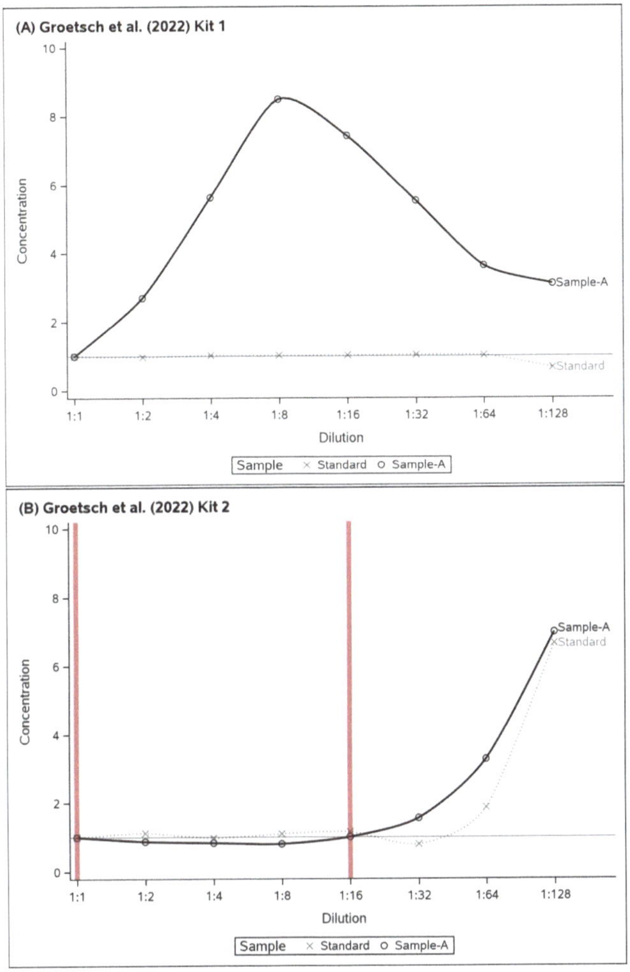

Figure 7. Literature example for allopregnanolone quantified by ELISA from saliva samples analysed [12]. (**A**) Lack of parallelism for Kit 1. The maximum error occurs at a dilution step of 1:8 with an ≈8-fold overestimation of the concentration of allopregnanolone (**B**) Partial parallelism between a dilution of 1:1 to 1:16 (red vertical reference lines). In this example, a horizontal black reference line is given at y = 1 which illustrates that there is also a problem with the standard in (**B**), most likely diluted beyond the analytical detection limit of the assay.

There are situations where it is desirable to quantify the same substance from different types of samples. Consequently, an Enzyme-Linked Immunosorbent Assay (ELISA) was developed for measurement of luteinizing hormone (LH) from whole blood, serum, cell extracts, cell culture medium, and pituitary gland extracts [13]. Averaged LH data on the parallelism experiments for these five different sample sources were provided in Tables 6 to 10. Based on these data, Figure 8 shows good partial parallelism for a dilution range from 1:1 to 1:4. After that, near perfect parallelism appears to be lost. The conclusion could be that the concentrations of LH cannot anymore be calculated reliably for comparison from different sources. But the deviation from one on the *y*-axis are only minimal (≤0.2 units). Therefore, in this example, partial parallelism persists up to a dilution of 1:32. After a dilution step of 1:64 the detection limit of the assay is reached for all sample sources.

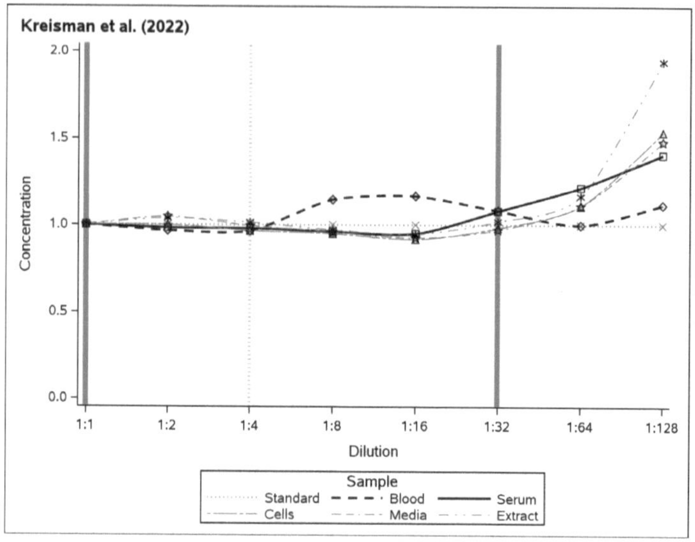

Figure 8. Literature example for luteinizing hormone quantified by ELISA from different sources [13]. At first glance, near perfect parallelism weakens after a dilution step of 1:4 (dotted vertical reference line). The error for partial parallelism is, however, minimal (≈0.2 units). Therefore, partial parallelism can be accepted for a dilution range of 1:1 to 1:32 (closed red vertical reference lines).

Lack of parallelism has also been reported explicitly [14]. These authors clarify in the abstract poor "dilution linearity" attributed to "presence of a matrix effect and/or different immunoreactivity of the antibodies to the recombinant standard and the endogenous analyte". The partial parallelism plot in Figure 9 is based on the raw data provided in Table 2 in [14]. Consistent with the author's conclusion, this graph convincingly shows absence of parallelism.

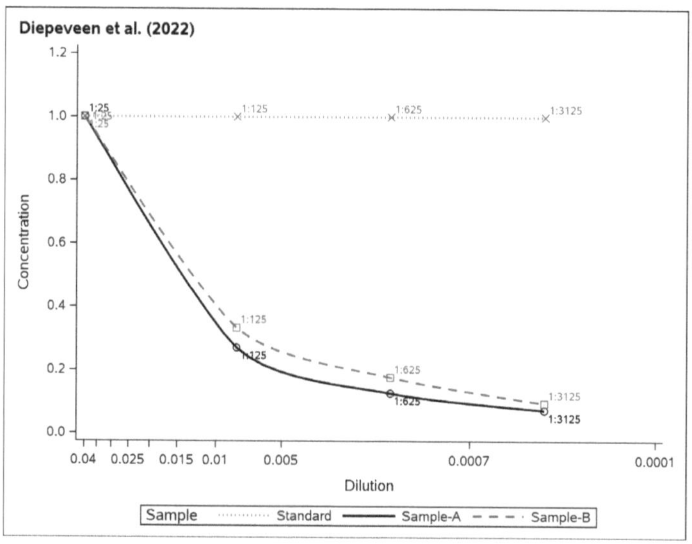

Figure 9. Literature example for reported lack of parallelism for erythroferrone quantified by ELISA from human serum samples [14]. Note that in this example an inverse logarithmic scale was used compared to what was presented in Figure 2B. This choice was based on the uneven dilution steps reported for this experiment. For clarity, each data point was labelled with the corresponding dilution step.

Finally, an example for perfect partial parallelism is shown in Figure 10. This example is based on an ELISA for quantification of human insulin (uU/mL) for a dilution range of 1:1 to 1:8 [15]. The data for the partial parallelism plot were taken from Table 4 in [15], which also details that Samples A to D were based on plasma samples with exogenous insulin added (dashed lines in Figure 10) and high endogenous insulin (dotted lines in Figure 10). Note that the range of the y-axis presented in Figure 10 is very narrow at only 1 uU/mL insulin (range 0.95 to 1.05 uU/mL). Both spiked (samples with exogenous insulin added) and native samples are perfectly parallel to the standard (solid horizontal reference line at y = 1).

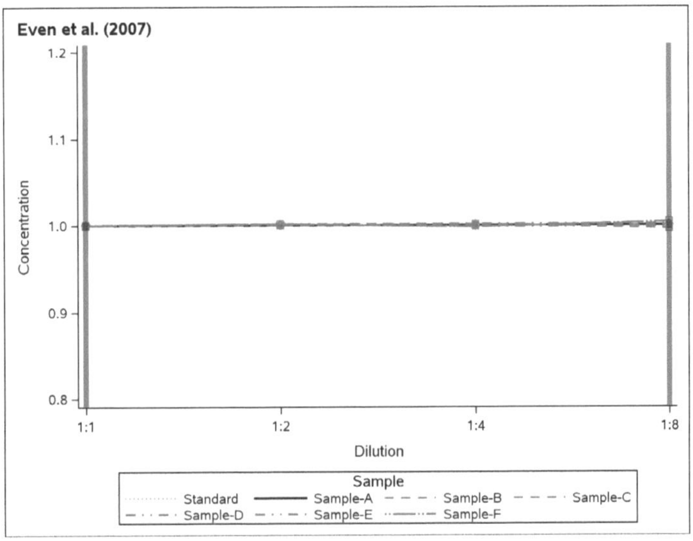

Figure 10. Literature example for perfect parallelism for insulin quantified by ELISA from human plasma samples [15]. The error for partial parallelism is negligible (<0.01 units).

Taken together, partial parallelism plots are a useful graphical method for evaluating the accuracy of calculating the biomarker concentration from a sample based on a biomarker standard curve for a defined range of concentrations. Therefore they can provide a more practical evaluation, which is also easy to understand, and can help identify potential sources of non-parallelism.

7. Relationships between Confounds and Parallelism

The relationship between confounds and parallelism in laboratory tests is an important topic, as confounds can have a significant impact on the accuracy and reliability of laboratory tests. By definition, confounds are variables that can affect the results of laboratory tests but are not directly related to the biomarker being measured. It has been noted that frequent examples of confounds include chemical stability of the biomarker, repeated freeze–thaw cycles, gender, height, weight, renal function, medication use, and co-morbidities such as diabetes mellitus [9].

Protein biomarker studies have shown that the presence of confounds, including sample preparation and storage, can impact the degree of parallelism between laboratory tests as earlier stated [5,10]. The relationship between confounds and parallelism can be expressed mathematically using regression equations as

$$y = \beta_0 + \beta_1 x + \epsilon \quad (10)$$

where y is the signal produced by the laboratory test, x is the concentration of the biomarker being measured, β_0 is the intercept, β_1 is the slope, and ϵ is the error term. Consequently, confounds can be added to equation (10) as additional independent variables:

$$y = \beta_0 + \beta_1 x + \beta_2 c + \epsilon \tag{11}$$

where c represents a confounding variable. The impact of the confound on the parallelism between laboratory tests can be assessed by comparing the slopes of the regression lines with and without the confound. The need for testing this has been highlighted in a recent white paper [3].

It is important for researchers and healthcare providers to be aware of the potential impact of confounds (c) on the accuracy and reliability of the biomarker tests. Laboratories should take steps to minimise the impact of confounds, including controlling for them in statistical analyses or by stratification of the analyses by confounding variables.

Overall, the need for research studies to include testing for confounds for their relationship with the degree of parallelism in biomarker tests has been recognised, but not yet been implemented systematically in the literature.

8. Discussion

The practical advantages of partial parallelism plots for biomarker tests has been illustrated statistically and graphically. Application of partial parallelism plots to real biomarker data has revealed the strength of the approach compared to alternatives which were reviewed and discussed with regard to their historical development. The assessment of partial parallelism is an essential step in the validation of laboratory tests, as it determines whether the assay produces accurate and reliable results. There are two primary methods for statistical and graphical assessment. Each method has its own benefits and limitations.

Statistical assessment, such as regression analysis and the calculation of parallelism and non-parallelism indexes, provides a quantitative measure of the degree of parallelism between two or more samples as discussed [1,2]. These methods are precise and objective, making them ideal for the assessment of large datasets or when a high level of accuracy is required. Additionally, statistical methods can detect subtle differences in parallelism that may be missed by graphical assessment.

However, statistical methods also have limitations. They assume that the assay follows a linear relationship between the concentration of the analyte and the response of the assay. This may not always be the case, as assays may exhibit non-linear responses at high or low concentrations. For example, the FDA and EMA guidelines state that "Parallelism is a performance characteristic that can detect potential matrix effects." [4]. A limitation of this definition is that it does not consider (i) the possibility of compound aggregate release or modifiable epitope masking in immunoassays [5]. Additionally, it was highlighted that statistical methods may not detect non-parallelism due to confounding factors, such as matrix effects or interference by endogenous substances, the CV, and the critical difference [16]. Statistical methods are also not necessarily easily comprehensible to many of the parties involved in appraisal of a biomarker test.

Visual assessment, on the other hand, relies on the interpretation of graphs or charts that depict the relationship between the concentration of the biomarker and the response of the assay. This method is very intuitive and can quickly identify gross deviations from partial parallelism, making it useful for screening biomarker assays for technicians, lay people, and regulatory authorities. Additionally, graphical assessment may detect non-parallelism due to confounding factors that are not detected by statistical methods. This includes, for example, a drop in the analytical sensitivity which affects a biomarker and test standard curve similarly (see Figure 7B). There are many chemical and biological reason to the lack of parallelism [17]. The major contributors to non-parallelism are related to interference or a mismatch with the capture antibody (or surface), the detection antibody, the surrogate reference material, the endogenous analyte, and specific and non-specific interactions. For optimal graphical presentation of the concentration range where parallelism applies in a test, a "raw signal" approach was proposed which includes a four parameter logistic regression curve fitting. The "raw signal" approach is similar to the Figure 3B. Present proposal of partial parallelism plots, as presented in Figures 5–9, should be interpreted as a further simplification of the "raw signal" approach. Individual researchers from all backgrounds and regulatory authorities may also find an advantage in the simplified pattern recognition of the partial parallelism plots. Importantly, both approaches emphasise that parallelism does not need to extend over the entire analytical range of a given test.

However, graphical assessment also has limitations. It is subjective and may vary depending on the experience and expertise of the assessor. Additionally, graphical assessment may not detect subtle deviations from partial parallelism that may affect the accuracy and reliability of the assay. One such factor relates to confounds and was expressed as the CV. In such a situation where the graph may not be clear cut, the visual approach can be improved by showing the parameters obtained from the fits, including the R-square values. Another improvement is the option to zoom into specific regions of the partial parallelism plots. One example was presented in Figure 8. After zooming in, it becomes visibly clearer that the data distribution is more random for the dilution steps 1:2 to 1:16 than for the following dilution steps which clearly demonstrate deviation from parallelism, even if the CV initially improves. Providing this level of interactivity will be a valuable improvement of the method for digital applications making use of proposed partial parallelism plots.

Taken together, both statistical and graphical assessment methods have their own benefits and limitations in the assessment of partial parallelism for biomarker tests. A combination of both methods may provide a comprehensive assessment of the degree of partial parallelism and the presence of non-parallelism due to analytical issues or confounding factors.

9. Conclusions

In conclusion, the introduction of partial parallelism plots as a tool for assessing parallelism in biomarker tests holds great promise. These plots offer a clear visualisation of the relationship between biomarker concentration and assay response for each sample, enabling the identification of non-parallelism arising from analytical challenges or confounding factors. Emphasising the importance of determining the optimal range of dilutions for each sample, these plots provide a language that is easily interpretable, ultimately leading to the attainment of accurate and reliable results. As such, incorporating partial parallelism plots into the validation process of quantitative laboratory tests is an essential step to ensure their appropriateness for clinical medicine, bolstering confidence in their utility.

Supplementary Materials: The following supporting information can be downloaded at https://www.mdpi.com/article/10.3390/app14020602/s1, Supplementary data are provided in form of an Excel sheet with two tabs for raw data and partial parallelism plot calculations. Table S1: Raw Data for Real-Life Examples: The first tab of this Excel sheet provides comprehensive details of the raw data used to generate the Figures based on real-life examples. Each row corresponds to an individual data point, and the first author's name is referenced for the source of the data. The columns are structured as follows: (a) Dilution Steps (String Variable): This column represents the dilution steps used in the experiments. (b) Dilution Steps (Numeric Variable): This column provides the numeric representation of the dilution steps. (c) Sample Description: This column provides information about the samples used in the experiments. (d) Numeric Values for *Y*-axis (Column 1): The first column containing numeric values used for the *Y*-axis in the Figures. (e) Numeric Values for *Y*-axis (Column 2—Optional): An optional second column containing additional numeric values for the *Y*-axis. Following the raw data presentation, each example is followed by a section detailing the conversion of the *Y*-axis values into values suitable for the partial parallelism plots (PPP plot calculations) presented in this paper. Table S2: Compact Format for PPP Plot Calculations: The second tab of the Excel sheet contains the data organised in a concise 5-column format, specifically designed for easy export into a comma-separated file (.csv). This format is suitable for generating graphical representations used in the present article. The columns are arranged as follows: (a) Author's first name and year of publication. (b) Dilution

Steps (String Variable): This column represents the dilution steps used in the experiments. (c) Dilutior Steps (Numeric Variable): This column provides the numeric representation of the dilution steps (d) Sample Description: This column offers a brief description of the samples. (e) Numeric Values for Y-axis (Column 1): The first column containing numeric values used for the Y-axis in the Figures (f) Numeric Values for Y-axis (Column 2—Optional): An optional second column containing additiona numeric values for the Y-axis. The section containing PPP plot calculations serves as a template fo readers to conduct their own calculations. It is crucial to verify the accuracy of the dilution steps and ensure that the reference for the "normalised" fields remains unchanged during the process. By presenting the raw data and providing a user-friendly template for PPP plot calculations, this Exce sheet aims to enhance reproducibility and facilitate further research in the field.

Funding: This research received no external funding.

Institutional Review Board Statement: Not applicable.

Informed Consent Statement: Not applicable.

Data Availability Statement: All data are provided in the Supplementary Materials.

Acknowledgments: Maria Grötsch is gratefully acknowledged for the sharing of raw data. The National Institute for Health Research (NIHR) Biomedical Research Centre based at Moorfields Eye Hospital NHS Foundation Trust and UCL Institute of Ophthalmology are also gratefully acknowledged. The views expressed are those of the author and not necessarily those of the NHS, the NIHR or the Department of Health.

Conflicts of Interest: The authors declare no conflict of interest.

Abbreviations

The following abbreviations are used in this manuscript:

EMA	European Medical Agency
ELISA	Enzyme-Linked Immunosorbent Assay
FDA	Food and Drug Adminsitration
GLM	General Linear Models
LH	luteinizing hormone
OD	Optical Density
SE	Standard Error

References

1. Plikaytis, B.D.; Holder, P.F.; Pais, L.B.; Maslanka, S.E.; Gheesling, L.L.; Carlone, G.M. Determination of parallelism and nonparallelism in bioassay dilution curves. *J. Clin. Microbiol.* **1994**, *32*, 2441–2447. [CrossRef]
2. Pum, J. A practical guide to validation and verification of analytical methods in the clinical laboratory. In *Advances in Clinical Chemistry*; Elsevier: Amsterdam, The Netherlands, 2019; pp. 215–281. [CrossRef]
3. Hersey, S.; Keller, S.; Mathews, J.; King, L.; Bandukwala, A.; Berisha, F.; Birchler, M.; Bower, J.; Clausen, V.; Duarte, J.; et al 2021 White Paper on Recent Issues in Bioanalysis: ISR for Biomarkers, Liquid Biopsies, Spectral Cytometry, Inhalation/Oral & Multispecific Biotherapeutics, Accuracy/LLOQ for Flow Cytometry (Part 2—Recommendations on Biomarkers/CDx Assays Development & Validation, Cytometry Validation & Innovation, Biotherapeutics PK LBA Regulated Bioanalysis, Critical Reagents & Positive Controls Generation). *Bioanalysis* **2022**, *14*, 627–692. [CrossRef] [PubMed]
4. ICH Guideline M10 on Bioanalytical Method Validation and Study Sample Analysis. Available online: https://www.ema europa.eu/en/documents/scientific-guideline/ich-guideline-m10-bioanalytical-method-validation-step-5_en.pdf (accessed on 27 July 2023).
5. Lu, C.H.; Kalmar, B.; Malaspina, A.; Greensmith, L.; Petzold, A. A method to solubilise protein aggregates for immunoassay quantification which overcomes the neurofilament hook effect. *J. Neurosci. Methods* **2011**, *195*, 143–150. [CrossRef] [PubMed]
6. Hardy-Sosa, A.; León-Arcia, K.; Llibre-Guerra, J.J.; Berlanga-Acosta, J.; de la C. Baez, S.; Guillen-Nieto, G.; Valdes-Sosa, P.A. Diagnostic Accuracy of Blood-Based Biomarker Panels: A Systematic Review. *Front. Aging Neurosci.* **2022**, *14*, 683689. [CrossRef] [PubMed]
7. Hauser, S.L.; Bar-Or, A.; Cohen, J.A.; Comi, G.; Correale, J.; Coyle, P.K.; Cross, A.H.; de Seze, J.; Leppert, D.; Montalban, X.; et al. Ofatumumab versus Teriflunomide in Multiple Sclerosis. *N. Engl. J. Med.* **2020**, *383*, 546–557. [CrossRef] [PubMed]
8. Miller, T.M.; Cudkowicz, M.E.; Genge, A.; Shaw, P.J.; Sobue, G.; Bucelli, R.C.; Chiò, A.; Damme, P.V.; Ludolph, A.C.; Glass, J.D.; et al. Trial of Antisense Oligonucleotide Tofersen for *SOD1* ALS. *N. Engl. J. Med.* **2022**, *387*, 1099–1110. [CrossRef] [PubMed]
9. Burtis, C.; Ashwood, E. (Eds.) *Tietz Textbook of Clinical Chemistry*, 2nd ed.; WB Saunders: Philadelphia, PA, USA, 1994.

10. Teunissen, C.E.; Petzold, A.; Bennett, J.L.; Berven, F.S.; Brundin, L.; Comabella, M.; Franciotta, D.; Frederiksen, J.L.; Fleming, J.O.; Furlan, R.; et al. A consensus protocol for the standardization of cerebrospinal fluid collection and biobanking. *Neurology* **2009**, *73*, 1914–1922. [CrossRef] [PubMed]
11. Petzold, A.; Altintas, A.; Andreoni, L.; Bartos, A.; Berthele, A.; Blankenstein, M.A.; Buee, L.; Castellazzi, M.; Cepok, S.; Comabella, M.; et al. Neurofilament ELISA validation. *J. Immunol. Methods* **2010**, *352*, 23–31. [CrossRef]
12. Grötsch, M.K.; Wietor, D.M.; Hettich, T.; Ehlert, U. Validation of a Commercial Enzyme-Linked Immunosorbent Assay for Allopregnanolone in the Saliva of Healthy Pregnant Women. *Biomolecules* **2022**, *12*, 1381. [CrossRef]
13. Kreisman, M.J.; McCosh, R.B.; Breen, K.M. A Modified Ultra-Sensitive ELISA for Measurement of LH in Mice. *Endocrinology* **2022**, *163*, bqac109. [CrossRef]
14. Diepeveen, L.; Roelofs, R.; Grebenchtchikov, N.; van Swelm, R.; Kautz, L.; Swinkels, D. Differentiating iron-loading anemias using a newly developed and analytically validated ELISA for human serum erythroferrone. *PLoS ONE* **2021**, *16*, e0254851. [CrossRef]
15. Even, M.S.; Sandusky, C.B.; Barnard, N.D.; Mistry, J.; Sinha, M.K. Development of a novel ELISA for human insulin using monoclonal antibodies produced in serum-free cell culture medium. *Clin. Biochem.* **2007**, *40*, 98–103. [CrossRef]
16. Jones, G.R.D. Critical difference calculations revised: inclusion of variation in standard deviation with analyte concentration. *Ann. Clin. Biochem.* **2009**, *46*, 517–519. [CrossRef] [PubMed]
17. Tu, J.; Bennett, P. Parallelism experiments to evaluate matrix effects, selectivity and sensitivity in ligand-binding assay method development: pros and cons. *Bioanalysis* **2017**, *9*, 1107–1122. [CrossRef] [PubMed]

applied sciences

MDPI

Article

Convolutional Neural Network-Based Classification of Steady-State Visually Evoked Potentials with Limited Training Data

Marcin Kołodziej *, Andrzej Majkowski, Remigiusz J. Rak and Przemysław Wiszniewski

Faculty of Electrical Engineering, Warsaw University of Technology, Pl. Politechniki 1, 00-661 Warsaw, Poland; andrzej.majkowski@pw.edu.pl (A.M.); remigiusz.rak@pw.edu.pl (R.J.R.); przemyslaw.wiszniewski.dokt@pw.edu.pl (P.W.)

* Correspondence: marcin.kolodziej@pw.edu.pl

Abstract: One approach employed in brain–computer interfaces (BCIs) involves the use of steady-state visual evoked potentials (SSVEPs). This article examines the capability of artificial intelligence, specifically convolutional neural networks (CNNs), to improve SSVEP detection in BCIs. Implementing CNNs for this task does not require specialized knowledge. The subsequent layers of the CNN extract valuable features and perform classification. Nevertheless, a significant number of training examples are typically required, which can pose challenges in the practical application of BCI. This article examines the possibility of using a CNN in combination with data augmentation to address the issue of a limited training dataset. The data augmentation method that we applied is based on the spectral analysis of the electroencephalographic signals (EEG). Initially, we constructed the spectral representation of the EEG signals. Subsequently, we generated new signals by applying random amplitude and phase variations, along with the addition of noise characterized by specific parameters. The method was tested on a set of real EEG signals containing SSVEPs, which were recorded during stimulation by light-emitting diodes (LEDs) at frequencies of 5, 6, 7, and 8 Hz. We compared the classification accuracy and information transfer rate (ITR) across various machine learning approaches using both real training data and data generated with our augmentation method. Our proposed augmentation method combined with a convolutional neural network achieved a high classification accuracy of 0.72. In contrast, the linear discriminant analysis (LDA) method resulted in an accuracy of 0.59, while the canonical correlation analysis (CCA) method yielded 0.57. Additionally, the proposed approach facilitates the training of CNNs to perform more effectively in the presence of various EEG artifacts.

Keywords: BCI; SSVEP; CNN; EEG; data augmentation; transfer-learning

Citation: Kołodziej, M.; Majkowski, A.; Rak, R.J.; Wiszniewski, P. Convolutional Neural Network-Based Classification of Steady-State Visually Evoked Potentials with Limited Training Data. *Appl. Sci.* **2023**, *13*, 13350. https://doi.org/10.3390/app132413350

Academic Editor: Alexander N. Pisarchik

Received: 20 November 2023
Revised: 16 December 2023
Accepted: 17 December 2023
Published: 18 December 2023

1. Introduction

Brain–computer interfaces (BCI) have been continuously developing over twelve years. They enable communication for completely paralyzed people, but at the same time they are increasingly being used by healthy individuals, for example in the entertainment industry [1–5]. BCI employs several EEG potentials. The most common are brain potentials associated with movement (ERD/ERS), P300 potentials, and steady-state visually evoked potentials (SSVEP) [6,7]. SSVEP-based BCIs are relatively common because they are easy to use. They require the user to observe flashing lights at a given frequency. The stimulators can be specially constructed panels with LEDs or LCD screens [8–10]. SSVEPs appear in the back of the head, where the visual cortex is located [11]. Many SSVEP-based interfaces utilize a limited number of electrodes, typically positioned over the visual cortex at the back of the head, with O_1, O_2, and O_z being the most commonly used [12]. We can observe the dominance of brain waves with the same frequencies as stimuli and their harmonics in the visual cortex. Power spectral density analysis (PSDA) methods [13] are the most widely used for feature extraction to distinguish stimulation frequencies. Dedicated methods have also been developed, such as canonical correlation analysis (CCA) [14] or simplified matching pursuit (sMP) [15].

Typically, a calibration session is performed in BCI systems to train the classifier to detect specific patterns. These patterns may differ for each person and each EEG signal registration. For example, each user may have slightly different SSVEPs (amplitudes). This may be due to anatomical and physiological differences (thickness of the skull, properties of head skin, structure of the cerebral cortex). Differences in the registration of SSVEPs may appear even for the same person (different electrode placements, skin contact surface with the electrode, stimulus power). In a calibration session, a user observes the known stimulation frequencies. The recorded EEG signal for a given stimulation frequency allows for the extraction of features to train the system. BCI can also run without a calibration session. In this case, we analyze the stimulation frequencies and their harmonics in the EEG signal. This simplification, however, results in lower efficiency of the system [16,17].

Features for SSVEP-based BCIs may encompass specific frequencies and their harmonics [18]. These features are utilized to train the classification and decision-making systems. For the SSVEP interface, numerous standard machine learning techniques are employed, including k-nearest neighbors (K-NN), linear discriminant analysis (LDA), quadratic discriminant analysis (QDA), support vector machines (SVM), and multilayer perceptron (MLP), among others [19]. Additionally, deep learning techniques are used for this purpose, with convolutional neural networks (CNN), long short-term memory networks (LSTM), and autoencoders (AE) being the most common structures [20]. Deep learning techniques offer several benefits, such as improved classification results and the capability for automatic feature extraction from signals and images, as seen with CNNs [21]. However, the disadvantages of deep learning are notable, including the necessity for a large dataset for training and the extensive time required for network training [22]. Given that deep learning techniques demand substantial training data, the development of effective methods for augmenting EEG data recorded during calibration sessions presents a significant challenge.

In recent years, numerous solutions employing convolutional neural networks (CNNs) for SSVEP detection have been developed. The study referenced in [23], discusses a machine learning approach for detecting SSVEP using a minimal number of channels. In [24], a proposed CNN model is compared with a standard neural network and other leading methods for SSVEP decoding—such as canonical correlation analysis (CCA), a CCA-based classifier, a multivariate synchronization index, and CCA combined with a k-nearest neighbors (K-NN) classifier—in an offline analysis. The research in [25] introduces a fusion algorithm (CCA-CWT-SVM) that integrates CCA, continuous wavelet transform (CWT), and support vector machine (SVM) to enhance classification accuracy for targetless stimuli when a single feature extraction method is used. In [26], a novel deep neural network (DNN) architecture is presented that processes multi-channel SSVEP signals by convolving across sub-bands of harmonics, channels, time, and classifies the signals at the fully connected layer. In [27], a classification method based on a convolutional neural network (CNN) was presented to enhance the detection accuracy of SSVEP amid competing stimuli. The method was evaluated using a seven-class SSVEP dataset from ten healthy participants. The study in [28] demonstrates the use of a compact convolutional neural network (Compact-CNN), which requires only raw EEG signals for automatic feature extraction, in decoding signals from a 12-class SSVEP dataset without user-specific calibration. In [29], a nonlinear model based on a convolutional neural network, named convolutional correlation analysis (Conv-CA), was introduced. Unlike pure deep learning models, Conv-CA combines a CNN with a unique correlation layer, where the CNN transforms multiple EEG channels into a single signal, and the correlation layer computes the correlation coefficients between this transformed signal and the reference signals. In [30], a complex-valued convolutional neural network (CVCNN) is proposed to overcome the limitation of SSVEP-based BCIs, which is the available stimulation frequency. The presented results demonstrate that the proposed method not only overcomes the limitation of the stimulation frequency but also outperforms conventional SSVEP feature extraction methods. Articles [31,32] introduce a convolutional neural network (CNN) specifically designed to learn the relationship between EEG signals and the templates corresponding to each stimulus frequency of SSVEPs. The effectiveness of the proposed method is validated by comparison with the standard canonical correlation analysis (CCA) and other state-of-the-art methods for decoding SSVEPs (i.e., CNN and task-related component analysis TRCA, Vaughan, ON, Canada) using actual SSVEP datasets. The study confirmed the efficiency of the proposed CNN-based network in decoding SSVEPs. A comprehensive list of various algorithms used for SSVEP classification, along with signal recording methods, number of channels, number of users, and classification accuracy, is available in the work cited as [23].

The analysis of the literature indicates that the use of convolutional neural networks allows for satisfactory SSVEP recognition accuracy. However, the practical application of CNNs has been investigated only on a limited basis. This limitation pertains to issues such as the small number of electrodes, extended training times for CNNs, the application of transfer learning techniques, and the effectiveness of CNNs for user-independent classification. A particularly significant challenge in practical CNN application for BCI systems is the limited size of training sets. Typically, the training (calibration) session is brief and includes only a few examples. While such a limited dataset suffices for classical machine learning algorithms, CNNs require many more training examples. Data augmentation (DA) strategies are beneficial in this context. There are numerous data augmentation techniques primarily developed for image processing, which include geometric transformations, flipping, cropping, rotation, photometric and color transformations, and noise injection [33]. However, techniques used for augmenting image data are not directly transferable to EEG data augmentation. Additionally, it is expected that not every data augmentation method will be applicable to all potentials (P300, ERD/ERS, SSVEP).

In recent years, deep-learning techniques have been employed for data augmentation with autoencoders (AE) and generative adversarial networks (GAN) being two common strategies. The impact of noise addition on time series is discussed in [34], where it was concluded that although noise can disrupt the amplitude and phase information, it does not change the spectral feature distribution. In [35], a data augmentation method based on graph empirical mode decomposition was introduced to generate EEG data, merging the benefits of the multiplex network model and the graph version of classical empirical mode decomposition. In [36], the authors explored the constraints of DA for EEG in emotion recognition. Direct geometric transforms and noise addition can impair the time domain features, potentially resulting in a negative DA impact. The issue of limited training data and a proposed solution are discussed in [37], where the authors employed the LST algorithm to transform SSVEP data across different users and devices to compile a larger dataset. In [38], a novel DNN model named FB-EEGNet for SSVEP target detection is introduced. This model integrates features from multiple neural networks to leverage information from various sub-bands and non-target stimulus data. Furthermore, it uses multiple labels for each sample and optimizes the parameters of FB-EEGNet across different stimuli to encompass information from non-target stimuli.

Aim of the Article

The aim of the article is to propose a CNN structure to classify SSVEPs for a significantly limited training dataset. An important element of our research was the development of an augmentation method dedicated to SSVEP detection. The data augmentation method that we applied is based on the spectral analysis of the EEG signal. Then we compared the efficiency of the proposed CNN with the methods commonly used for SSVEP detection, such as: CCA, MLP, sMP, LDA, and QDA. All comparisons were made under the same conditions: window width, number of testing examples, etc. The idea of our research is presented in Figure 1.

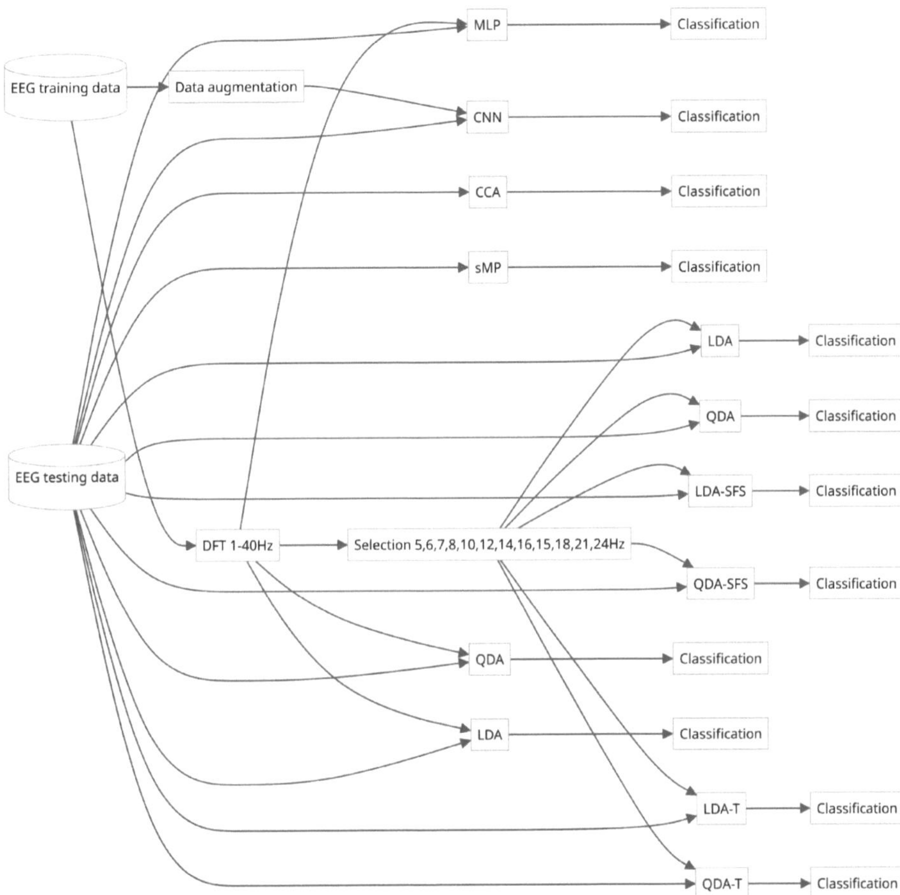

Figure 1. Diagram of the conducted research.

2. Materials

Five users aged 23, 25, 31, 42, and 46 participated in the experiment. The users sat comfortably in a chair. A green LED of 1 cm diameter was placed at a distance of 1 m from a person's eyes. The brightness of the LED light was set based on user evaluation, to be bright enough but not to cause discomfort.

The EEG signals were recorded using a g.USBAmp 2.0 (g.tec Guger Technologies, Graz, Austria) with three active electrodes. Participants were exposed to flickering LED lights at frequencies of 5 Hz, 6 Hz, 7 Hz, and 8 Hz. Research outlined in [39] examined the impact of stimulation frequency and color on the signal-to-noise ratio (SNR) of the recorded SSVEP responses, revealing that frequencies below 10 Hz are adequate for eliciting robust SSVEP responses. Additionally, such stimulation frequencies were found to influence the power of SSVEP responses. We chose frequencies of 5, 6, 7, and 8 Hz to ensure the stability of the generated signals and to distinguish between stimulations with similar frequencies, spaced 1 Hz apart. To generate stable frequencies, a Siglent SDG1062X function generator was utilized. The LED was wired in series with a 220 ohm resistor, and the LED brightness was regulated by altering the voltage at the function generator's output.

The stimulation lasted for 20 s in the training sessions and for 10 s in the testing sessions. To minimize circadian influences on the measurements, all sessions were conducted at the same time each day. For the recordings, three measurement electrodes (O_2, O_z, O_1), a reference electrode (Ref), and a ground electrode (Gnd) to balance the amplifier potential, were employed. The EEG signals were sampled at a frequency of 256 Hz. They were processed using a Butterworth bandpass filter with a range of 0.1–100 Hz and a notch filter set between 48–52 Hz to eliminate power network artifacts. The recorded database has been made available on the Internet.

3. Methods

3.1. Data Augmentation

The method of data augmentation applied by us is based on the spectral analysis of the EEG signal. First, the spectral representation of EEG signals is built on the basis of previously recorded signals for stimulations with frequencies 5, 6, 7, and 8 Hz. The data augmentation procedure for each of the EEG signal channels is independent. The S signal recorded for each electrode is split into 1 s S_m windows. For the data recorded in the experiment, the sampling frequency is fs = 256 Hz. A window width of N = 256 samples was used, and the window was shifted with a small overlap of o = 10 samples. This made it possible to create a large number of M time windows. Then, for each S_m window, a spectral analysis was performed using discrete Fourier transform (DFT) [40]:

$$X_k = \sum_{n=0}^{N-1} (S_m)_n e^{-\frac{i2\pi}{N}kn} \tag{1}$$

The spectral analysis enables the determination of the amplitudes of individual frequencies, which range from 0 to $fs/2$ Hz. The number of samples, N = 256, allowed for the acquisition of a frequency resolution of the signal equal to 1 Hz. As a result of the DFT analysis, we obtained the sets $P_k = \{X_{k1}, X_{k2}, X_{k3}, \ldots, X_{kM}\}$ representing the amplitude values for the frequencies $k = 0 \ldots fs/2$ Hz. The P_k sets were used to generate new EEG signals. The augmentation algorithm enables the creation of a new artificial EEG signal with any number of L samples. The algorithm to create an artificial EEG signal is as follows:

1. Create a new zero-time vector S_a of length L. This vector corresponds to the newly generated EEG signal for time samples from 0 to $L \times T - T$ (with step T).
2. In a loop, for each value of frequency k = 0 to $fs/2$, perform the following:
    a. Choose an A_r value randomly from the range $\langle -0.82; 0.82 \rangle$,
    b. Choose a φ_r value randomly from the range $\langle -2\pi; 2\pi \rangle$
    c. Choose a P_{kr} element randomly from the P_k set
    d. Update vector S_a according to the formula:

$$S_a = S_a + (P_{kr} + A_r)\sin(2\pi kt + \varphi_r),$$

where t is a vector of time samples

3. Add a vector R of length L to the vector Sa containing values chosen randomly from the range $-\varepsilon$ to ε, where $\varepsilon = \langle -1.84 \times 10^{-8}; 1.84 \times 10^{-8} \rangle$

The result is a vector S_a corresponding to the newly generated EEG signal. Particular attention should be paid to the ranges from which the values A_r, φ_r, and ε are to be chosen. The typical values of the parameters were selected based on observations and were $A_r = \langle -0.82; 0.82 \rangle$, $\varphi_r = \langle -2\pi; 2\pi \rangle$, and $\varepsilon = \langle -1.84 \times 10^{-8}; 1.84 \times 10^{-8} \rangle$, respectively.

To obtain the augmented signal, the first 20 s of the recorded real EEG signals were used. As a result of data augmentation, we obtained 90,000 examples per class for each user (S01–S05), totaling 360,000 examples. Out of these, 10% were designated as validation data. Consequently, the CNN training set comprised 324,000 examples, while the validation set included 36,000 examples. Only the generated data were utilized for training the CNN. However, to evaluate the network's performance, the last 10 s of the real recorded EEG signals were used. The method itself does not limit the number of examples that can be generated. From several hundred real EEG examples, it is possible to generate several thousand artificial examples. The morphology of the generated EEG signals is distinct, exhibiting completely new signal characteristics in the time domain. Nevertheless, the generated EEG signal maintains the same statistical parameters as the real one. Moreover, the spectrum of the generated signal closely resembles that of the real one. An illustration of one second of the real EEG signal (in blue) and the generated signal (in red) is presented in Figure 2. Figure 3 displays a histogram comparing samples of one second from the real EEG signal (in blue) and the generated signal (in red), highlighting their strong similarity. The spectra of the real EEG signal (in blue) and the generated one (in red) are depicted in Figure 4.

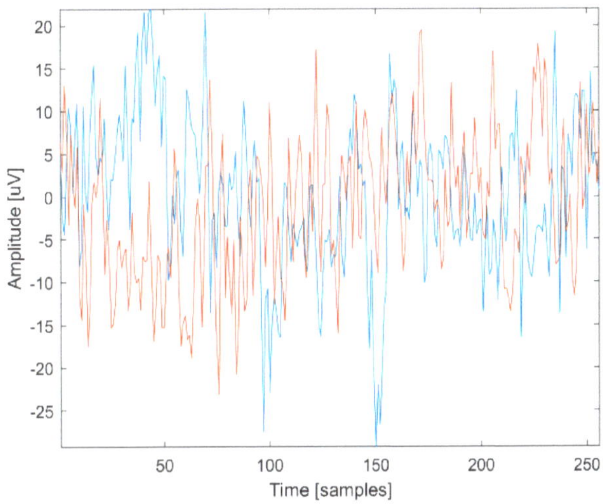

Figure 2. Example of one second of the real EEG signal (blue) and the generated signal (red).

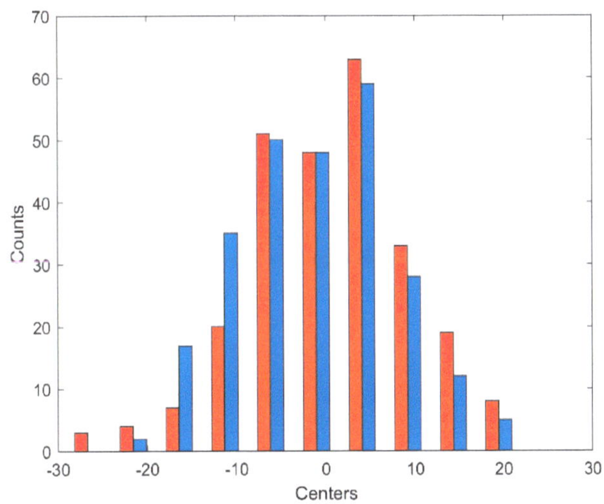

Figure 3. Histogram of samples of one second of the real EEG signal (blue) and the generated signal (red).

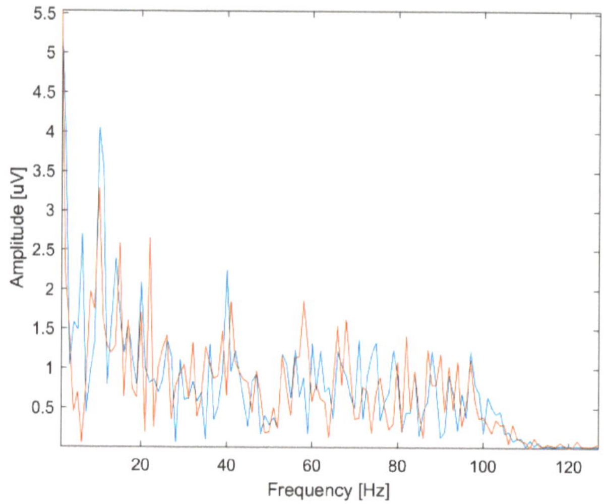

Figure 4. Spectrum of the real EEG signal (blue) and the generated signal (red).

3.2. Convolutional Neural Network

The operation of CNNs is based on convolutional filters. When a signal passes through these filters, it is transformed into a vast array of features that are then classified by a fully connected layer. In the search for the optimal CNN architecture, the impact of varying the number of convolutional layers (ranging from 2 to 5) was examined. Additionally, the effect of the number of filters with values of 2, 4, 8, 16, 32, 64, 128, and 256 was evaluated. Subsequently, the influence of different filter sizes—2, 4, 8, 16, 32, and 64—was investigated. The selection of the network structure was derived from an automated search for the optimal combinations of layer count, filter count, and filter size. During our research, we did not consider the impact of the dropout layer on the CNN training process. During the selection of the best parameters, different optimizer algorithms (ADAM, SGD) and a range of values for InitialLearnRate (0.0001, 0.001, 0.01) and L2Regularization (0.01, 0.001 0.0001) were evaluated. The search for the optimal combination of network structure and learning parameters spanned several days. The best network structure and parameters were determined based on the classification accuracy obtained for the validation data. The accuracies for the validation set for the considered structures ranged on average for all users from 0.65 to 0.72. The best results were achieved for the CNN network structure, which consisted of four convolutional layers, applying a ReLU activation function after each. The final convolutional layer, along with the subsequent ReLU layer, consists of 128 filters, resulting in a considerable number of features fed into the SoftMax classifier. The ADAM optimizer [37] was employed to train the CNN network, with an InitialLearnRate set at 0.001. Training was conducted over a maximum of 50 epochs, with a MiniBatchSize of 128 and an L2Regularization factor of 0.0001. The architecture of the CNN used in this study is detailed in Table 1. During the training of the network, the learning curve and error for the validation data were observed. No signs of overfitting in the CNN were noticed.

To train a CNN, a large number of training examples are needed. During training, we utilized a dataset obtained through the proposed augmentation method. However, for testing the performance of the CNN, we employed EEG signals recorded during the test session. The schematic for CNN application is presented in Figure 5. In the Supplementary Materials, there is the source code of our developed method for EEG data augmentation, as well as the code for the implementation of a CNN network that enables the classification of SSVEP.

Table 1. CNN structure.

No.	Name of Layer	Parameters
1	Input Layer	256 × 3 × 1 signals with zero-center normalization
2	Convolution_1	32 filters of size 8 × 3 with stride [1 1] and padding 'same'
3	Batch Normalization_1	Batch normalization with 32 channels
4	ReLU_1	ReLU
5	Convolution_2	64 filters of size 16 × 3 with stride [1 1] and padding
6	Batch Normalization_2	Batch normalization with 64 channels
7	ReLU_2	ReLU
8	Convolution_3	128 filters of size 32 × 1 with stride [1 1] and padding
9	Batch Normalization_3	Batch normalization with 128 channels
10	ReLU_3	ReLU
11	Convolution_4	128 filters of size 64 × 1 with stride [1 1] and padding
12	Batch Normalization_4	Batch normalization with 128 channels
13	ReLU_4	ReLU
14	Fully Connected	4 fully connected layer
15	Softmax	Softmax
16	Classification Output	Crossentropyex

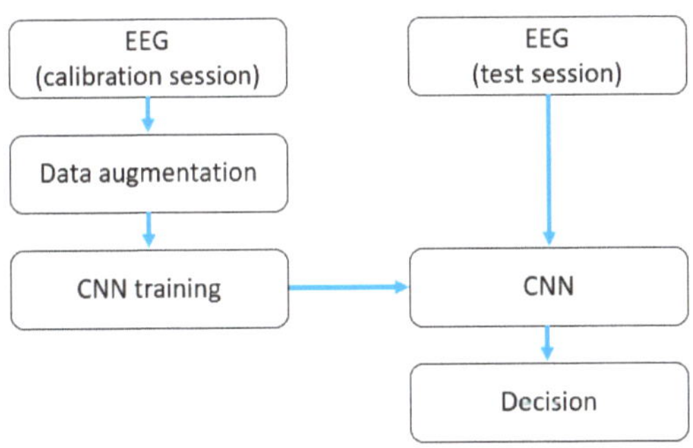

Figure 5. Schematic illustration of using a CNN to classify SSVEPs.

3.3. Classical SSVEP Detection Methods

The proposed CNN algorithm has been compared with a number of classical methods traditionally used for SSVEP detection. The concepts of utilizing classical and dedicated algorithms for SSVEP detection are illustrated in Figures 6 and 7. Figure 6 demonstrates the application of typical machine learning methods, employing classifiers such as LDA, QDA, SVM, or MLP. The initial step involves training the classifier with data from a calibration session. Only after this step can the test data be classified. The classification process begins with the extraction of features from the EEG signal, followed by the selection of the most effective features. Figure 7 delineates the application of typical dedicated methods (such as CCA and sMP) for analyzing SSVEP. These methods do not necessitate a training session, but they do require knowledge of the frequencies of the stimuli. The aim is to find base signals that most closely correspond to stimuli at frequencies of 5 Hz, 6 Hz, 7 Hz, and 8 Hz. Canonical correlation analysis (CCA) seeks a linear combination between EEG signals and sinusoidal signals at the stimulation frequency and its harmonics. The frequency sought is the one for which the maximum correlation between EEG signals and sinusoidal signals either at the stimulation frequency or its harmonics, is observed to be the largest [41]. Another method tailored for SSVEP detection is the sMP algorithm, which is derived from the well-known matching pursuit (MP) algorithm. However, the set of base functions in sMP is drastically narrowed down to sinusoidal signals at frequencies specifically chosen for visual stimulation.

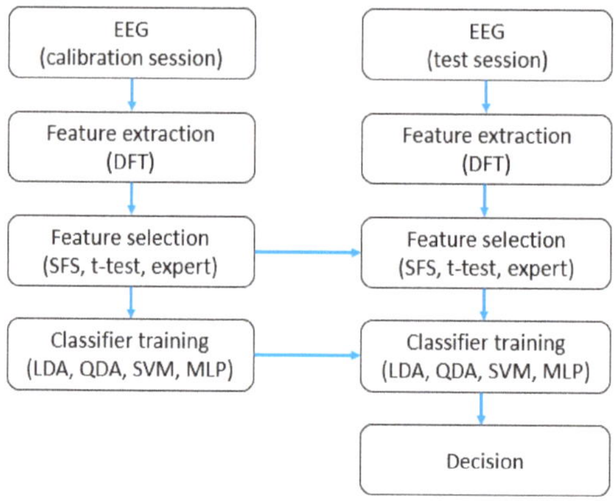

Figure 6. Schematic illustration of using classical methods for SSVEP detection.

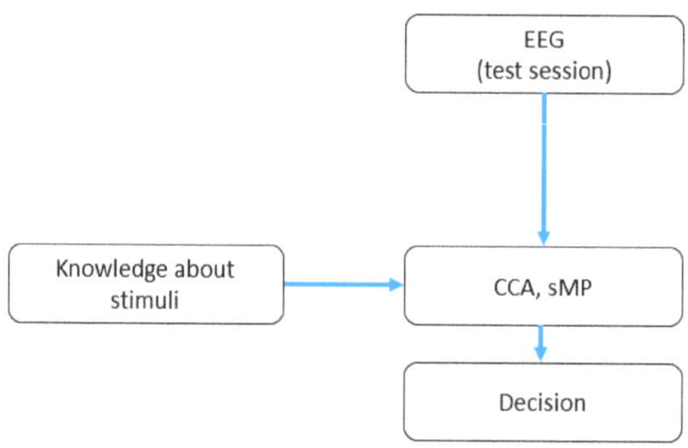

Figure 7. Schematic illustration of using dedicated methods (CCA, sMP) for SSVEP detection.

Traditional machine learning methods require a feature extraction stage. Frequency analysis is often used to extract features from the EEG signal to detect SSVEP [42]. This is because, during user stimulation with frequency k, we expect an increased amplitude of the EEG signal in the visual cortex for the stimulation frequency k and its harmonics $2k$, $3k$. In our experiments, feature extraction was performed using DFT analysis. The spectrum was calculated from each second of the EEG signal, with a frequency resolution of 1 Hz. Such resolution should be sufficient to distinguish between the SSVEPs at stimulating frequencies of 5, 6, 7, and 8 Hz.

Feature vectors were constructed for each channel, corresponding to one of the two cases:

1. All frequencies between 1 and 40 Hz were extracted.
2. Only the frequencies of possible stimulations and their second and third harmonics were extracted. For the frequencies of 5, 6, 7, 8 Hz, these were, respectively: 5, 6, 7, 8, 10, 12, 14, 16, 15, 18, 21, and 24 Hz.

Feature extraction was performed for each channel separately. The use of three EEG signal channels triples the number of features. For case 1, there are 120 features, and for case 2, there are 48 features in total.

To ensure that the classifier is correctly trained using standard machine learning techniques, only the most useful features should be utilized, necessitating a feature selection stage. Various methods are employed to select the best features, with filter and wrapper approaches being the most common [43]. A typical filter method that is widely used is the *t*-test, which assumes a normal distribution of features. We implemented the absolute value two-sample *t*-test with a pooled variance estimate [44]. However, the *t*-test selection is typically designed for two groups, and in our case, there are four classes (5, 6, 7, and 8 Hz). Consequently, we adopted a strategy of selecting the best features for one class in contrast to all other combined classes. This approach allowed us to select the most distinctive features for the groups: 5 Hz versus (6, 7, 8 Hz), 6 Hz versus (5, 7, 8 Hz), 7 Hz versus (5, 6, 8 Hz), and 8 Hz versus (5, 6, 7 Hz). Subsequently, a subset of 14 features was chosen, which yielded the best classification performance for the training set. The number 14 was determined experimentally.

Unfortunately, feature selection methods do not always yield the best results because they do not consider the interdependencies between features [45]. A method that accounts for these types of dependencies is sequential forward selection (SFS) [46]. This method operates by selecting an initial feature, assessing the classification accuracy, and then incrementally adding the feature that most improves classification. For feature selection using SFS, the LDA and QDA classifiers were employed [47]. During the experiments, it was observed that the optimal number of features for achieving the highest classification accuracy was 25 for both LDA and QDA methods.

All experiments were conducted using MATLAB R2021a software on a computer equipped with an Intel Core i7-9800X processor, 128 GB of RAM, and an NVIDIA GeForce RTX 2080 Ti graphics card. The time required to execute various algorithms on the applied dataset, considering the established training parameters, varied significantly. Table 2 illustrates the time required to create the augmentation set, train the different classification methods, conduct feature selection, and train the CNN. However, it is important to note that both the CNN and MLP algorithms used a GPU for their calculations.

Table 2. Execution times of the individual algorithms.

Algorithm	Execution Time
Data set augmentation	12.115 s
CCA algorithm for classification (does not require training)	0.112 s
sMP algorithm for classification (does not require training)	0.311 s
Training the CNN for 50 epochs	145 min 8 s
Training the MLP	11.1 s
Training the LDA	14.3 s
Training the QDA	17.2 s
Training LDA with SFS/*t*-test feature selection	31.4 s/19.2 s
Training QDA with SFS/*t*-test feature selection	41.8 s/22.5 s

4. Results

Classification accuracy was used to evaluate the performance of individual classification methods. This measure is commonly used to assess classifiers and the effectiveness of BCI systems. The classification accuracy for each classifier was determined based on the last 10 s of real recorded EEG signals. During testing, 1 s windows overlapping by 0.5 s were employed, resulting in 72 windows for four stimulation frequency classes: 5, 6, 7, and 8 Hz. Table 3 shows the accuracy and macro average F1-score results obtained for the individual classification methods on the test set. Macro average F1-score provides a balanced assessment of precision and sensitivity. In addition to the methods' symbolic names (CNN, CCA, sMP, MLP, QDA, LDA, QDA-SFS, LDA-SFS, QDA-T, LDA-T), details about the data used at the classifiers' input (EEG raw, DFT 1–40 Hz, DFT specific frequencies) and selection methods (SFS with 25 features, *t*-test with 14 features) are included. The table also indicates whether data augmentation was used for training (Y) or if it was the first 20 s of the recorded EEG signal (N).

Table 3. Comparison of the classification accuracies for the tested methods.

Method	CNN	CCA	sMP	MLP	QDA	LDA	QDA	LDA	QDA-SFS	LDA-SFS	QDA-T	LDA-T
Input	EEG raw			DFT 1–40 Hz	DFT 1–40 Hz	DFT 1–40 Hz	DFT 5 Hz, 6 Hz, 7 Hz, 8 Hz, 10 Hz, 12 Hz, 14 Hz, 16 Hz, 15 Hz, 18 Hz, 21 Hz, 24 Hz					
Training the classifier on the generated data	Y	N	N	Y	Y	Y	N	N	N	N	N	N
Feature selection	-	-	-	-	-	-	-	-	SFS 25 features	SFS 25 features	*t*-test 14 features	*t*-test 14 features
						Accuracy						
User S01	0.81	0.75	0.58	0.62	0.62	0.75	0.65	0.66	0.62	0.63	0.68	0.76
User S02	0.88	0.54	0.51	0.61	0.61	0.59	0.56	0.40	0.48	0.47	0.61	0.50
User S03	0.42	0.40	0.29	0.30	0.27	0.31	0.23	0.33	0.26	0.33	0.18	0.22
User S04	0.75	0.54	0.54	0.70	0.65	0.68	0.58	0.65	0.56	0.65	0.59	0.56
User S05	0.75	0.63	0.61	0.63	0.63	0.65	0.62	0.58	0.61	0.59	0.66	0.62
Mean value	0.72	0.57	0.51	0.57	0.55	0.59	0.53	0.52	0.51	0.53	0.54	0.53
						F1-score						
User S01	0.79	0.59	0.46	0.51	0.48	0.59	0.56	0.53	0.51	0.51	0.55	0.60
User S02	0.87	0.41	0.33	0.49	0.50	0.48	0.45	0.28	0.31	0.33	0.48	0.33
User S03	0.29	0.30	0.19	0.18	0.18	0.23	0.15	0.24	0.12	0.25	0.11	0.14
User S04	0.60	0.40	0.42	0.56	0.54	0.55	0.46	0.55	0.43	0.54	0.47	0.45
User S05	0.60	0.51	0.50	0.50	0.52	0.51	0.50	0.46	0.50	0.48	0.53	0.49
Mean value	0.63	0.44	0.38	0.44	0.44	0.47	0.42	0.41	0.37	0.42	0.42	0.40

The highest mean classification accuracy was achieved with the CNN at 0.72. A lower average accuracy of 0.57 was observed for both CCA and MLP. The sMP method yielded slightly inferior results, with an average classification accuracy of 0.51. Standard machine learning methods that employ spectral features and feature selection achieved classification accuracies ranging from 0.51 to 0.54. It is important to note that the classification pertained to 1 s windows across four classes. The random operation of a four-class classifier would result in a classification accuracy of 0.25. Therefore, it can be concluded that the methods under consideration deliver satisfactory results that are practically applicable. Attention should also be given to the variations in classifier accuracies among individual users. These differences can be attributed to the psychophysical characteristics of the person being tested and are a normal phenomenon. Additionally, some individuals are more naturally inclined to generate SSVEP responses to visual stimuli. To determine if the comparison of classification algorithms across five subjects (S01–S05) is reliable, statistical tests were conducted. Given the small sample size and the uncertain distribution of results, the non-parametric Wilcoxon–Mann–Whitney test was utilized [48]. *p*-values were calculated from a two-sided Wilcoxon signed-rank test. The classification accuracy results for the CNN method compared with other methods used by us (QDA, LDA, QDA-SFS, LDA-SFS, QDA-T, LDA-T, CCA and sMP) were found to be statistically significant at

$p = 0.0625$. The improved classification accuracy of the CNN network may be due to its ability to automatically generate features. In contrast, other algorithms—whether specialized for SSVEP interfaces (like CCA and sMP) or standard machine learning methods (such as QDA, LDA, QDA-SFS, LDA-SFS, QDA-T, LDA-T)—relied on features derived from frequency analysis.

Table 3 presents the calculated F1-scores for various SSVEP potential classification methods. Among these, the CNN method attains the highest average F1-score of 0.63. Other methods, including CCA, MLP, QDA, and LDA, exhibit comparable results, with their average F1-scores ranging approximately from 0.44 to 0.47. This range indicates a moderate level of effectiveness for these techniques in SSVEPs classification. The sMP method recorded the lowest F1-score at 0.38, suggesting its comparatively limited utility. Meanwhile, the QDA and LDA methods, after incorporating SFS feature selection and the *t*-test, achieved F1-scores between 0.37 and 0.42. Overall, these findings imply that the CNN method is the most effective for SSVEP classification, whereas the other techniques demonstrate similar yet generally lower levels of effectiveness.

Future research should consider expanding the training dataset with EEG recordings from a greater number of individuals and employing different methods of stimulation, as well as various EEG signal acquisition systems.

5. Discussion

The results obtained can be converted into the information transfer rate (ITR), which are commonly used to compare brain–computer interface (BCI) systems. Table 4 compares the ITR results for individual users using both the CCA and CNN methods. The calculations reveal significant variations in the practical usability of the BCI interface among different individuals. It is important to note that the ITR was calculated based on the classification of one second continuous EEG signal segments. The decision-making time and classification accuracy substantially influence the information transfer rate. In practice, the actual ITR would be lower than the estimated values. Nonetheless, we can approximate the disparity in ITR by comparing the CNN and CCA methods for EEG signal classification. The largest difference, favoring the CNN method, is observed for user S02, at approximately 60.3 bits per minute, and the smallest for user S03, at 1.3 bits per minute.

Table 4. ITR comparison for classifiers [bit/min].

Subject	CNN	CCA
S01	59.8	47.5
S02	76.8	16.5
S03	5.95	4.6
S04	47.5	16.5
S05	47.5	27.7
Mean	42.0	19.9

It is important to consider that the analyses were conducted on SSVEP signals recorded under specific conditions and with individuals who had no previous experience with SSVEP interfaces, utilizing only three EEG signal electrodes. Various types of amplifiers, stimulation methods (such as stimulus brightness and LED size), and numbers of stimuli have been employed for recording SSVEP signals in the literature, complicating the comparison of classification results and ITR values across studies. In publication [38], EEG signals recorded using 8 channels and 12 stimulations were utilized, and the FB-EEGNet algorithm applied for classification yielded an ITR of 70.45 bits/min. In publication [49], a method based on task-related component analysis (TRCA) and an extended method based on canonical correlation analysis (CCA) for a 40-class SSVEP were implemented, with the online BCI speller achieving an average ITR of 325.33 $\pm$ 38.17 bits/min. Lastly, in publication [50], EEG data were recorded from 32 active electrodes, and by employing a spatially-coded BCI, the classification method reached an ITR of 31 $\pm$ 17 bits/min in novice users completing the task for the first time.

CNN delivers significantly better results for classification accuracy compared to other methods. During CNN training, filter weights are optimized to select useful features. The number of features processed through the fully connected layer is considerable: 128 filters $\times$ 3 EEG channels $\times$ 256 features per filter. This exceeds the number of features derived from selecting the 1–40 Hz frequency band, which is common in other methods. However interpreting the function of these filters can be challenging. We can visualize the effects of these filters on the signals. Figure 8 displays a one second segment of the EEG signal from the O1 channel during a 5 Hz stimulus. Figure 9 illustrates the same signal after processing through a chosen filter from the fourth convolutional layer. Additionally, the spectra of these signals are shown, allowing for the analysis of the filter's effect. In Figure 8, the original input signal to the filter has a broad frequency spectrum, but frequencies at 5 Hz, 10 Hz, and 15 Hz are not readily distinguishable. In contrast, Figure 9 reveals that the output signal from the filter predominantly features frequencies around 5, 10, and 15 Hz, which correspond to the stimulation frequency and its harmonics. Therefore, the signal post-filtering contains frequencies potentially beneficial for the classification of SSVEPs.

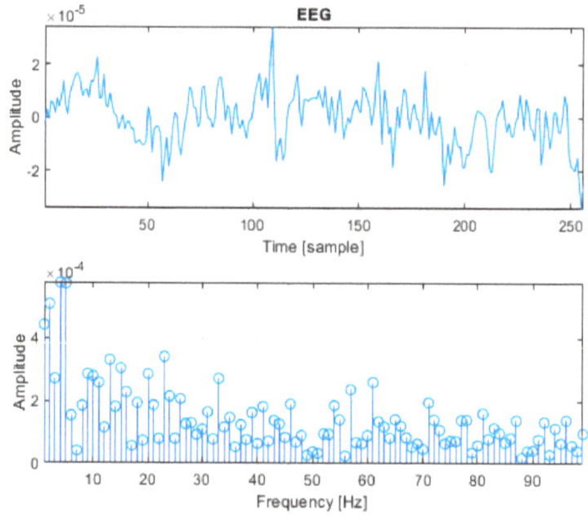

Figure 8. One second fragment of the EEG signal fed to the network input and the spectrum of this signal.

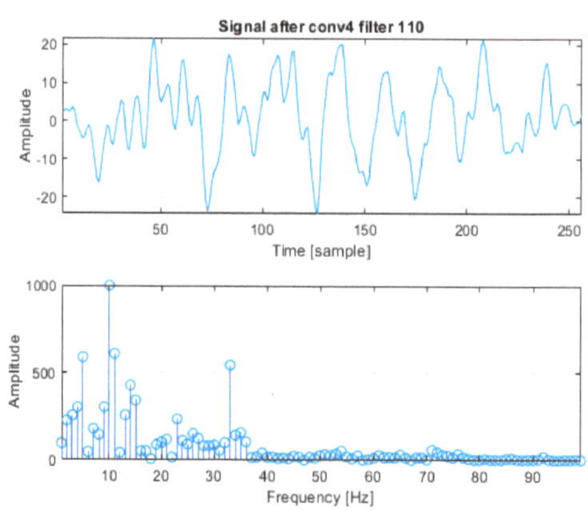

Figure 9. One second fragment of the EEG signal after applying the exemplary convolutional filter (no 110) in the 4th layer and spectrum of this signal.

Several studies on CNNs indicate that the network is more robust to artifacts [51,52]. To determine whether the CNN approach is more effective in classifying SSVEPs with artifacts, we introduced Gaussian noise into the test signal. Gaussian noise closely approximates EMG artifacts resulting from muscular activities like jaw clenching, tongue movement, and swallowing [53]. We then attempted to classify sections of the noisy EEG signals for stimuli at 5, 6, 7, and 8 Hz. The classification accuracies for the CNN and CCA methods are listed in Table 5. Case I presents the classification accuracies (0.81 for CNN and 0.75 for CCA) obtained with the originally recorded EEG signal, which had a standard deviation of 0.87×10^{-5}. In Case II, Gaussian noise was added to the EEG signal with a standard deviation of 5.99×10^{-6}, leading to a decrease in classification accuracy (0.69 for CNN and 0.54 for CCA). For Case III, the noise standard deviation was significantly increased to 1.60×10^{-5}, which resulted in a further reduction in classification accuracy to 0.59 for CNN and 0.45 for CCA.

Table 5. Comparison of classification accuracy for a noisy signal.

	EEG Signal (Std)	Noise (Std)	CNN	CCA
I	EEG (0.87×10^{-5})	-	0.81	0.75
II	EEG (0.87×10^{-5})	5.99×10^{-6}	0.69	0.54
III	EEG (0.87×10^{-5})	1.60×10^{-5}	0.59	0.45

The augmentation of EEG data using the proposed method proved to be effective for SSVEP. This technique enables the creation of any number of training examples. However, the data augmentation method does not account for inter-channel relationships. If there are significant dependencies between channels O_1, O_2, and O_z—related to phase, frequencies, or amplitudes, for instance—the method may not generate accurate data for network training. Therefore, caution is advised when applying this technique to other potentials used in BCI, such as P300 or ERD/ERS.

The results we obtained align with those of other researchers who have applied CNN and deep learning to classification tasks in BCI systems. The experiment detailed in [54] involved nine flicker stimuli of different frequencies, and a CNN-based multitarget rapid classification method was constructed for nine classification tasks. The average accuracy of AR-BCI using the CNN model at a 1 s stimulus duration was about 81.83%. In [55], to enhance the classification accuracy of SSVEP signals during movement, SSVEP data were collected from five targets moving at speeds of 0 km/h, 2.5 km/h, and 5 km/h. A convolutional neural network (CNN) was developed to discern the relationship between the EEG signal and the pattern corresponding to each stimulus frequency. The proposed method outperformed traditional methods (i.e., CCA, FBCCA, and SVM) at all speeds, with CNN accuracies of 86.08%, 71.53%, and 60.63% from the lowest to highest walking speeds, respectively. In [26], the use of 64 channels yielded excellent results; however, when reduced to three channels, the classification accuracy was approximately 51% and 42% for sets of EEG signals. In [56], a BCI was utilized in an online experiment to spell the word 'SPELLER' using a 2 s time window. The system attained an average accuracy of 97.4% and an information transfer rate of 49 bpm, demonstrating the practicality and feasibility of implementing a reliable single-channel SSVEP-based speller using a 1D CNN. The study in [57] introduced a filter bank convolutional neural network (FBCNN) approach to optimize SSVEP classification. Three filters, each covering a harmonic of the SSVEP signals, were used to extract and differentiate the relevant components, with their information transformed into the frequency domain. Experimental results indicated that FBCNN enhances the performance of CNN-based SSVEP classification methods and holds significant potential for SSVEP-based BCIs. FBCNN results were approximately 2% better than those of traditional CNNs, though a wide dispersion of results was observed for both methods, varying by individual.

When attempting to implement CNNs in practical applications, certain challenges may arise. In our study, the classification time for 1 s of EEG signal was a rapid 3.7 ms. However, the training time required for the CNN poses a challenge. Here, transfer learning techniques could be vitally important. Utilizing transfer learning may necessitate adjustments to the signal sampling frequency and the number of network inputs, which must align with the number of recorded EEG channels. Additionally, it is crucial to retrain the network using a relatively large dataset.

We implemented the CNN proposed in article [26] to explore the potential of using transfer learning. The proposed network yields impressive results, achieving close to 98% accuracy for 1 s segments of the signal across all 64 channels. Its architecture reflects an understanding of EEG signal processing and analysis methods. The network was originally trained on data from 70 healthy individuals and 40 target characters, which flickered at frequencies ranging from 8 to 15.8 Hz in 0.2 Hz increments. This training used EEG data recorded at 250 Hz. We adapted this network structure for the data recorded from users S01–S05. The adaptation involved modifying the first and last layers of the CNN to accommodate three input channels (O_1, O_2, O_z) and four SSVEP frequencies (5 Hz, 6 Hz, 7 Hz, 8 Hz). We then retrained the network with the EEG training data, using the initial 20 s of the actual recorded EEG signal for S01–S05 users, after resampling the signals from 256 Hz to 250 Hz. Subsequently, we calculated the classification accuracy for SSVEP recognition on the training data (last 10 s) for each user. The classification results obtained for the adapted CNN [26] using transfer learning techniques are summarized in Table 6. The table also includes comparative results from the CNN network that we developed as well as the CCA method.

Table 6. Comparison of classification accuracy for CNN [26].

Subject	CNN [26]	Our CNN	CCA
S01	0.85	0.81	0.75
S02	0.55	0.88	0.54
S03	0.30	0.42	0.40
S04	0.52	0.75	0.54
S05	0.82	0.75	0.63
Mean	0.61	0.72	0.57

The average recognition accuracy for the CNN [26] method is 61%, for the CCA method it is 57%, and for the CNN that we proposed, which includes data augmentation, it is 72%. These results suggest that the application of transfer learning techniques yields better outcomes than the use of standard machine learning methods like CCA. Nonetheless, our specialized approach achieved an 11% higher accuracy.

6. Conclusions

The results presented demonstrate that the use of CNN can significantly enhance the efficiency of SSVEP-based BCIs. Compared to traditional machine learning methods, CNN can provide up to 20% better results. This improvement leads to a substantially higher ITR and more effective BCI system operations. A CNN classifier trained for this purpose is more resistant to artifacts in the EEG signal than other SSVEP detection methods. The data augmentation method proposed for calibration sessions enables effective CNN training. Unfortunately, the use of CNN is not without practical limitations. One drawback is the extensive training time required, which may span several hours. Additionally, high classification accuracy is typically achieved only when the data from a specific individual's calibration session are used for training. Furthermore, the same network structure cannot be directly applied to different databases. The CNN structure must be modified for signals recorded with varying equipment and different sampling frequencies.

Supplementary Materials: The following supporting information can be downloaded at: https://github.com/kolodzima/CNN_limited_SSVEP_dataset (accessed on 16 Decmeber 2023). The source code, which presents the proposed augmentation method, along with the structure of the CNN and methods for training and testing, has been placed at the link.

Author Contributions: Conceptualization, M.K., A.M., R.J.R. and P.W.; methodology, M.K., R.J.R. and A.M.; software, M.K. and P.W.; validation, M.K. and A.M.; formal analysis, A.M. and R.J.R.; investigation, A.M. and M.K.; resources, M.K., P.W., A.M. and R.J.R.; data curation, A.M., R.J.R. and M.K.; writing—original draft preparation, A.M., R.J.R., P.W. and M.K.; writing—review and editing, A.M., M.K., R.J.R. and P.W.; visualization, M.K.; supervision, A.M. All authors have read and agreed to the published version of the manuscript.

Funding: This research received no external funding.

Institutional Review Board Statement: Not applicable.

Informed Consent Statement: Not applicable.

Data Availability Statement: Marcin Kołodziej, Andrzej Majkowski. (2022). SSVEP classification with a limited training dataset. IEEE Dataport. https://dx.doi.org/10.21227/kc6z-8447 (accessed on 16 December 2023).

Conflicts of Interest: The authors declare no conflict of interest.

Abbreviations

ADAM	Adaptive Moment Estimation
AR-BCI	Augmented Reality based Brain–Computer Interface
BCI	Brain–Computer Interface
CCA	Canonical Correlation Analysis
CNN	Convolutional Neural Network
DFT	Discrete Fourier Transform
DNN	Deep Neural Network
EEG	Electroencephalography
EMG	Electromyography
ERD/ERS	Event-Related Desynchronization/Event-Related Synchronization
FBCCA	Filter Bank Canonical Correlation Analysis
FBCNN	Filter Bank Convolutional Neural Network
FB-EEGNet	Filter Bank EEGNet
GAN	Generative Adversarial Network
ITR	Information Transfer Rate
K-NN	K-Nearest Neighbors
LDA	Linear Discriminant Analysis
LDA-SFS	Linear Discriminant Analysis with Sequential Feature Selection
LDA-T	Linear Discriminant Analysis with *t*-test
MPL	Multi-layer Perceptron
P300	P300 Wave
QDA	Quadratic Discriminant Analysis
QDA-SFS	Quadratic Discriminant Analysis with Sequential Feature Selection
QDA-T	Quadratic Discriminant Analysis with *t*-test
SDG	Stochastic Gradient Descent
SSVEP	Steady State Visually Evoked Potentials
SVM	Support Vector Machine
MLP	Multilayer perceptron
sMP	simplified Matching Pursuit

References

1. Gu, X.; Cao, Z.; Jolfaei, A.; Xu, P.; Wu, D.; Jung, T.-P.; Lin, C.-T. EEG-Based Brain-Computer Interfaces (BCIs): A Survey of Recent Studies on Signal Sensing Technologies and Computational Intelligence Approaches and Their Applications. *IEEE/ACM Trans Comput. Biol. Bioinform.* **2021**, *18*, 1645–1666. [CrossRef] [PubMed]
2. Li, M.; He, D.; Li, C.; Qi, S. Brain–Computer Interface Speller Based on Steady-State Visual Evoked Potential: A Review Focusing on the Stimulus Paradigm and Performance. *Brain Sci.* **2021**, *11*, 450. [CrossRef] [PubMed]
3. Norizadeh Cherloo, M.; Mijani, A.M.; Zhan, L.; Daliri, M.R. A Novel Multiclass-Based Framework for P300 Detection in BCI Matrix Speller: Temporal EEG Patterns of Non-Target Trials Vary Based on Their Position to Previous Target Stimuli. *Eng. Appl. Artif. Intell.* **2023**, *123*, 106381. [CrossRef]
4. Ramkumar, S.; Amutharaj, J.; Gayathri, N.; Mathupriya, S. A Review on Brain Computer Interface for Locked in State Patients. *Mater. Today Proc.* **2021**. SSN 2214-7853. [CrossRef]
5. Choi, W.-S.; Yeom, H.-G. Studies to Overcome Brain–Computer Interface Challenges. *Appl. Sci.* **2022**, *12*, 2598. [CrossRef]
6. Abibullaev, B.; Kunanbayev, K.; Zollanvari, A. Subject-Independent Classification of P300 Event-Related Potentials Using a Small Number of Training Subjects. *IEEE Trans. Hum.-Mach. Syst.* **2022**, *52*, 843–854. [CrossRef]
7. Edlinger, G.; Allison, B.Z.; Guger, C. How Many People Can Use a BCI System? In *Clinical Systems Neuroscience*; Kansaku, K., Cohen, L.G., Birbaumer, N., Eds.; Springer: Tokyo, Japan, 2015; pp. 33–66; ISBN 978-4-431-55037-2.
8. Mu, J.; Grayden, D.B.; Tan, Y.; Oetomo, D. Comparison of Steady-State Visual Evoked Potential (SSVEP) with LCD vs. LED Stimulation. In Proceedings of the 2020 42nd Annual International Conference of the IEEE Engineering in Medicine Biology Society (EMBC), Montreal, QC, Canada, 20–24 July 2020; pp. 2946–2949.
9. Wang, J.; Bi, L.; Fei, W. Multitask-Oriented Brain-Controlled Intelligent Vehicle Based on Human–Machine Intelligence Integration *IEEE Trans. Syst. Man. Cybern. Syst.* **2022**, *53*, 2510–2521. [CrossRef]
10. Waytowich, N.R.; Krusienski, D.J. Multiclass Steady-State Visual Evoked Potential Frequency Evaluation Using Chirp-Modulated Stimuli. *IEEE Trans. Hum.-Mach. Syst.* **2016**, *46*, 593–600. [CrossRef]
11. Lin, B.-S.; Wang, H.-A.; Huang, Y.-K.; Wang, Y.-L.; Lin, B.-S. Design of SSVEP Enhancement-Based Brain Computer Interface. *IEEE Sens. J.* **2021**, *21*, 14330–14338. [CrossRef]
12. Brennan, C.; McCullagh, P.; Lightbody, G.; Galway, L.; McClean, S.; Stawicki, P.; Gembler, F.; Volosyak, I.; Armstrong, E.; Thompson, E. Performance of a Steady-State Visual Evoked Potential and Eye Gaze Hybrid Brain-Computer Interface on Participants With and Without a Brain Injury. *IEEE Trans. Hum.-Mach. Syst.* **2020**, *50*, 277–286. [CrossRef]
13. Castillo, J.; Müller, S.; Caicedo, E.; Bastos, T. Feature Extraction Techniques Based on Power Spectrum for a SSVEP-BCI. In Proceedings of the 2014 IEEE 23rd International Symposium on Industrial Electronics (ISIE), Istanbul, Turkey, 1–4 June 2014; pp. 1051–1055.
14. Shao, X.; Lin, M. Filter Bank Temporally Local Canonical Correlation Analysis for Short Time Window SSVEPs Classification. *Cogn. Neurodyn.* **2020**, *14*, 689–696. [CrossRef] [PubMed]
15. Kołodziej, M.; Majkowski, A.; Rak, R.J. Simplified Matching Pursuit as a New Method for SSVEP Recognition. In Proceedings of the 2016 39th International Conference on Telecommunications and Signal Processing (TSP), Vienna, Austria, 27–29 June 2016; pp. 346–349.
16. Waytowich, N.R.; Faller, J.; Garcia, J.O.; Vettel, J.M.; Sajda, P. Unsupervised Adaptive Transfer Learning for Steady-State Visual Evoked Potential Brain-Computer Interfaces. In Proceedings of the 2016 IEEE International Conference on Systems, Man, and Cybernetics (SMC), Budapest, Hungary, 9–12 October 2016; pp. 004135–004140.
17. Müller-Putz, G.R.; Scherer, R.; Brauneis, C.; Pfurtscheller, G. Steady-State Visual Evoked Potential (SSVEP)-Based Communication: Impact of Harmonic Frequency Components. *J. Neural Eng.* **2005**, *2*, 123–130. [CrossRef] [PubMed]
18. Liu, Q.; Jiao, Y.; Miao, Y.; Zuo, C.; Wang, X.; Cichocki, A.; Jin, J. Efficient Representations of EEG Signals for SSVEP Frequency Recognition Based on Deep Multiset CCA. *Neurocomputing* **2020**, *378*, 36–44. [CrossRef]
19. Lahane, P.; Jagtap, J.; Inamdar, A.; Karne, N.; Dev, R. A Review of Recent Trends in EEG Based Brain-Computer Interface. In Proceedings of the 2019 International Conference on Computational Intelligence in Data Science (ICCIDS), Chennai, India, 21–23 February 2019; pp. 1–6.
20. Osowski, S.; Cichocki, A.; Lempitsky, V.; Poggio, T. Deep Learning: Theory and Practice. *Bull. Pol. Acad. Sci. Tech. Sci.* **2018**, *66*, 757–759.
21. Shen, C.; Nguyen, D.; Zhou, Z.; Jiang, S.B.; Dong, B.; Jia, X. An Introduction to Deep Learning in Medical Physics: Advantages, Potential, and Challenges. *Phys. Med. Biol.* **2020**, *65*, 05TR01. [CrossRef]
22. Voulodimos, A.; Doulamis, N.; Doulamis, A.; Protopapadakis, E. Deep Learning for Computer Vision: A Brief Review. *Comput. Intell. Neurosci.* **2018**, *2018*, e7068349. [CrossRef]
23. Israsena, P.; Pan-Ngum, S. A CNN-Based Deep Learning Approach for SSVEP Detection Targeting Binaural Ear-EEG. *Front. Comput. Neurosci.* **2022**, *16*, 868642. [CrossRef]
24. Kwak, N.-S.; Müller, K.-R.; Lee, S.-W. A Convolutional Neural Network for Steady State Visual Evoked Potential Classification under Ambulatory Environment. *PLoS ONE* **2017**, *12*, e0172578. [CrossRef]
25. Ma, P.; Dong, C.; Lin, R.; Ma, S.; Jia, T.; Chen, X.; Xiao, Z.; Qi, Y. A Classification Algorithm of an SSVEP Brain-Computer Interface Based on CCA Fusion Wavelet Coefficients. *J. Neurosci. Methods* **2022**, *371*, 109502. [CrossRef]

26. Guney, O.B.; Oblokulov, M.; Ozkan, H. A Deep Neural Network for SSVEP-Based Brain-Computer Interfaces. *IEEE Trans. Biomed. Eng.* **2022**, *69*, 932–944. [CrossRef]
27. Ravi, A.; Manuel, J.; Heydari, N.; Jiang, N. A Convolutional Neural Network for Enhancing the Detection of SSVEP in the Presence of Competing Stimuli. In Proceedings of the 2019 41st Annual International Conference of the IEEE Engineering in Medicine and Biology Society (EMBC), Berlin, Germany, 23–27 July 2019; pp. 6323–6326.
28. Waytowich, N.; Lawhern, V.J.; Garcia, J.O.; Cummings, J.; Faller, J.; Sajda, P.; Vettel, J.M. Compact Convolutional Neural Networks for Classification of Asynchronous Steady-State Visual Evoked Potentials. *J. Neural Eng.* **2018**, *15*, 066031. [CrossRef] [PubMed]
29. Li, Y.; Xiang, J.; Kesavadas, T. Convolutional Correlation Analysis for Enhancing the Performance of SSVEP-Based Brain-Computer Interface. *IEEE Trans. Neural Syst. Rehabil. Eng.* **2020**, *28*, 2681–2690. [CrossRef] [PubMed]
30. Ikeda, A.; Washizawa, Y. Steady-State Visual Evoked Potential Classification Using Complex Valued Convolutional Neural Networks. *Sensors* **2021**, *21*, 5309. [CrossRef] [PubMed]
31. Xing, J.; Qiu, S.; Ma, X.; Wu, C.; Li, J.; Wang, S.; He, H. A CNN-Based Comparing Network for the Detection of Steady-State Visual Evoked Potential Responses. *Neurocomputing* **2020**, *403*, 452–461. [CrossRef]
32. Xing, J.; Qiu, S.; Wu, C.; Ma, X.; Li, J.; He, H. A Comparing Network for the Classification of Steady-State Visual Evoked Potential Responses Based on Convolutional Neural Network. In Proceedings of the 2019 IEEE International Conference on Computational Intelligence and Virtual Environments for Measurement Systems and Applications (CIVEMSA), Tianjin, China, 14–16 June 2019; pp. 1–6.
33. Maharana, K.; Mondal, S.; Nemade, B. A Review: Data Pre-Processing and Data Augmentation Techniques. *Glob. Transit. Proc.* **2022**, *3*, 91–99. [CrossRef]
34. He, C.; Liu, J.; Zhu, Y.; Du, W. Data Augmentation for Deep Neural Networks Model in EEG Classification Task: A Review. *Front. Hum. Neurosci.* **2021**, *15*, 765525. [CrossRef] [PubMed]
35. Kalaganis, F.P.; Laskaris, N.A.; Chatzilari, E.; Nikolopoulos, S.; Kompatsiaris, I. A Data Augmentation Scheme for Geometric Deep Learning in Personalized Brain–Computer Interfaces. *IEEE Access* **2020**, *8*, 162218–162229. [CrossRef]
36. Wang, F.; Zhong, S.; Peng, J.; Jiang, J.; Liu, Y. Data Augmentation for EEG-Based Emotion Recognition with Deep Convolutional Neural Networks. In *Proceedings of the MultiMedia Modeling*; Schoeffmann, K., Chalidabhongse, T.H., Ngo, C.W., Aramvith, S., O'Connor, N.E., Ho, Y.-S., Gabbouj, M., Elgammal, A., Eds.; Springer International Publishing: Cham, Switzerland, 2018; pp. 82–93.
37. Chiang, K.-J.; Wei, C.-S.; Nakanishi, M.; Jung, T.-P. Boosting Template-Based SSVEP Decoding by Cross-Domain Transfer Learning. *J. Neural Eng.* **2021**, *18*, 016002. [CrossRef]
38. Yao, H.; Liu, K.; Deng, X.; Tang, X.; Yu, H. FB-EEGNet: A Fusion Neural Network across Multi-Stimulus for SSVEP Target Detection. *J. Neurosci. Methods* **2022**, *379*, 109674. [CrossRef]
39. Duart, X.; Quiles, E.; Suay, F.; Chio, N.; García, E.; Morant, F. Evaluating the Effect of Stimuli Color and Frequency on SSVEP. *Sens.* **2020**, *21*, 117. [CrossRef]
40. Hui, S.; Żak, S.H. Discrete Fourier Transform and Permutations. *Bull. Pol. Acad. Sciences. Tech. Sci.* **2019**, *675*, 130874. [CrossRef]
41. Bin, G.; Gao, X.; Yan, Z.; Hong, B.; Gao, S. An Online Multi-Channel SSVEP-Based Brain-Computer Interface Using a Canonical Correlation Analysis Method. *J. Neural Eng.* **2009**, *6*, 046002. [CrossRef] [PubMed]
42. Tanaka, T.; Zhang, C.; Higashi, H. SSVEP Frequency Detection Methods Considering Background EEG. In Proceedings of the The 6th International Conference on Soft Computing and Intelligent Systems, and The 13th International Symposium on Advanced Intelligence Systems, Kobe, Japan, 20–24 November 2012; pp. 1138–1143.
43. Jović, A.; Brkić, K.; Bogunović, N. A Review of Feature Selection Methods with Applications. In Proceedings of the 2015 38th International Convention on Information and Communication Technology, Electronics and Microelectronics (MIPRO), Opatija, Croatia, 25–29 May 2015; pp. 1200–1205.
44. Wang, M.; Liu, G. A Simple Two-Sample Bayesian t-Test for Hypothesis Testing. *Am. Stat.* **2016**, *70*, 195–201. [CrossRef]
45. Zhou, X.; Wang, J. Feature Selection for Image Classification Based on a New Ranking Criterion. *J. Comput. Commun.* **2015**, *3*, 74–79. [CrossRef]
46. Tahir, M.A.; Bouridane, A.; Kurugollu, F. Simultaneous Feature Selection and Feature Weighting Using Hybrid Tabu Search/K-Nearest Neighbor Classifier. *Pattern Recognit. Lett.* **2007**, *28*, 438–446. [CrossRef]
47. Lotte, F.; Bougrain, L.; Cichocki, A.; Clerc, M.; Congedo, M.; Rakotomamonjy, A.; Yger, F. A Review of Classification Algorithms for EEG-Based Brain–Computer Interfaces: A 10 Year Update. *J. Neural Eng.* **2018**, *15*, 031005. [CrossRef] [PubMed]
48. Howard, C.W.; Zou, G.; Morrow, S.A.; Fridman, S.; Racosta, J.M. Wilcoxon-Mann-Whitney Odds Ratio: A Statistical Measure for Ordinal Outcomes Such as EDSS. *Mult. Scler. Relat. Disord.* **2022**, *59*, 103516. [CrossRef]
49. Nakanishi, M.; Wang, Y.; Chen, X.; Wang, Y.-T.; Gao, X.; Jung, T.-P. Enhancing Detection of SSVEPs for a High-Speed Brain Speller Using Task-Related Component Analysis. *IEEE Trans. Biomed. Eng.* **2018**, *65*, 104–112. [CrossRef]
50. Maÿe, A.; Mutz, M.; Engel, A.K. Training the Spatially-Coded SSVEP BCI on the Fly. *J. Neurosci. Methods* **2022**, *378*, 109652. [CrossRef]
51. Kołodziej, M.; Majkowski, A.; Tarnowski, P.; Rak, R.J.; Rysz, A. A New Method of Cardiac Sympathetic Index Estimation Using a 1D-Convolutional Neural Network. *Bull. Pol. Acad. Sciences. Tech. Sci.* **2021**, *69*, 136921. [CrossRef]
52. Zhang, Q.; Zhou, D.; Zeng, X. HeartID: A Multiresolution Convolutional Neural Network for ECG-Based Biometric Human Identification in Smart Health Applications. *IEEE Access* **2017**, *5*, 11805–11816. [CrossRef]

53. Furui, A.; Hayashi, H.; Nakamura, G.; Chin, T.; Tsuji, T. An Artificial EMG Generation Model Based on Signal-Dependent Noise and Related Application to Motion Classification. *PLoS ONE* **2017**, *12*, e0180112. [CrossRef] [PubMed]
54. Zhao, X.; Du, Y.; Zhang, R. A CNN-Based Multi-Target Fast Classification Method for AR-SSVEP. *Comput. Biol. Med.* **2022**, *141*, 105042. [CrossRef] [PubMed]
55. Wu, C.; Qiu, S.; Xing, J.; He, H. A CNN-Based Compare Network for Classification of SSVEPs in Human Walking. In Proceedings of the 2020 42nd Annual International Conference of the IEEE Engineering in Medicine Biology Society (EMBC), Montreal, QC, Canada, 20–24 July 2020; pp. 2986–2990.
56. Nguyen, T.-H.; Chung, W.-Y. A Single-Channel SSVEP-Based BCI Speller Using Deep Learning. *IEEE Access* **2019**, *7*, 1752–1763. [CrossRef]
57. Zhao, D.; Wang, T.; Tian, Y.; Jiang, X. Filter Bank Convolutional Neural Network for SSVEP Classification. *IEEE Access* **2021**, *9*, 147129–147141. [CrossRef]

applied sciences

MDPI

Article

Novel Deep-Learning Approach for Automatic Diagnosis of Alzheimer's Disease from MRI

Omar Altwijri [1], Reem Alanazi [2], Adham Aleid [1], Khalid Alhussaini [1], Ziyad Aloqalaa [1], Mohammed Almijalli [1] and Ali Saad [1,*]

1 Department of Biomedical Technology, College of Applied Medical Sciences, King Saud University, Riyadh 11433, Saudi Arabia
2 Department of Physics and Astronomy, College of Science, King Saud University, Riyadh 11451, Saudi Arabia
* Correspondence: alisaad@ksu.edu.sa

Abstract: This study introduces a novel deep-learning methodology that is customized to automatically diagnose Alzheimer's disease (AD) through the analysis of MRI datasets. The process of diagnosing AD via the visual examination of magnetic resonance imaging (MRI) presents considerable challenges. The visual diagnosis of mild to very mild stages of AD is challenging due to the MRI similarities observed between a brain that is aging normally and one that has AD. The detection of AD with extreme precision is critical during its early stages. Deep-learning techniques have recently been shown to be significantly more effective than human detection in identifying various stages of AD, enabling early-stage diagnosis. The aim of this research is to develop a deep-learning approach that utilizes pre-trained convolutional neural networks (CNNs) to accurately detect the severity levels of AD, particularly in situations where the quantity and quality of available datasets are limited. In this approach, the AD dataset is preprocessed via a refined image processing module prior to the training phase. The proposed method was compared to two well-known deep-learning algorithms (VGG16 and ResNet50) using four Kaggle AD datasets: one for the normal stage of the disease and three for the mild, very mild, and moderate stages, respectively. This allowed us to evaluate the effectiveness of the classification results. The three models were compared using six performance metrics. The results achieved with our approach indicate an overall detection accuracy of 99.3%, which is superior to the other existing models.

Keywords: Alzheimer's disease; image processing; deep learning; transfer learning; classification

Citation: Altwijri, O.; Alanazi, R.; Aleid, A.; Alhussaini, K.; Aloqalaa, Z.; Almijalli, M.; Saad, A. Novel Deep-Learning Approach for Automatic Diagnosis of Alzheimer's Disease from MRI. *Appl. Sci.* **2023**, *13*, 13051. https://doi.org/10.3390/app132413051

Academic Editor: Alexander N. Pisarchik

Received: 27 October 2023
Revised: 2 December 2023
Accepted: 3 December 2023
Published: 7 December 2023

1. Introduction

Alzheimer's disease (AD) is a degenerative neurological disorder that causes permanent brain cell loss and long-term cognitive impairment [1]. Alzheimer's disease (AD) causes cognitive and mental deterioration, behavioral issues, language problems, and difficulty doing fundamental tasks. AD is a sixth-order death that destroys the brain area that controls breathing and cardiac function. There is no treatment to stop or slow the progression of Alzheimer's disease [2], and its cause is unknown. Defects in the hippocampus, cerebral cortex, and ventricles are signs of Alzheimer's disease. These areas control memory, planning, reasoning, and judgment [3]. Alzheimer's disease (AD) progresses to varying degrees of severity. It is challenging to diagnose AD in its early and late stages because of MRI similarities between a normal aging brain and an AD brain. As a result, analyzing and assessing these pictures is challenging [4,5]. Until patients reach a moderate stage of AD, detection accuracy is low. Thus, it is crucial for AD diagnosis to detect changes in specific brain regions early on so that the disease can be halted in its tracks [6]. The ability of machine learning algorithms to detect AD has recently been demonstrated in studies [7,8]. MRI scans are frequently used in medical diagnoses. MRI scans may have varying meanings, depending on the reader. Supervised systems are trained using feature vectors extracted from medical imaging data. To extract these characteristics, human experts must expend significant time, resources, and energy. Rapid patient screening and diagnosis may be aided by deep-learning models. This technology can instantly analyze photographs without the need for skilled manual extraction. Features from MRI brain images are extracted using deep-learning-based methods, allowing for the early detection of AD. To mimic the performance of biological neural networks, scientists created ANNs [4,9]. Computers can learn from data at varying granularities due to their multi-layer processing architecture [10]. Deep learning is a subfield of machine learning (ML), which is a core component of artificial intelligence (AI). AI is used in many areas, including neuroscience. Predicting and diagnosing brain diseases, such as Alzheimer's, has become much simpler thanks to AI. Deep learning has many types, such as the feed-forward deep neural network, the convolutional neural network (CNN), the auto-encoder (AE), the recurrent neural network (RNN), the deep belief network (DBN), and the generative adversarial network (GAN) [11].

CNN is a feed-forward neural network that makes use of convolutional features [12,13]. CNN, unlike other methods, does not require manual feature extraction. CNN kernels are analogous to various sensors that can respond to a wide variety of stimuli. Activation functions are similar to the way in which neurons send electric impulses to the next cell when a certain threshold is reached. CNN is better than most artificial neural networks in three ways: First, local connections are used between neurons in the same layer instead of between all neurons in the layer below; this lowers the parameters and speeds up convergence. Second, sharing the weight of links may reduce the total number of parameters if we combine link weights. Third, because convolution makes feature maps with a lot of features, the chance of overfitting goes up. Maximum and average pooling are two types of pooling that are recommended for reducing redundant work. The downsampling of dimensions: a pooling layer uses the idea of local correlation to downscale an image while preserving its essential details.

The following study demonstrates the efficacy of CNN in identifying AD. Ref. [14] proposed a CNN-Sparse Regression Network combination model for AD diagnosis. The model generated numerous representations at the target level using sparse regression networks. CNN was used to combine these representations at the target level to enhance output label recognition. A 16-layer VGGNet was used by the authors of reference [15] to effectively divide structural MRI scans into three groups: Alzheimer's disease (AD), mild cognitive impairment (MCI), and normal cognitive (NC). The authors reported that segmentation was not conducted on the magnetic resonance (MR) images. Using functional MRI, ref. [16,17] applied the LeNet architecture to classify patients with Alzheimer's disease from healthy controls. They came up with a technique for structural MRI that uses CNNs. The research demonstrated that CNN outperformed SVM. Future studies will likely include axial and sagittal MRI scans in addition to the standard coronal ones. The current result is 98.84% accurate, which is quite good. The use of structural MR images allowed for the development of a CNN-based AD diagnosis model [18]. Researchers found that by using both data augmentation and transfer learning together, overfitting could be lessened, and the models could use less computing power. Previous studies relied on smaller, regional datasets, but the authors of this one claim their work can be applied much more widely. In [19], an 8-layer CNN model was created specifically for AD diagnosis. To find the best model setup, the authors looked at many different activation function combinations, such as stochastic, max, and average pooling with ReLU, sigmoid, and leaky ReLU. A leaky ReLU activation function and a max pooling function were used in the most effective CNN models. A 3D-CNN model trained on MR images was proposed for AD diagnosis in [20]. They propose a 3D-CNN using the ResNet framework. Convolutional, dropout, pooling, and fully connected layers are some of the 36 it contains. The model outperformed expectations on a variety of performance metrics in experimental testing. Another method in [21] using 3D-CNN to examine MR images for AD signs was proposed. The authors deployed a Sobolev gradient optimizer, a leaky ReLU activation function, and a Max

Pooling function. The three functions worked better together than separately. To aid in the diagnosis of AD, ref. [22] developed a 3D-FCNN-based model using MR images. The authors revealed that their proposed model outperformed several industry standards for both accuracy and robustness. The 3D-FCNN model outperformed the 2D-CNN on binary and multi-class classification tasks. An approach to diagnosing AD that uses structural MRI, genetic testing, and clinical evaluation was suggested in [23]. It is based on convolutional neural networks. The framework required fewer parameters than rivals like VGGNet and AlexNet for building CNN models. The method was quicker and less prone to overfitting in scenarios with sparse data. In [2], the authors propose using a CNN-based MR image model for the early detection of AD. The OASIS dataset, which is notoriously skewed, was used to train the model. To address the discrepancies present in the OASIS data set, data augmentation was implemented. According to the results of the experiments, the proposed model is superior to several state-of-the-art models.

A CNN-based MR image AD diagnosis model was proposed in [24]. It all started with voxelizing MR scans. Skull stripping was used to get rid of extra voxels, and the quality of the remaining ones was enhanced with a Gaussian filter. Independent component analysis was used to separate out different regions of the brain. In the end, the gray matter in the model was segmented. According to the results of the experiments, the proposed model is superior to other state-of-the-art models. The eight-layer CNN AD diagnosis model in [25] used drop-out regularization, data augmentation, and batch normalization to ensure excellent precision. The Siamese Convolutional Neural Network (SCNN) was introduced in [26] as a CNN-based model for dividing dementia into four distinct stages: moderate Alzheimer's disease (MAD), mild dementia (MD), very mild dementia (VMD), and no dementia. Despite just having a small sample size to train on, the model's results were reliable. The proposed model was shown to be superior to five other state-of-the-art studies. Ref. [27] proposed a cascaded 3D-CNN for AD diagnosis using structural MR images. To classify the input, the CNN model first retrieved features from it.

In order to diagnose Alzheimer's disease using 3D-MR brain pictures, the authors in [28] modified V-Net to partition the bilateral hippocampus. They demonstrate the need for accurate hippocampi segmentation for an accurate AD diagnosis model. Compared to other segmentation and classification approaches, they state that the proposed design performed better. Ref. [29] developed a CNN model for AD diagnosis using MR images. To put the method to the test in real-world settings, the researchers looked at the correlation between relevance score and hippocampus volume. The 3D-CNN-SVM model for AD diagnosis was proposed in [30] based on MR images. It combines the 3D-CNN model to obtain features from MR images and SVM to classify the features. The 3D-CNN-SVM model provides a significant improvement over both 2D-CNN and classic 3D-CNN. Ref. [31] created a CNN using the DenseNet Bottleneck-Compressed architecture for the diagnosis of AD using MR images. The proposed model correctly classified the input 86% of the time. The EfficientNet models [32] are developed through the implementation of uncomplicated and exceptionally effective compound scaling methods. EfficientNet models demonstrate an enhanced level of precision and effectiveness in comparison to modern CNNs, including MobileNetV2, AlexNet, ImageNet, and GoogleNet [33]. EfficientNets show better accuracy through their compactness, computational efficiency, and generalization capabilities. Comparing eight different convolutional neural networks for early detection of Alzheimer's disease, the EfficientNetB0 model has better evaluation metrics and needs fewer model parameters [34]. In a recent study [35], an EfficientNetB0 model was employed to diagnose AD. The results obtained for all performance metrics varied from 87% to 95%. The EfficientNetB0 model is very good at finding COVID-19 patterns in X-ray images while using a small amount of computing power compared to other popular architectures like ResNets and VGGs [36]. Subsequently, EfficientNets demonstrated high efficiency in numerous applications, including the detection of malaria parasites from blood smears and various COVID-19 detection applications [37]. Thus, the results of the EfficientNet series on different medical applications inspired us to develop an approach based on the EfficientNetB0 structure for AD.

This study introduces a comprehensive methodology for evaluating the severity and progression of Alzheimer's disease (AD) from start to finish. Deep-learning techniques were utilized to distinguish between four stages of Alzheimer's disease, specifically normal control, very mild, mild, and moderate dementia. This research endeavor aims to improve the performance of efficientNetB0 by implementing a three-step data processing approach. These steps involve the use of an alpha-trimmed filter for low-pass filtering, histogram equalization, and the application of transfer-learning techniques. In order to evaluate our approach performance, we compared it to two commonly utilized models, specifically ResNet50 and VGG16. The evaluation metrics employed for comparison encompassed precision, recall, accuracy, F1 score, confusion matrices, and receiver operating characteristic (ROC). In the following sections, the materials and methods are described. The results from experimental data, using datasets from official GitHub and Kaggle repositories, of three DL models are compared using six performance metrics. The proposed method was compared with two other well-performing deep-learning algorithms. Then, the results were discussed, conclusions drawn, and future directives were suggested.

2. Materials and Methods

2.1. Dataset Description

The benchmark dataset research on the "Kaggle" website is available online: https://www.kaggle.com/datasets/tourist55/alzheimers-dataset-4-class-of-images (accessed on 1[illegible] February 2023). It provided MRI pictures of Alzheimer's disease for this study [38]. Kaggle serves as a platform for providing online datasets for research and analysis in various fields. To expedite the development of enhanced algorithms for diagnosing and treating Alzheimer's disease, we chose this dataset owing to its complete freedom, availability in diverse categories, and relatively small hard disk size, setting it apart from other popular datasets in the field. This manually collected dataset comprises MRI images verified and classified by Sarvesh Dubey [38]. Serving as valuable resources for training and testing deep-learning models with the objective of accurately predicting the stage of Alzheimer's disease. By affording researchers and practitioners an opportunity to create algorithms for precise Alzheimer's disease diagnosis, this dataset assumes a crucial role. Additionally, it contributes to the development of effective treatments. As the global burden of Alzheimer's disease escalates, this dataset gains significance in advancing our understanding of the disease and improving patient outcomes [36].

A total of 6400 photos make up the Kaggle Alzheimer's classification dataset (KACD). The dataset was divided into four groups: 896 mild AD, 64 moderate AD, 3200 normal, and 2240 very mild AD. To test the models, 20% of the dataset was used for testing, while 80% was used for training and analysis. The sample included 2560 normal controls, 717 participants with mild AD, 52 with moderate AD, and 1792 with very mild AD. A typical sample from each dataset class is shown in Figure 1. Doctors use the Hippocampal area as a biomarker to diagnose Alzheimer's disease (AD) with great accuracy, making it a significant factor. However, hippocampus volume alone cannot predict early stages. According to prior study [4], cortical regions and thickness affect the illness's progression. Due to its high resolution and contrast for soft tissues, structural MRI is used to evaluate the parietal, temporal, hippocampal, entorhinal cortex, and ventricular atrophy [6]. Different brain regions are modified depending on illness progression [4].

This study's objective, as depicted in Figure 1, is to demonstrate how cognitive decline manifests differently in areas that have not experienced any disease and areas that have experienced the worst cases of the same disease [10]. In contrast to the moderate stage, individuals in the very mild and light stages of dementia exhibit a somewhat better level of independence in their functioning. However, due to notable memory impairment, they often require some level of support with various everyday activities. The severity stage is characterized by a prolonged duration compared to the very mild and mild stages. During this stage, the patient has a progressive deterioration of their physical condition, ultimately leading to mortality [39].

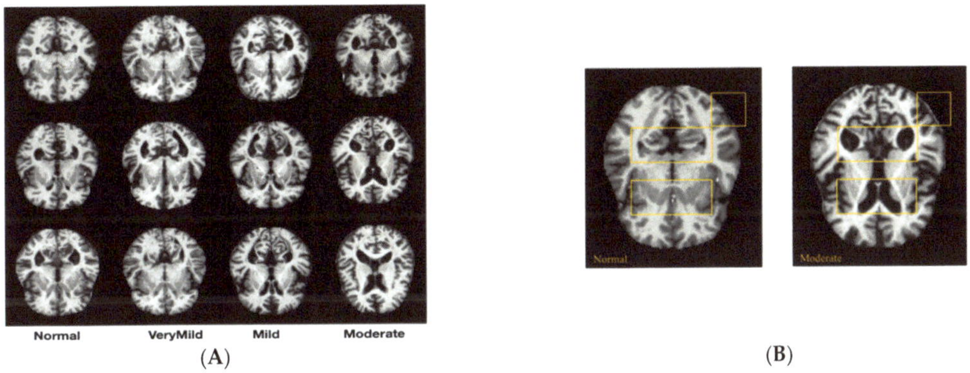

Figure 1. (**A**) Magnetic resonance imaging (MRI) images (2D) from the KACD dataset with four of Alzheimer's dementia's stages. (**B**) The images highlight major regions of healthy brain (**left**) and an Alzheimer's brain (**right**), as indicated in yellow.

2.2. Proposed Method for AD Diagnosis

Figure 2 presents the general deep-learning workflow, where the first step is to choose the datasets for training and validation. Then, the selection of the hyper-parameters for the neural network model follows. In the third step, the choice of the CNN model and framework is determined by the related parameters, including the loss rate, learning function, and optimizer. The fourth step is the training and validation phase of the model. The fifth step is the testing and prediction phase using new input datasets. The final step is the assessment of the performance of the model.

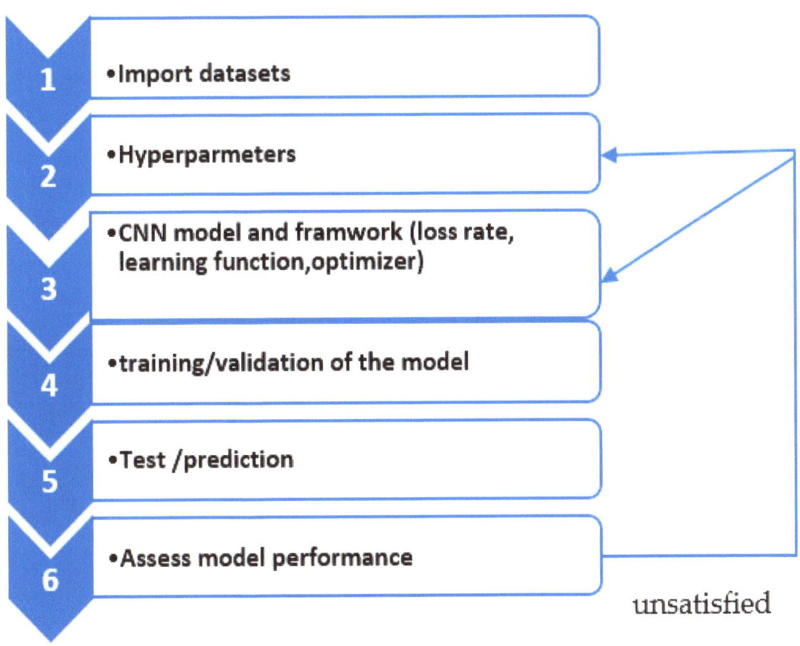

Figure 2. Deep-learning workflow.

Figure 3 presents the proposed approach for the classification of the AD images. The AD dataset is pre-processed first using pre-processing techniques, including skull removal and spatial registration, then another processing phase, including histogram equalization slicing, and image resizing, followed by low-pass alpha-trimmed filtering. Skull removal is used to remove bones from the image. Histogram equalization is the normalization process of the gray levels in the images from various subjects and maps the pixel intensity values to a wide range.

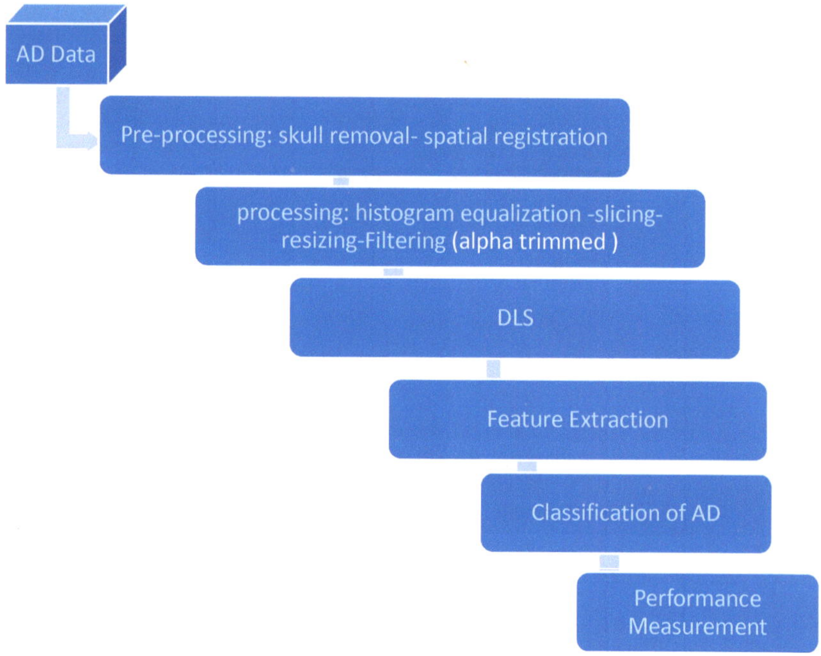

Figure 3. Proposed DL (PDL)-based classification approach.

In order to reduce the impact of orientation and spatial differences among scanner users, registration is performed. Registration improves the precision of the classification. The MNI152 brain template [40] was used by averaging 152 structural pictures into a single high-resolution image using non-linear registration. Slicing divides the image into multiple logical images. Resizing is carried out in order to get the desired image size (224 × 224).

Filtering improves the quality of the images by removing noise and artifacts. An alpha-trimmed filter was used to filter images from noise and artifacts. It first ranked the values of pixels in the neighbor window (5 × 5 pixels) centered on the pixel under processing. It ranks the pixels from the smallest to the biggest, eliminates the extremities according to the dimension of the parameter 'd' (Equation (1)), and then calculates the average of the remaining pixels. The resultant average will be placed at the same location (x,y) as the central pixel under processing in a new image.

The alpha-trimmed filter equation is

$$F(x,y) = (1/(mn - d))\sum gr(s,t), \quad (1)$$

where

F(x,y): F represents the filtered image, and x,y are the coordinate of the pixel processed.

gr is the set of pixels left after excluding the d/2 extreme pixels.

mn is the dimensions of the filter.

d represents the number of pixels that will be excluded from the averaging.

(s,t) represents the set of coordinates of the remaining pixels.

In our implementation, the filter dimensions' mn is 5 × 5, and d was set to 8. It means that we exclude 8 pixels out of 25, 4 from the beginning and 4 from the end of the ranked values of the pixels.

Then, the pre-processed data are fed as input to the DLS model using CNN model that performs feature extraction and classification of the input data. Finally, the model is evaluated using performance metrics such as F1 score, area under curve (AUC), recall, and precision.

Due to the small dataset, training big convolutional neural networks (CNNs) from the beginning proved difficult. Neural networks need a lot of data to train well, which may not be available. Instead of starting from scratch, using an existing model for a comparable job can save training time and improve outcomes [41]. CNN structure is presented in Figure 4.

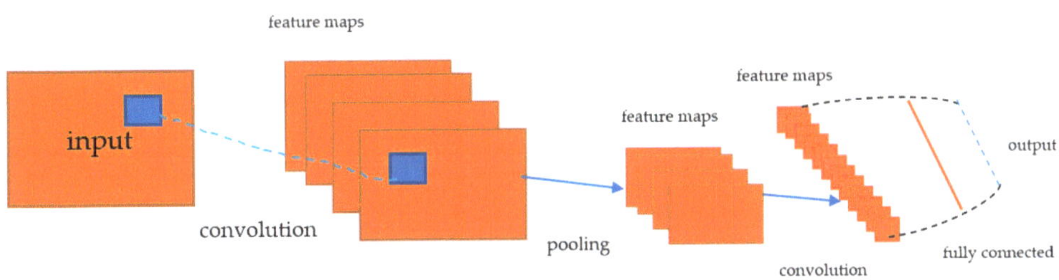

Figure 4. CNN architecture.

As stated in reference [4], transfer learning facilitates accelerated training and compensates for the limited data set. It has been demonstrated that transfer learning is a dependable and effective initial method for developing interpretable deep-learning models. Transfer learning is an approach that applies pre-trained networks to novel tasks by modifying them, thereby efficiently classifying diverse datasets. Classifying medical images, such as brain MRI scans, with models initially trained on natural images from ImageNet [42] proves to be a particularly advantageous application. During training, this study utilizes pre-trained weights obtained from ImageNet [42]. Then, retraining networks on a new dataset through the modification of the final fully connected layers, except for the final fully connected layer, pre-trained layers are frozen, and each model is retrained using the dataset in this scenario. In order to generate class prediction probabilities, the output layer incorporates a fully connected layer with Softmax activation and a global pooling layer (Global Average Pooling).

The selection of ResNet50 and VGG16 for our research was predicated on their distinct attributes and benefits within the domain of medical image classification, specifically in the context of transfer-learning-based Alzheimer's disease (AD) diagnosis. ResNet50 and VGG16 are pre-trained models, as described in [42]. By utilizing deep residual learning, ResNet50 overcomes the difficulty associated with training extremely deep networks. Additionally, ResNet50's skip connection lets each layer make a link between its input and output, which makes it easier for the model to understand complicated features in medical images and speeds up the flow of data. The VGG16 architecture is selected due to its simple design, which incorporates deeper networks utilizing smaller convolutional filters (3 × 3). The simplicity of this approach facilitates the comprehension of acquired features, which is particularly critical in medical situations where it is vital to grasp the model's reasoning process. Moreover, every layer in VGG16 represents a distinct level of abstraction, thereby establishing a distinct hierarchy of features. The hierarchical structure of this representation proves to be highly advantageous in the field of medical image analysis, wherein the significance of multiple levels of detail varies.

The configuration parameters for the three tested models are summarized in Table 1. Under identical conditions, the objective of this configuration is to compare the F1 score, recall, and precision of these various neural network architectures. This guarantees that the hyperparameters for all three models are identical.

Table 1. Summarize the hyper-parameters used for training.

Parameters	Proposed Method (PDL)	VGG16	ResNet50
Number of epochs	30	30	30
Batch Size	34	34	34
Optimiser	Adam	Adam	Adam
Learning Rate	0.0001	0.0001	0.0001
Loss Function	Categorical cross-entropy	Categorical cross-entropy	Categorical cross-entropy

Throughout the experimental procedure, data preprocessing was conducted in a consistent manner for all models, thereby guaranteeing consistent data partitions for the purposes of training, validation, and testing as follows:

- For a consistent starting point, identical weights were assigned to each model during initialization.
- During training, the designated hyper-parameters were applied to the training set while the progress of training was consistently monitored and assessed on the validation set.
- Using identical test set, performance metrics were computed for every model undergoing evaluation.
- To determine whether or not there were significant differences in performance metrics between models, statistical tests were employed, including ROC curve tests.

The results were graphically represented, incorporating precise recall curves or confusion matrices, which offered valuable insights into the merits and demerits of every model.

Proposed deep learning (PDL) is a CNN framework using EfficientNetB0, where architecture is presented in Table 2, and a few other compounds explained in the next section. EfficientNetB0 has been purposefully engineered to attain competitive performance while minimizing computational demands. This characteristic renders it a highly suitable option for situations in which there are limitations on resources, such as those encountered in medical environments where computational resources may be limited, where EfficientNetB0 achieves a balance between model complexity and efficiency by uniformly scaling network dimensions. This is beneficial in the context of medical applications where optimizing resource utilization is a critical factor in achieving high predictive performance. EfficientNetB0 is famous for its efficiency and low cost. Due to compound scaling, the network's depth, breadth, and resolution are equal. The EffifientNetB0 model was pre-trained using ImageNet, a large labeled dataset. Pre-trained weights from the ImageNet dataset were used during training. The transfer-learning technique was used to repair the pre-existing layers and then retrain the PDL model using Kaggle datasets.

Table 2. EfficientNetB0 architecture used in our model.

Steps	Operator	Resolution	Channels	Layers
1	Conv 3 × 3	224 × 224	32	1
2	MBconv1, 3 × 3	112 × 112	16	1
3	MBconv1, 3 × 3	112 × 112	24	2
4	MBconv1, 5 × 5	56 × 56	40	2
5	MBconv1, 3 × 3	28 × 28	80	3
6	MBconv1, 5 × 5	14 × 14	112	3
7	MBconv1, 5 × 5	14 × 14	192	4
8	MBconv1, 3 × 3	7 × 7	320	1
9	Conv 1 × 1 pooling	7 × 7	1280	1

PDL is composed of several blocks. Convolutional and pooling layers precede fully connected classification layers in each block. The PDL design included a batch normalizing layer before the fully connected layer, and a global average pooling (GAP) was used to build the model's output layer. Layers were added after the fully connected layer flattened the model. The rectified linear unit (ReLU) activation function, global average pooling (GAP), 0.5 dropout layer, 4-unit dense layer, and Softmax activation function were used in these layers. After flattening the fully connected layer, one dense layer was applied; this layer was activated using ReLU, a dropout layer with a 0.5 dropout rate.

Mobile inverted residual bottleneck convolution (MBConv) [8] is a key characteristic used in several building blocks. Two pointwise convolution layers have a bottleneck layer. Pointwise convolutions increase output channels, while the bottleneck layer reduces input channels. To balance accuracy and performance, DLS uses compound scaling to customize its stack of MBConv layers. The compound scaling approach simultaneously modifies network depth, breadth, and resolution; this optimizes computing resources.

The DLS underwent training using an adaptive moment estimation (ADAM) optimizer, with a learning rate of 0.0001 and a batch size of 34. The training process consisted of a minimum of 20 epochs, during which a dropout rate of 0.5 was applied to the dropout layer.

In PDL, Softmax activation [43] was used to calculate class prediction probabilities using the dense function [44].

3. Results

The proposed method achieves its highest training and validation accuracy at epoch 20, reaching 99.8% and 99.0%, respectively. The corresponding losses for the training and validation sets are 0.006 and 0.02. The VGG16 architecture demonstrated lower performance in terms of training and validation accuracy at epoch 20, achieving rates of 99.4% and 98.2%, respectively. The corresponding losses were recorded as 0.025 and 0.05. In contrast, the ResNet50 network has a training accuracy of 98.0% and a validation accuracy of 96.5%, with corresponding loss values of 0.04 and 0.15. On the other hand, the proposed method has the advantage of requiring the least amount of time per iteration. Upon assessing the loss curve, it becomes apparent that the loss values of PDL exhibit a more rapid fall and tend towards zero in comparison to other networks. The VGG16 model has a higher iteration time compared to the PDL model, with the former taking around twice as long. On the other hand, the ResNet50 model demonstrates the longest training duration among the three models. All three models eventually converge; however, the PDL and VGG16 models have the fastest convergence rates.

The PDL model demonstrates a classification accuracy of over 98% and an error rate below 2% after five iterations. Both the ResNet50 and VGG16 models need more than 10 iterations. As a result, adversarial pictures only exhibit a minimal level of resilience. Consequently, the PDL and VGG16 models exhibit notable efficacy and robust convergence in the context of Alzheimer's disease identification. Figure 5 presents the accuracy and loss metrics obtained from trained and validated databases over a span of 20 iterations using mixed data sets. Consequently, the PDL exhibits superior efficiency and accuracy in recognizing Alzheimer's disease. The results prove that among the selected methods, the PDL has a notable capacity for generalization in Alzheimer's disease recognition and is well suited for a broader range of diagnostic situations related to Alzheimer's disease.

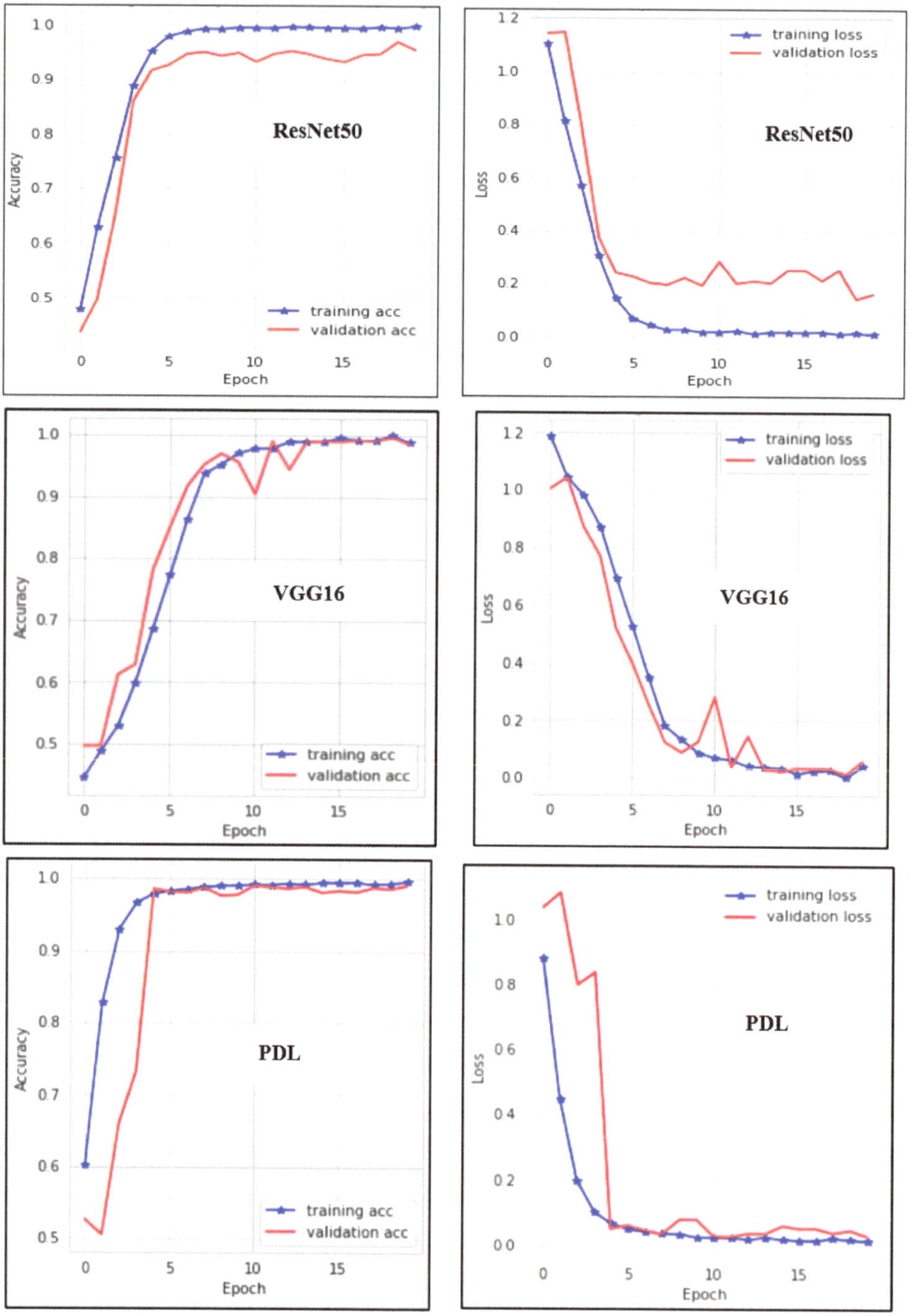

Figure 5. The accuracy and loss curves in the training and validation stages for the 3 models; blue lines represent training, and red lines represent validation. The first raw images, from the top, represent the accuracy and loss of Resnet50; the second raw images represent the accuracy and loss of VGG16; and the third raw images represent the accuracy and loss of PDL.

3.1. Prediction Performance

Figure 5 shows the accuracy and loss curves in the training and validation stages, where a blue line indicates training loss and a red line for validation loss indicates the three convolutional models we experimented with during this work, which were trained on four class datasets for 25 epochs.

Performance metrics are applied to test data by considering normal, very mild, mild, and moderate AD cases.

As shown in Figure 6, in terms of precision, recall, and F1 score, when comparing all methods, it is observed that PDL achieved the lowest loss value of 0.02 and performed the best accuracy of 99%. The lowest accuracy is obtained with ResNet50 (96.5%). For further in-depth evaluation of performance, the results are reported in Table 3.

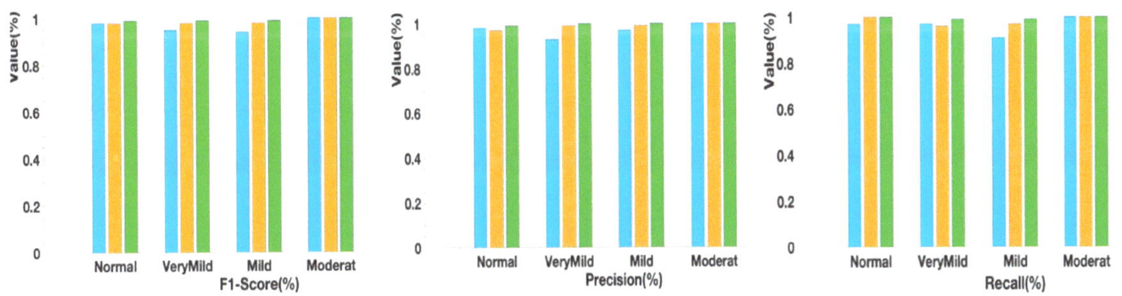

(**A**) F1 Score (**B**) Precision (**C**) Recall

Figure 6. (**A**) The performance metric of F1 score, where ResNet50 results are shown in blue, VGG16 results are shown in yellow, and PDL results are shown in green. (**B**) The performance metric of precision, where ResNet50 results are shown in blue, VGG16 results are shown in yellow, and PDL results are shown in green. (**C**) The performance metric of recall, where ResNet50 results are shown in blue, VGG16 results are shown in yellow, and PDL results are shown in green.

Table 3. Performance measures: comparison between pertained models architecture based on AD patients.

Models	Class Label	Precision (%)	Recall (%)	F1-Score (%)	Average Score
ResNet50	Normal	98	97	98	97.6%
	Very Mild	93	97	95	95%
	Mild	97	91	94	94%
	Moderate	100	100	100	100%
VGG16	Normal	98	99	99	98.6%
	Very Mild	98	97	97	97.3%
	Mild	100	98	99	99%
	Moderate	100	100	100	100%
Proposed Method	Normal	99	99	98	98.6%
	Very Mild	100	99	99	99.3%
	Mild	100	99	99	99.3%
	Moderate	100	100	100	100%

The prediction of the early stage of a very mild class is intriguing due to the inherent challenges associated with its diagnosis. In contrast, the classification of the late stage, which falls under the moderate category, is rather straightforward since all algorithms consistently yield a 100% accuracy rate in their results. In the early stages of AD, the PDL demonstrates superior performance with a precision of 100%, recall of 99%, and F1 score of 99%. In contrast, the ResNet50 and VGG16 models exhibit lower predictions, as shown in

Table 3. Consequently, it is evident that PDL has superior performance in comparison t the other two models.

Additionally, in order to conduct a more comprehensive assessment of the classificatio models, Figure 7 presents a confusion matrix diagram that serves as a concise representatio of the prediction outcomes during the evaluation of classification models on test data. Th PDL model demonstrates remarkable performance in detecting both normal and earl disease cases. In particular, the PDL yields 2–3% superior results compared to the VGG1 and ResNet50 models. It systematically summarizes the number of correctly or incorrectl predicted images. The vertical axis corresponds to the predicted class (output class), whil the horizontal axis represents the true class (target class). Each confusion matrix is visuall depicted as a heat map, utilizing color-coding techniques. The presence of darker pixel representing the diagonal elements is observable in all of the confusion matrices that hav been displayed. This observation suggests that a substantial quantity of data is accuratel classified in its corresponding category. In contrast, bright hues show instances of mode misclassifications. The PDL achieved accurate classification for 1044 out of 1047 norma images, 698 out of 703 very mild images, and 269 out of 274 mild images. In contrast, it wa shown that all algorithms exhibited accurate classification of the moderate AD group, whil the very mild class demonstrated the lowest accuracy in classification. In the mild class the VGG-16 model accurately predicted 267 out of the total mild images (274), whereas th ResNet-50 model properly identified 249 out of the total 274 AD images.

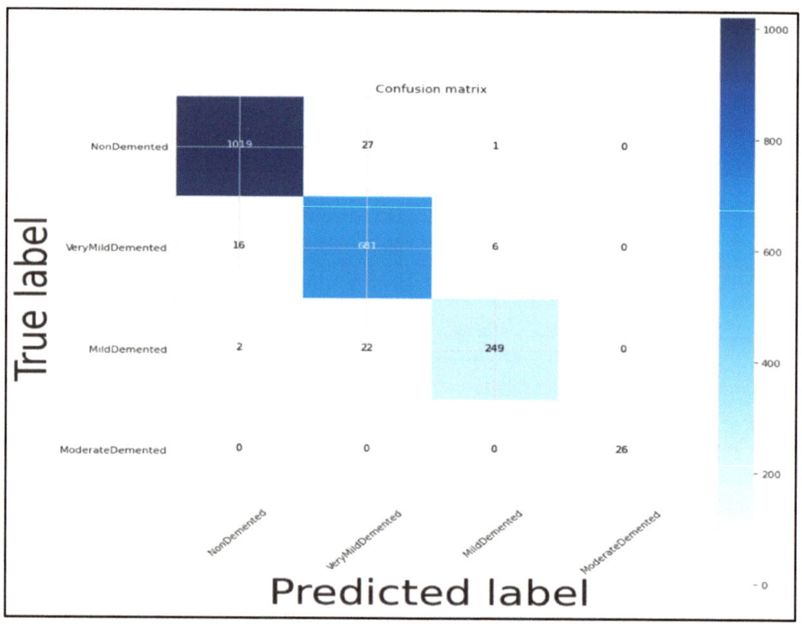

(**A**) Confusion matrix of ResNet50

Figure 7. *Cont.*

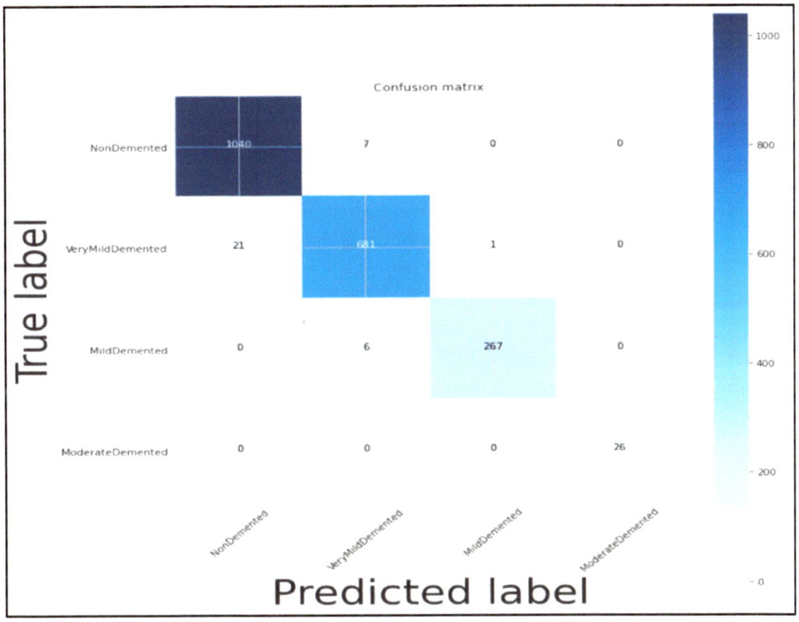

(**B**) Confusion matrix of VGG16

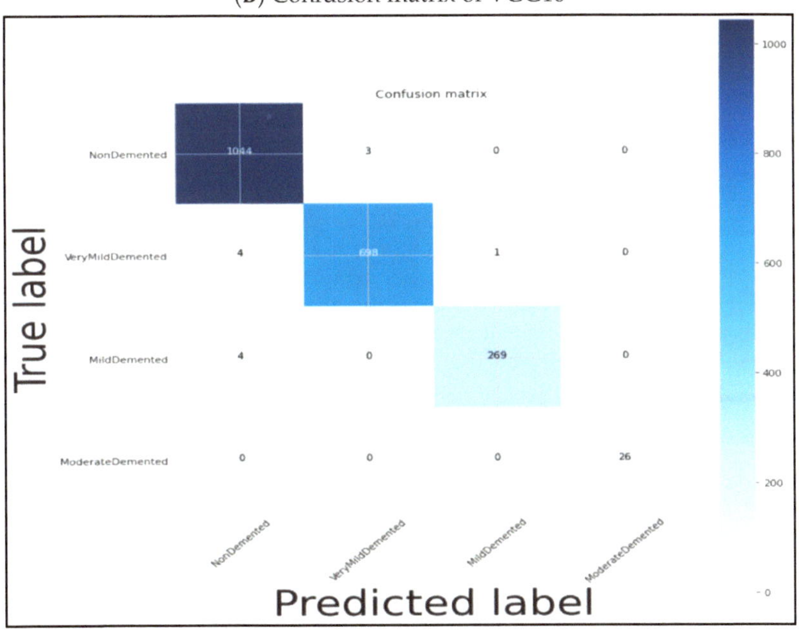

(**C**) Confusion matrix of the proposed method (PDL)

Figure 7. The confusion matrices of ResNet50, VGG16, and PDL are presented in (**A**–**C**), respectively. The dark blue square represents normal (non-AD), the sky blue represents very mild, the gray represents mild, and the non-colored square represents moderate. The horizontal axis represents the predicted label, and the vertical axis represents the true label. The number in the center of each square represents the number of images that were classified correctly. The other numbers represent the misclassification.

3.2. ROC Curves

The receiver operating characteristic (ROC) curve, which stands for "true positive rate vs. false positive rate", is a graph that shows how well a classification model works with different types of classification criteria. The ROC curves, specifically the AUC (area under the ROC curve) values, for the proposed approach (PDL), VGG16, and ResNet50 models are compared in Figure 8 in relation to the four cases of Alzheimer's disease. The ROC curves illustrate the individual AUC scores for each class as generated by three models, i.e. the classifiers that underwent training using EfficientNetB0 exhibited superior performance in comparison to those that were trained using ResNet50 and VGG16.

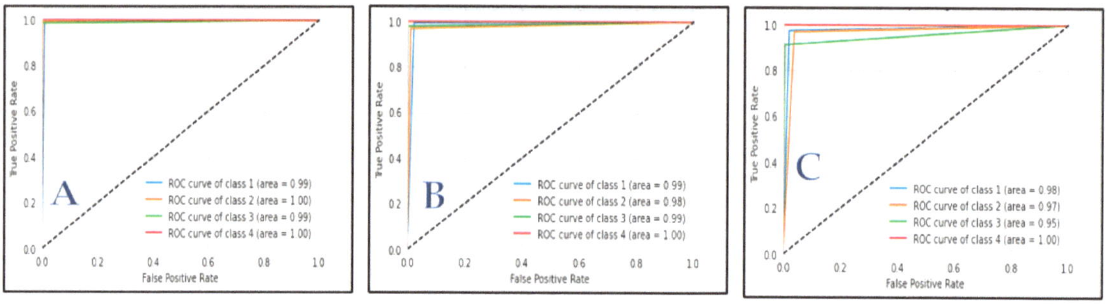

Figure 8. Representing the ROC curves, the vertical axis represents the true positive rate, and the horizontal axis represents the false positive rate. It shows the class-wise AUC scores obtained by the three models: (**A**) represents the proposed method; (**B**) represents VGG16; and (**C**) represents ResNet50. Blue represents the normal case (class 1), orange represents very mild AD (class 2), green represents mild AD (class 3), and red represents moderate AD (class 4).

4. Discussion

This study tested transfer learning to train our proposed deep-learning algorithm with the objective of accurately classifying different stages of Alzheimer's disease. This study compared our proposed method with two other well-known methods (VGG16 and ResNet50) on Kaggle AD datasets. The findings of our study exhibited a higher level of accuracy and yielded favorable results in comparison to other investigations [45]. The results of our work indicate that pre-trained models achieved high levels of accuracy without requiring data augmentation or extended training epochs. The utilization of hyper-parameters, as informed by previous research [46], involved the adjustment of batch size and learning rate to enhance the learning process and improve generalization accuracy. This enabled us to train CNNs that exhibit effective picture classification capabilities even when using less precise hyper-parameter values. Based on our results, it has been shown that a trained model has the ability to effectively classify AD into distinct phases with a high degree of accuracy. The classification results produced by the three distinct models are presented in Table 3, along with the corresponding values for four performance indicators. The proposed model demonstrated higher performance than VGG16 and Resnet50 for normal cases in terms of overall precision, with scores of 99.00%, 98.00%, and 98%, respectively. The evaluation of F1 score, precision, and recall performance metrics indicates that the proposed framework outperforms VGG16 in most scenarios. Equally, the ResNet50 model exhibits comparatively inferior performance outcomes compared to the other models. The consistency of the AUC values across all categories indicates that the predictions made by the proposed model are stable. Furthermore, the findings indicate that the prediction accuracy of both the VGG16 and ResNet50 models was comparatively lower for the mild and very mild stages. The results also indicated that the utilization of AD improved the ability to classify across all categories. A notable observation is that a significant proportion of the receiver operating characteristic (ROC) curves are situated above the linear reference line that connects the points (0,0) and (1,1). Nevertheless, the curves in question do not demonstrate a significant closeness to the upper-left corner, mostly because a restricted dataset was used for testing. Based on the confusion matrices depicted in Figure 7, it is apparent that the two highest-performing models exhibited nearly identical levels of accuracy when identifying MRI pictures associated with Alzheimer's disease. The proposed model demonstrates accurate classification, with a success average rate of 98.6% for normal brain pictures and 99.3% for mild and very mild AD images. The VGG16-based model has an accuracy rate of 98.6% for correctly classifying normal cases, 99% for mild AD, and 97.3% for very mild AD images. The ResNet50 model demonstrates an accuracy rate of 97.3% for properly classifying normal brain images, 95% for accurately identifying very mild AD, and 94% for correctly classifying mild AD images. Based on the results of the analysis, it can be concluded that the performance of the proposed model surpassed that of the VGG16 and ResNet50 models for mild and very mild AD cases.

5. Conclusions

This paper describes an automated new method for diagnosing Alzheimer's disease (AD) that uses image processing and novel deep transfer learning to figure out how bad the disease is and find important brain areas linked to it. The models use limited training sets of brain MRI scans. The empirical evaluations conducted in our study demonstrate that our proposed approach exhibited superior performance in handling the classification method compared to other popular state-of-the-art models. It attained an impressive overall average classification accuracy of 99.3%.

The findings of this work indicate that the proposed approach shows excellent performance in properly classifying Alzheimer's disease (AD) and its various stages within a limited and restricted dataset. The findings underscore the capacity of computers to aid physicians in the process of diagnosing AD conditions. The proposed method demonstrated remarkable efficacy in extracting valuable information from pictures and accurately predicting prognostic indicators of the disease. Notably, this was achieved without requiring extensive image processing, optimization, or data augmentation techniques. Further research will be conducted to examine the impact of data augmentation techniques on the outcomes of various AD datasets.

Author Contributions: Conceptualization, A.S.; methodology, A.S., R.A. and O.A.; software, R.A. and O.A.; validation, A.A., K.A. and A.S.; formal analysis, A.S., A.A. and M.A.; investigation, O.A., R.A. and A.S.; resources, Z.A. and M.A.; data curation, O.A. and R.A.; writing original draft, R.A. and A.S.; writing review and editing, all authors; visualization, R.A.; supervision, A.S., A.A. and K.A.; funding acquisition, O.A., M.A. and Z.A. All authors have read and agreed to the published version of the manuscript.

Funding: The Deputyship for Research and Innovation, Ministry of Education in Saudi Arabia, project no. (IFKSUOR–3–206).

Institutional Review Board Statement: Not applicable.

Informed Consent Statement: Not applicable.

Data Availability Statement: The data presented in this study are available on request from the corresponding author. The data are not publicly available due to privacy.

Acknowledgments: The authors extend their appreciation to the Deputyship for Research and Innovation, Ministry of Education in Saudi Arabia for funding this research through project no. (IFKSUOR–3–206).

Conflicts of Interest: The authors declare no conflict of interest.

References

1. Helaly, H.A.; Badawy, M.; Haikal, A.Y. Deep Learning Approach for Early Detection of Alzheimer's Disease. *Cogn. Comput.* **2021**, *14*, 1711–1727. [CrossRef] [PubMed]
2. Islam, J.; Zhang, Y. Brain MRI Analysis for Alzheimer's Disease Diagnosis Using an Ensemble System of Deep Convolutional Neural Networks. *Brain Inform.* **2018**, *5*, 2. [CrossRef] [PubMed]

3. Andrushia, A.D.; Sagayam, K.M.; Dang, H.; Pomplun, M.; Quach, L. Visual-Saliency-Based Abnormality Detection for MRI Brain Images—Alzheimer's Disease Analysis. *Appl. Sci.* **2021**, *11*, 9199. [CrossRef]
4. Raju, M.; Thirupalani, M.; Vidhyabharathi, S.; Thilagavathi, S. Deep Learning Based Multilevel Classification of Alzheimer's Disease Using MRI Scans. *IOP Conf. Ser. Mater. Sci. Eng.* **2021**, *1084*, 012017. [CrossRef]
5. Sethi, M.; Ahuja, S.; Rani, S.; Koundal, D.; Zaguia, A.; Enbeyle, W. An Exploration: Alzheimer's Disease Classification Based on Convolutional Neural Network. *BioMed Res. Int.* **2022**, *2022*, 8739960. [CrossRef] [PubMed]
6. Pushpa, B.R.; Amal, P.S.; Kamal, N.P. Detection and stage wise classification of Alzheimer disease using deep learning methods. *Int. J. Recent Technol. Eng. (IJRTE)* **2019**, *7*, 206–212.
7. Sadat, S.U.; Shomee, H.H.; Awwal, A.; Amin, S.N.; Reza, T.; Parvez, M.Z. Alzheimer's disease detection and classification using transfer learning technique and ensemble on Convolutional Neural Networks. In Proceedings of the 2021 IEEE Interna-tional Conference on Systems, Man, and Cybernetics (SMC), Melbourne, Australia, 17–20 October 2021. [CrossRef]
8. Vrahatis, A.G.; Skolariki, K.; Krokidis, M.G.; Lazaros, K.; Exarchos, T.P.; Vlamos, P. Revolutionizing the Early Detection of Alzheimer's Disease through Non-Invasive Biomarkers: The Role of Artificial Intelligence and Deep Learning. *Sensors* **2023**, *23*, 4184. [CrossRef]
9. Lecun, Y.; Bengio, Y.; Hinton, G. Deep learning. *Nature* **2015**, *521*, 436–444. [CrossRef]
10. Danker, A.; Wirgård Wiklund, J. *Using Transfer Learning to Classify Different Stages of Alzheimer's Disease*; KTH Royal INSTITUTE of Technology: Stockholm, Sweden, 2021; p. 33.
11. Saleem, T.J.; Zahra, S.R.; Wu, F.; Alwakeel, A.; Alwakeel, M.; Jeribi, F.; Hijji, M. Deep Learning-Based Diagnosis of Alzheimer's Disease. *J. Pers. Med.* **2022**, *12*, 815. [CrossRef]
12. Saleem, T.J.; Chishti, M.A. Deep learning for Internet of Things data analytics. *Procedia Comput. Sci.* **2019**, *163*, 381–390. [CrossRef]
13. Suk, H.I.; Shen, D. Deep ensemble sparse regression network for Alzheimer's disease diagnosis. In *International Workshop on Machine Learning in Medical Imaging*; Springer: Cham, Switzerland, 2016; pp. 113–121.
14. Billones, C.D.; Demetria, O.J.L.D.; Hostallero, D.E.D.; Naval, P.C. DemNet: A convolutional neural network for the detection of Alzheimer's disease and mild cognitive impairment. In Proceedings of the 2016 IEEE Region 10 Conference (TENCON), Singapore, 22–25 November 2016; pp. 3724–3727.
15. Sarraf, S.; Tofighi, G. Classification of alzheimer's disease using fmri data and deep learning convolutional neural networks. *arXiv* **2016**, arXiv:1603.08631.
16. Sarraf, S.; Tofighi, G. Classification of Alzheimer's disease structural MRI data by deep learning convolutional neural networks. *arXiv* **2016**, arXiv:1607.06583.
17. Gunawardena, K.A.N.N.P.; Rajapakse, R.N.; Kodikara, N.D. Applying convolutional neural networks for pre-detection of alzheimer's disease from structural MRI data. In Proceedings of the 2017 24th International Conference on Mechatronics and Machine Vision in Practice (M2VIP), Auckland, New Zealand, 21–23 November 2017; pp. 1–7.
18. Basaia, S.; Agosta, F.; Wagner, L.; Canu, E.; Magnani, G.; Santangelo, R.; Filippi, M.; Alzheimer's Disease Neuroimaging Initiative. Automated classification of Alzheimer's disease and mild cognitive impairment using a single MRI and deep neural networks. *NeuroImage Clin.* **2019**, *21*, 101645. [CrossRef] [PubMed]
19. Wang, S.-H.; Phillips, P.; Sui, Y.; Liu, B.; Yang, M.; Cheng, H. Classification of Alzheimer's disease based on eight-layer convolutional neural network with leaky rectified linear unit and max pooling. *J. Med. Syst.* **2018**, *42*, 85. [CrossRef]
20. Karasawa, H.; Liu, C.-L.; Ohwada, H. Deep 3d convolutional neural network architectures for alzheimer's disease diagnosis. In *Asian Conference on Intelligent Information and Database Systems*; Springer: Cham, Switzerland, 2018; pp. 287–296.
21. Goceri, E. Diagnosis of Alzheimer's disease with Sobolev gradient-based optimization and 3D convolutional neural network. *Int. J. Numer. Methods Biomed. Eng.* **2019**, *35*, e3225. [CrossRef]
22. Tang, H.; Yao, E.; Tan, G.; Guo, X. A fast and accurate 3D fine-tuning convolutional neural network for Alzheimer's disease diagnosis. In *International CCF Conference on Artificial Intelligence*; Springer: Singapore, 2018; pp. 115–126.
23. Spasov, S.E.; Passamonti, L.; Duggento, A.; Liò, P.; Toschi, N. A multi-modal convolutional neural network framework for the prediction of Alzheimer's disease. In Proceedings of the 2018 40th Annual International Conference of the IEEE Engineering in Medicine and Biology Society (EMBC), Honolulu, HI, USA, 18–21 July 2018; pp. 1271–1274.
24. Basheera, S.; Ram, M.S.S. Convolution neural network–based Alzheimer's disease classification using hybrid enhanced independent component analysis based segmented gray matter of T2 weighted magnetic resonance imaging with clinical valuation. *Alzheimer's Dement. Transl. Res. Clin. Interv.* **2019**, *5*, 974–986. [CrossRef]
25. Jiang, X.; Chang, L.; Zhang, Y.-D. Classification of Alzheimer's disease via eight-layer convolutional neural network with batch normalization and dropout techniques. *J. Med. Imaging Health Inform.* **2020**, *10*, 1040–1048. [CrossRef]
26. Mehmood, A.; Maqsood, M.; Bashir, M.; Shuyuan, Y. A deep siamese convolution neural network for multi-class classification of alzheimer disease. *Brain Sci.* **2020**, *10*, 84. [CrossRef]
27. Raju, M.; Gopi, V.P.; Anitha, V.S.; Wahid, K.A. Multi-class diagnosis of Alzheimer's disease using cascaded three dimensional convolutional neural network. *Phys. Eng. Sci. Med.* **2020**, *43*, 1219–1228. [CrossRef]
28. Sun, J.; Yan, S.; Song, C.; Han, B. Dual-functional neural network for bilateral hippocampi segmentation and diagnosis of Alzheimer's disease. *Int. J. Comput. Assist. Radiol. Surg.* **2020**, *15*, 445–455. [CrossRef]

29. Dyrba, M.; Hanzig, M.; Altenstein, S.; Bader, S.; Ballarini, T.; Brosseron, F.; Buerger, K.; Cantré, D.; Dechent, P.; Dobisch, L.; et al. Improving 3D convolutional neural network comprehensibility via interactive visualization of relevance maps: Evaluation in Alzheimer's disease. *arXiv* **2020**, arXiv:2012.10294. [CrossRef] [PubMed]
30. Feng, W.; Halm-Lutterodt, N.V.; Tang, H.; Mecum, A.; Mesregah, M.K.; Ma, Y.; Li, H.; Zhang, F.; Wu, Z.; Yao, E.; et al. Automated MRI-based deep learning model for detection of Alzheimer's disease process. *Int. J. Neural Syst.* **2020**, *30*, 2050032. [CrossRef] [PubMed]
31. Solano-Rojas, B.; Villalón-Fonseca, R. A Low-Cost Three-Dimensional DenseNet Neural Network for Alzheimer's Disease Early Discovery. *Sensors* **2021**, *21*, 1302. [CrossRef] [PubMed]
32. Tan, M.; Le, Q.V. EfficientNet: Rethinking Model Scaling for Convolutional Neural Networks. In Proceedings of the 36th International Conference on Machine Learning, Long Beach, CA, USA, 10–15 June 2019.
33. Marques, G.; Agarwal, D.; de la Torre Díez, I. Automated medical diagnosis of COVID-19 through EfficientNet convolutional neural network. *Appl. Soft Comput.* **2020**, *96*, 106691. [CrossRef] [PubMed]
34. Agarwal, D.; Berbis, M.A.; Martín-Noguerol, T.; Luna, A.; Garcia, S.C.P.; de la Torre-Díez, I. End-to-end deep learning architectures using 3D neuroimaging biomarkers for early Alzheimer's diagnosis. *Mathematics* **2022**, *10*, 2575. [CrossRef]
35. Agarwal, D.; Berbís, M.Á.; Luna, A.; Lipari, V.; Ballester, J.B.; de la Torre-Díez, I. Automated Medical Diagnosis of Alzheimer's Disease Using an Efficient Net Convolutional Neural Network. *J. Med. Syst.* **2023**, *47*, 57. [CrossRef] [PubMed]
36. Luz, E.; Silva, P.L.; Silva, R.; Silva, L.; Guimarães, J.; Miozzo, G.; Moreira, G.; Menotti, D. Towards an effective and efficient deep learning model for COVID-19 patterns detection in X-ray images. *arXiv* **2021**, arXiv:2004.05717. [CrossRef]
37. Zebin, T.; Rezvy, S. COVID-19 Detection and Disease Progression Visualization: Deep Learning on Chest X-rays for Classification and Coarse Localization. *Appl. Intell.* **2021**, *51*, 1010–1021. [CrossRef]
38. Dubey, S. Alzheimer's Dataset (4 Class of Images). Available online: https://www.kaggle.com/tourist55/alzheimers-dataset-4-class-of-images (accessed on 29 November 2020).
39. Fu'adah, Y.N.; Wijayanto, I.; Pratiwi, N.K.C.; Taliningsih, F.F.; Rizal, S.; Pramudito, M.A. Automated Classification of Alzheimer's Disease Based on MRI Image Processing Using Convolutional Neural Network (CNN) With AlexNet Architecture. *J. Physics Conf. Ser.* **2021**, *1844*, 012020. [CrossRef]
40. Atlases—NIST. Available online: https://nist.mni.mcgill.ca/atlases/ (accessed on 20 March 2023).
41. Kang, H.; Park, H.-M.; Ahn, Y.; Van Messem, A.; De Neve, W. Towards a Quantitative Analysis of Class Activation Mapping for Deep Learning-Based Computer-Aided Diagnosis. In Proceedings of the Medical Imaging 2021: Image Perception, Observer Performance, and Technology Assessment, Online, 15–19 February 2021; Samuelson, F.W., Taylor-Phillips, S., Eds.; SPIE: Bellingham, WA, USA, 2021; Volume 11599. [CrossRef]
42. Lim, B.Y.; Lai, K.W.; Haiskin, K.; Kulathilake, K.A.S.H.; Ong, Z.C.; Hum, Y.C.; Dhanalakshmi, S.; Wu, X.; Zuo, X. Deep Learning Model for Prediction of Progressive Mild Cognitive Impairment to Alzheimer's Disease Using Structural MRI. *Front. Aging Neurosci.* **2022**, *14*, 876202. [CrossRef]
43. Kabani, A.; El-Sakka, M.R. Object Detection and Localization Using Deep Convolutional Networks with Softmax Activation and Multi-class Log Loss. In Proceedings of the 2016 International Conference on Image Analysis and Recognition (ICIAR 2016), Póvoa de Varzim, Portugal, 13–15 July 2016; pp. 358–366.
44. Deng, J.; Dong, W.; Socher, R.; Li, L.-J.; Li, K.; Fei-Fei, L. ImageNet: A large-scale hierarchical image database. In Proceedings of the 2009 IEEE Conference on Computer Vision and Pattern Recognition, Miami, FL, USA, 20–25 June 2009; pp. 248–255. [CrossRef]
45. Prakash, D.; Madusanka, N.; Bhattacharjee, S.; Park, H.-G.; Kim, C.-H.; Choi, H.-K. A Comparative Study of Alzheimer's Disease Classification using Multiple Transfer Learning Models. *J. Multimedia Inf. Syst.* **2019**, *6*, 209–216. [CrossRef]
46. Kandel, I.; Castelli, M. The Effect of Batch Size on the Generalizability of the Convolutional Neural Networks on a Histopathology Dataset. *ICT Express* **2020**, *6*, 312–315. [CrossRef]

applied sciences

MDPI

Review

Design, Synthesis and Molecular Modeling Study of Radiotracers Based on Tacrine and Its Derivatives for Study on Alzheimer's Disease and Its Early Diagnosis

Przemysław Koźmiński * and Ewa Gniazdowska *

Centre of Radiochemistry and Nuclear Chemistry, Institute of Nuclear Chemistry and Technology, 03-195 Warsaw, Poland
* Correspondence: p.kozminski@ichtj.waw.pl (P.K.); e.gniazdowska@ichtj.waw.pl (E.G.)

Abstract: From 1993 to 2013, tacrine was an approved drug for Alzheimer's disease. Due to its strong inhibitory properties towards cholinesterase, tacrine causes an increase in the level of the neurotransmitter acetylcholine in the cholinergic system of the central nervous system. This work presents a review of articles in which tacrine or its derivatives labeled with the radionuclides ^{3}H, ^{11}C, ^{14}C, ^{123}I, ^{99m}Tc and ^{68}Ga were used as vectors in radiotracers dedicated to the diagnosis of Alzheimer's disease. The possibility of clinical applications of the obtained radiopreparations was assessed by analyzing their physicochemical properties, ability to cross the blood–brain barrier and the level of uptake in the brain. Based on these data, it was shown that radiopreparations based on the tacrine molecule or its very close analogues retain the ability to cross the blood–brain barrier, while radiopreparations containing a more modified tacrine molecule (connected via a linker to a radionuclide chelator) lose this ability. This is probably the result of the addition of a chelator, which significantly increases the size of the radiopreparation and reduces its lipophilicity. Computer docking studies of tacrine derivatives and/or radiopreparations showed how these compounds bind to the active sites of acetyl- and butyrylcholinesterase.

Keywords: Alzheimer's disease; tacrine; radiopharmaceuticals; molecular modeling; PET; SPECT

Citation: Koźmiński, P.; Gniazdowska, E. Design, Synthesis and Molecular Modeling Study of Radiotracers Based on Tacrine and Its Derivatives for Study on Alzheimer's Disease and Its Early Diagnosis. *Appl. Sci.* **2024**, *14*, 2827. https://doi.org/10.3390/app14072827

Academic Editor: Qi-Huang Zheng

Received: 30 January 2024
Revised: 19 March 2024
Accepted: 25 March 2024
Published: 27 March 2024

1. Introduction

Alzheimer's disease (AD) is a progressive central nervous system (CNS) disease leading to the loss of cognitive abilities, the initial symptoms of which are often attributed to the normal aging process [1–3]. The initial stage of the disease may last for many years and is often latent. In the final stage of the disease, the patient is unable to perform basic everyday activities. Alzheimer's disease cannot be cured. Nevertheless, early symptomatic treatment helps alleviate the symptoms and delay the progression of the disease. However, this requires an early diagnosis, which is usually unattainable due to the long latent period of the disease and the lack of morphological symptoms. Such pathophysiological symptoms in everyday functioning, such as dementia, loss of memory and orientation and loss of daily physical activities, which are often similar to those of other diseases, are already visible at the stage of very advanced AD. The causes of Alzheimer's disease are not clearly defined. Many risk factors, both environmental and genetic, are considered here. These may be, for example, a head injury, clinical depression, high blood pressure [3] or a genetic factor [4,5]. In the course of Alzheimer's disease, increased amounts of beta-amyloid (Aβ) accumulate in the brain, which can accumulate extracellularly in the form of amyloid plaques and tau proteins or intracellularly in the form of neurofibrillary tangles—both of these phenomena cause impairment of neuronal transmission, leading to the loss of proper brain function [6]. AD cannot be cured, and already existing pathophysiological symptoms cannot be reversed. However, it is possible to slow down the course of this disease and reduce its cognitive symptoms. An early pathophysiological feature of the mild to moderate stages of AD is loss of memory and cognitive function, caused by a deficiency of the neurotransmitter acetylcholine (ACh). ACh deficiency results from its hydrolysis, induced by the enzyme acetylcholinesterase (AChE), and leads to the selective loss of cholinergic neurons in the cerebral cortex, basal ganglia and hippocampus [7]. The two main therapeutic strategies in AD are influencing the processing of amyloid precursor protein (APP) and slowing the decline of neuronal degeneration and improving cholinergic neurotransmission [1,7–10]. As therapeutic agents in the mild and moderate stages of Alzheimer's disease, AChE inhibitors, e.g., tacrine, rivastigmine, galantamine and donepezil, have been used [11–14]. One of the known and tested inhibitors is tacrine (1,2,3,4-tetrahydro-9-amino acridine, THA)—the active substance of the drug Cognex [15,16], approved in 1993 by the US Food and Drug Administration (FDA) for use in treating the symptoms of AD. The therapeutic effect of THA is achieved by reversible binding to AChE and its inactivation, which results in an increase in the concentration of ACh at cholinergic synapses, thanks to which a greater number of cholinergic neurons remain intact, and the progression of the disease is slowed down [10,12,17,18]. Tacrine is characterized by high biological activity towards AChE, but unfortunately, it causes a number of common side effects (nausea, indigestion, vomiting, anorexia, abdominal pain, diarrhea, skin rash) and a high risk of liver damage, especially when using large doses of the drug [11–14,19]. In 2013, it was withdrawn due to hepatotoxicity and cardiovascular problems occurring in patients.

Nevertheless, both tacrine and its derivatives have been used as biologically active molecules in potential radiopharmaceuticals dedicated to the early diagnosis of Alzheimer's disease [1]. Diagnostic radiopharmaceuticals are compounds that use a biologically active molecule as a vector and contain a diagnostic radionuclide emitting gamma or beta plus radiation. They are administered to the patient in nanomolar amounts, so they do not cause any morphological changes in the body. At the same time, registration of the emitted radiation allows for the precise location of the radiopharmaceutical in the patient's body and thus the location of the disease lesion.

In the presented review, we collected and discussed data on radioactively labeled tacrine and its derivatives indicated to the early diagnosis of Alzheimer's disease. Specific consideration has been placed on the role of computational molecular modeling in the visualization of the interaction of tacrine with cholinesterase.

2. Radiolabeled Tacrine and Its Derivatives Used in Alzheimer's Disease

There are many papers on tacrine and its use in treating the symptoms of Alzheimer's disease but only a dozen papers on the use of tacrine and its derivatives as a vector in described radiotracers.

2.1. Radiotracers Based on Tacrine

One of the first tacrine-based radiopreparations was the [9-^{14}C]tacrine radiotracer (Figure 1A) [20,21]. The authors examined the distribution of this radiotracer in the rat body after both intravenous and oral administration using a quantitative whole-body autoradiographic method [20]. Based on the results obtained, the authors concluded that, in both cases, [^{14}C]tacrine ([^{14}C]THA) crosses the blood—brain barrier; the biodistribution of the radiotracer is similar, although after oral administration, the absorption of the radiotracer persists in the organs noticeably longer. Due to the potential use of ([^{14}C]THA) in the diagnosis of neurological diseases, the authors examined the regional distribution of the radiotracer in the brain—the highest levels of radioactivity were detected in the cortex, hippocampus, cerebellum and striatum. Analyzing the distribution of the radiotracer in individual parts of the brain, the authors found that it did not correlate consistently with the distribution of acetylcholinesterase (AChE), which may suggest that the effect of tacrine in the treatment of senile dementia may occur in a way other than by inhibiting the enzyme. Based on the high level of activity detected in the kidneys and ureters, the authors also concluded that the radiotracer is excreted primarily in the urine, although they also noted that there are indications of excretion of the radiotracer and its metabolites into the intestines.

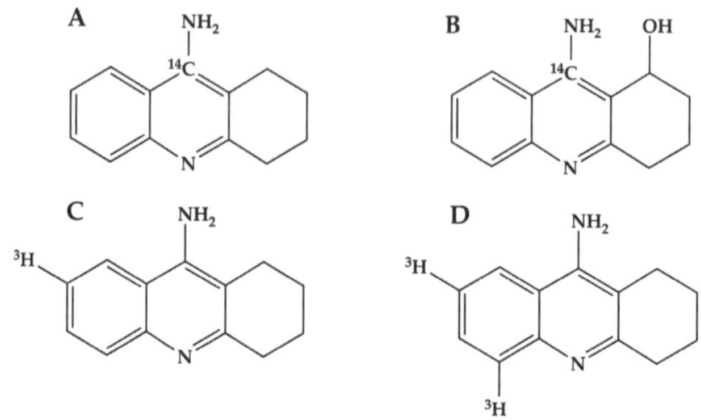

Figure 1. Radiotracers based on tacrine: (**A**) [9-^{14}C]tacrine [20,21]; (**B**) [9-^{14}C]1-OH-THA [21]; (**C**) [7-^{3}H]tacrine [22,23]; (**D**) [5,7-^{3}H]tacrine [23].

Further research by McNally et al. on the application of the radiotracer [9-^{14}C]tacrine (Figure 1A) in the study of THA distribution in the brain is presented in [21]. Using oral administration (by oral gavage) and single or multiple (twice daily for 3 days) doses (SD or MD, respectively), the authors performed in vivo studies of the distribution of [9-^{14}C]THA and its main metabolite [9-^{14}C]1-OH-tacrine ([9-^{14}C]1-OH-THA, Figure 1B) in rats. At selected time intervals, ranging from 0.5 h to 96 h, they examined the level of radioactivity in blood, plasma, the heart, the lungs, the kidneys, the liver, the pancreas and various brain regions (brainstem, cerebellum, cortex, hippocampus, striatum and thalamus) using the autoradiography method. The authors also performed in vitro studies of the process of tacrine metabolism in the brain. The rat brain homogenate was incubated with the [9-^{14}C]THA radiotracer and its metabolite [9-^{14}C]1-OH-THA. Then, after the incubation, the products of tacrine metabolism were identified using the HPLC method in specially prepared samples. For this purpose, various products of tacrine metabolism ([9-^{14}C]1-OH-THA, [9-^{14}C]2-OH-THA and [9-^{14}C]4-OH-THA) and their reference compounds (1-OH-THA, 2-OH-THA and 4-OH-THA) were previously characterized by HPLC. Based on the results obtained, McNally et al. drew a number of conclusions. They found that for both doses used (SD and MD), the distribution of [9-^{14}C]THA in the brain was similar although in the case of MD, the radioactivity levels in the tested organs were visibly higher. After oral administration of [9-^{14}C]THA, the radiotracer penetrates very quickly into the brain and accumulates in the cortex and hippocampus, i.e., in the areas responsible for the cognitive functions of the brain. In vitro studies of tacrine metabolism in the brain performed using rat brain homogenate showed that this process practically does not occur in the brain and that metabolite [9-^{14}C]1-OH-THA has a very limited ability to cross the blood–brain barrier. This difference in the ability to cross the blood–brain barrier correlates well with the log p values calculated by the authors for THA and its metabolite 1-OH-THA (3.30 and 1.66, respectively). The lack of transformation of tacrine into its metabolites in the brain and low ability of metabolites to cross the blood-brain barrier correlate perfectly with (observed in radioautographic studies in rats) a slight accumulation of [9-^{14}C]1-OH-THA in the brain and a high level of [9-^{14}C]1-OH-THA in the blood, which explains the significantly higher brain-to-plasma ratio determined for tacrine than for its 1-OH-THA metabolite. The results of the experiments conducted by the authors also indicate that the transport of THA and its metabolites to the brain tissue takes place through a simple passive process, and the excretion of these compounds occurs through the biliary and urinary tracts.

Tritium-labelled tacrine, [7-^{3}H]tacrine ([^{3}H]THA, a very close analogue of tacrine, custom synthesized from 7-bromo-9-amino-1,2,3,4-tetrahydroacridine, Figure 1C), was used in the study by Mena et al. to locate tacrine binding sites in the rat brain [22]. Using P2 membrane fractions prepared from rat brain, Mena and Desai performed a number of tests: AChE enzymatic test, localization of [^{3}H]THA binding sites, kinetic parameters of [^{3}H]THA binding using various concentrations of inactive THA and the ability of THA to inhibit AChE. In order to be able to compare the results obtained, experimental conditions (buffer, pH, concentration of the tested compound) as similar as possible were used in all studies. Based on the results obtained, the authors concluded that [^{3}H]THA binds to the membrane almost rapidly but in a reversible manner—adding an inactive ligand completely removed [^{3}H]THA from the membrane. They also showed that the binding of [^{3}H]THA in the rat brain is not blocked by a number of other neurotransmitters/neuromodulators, so the binding site of [^{3}H]THA is different from the sites of action (sites of receptors) of these compounds. Moreover, their research (autoradiography) also showed that [^{3}H]THA binding sites are not located together with the activity of acetylcholine (and other acetylcholinesterase inhibitors); therefore, it can be concluded that the clinical effect of THA may also result from an action other than through the cholinergic neuronal system.

The procedure for the synthesis of tritium-labeled tacrine was described by Egan et al. [23]. The authors synthesized two radiopreparations, [7-^{3}H]tacrine (Figure 1C) and [5,7-^{3}H]tacrine (Figure 1D), using the catalytic tritium dehalogenation of 7-bromo-tacrine and 5,7-dibromo-tacrine compounds, respectively. The products of the individual stages of synthesis were analyzed using the proton and tritium NMR method and TLC and HPLC methods equipped with UV and/or liquid scintillation detectors.

In 2007, Jogani et al. presented the results of studies on the intravenous and intranasal administration of tacrine labeled with technetium-99m [24]. The labeling reaction was performed directly using a solution of tacrine in propylene glycol, reducing agent $SnCl_2$ and pertechnetate solution [^{99m}Tc]TcO_4^- (eluate from the ^{99}Mo/^{99m}Tc generator). The authors demonstrated the stability of the radiopreparation in normal saline solution and in mouse serum as well as in a challenge experiment with DTPA (diethylenetriamine pentaacetic acid). Unfortunately, the obtained radiopreparation [^{99m}Tc]Tacrine solution ([^{99m}Tc]Tc-TS) was tested only by thin-layer chromatography (TLC) using silica gel-coated fiberglass sheets as a stationary phase and two different mobile phases: acetone and pyridine:acetic acid:water (3:5:1.5, v/v). The structure of [^{99m}Tc]Tc-TS is also unknown. The paper presents the results of biodistribution studies in mice and γ-scintigraphy imaging studies (performed using single photon emission computerized tomography, SPECT) in rabbits. In both studies, intravenous (IV) and intranasal (IN) administrations of [^{99m}Tc]Tc-TS were used. The results of these studies showed that the tacrine concentration in the brain after intranasal administration was significantly higher than after intravenous administration. The authors concluded that nasal-to-brain administration of tacrine may provide an alternative route to the currently used oral route, which is limited by tacrine's low bioavailability and pronounced side effects.

Jogani et al. continued to study tacrine delivery to the brain via the nasal route using other tacrine formulations [25]. It is known that, in nasal drug administration, the application of microemulsion and mucoadhesive agent, due to the small size of the globules and their lipophilicity, effectively improves the delivery process of a drug by increasing the retention of formulations at the absorption site. The authors prepared and characterized tacrine solution (TS), tacrine microemulsion (TME) and tacrine mucoadhesive microemulsion (TMME) and assessed their pharmacokinetic and pharmacodynamic properties in the process of tacrine delivery to the brain. Then, similarly to previous studies [24], all three tacrine formulations were labeled with ^{99m}Tc radionuclide to obtain the radiopreparations [^{99m}Tc]Tc-TS, [^{99m}Tc]Tc-TME and [^{99m}Tc]Tc-TMME. As before, the stability of all three radiotracers was tested in normal saline solution and in mouse serum as well as in a challenge experiment with DTPA. Using different routes of administration (intranasal (IN) and intravenous (IV)) and all three radiotracers, the authors performed biodistribution studies in mice and γ-scintigraphy imaging studies in rabbits. Analyzing the results obtained in all studies (for TS, TME, TMME, [^{99m}Tc]Tc-TS, [^{99m}Tc]Tc-TME and [^{99m}Tc]Tc-TMME and IN or IV administration), the authors showed that, in the case of intranasal administration, the accumulation of a given tacrine formulation (TS, TME, TMME, [^{99m}Tc]Tc-TS, [^{99m}Tc]Tc-TME and [^{99m}Tc]Tc-TMME) in the brain and the brain/blood ratio was higher than in the case of intravenous administration. They also showed that the most efficient transport of tacrine to the brain was observed in the case of TMME-based preparations, followed by TME-based preparations, and the lowest was observed in the case of TS-based preparations. The results of the study by Jogani et al. suggest that the intranasal administration of an appropriate tacrine formulation may minimize gastrointestinal and hepatic side effects and may play an important role in the treatment of patients with Alzheimer's disease.

Concise information concerning radiotracers based on tacrine is presented in Table 1.

Table 1. Radiotracers based on tacrine.

Tacrine-Based Radiotracers	Research Purpose and Conclusions	References
[9-^{14}C]tacrine ([^{14}C]THA)	Distribution of the radiopreparation in the rat body after both intravenous and oral administration, studies of the process of tacrine metabolism in the brain • In both cases, [^{14}C]THA crosses the blood–brain barrier; • Biodistribution of the radiotracer is similar, although after oral administration, the absorption of the radiotracer persists noticeably longer in the organs; • The highest levels of radioactivity were detected in the cortex, hippocampus, cerebellum and striatum; • The radiotracer is excreted mainly in the urine and partly through the intestines; • Tacrine metabolism practically does not occur in the brain; • The main metabolite of tacrine is [9-^{14}C]1-OH-THA, and it has a very limited ability to cross the blood–brain barrier.	[20,21]
[7-^{3}H]tacrine ([^{3}H]THA)	AChE enzymatic test, localization of [^{3}H]THA binding sites in the rat brain, kinetic parameters of [^{3}H]THA binding, ability of THA to inhibit AChE • [^{3}H]THA binds to the membrane almost rapidly but in a reversible manner; • Binding of [^{3}H]THA in the rat brain is not blocked by a number of other neurotransmitters/neuromodulators, so the binding site of [^{3}H]THA is different from the sites of action of these compounds; • [^{3}H]THA binding sites are not located together with the activity of acetylcholine (and other acetylcholinesterase inhibitors), so it can be concluded that the clinical effect of THA may also result from an action other than through the cholinergic neuronal system.	[22]
[7-^{3}H]tacrine and [5,7-^{3}H]tacrine	Procedure for the synthesis of tritium-labeled tacrine, analysis of products using TLC, HPLC and NMR methods	[23]
[^{99m}Tc]Tacrine solution ([^{99m}Tc]Tc-TS)	Studies on the intravenous (IV) and intranasal (IN) administration of tacrine labeled with technetium-99m, biodistribution studies in mice • The tacrine concentration in the brain after IN administration is significantly higher than after IV administration; • Nasal-to-brain administration of tacrine may provide an alternative route to the currently used oral route, which is limited by tacrine's low bioavailability and pronounced side effects.	[24]
[^{99m}Tc]Tc-TS, [^{99m}Tc]Tc-TME, [^{99m}Tc]Tc-TMME	Syntheses and characterization of tacrine solution (TS), tacrine microemulsion (TME) and tacrine mucoadhesive microemulsion (TMME) and assessment of their pharmacokinetic and pharmacodynamic properties in the process of tacrine delivery to the brain, radiolabeling of these tacrine formulations with ^{99m}Tc, biodistribution studies in mice after intravenous (IV) and intranasal (IN) administration • In the case of IN administration, the accumulation of a given tacrine formulation in the brain and the brain/blood ratio are higher than in the case of IV administration; • The most efficient transport of tacrine to the brain was observed in the case of TMME-based preparations, followed by TME-based preparations, and the lowest was observed in the case of TS-based preparations • IN administration of an appropriate tacrine formulation may minimize gastrointestinal and hepatic side effects and may play an important role in the treatment of patients with AD	[25]

2.2. Radiotracers Based on Tacrine Derivatives

Some of the first reports on radionuclide-labeled tacrine derivatives are the papers presenting 1,2,3,4-tetrahydro-9-methyl-amino acridine (N-methyl-THA, MTHA) labeled with ^{11}C radionuclide, synthesized by Bonnot et al. [26] and studied in vivo in non-human primates by Tavitian et al. [27] and in healthy human volunteers by Traykov et al. [28]. In vitro testing of tacrine (THA) and its N-methyl derivative MTHA showed that both compounds have very similar inhibitory properties towards acetylcholinesterase (AChE) [27]. The authors conducted a study of the distribution of the [methyl-^{11}C]1,2,3,4-tetrahydro-9-methyl-amino acridine radiotracer ([^{11}C]MTHA, Figure 2A) in rats (radioactivity was measured in blood, plasma, the heart, the liver, the kidneys, the lungs, skeletal muscles and the brain (separately in the pons, cerebellum, colliculi, hypothalamus, hippocampus, striatum and anterior and posterior cortices)), and positron emission tomography (PET) imaging was performed on two male adult baboons. In these experiments (distribution study in rats and PET imaging in baboons), both [^{11}C]MTHA radiotracer alone and [^{11}C]MTHA radiotracer together with unlabeled THA administered simultaneously or 20 min before tracer injection were used. The results, in the case of using the [^{11}C]MTHA radiopreparation and in the case of administering the [^{11}C]MTHA radiopreparation and then unlabeled THA, showed significantly different amounts of radioactivity accumulated in the examined organs, significantly lower in the case of using unlabeled THA in the experiment, e.g., radioactivity accumulated in all brain regions studied in the case of coinjection of THA together with [^{11}C]MTHA was 40–50% lower. This allows for us to conclude that these two molecules ([^{11}C]MTHA and THA) compete for the same binding sites. Based on these results, the authors concluded that the [^{11}C]MTHA radiotracer could be considered a promising PET ligand for studying THA binding in the brain.

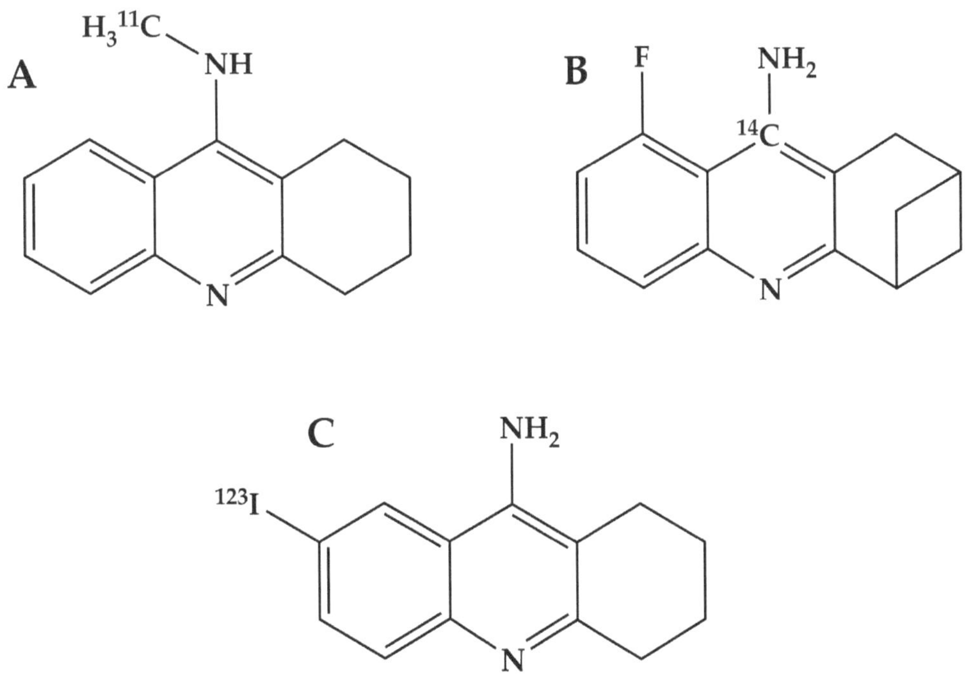

Figure 2. Radiotracers based on tacrine derivatives: (**A**) [methyl-^{11}C]1,2,3,4-tetrahydro-9-methyl-amino acridine [26–28]; (**B**) 9-amino-8-fluoro-2,4-methane-1,2,3,4-[9-^{14}C]tetrahydroacridine [29]; (**C**) 7-[^{123}I]iodotacrine [30].

The first studies of the [^{11}C]MTHA radiotracer (Figure 2A) in position emission tomography (PET) imaging in healthy volunteers are presented in the publication by Traykov et al. [28]. The study involved four healthy men who had never been diagnosed with Alzheimer's disease or other chronic diseases. The [^{11}C]MTHA radiotracer was used in PET and magnetic resonance imaging (MRI) methods in order to obtain individual cerebral anatomy. At specific time intervals (up to 70 min after intravenous administration of the tracer), the authors determined the level of radioactivity in the blood and in the selected areas of the brain (white matter, putamen, thalamus, brainstem, cerebellum cortex). The determined radioactivity in the brain was relatively high and amounted to approximately 6% of the administered dose. Radioactivity in the blood, after a rapid increase (approximately 1.5 min after administration) and then an equally rapid decline (approximately 5 min after administration), increased very slightly; this is related to the circulation of radiotracer metabolites ([^{11}C]1-hydroxy-MTHA) in the blood [21], which are, however, capable of crossing the blood–brain barrier to a significantly lesser extent than the administered radiotracer [21]. Traykov's study also showed that brain accumulation of the radiotracer [^{11}C]MTHA differs from the sites of action and/or sites of elevated AChE concentrations detected post-mortem in human brains. This allows for us to assume that the cerebral distribution of the [^{11}C]MTHA radiotracer in the human nervous system is not parallel with AChE, so the clinical effect of THA may also occur in a way other than through the cholinergic neuronal system, which has already been discussed in previous works [20,22].

Another radionuclide-labeled tacrine derivative is the radiopreparation based on the cholinesterase inhibitor 9-amino-8-fluoro-2,4-methane-1,2,3,4-tetrahydroacridine in which ^{14}C radionuclide is located in position 9 of the tetrahydroacridine ring (Figure 2B) [29]. Practically in every step of the synthesis of this radiotracer, the obtained intermediates were tested using many analytical methods (radio-thin layer chromatography (RTLC), radio-high performance liquid chromatography (RHPLC), infrared spectrum (IR), proton nuclear magnetic resonance (NMR) and mass spectrum (MS)). According to the authors' intention, this radiopreparation was to be used in metabolic studies, but due to the presence of a tacrine derivative and therapeutic radionuclide C-14, it could also be considered as a potential therapeutic radiopharmaceutical in Alzheimer's disease. However, this would also require physicochemical tests of the compound (lipophilicity, stability in body fluids and serum), which were not tested in this work.

Akula et al. developed a four-step synthesis procedure of 7-[^{123}I]iodotacrine ([^{123}I]7-I-THA, Figure 2C), a potential imaging agent for single photon emission computer tomography (SPECT), to map acetylcholinesterase (AChE) receptor sites in living organisms [30]. In each step of the synthesis of the [^{123}I]7-I-THA radiopreparation, the obtained intermediates were tested using analytical methods: melting points, elemental analysis and ^{1}H and ^{13}C-NMR analyses. The radiochemical purity of the [^{123}I]7-I-THA radiopreparation was tested by thin-layer chromatography (TLC) using an aluminum silica gel plate and a chloroform/methanol mixture (4:1, v/v) as a developing solvent. However, there are no reports in the literature of further studies using this radiotracer.

In the years 2017–2022, several works were published in which the labeling of tacrine derivatives with the diagnostic radionuclides ^{99m}Tc and ^{68}Ga was presented. The procedures for synthesizing and testing the physicochemical properties of the tacrine derivatives used here were designed and described by Szymański et al. [31,32]. The structural modification of tacrine consisted of attaching to the amino group of tacrine an aliphatic hydrocarbon chain composed of -$(CH_2)_n$- groups (where the number of carbon atoms n was from two to nine) and a bifunctional coupling agent (BFCA) capable of forming complexes with a given radionuclide. It was a series of compounds that differed only in the number of methylene groups (affecting the lipophilicity parameter of the compound) between tetrahydroacridine and the radionuclide complexing moiety.

The work by Gniazdowska et al. presents the syntheses and physicochemical and biological studies of eight [^{99m}Tc]Tc(NS$_3$)(CN-NH$(CH_2)_n$Tac radioconjugates (Figure 3A) consisting of ^{99m}Tc radionuclide coordinated by the tetradentate tripodal chelator (NS_3, tris(2-mercaptoethyl)-amine) and a monodentate isocyanide ligand (CN-BFCA, succinimidyl isocyanobutyric ester) previously conjugated to the tacrine molecule (Tac) [33]. All radioconjugates turned out to be completely stable in a challenge experiment with cysteine and histidine solutions and in human serum. For the [^{99m}Tc]Tc(NS_3)(CN-NH(CH_2)$_7$Tac radioconjugate characterized by the highest lipophilicity, stability tests in cerebrospinal fluid, biological activity towards acetylcholinesterase using Ellman's method and a multiorgan biodistribution study in normal mice were performed. For this radioconjugate and its parent tacrine-based complexing agent CN-NH(CH_2)$_7$Tac, computer docking studies were also performed. The biodistribution study showed a higher uptake of [^{99m}Tc]Tc(NS_3)(CN-NH(CH_2)$_7$Tac in the liver than in the kidney, indicating the clearance of the radioconjugate mainly through the hepatic route. High uptake was also observed in the lung, but uptake in the brain was relatively low, nevertheless demonstrating the ability of the [^{99m}Tc]Tc(NS_3)(CN-NH(CH_2)$_7$Tac radioconjugate to cross the blood–brain barrier.

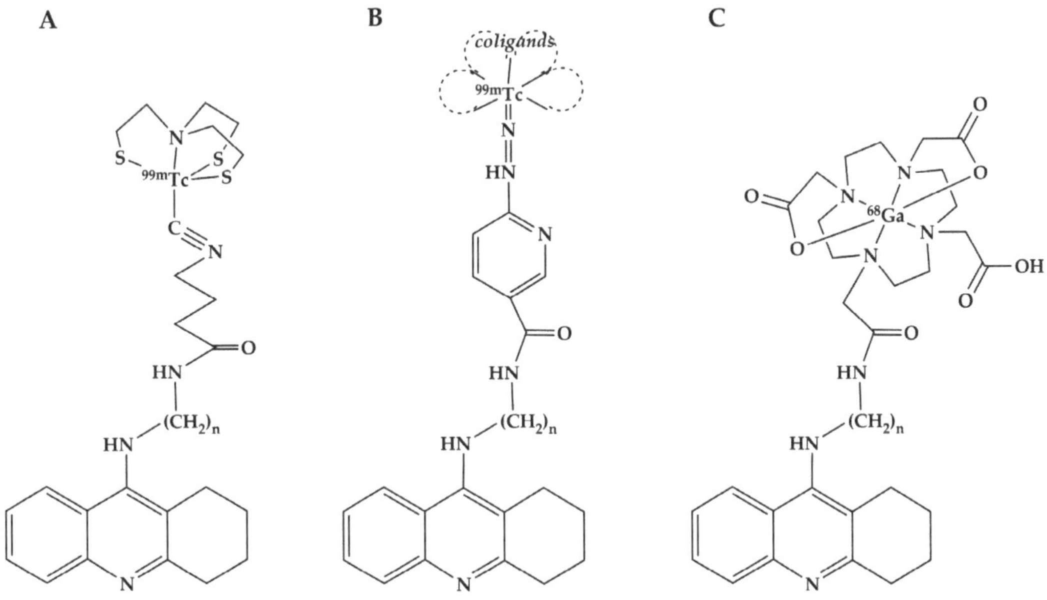

Figure 3. Radioconjugates based on tacrine derivatives: (**A**) [^{99m}Tc]Tc(NS_3)(CN-NH(CH_2)$_7$tacrine [33]; (**B**) [^{99m}Tc]Tc-Hynic-(tricine)$_2$NH(CH_2)$_n$tacrine [34]; (**C**) [^{68}Ga]Ga-DOTA-NH(CH_2)$_n$tacrine [34].

Radioconjugates based on the same tacrine derivatives, containing the radionuclide ^{99m}Tc complexed by 6-hydrazinonicotinamide (HYNIC) and ^{68}Ga complexed by macrocyclic ligand 1,4,7,10-tetraazacyclododecane-1,4,7,10-tetraacetic acid (DOTA), are presented in another work by Gniazdowska et al. [34]. All synthesized [^{99m}Tc]Tc-Hynic-(tricine)$_2$NH(CH_2)$_n$tacrine ([^{99m}Tc]Tc-Hynic-NH(CH_2)$_n$Tac) (Figure 3B), where n was in the range from two to nine, and [^{68}Ga]Ga-DOTA-NH(CH_2)$_n$tacrine ([^{68}Ga]Ga-DOTA-NH(CH_2)$_n$Tac), where n was seven, eight or nine (Figure 3C), turned out to be completely stable in cysteine and histidine solutions (challenge experiments), in human serum and in cerebrospinal fluid. The lipophilicity parameter was determined for all radioconjugates, and the determined log D parameters showed that all [^{99m}Tc]Tc-Hynic-NH(CH_2)$_n$Tac and [^{68}Ga]Ga-DOTA-NH(CH_2)$_n$Tac radioconjugates are definitely hydrophilic compounds. Furthermore, the radioconjugates containing the [^{68}Ga]Ga-DOTA complex were significantly more hydrophilic than the radioconjugates containing the [^{99m}Tc]Tc-Hynic complex. For the two radioconjugates (one from each series, [^{99m}Tc]Tc-Hynic-NH(CH_2)$_9$Tac and [^{68}Ga]Ga-

DOTA-NH(CH_2)$_9$Tac, containing nine methylene CH_2 groups in the aliphatic chain) with the most appropriate physicochemical properties (the highest possible lipophilicity parameter), cholinesterase inhibitory activity tests, biodistribution studies in mice and molecular modelling studies were carried out. Studies of the biological activity of the radioconjugates [^{99m}Tc]Tc-Hynic-NH(CH_2)$_9$Tac and [^{68}Ga]Ga-DOTA-NH(CH_2)$_9$Tac showed that these compounds have equally strong inhibitory properties against acetylcholinesterase (AChE) and butyrylcholinesterase (BuChE) as the reference compound tacrine. The in vivo biodistribution study showed the uptake of both radioconjugates in the brain, spleen, lungs, heart kidneys and liver; however, the uptake of [^{68}Ga]Ga-DOTA-NH(CH_2)$_9$Tac in the brain was very low. In general, the uptake of the [^{99m}Tc]Tc-Hynic-NH(CH_2)$_9$Tac radioconjugate was significantly higher in all analyzed organs (its uptake in the brain was four times greater than that of the gallium radioconjugate). According to the authors, the low efficiency of crossing the blood–brain barrier of the tested radioconjugates may be due to their hydrophilic nature, and, for example, in the case of the gallium radioconjugates, the use of a chelator less hydrophilic than DOTA could increase the lipophilicity of the radioconjugate.

Studies on radioconjugates, still based on the same tacrine derivatives, in which five various chelators were used to complex the ^{68}Ga radionuclide, were presented in the work of Koźmiński et al. [35]. The chelators used for the synthesis of the ^{68}Ga-radioconjugates were 2,2′-(7-(1-carboxy-4-((2,5-dioxopyrrolidin-1-yl)oxy)-4-oxobutyl)-1,4,7-triazonane-1,4-diyl)diacetic acid (NODAGA-NHS), 2,2′-(7-(1-carboxy-4-((4-isothiocyanatobenzyl)amino)-4-oxobutyl)-1,4,7-triazonane-1,4-diyl)diacetic acid (NODAGA-Bn-NCS), 2,2′,2″-(10-(1-carboxy-4-((4-isothiocyanatobenzyl)amino)-4-oxobutyl)-1,4,7,10-tetraazacyclododecane-1,4,7-triyl)triacetic acid (DOTAGA-Bn-NCS), [(R)-2-Amino-3-(4-isothiocyanatophenyl)propyl]-trans-(S,S)-cyclohexane-1,2-diamine-pentaacetic acid (DTPA-CHX-Bn-NCS) and N1,N7-bis((3-hydroxy-1,6-dimethyl-4-oxo-1,4-dihydropyridin-2-yl)methyl)-4-(3-(((3-hydroxy-1,6-dimethyl-4-oxo-1,4-dihydropyridin-2-yl)methyl)amino)-3-oxopropyl)-4-(3-(3-(4-isothiocyanatophenyl)thioureido)propanamido)heptanediamide (THP-Bn-NCS). In order to obtain the highest possible lipophilicity of the synthesized radioconjugates (recommended for radiopreparations capable of crossing the blood–tissue and blood–brain barriers), a tacrine derivative containing nine methylene groups in the aliphatic hydrocarbon chain was used for the synthesis of the ^{68}Ga-radioconjugates. All obtained radioconjugates (Figure 4) met the physicochemical properties required for radiopharmaceuticals. The tested inhibitory properties of the radioconjugates turned out to be no less than those of tacrine, which confirmed that the attachment of the radionuclide complex to tacrine through an appropriate linker does not change the biological properties of tacrine. Moreover, the [^{68}Ga]Ga-THP-NH(CH_2)$_9$Tac radioconjugate showed a much stronger activity towards both AChE and BuChE than the parent tacrine compound. Lipophilicity tests showed that all newly synthesized radioconjugates were significantly less hydrophilic, and two of them, [^{68}Ga]Ga-NODAGA-Bn-NH(CH_2)$_9$Tac and [^{68}Ga]Ga-THP-NH(CH_2)$_9$Tac, were already hydrophobic. These radioconjugates were selected for the biodistribution studies and molecular docking studies. The in vivo biodistribution studies in rats showed clearly different profiles. The [^{68}Ga]Ga-NODAGA-Bn-NH(CH_2)$_9$Tac compound, apart from a relatively large accumulation in the excretory organs (kidneys and liver), accumulated in other organs (lungs, blood, heart and spleen) in small amounts. The compound [^{68}Ga]Ga-THP-NH(CH_2)$_9$Tac circulated in the blood in large quantities and accumulated in comparable amounts in the excretory organs (kidneys, liver and spleen) as well as in the lungs and heart. Unfortunately, the uptake of both radioconjugates in the brain was low and insufficient from the point of view of potential application of these radioconjugates as a tool for the early diagnosis of Alzheimer's disease. Particularly noteworthy is the high uptake of the radioconjugate [^{68}Ga]Ga-THP-NH(CH_2)$_9$Tac in the lungs, which indicates its specificity for this organ, and due to the presence of cholinesterase in the glial tissue of the lungs, it allows for the use of this radioconjugate as a tool for imaging pathological conditions of the lungs.

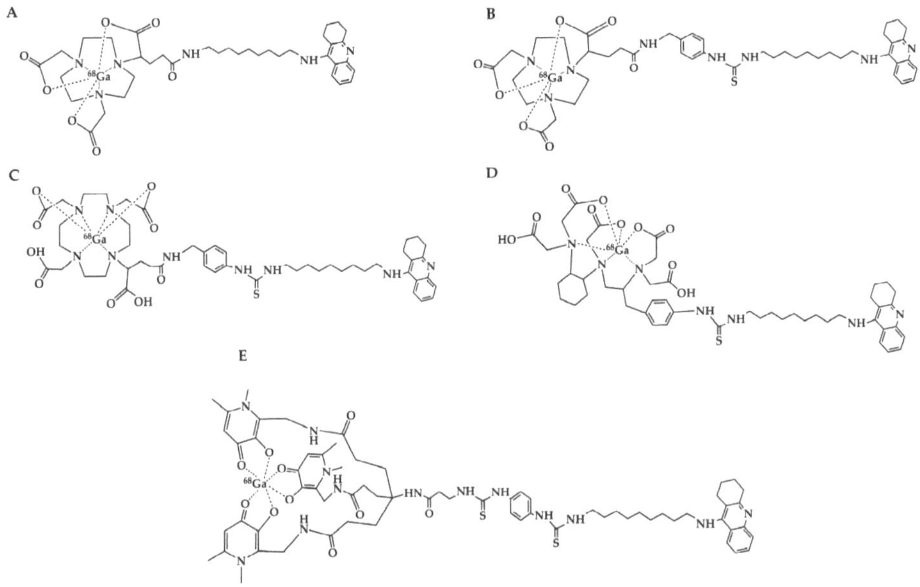

Figure 4. Radioconjugates based on tacrine derivatives [35]: (**A**) [^{68}Ga]Ga-NODAGA-NH(CH_2)$_9$tacrine; (**B**) [^{68}Ga]Ga-NODAGA-Bn-NH(CH_2)$_9$tacrine; (**C**) [^{68}Ga]Ga-DOTAGA-Bn-NH(CH_2)$_9$tacrine; (**D**) [^{68}Ga]Ga-DTPA-CHX-NH(CH_2)$_9$tacrine; (**E**) [^{68}Ga]Ga-THP-NH(CH_2)$_9$tacrine.

In a review article focusing on potential radiotracers based on tacrine and its analogues, it is worth presenting the radioconjugates synthesized and tested by Szymański et al. [36]. Compared to previously used tacrine derivatives [33–35], the structural modification of tacrine additionally consisted of replacing the six-membered tetrahydroacridine ring with a five-membered ring. As previously reported [33–35], 6-hydrazinonicotinamide (HYNIC, radionuclide ^{99m}Tc complexing agent) was attached to the amino group of tacrine through a hydrocarbon chain with a different number of (CH_2)$_n$ groups, where n = 2 ÷ 9. Biochemical tests performed spectrophotometrically using the Ellman method showed that tacrine derivatives with the long hydrocarbon chain (n = 7 ÷ 9) have higher inhibition activity and are more selective towards acetylcholinesterase (AChE) than tacrine, while all compounds showed less selectivity for butyrylcholinesterase (BChE) compared to tacrine. The docking studies of the new tacrine derivatives to AChE and BChE showed that all tacrine derivatives bound to AChE in a similar way—they extended along the active gorge of the enzyme and interacted with the catalytic and peripheral sites. In the case of BChE, the binding method of the tacrine derivatives to the enzyme was similar with a slight difference regarding the location of the hydrazinnicotin fragment in the reduced peripheral anionic site of BChE. Among these tacrine derivatives, for the synthesis of the ^{99m}Tc-radioconjugate, the authors chose the derivative that contained two methylene groups in the hydrocarbon chain and was characterized by the highest activity towards BChE (the level of this enzyme varies in different stages of Alzheimer's disease). The spectrophotometric test of this compound (6-Hydrazino-*N*-[2-(2,3-dihydro-1*H*-cyclopenta[*b*]quinolin-9-ylamino)Ethyl]nicotinamide) showed its stability in water. After labeling this compound with ^{99m}Tc radionuclide, the obtained radioconjugate [^{99m}Tc]Tc-Hynic-2,3-dihydro-1H-cyclopenta[b]quinolone (Figure 5) was used to study biodistribution in rats. The greatest accumulation of radioactivity was observed in the liver, followed by the kidneys, lungs and gastrointestinal tract. Unfortunately, the uptake of the radioconjugate in the brain was very low (probably due to the hydrophilic nature of the radioconjugate), which indicates that the tested radioconjugate does not have sufficient ability to cross the blood–brain barrier and cannot be considered as a potential agent in the diagnosis of Alzheimer's disease.

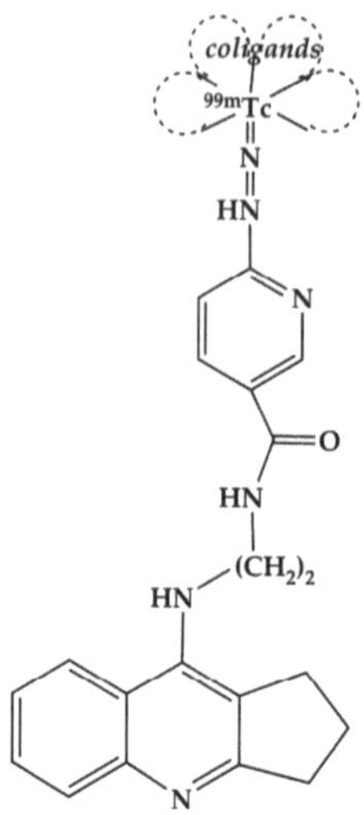

Figure 5. ^{99m}Tc-radioconjugate based on 2,3-dihydro-1H-cyclopenta[b]quinolone [36].

Concise information concerning radioconjugates based on tacrine derivatives is presented in Table 2.

Table 2. Radioconjugates based on tacrine derivatives.

Tacrine Derivative-Based Radioconjugates	Research Purpose and Conclusions	References
[^{11}C]MTHA	Synthesis procedure	[26]
	Studies in vivo in non-human primates: in rats (radioactivity was measured in blood, plasma, heart, liver, kidneys, lungs, skeletal muscles and brain (separately in the pons, cerebellum, colliculi, hypothalamus, hippocampus, striatum and anterior and posterior cortices)), and PET imaging on two male adult baboons using both [^{11}C]MTHA radiotracer alone and [^{11}C]MTHA radiotracer together with unlabeled THA • Radioactivity accumulated in all brain regions studied; in the case of co-injection of THA together with [^{11}C]MTHA, accumulation was significantly lower; • These two molecules, [^{11}C]MTHA and THA, compete for the same binding sites; • The [^{11}C]MTHA radiotracer could be considered a promising PET ligand for studying THA binding in the brain.	[27]
	Studies in vivo in healthy human volunteers and measurement of radioactivity accumulated in all brain regions studied in the case of co-injection of THA together with [^{11}C]MTHA • Determined radioactivity in the brain was relatively high and amounted to approximately 6% of the administered dose; • Brain accumulation of the radiotracer [^{11}C]MTHA differs from the sites of action and/or sites of elevated AChE concentrations detected post-mortem in human brains; • Cerebral distribution of the [^{11}C]MTHA radiotracer in the human nervous system is not parallel with AChE, so the clinical effect of THA may also occur in a way other than through the cholinergic neuronal system.	[28]

Table 2. *Cont.*

Tacrine Derivative-Based Radioconjugates	Research Purpose and Conclusions	References
9-amino-8-fluoro-2,4-methane-1,2,3,4-[9-^{14}C]tetrahydroacridine	Synthesis procedure of radiopreparation dedicated for metabolic studies, potential therapeutic radiopharmaceutical in AD	[29]
[^{123}I]7-I-THA	Procedure for the synthesis of a potential agent for imaging the map of acetylcholinesterase (AChE) receptor sites in living organisms	[30]
[^{99m}Tc]Tc(NS$_3$)(CN-NH(CH$_2$)$_n$Tac	Synthesis and physicochemical properties of radioconjugates, multiorgan biodistribution study of [^{99m}Tc]Tc(NS$_3$)(CN-NH(CH$_2$)$_7$Tac in normal mice, computer docking studies of [^{99m}Tc]Tc(NS$_3$)(CN-NH(CH$_2$)$_7$Tac and its parent tacrine-based complexing agent CN-NH(CH$_2$)$_7$Tac • A higher uptake in liver than in kidney indicates the clearance of the radioconjugate mainly through the hepatic route; • A high uptake in the lung and relatively low uptake in the brain	[33]
[^{99m}Tc]Tc-Hynic-NH(CH$_2$)$_n$Tac, [^{68}Ga]Ga-DOTA-NH(CH$_2$)$_n$Tac	Synthesis and physicochemical properties of radioconjugates, biodistribution studies in mice and molecular modelling studies of [^{99m}Tc]Tc-Hynic-NH(CH$_2$)$_9$Tac and [^{68}Ga]Ga-DOTA-NH(CH$_2$)$_9$Tac • Radioconjugates containing the [^{68}Ga]Ga-DOTA complex are significantly more hydrophilic than radioconjugates containing the [^{99m}Tc]Tc-Hynic complex; • The uptake of the [^{99m}Tc]Tc-Hynic-NH(CH$_2$)$_9$Tac radioconjugate was significantly higher in all analyzed organs than that of [^{68}Ga]Ga-DOTA-NH(CH$_2$)$_9$Tac.	[34]
[^{68}Ga]Ga-NODAGA-NH(CH$_2$)$_9$Tac, [^{68}Ga]Ga-NODAGA-Bn-NH(CH$_2$)$_9$Tac, [^{68}Ga]Ga-DOTAGA-Bn-NH (CH$_2$)$_9$Tac, [^{68}Ga]Ga-DTPA-CHX-NH (CH$_2$)$_9$Tac, [^{68}Ga]Ga-THP-NH(CH$_2$)$_9$Tac	Synthesis and physicochemical properties of ^{68}Ga-radioconjugates using the five various chelators NODAGA-NHS, NODAGA-Bn-NCS, DOTAGA-Bn-NCS, DTPA-CHX-Bn-NCS-THP-Bn-NCS, biodistribution studies in mice and molecular modelling studies of [^{68}Ga]Ga-NODAGA-Bn-NH(CH$_2$)$_9$Tac and [^{68}Ga]Ga-THP-NH(CH$_2$)$_9$Tac • Radioconjugates [^{68}Ga]Ga-NODAGA-NH(CH$_2$)$_9$Tac, [^{68}Ga]Ga-DOTAGA-Bn-NH(CH$_2$)$_9$Tac and [^{68}Ga]Ga-DTPA-CHX-NH(CH$_2$)$_9$Tac were hydrophilic; • [^{68}Ga]Ga-NODAGA-Bn-NH(CH$_2$)$_9$Tac and [^{68}Ga]Ga-THP-NH(CH$_2$)$_9$Tac were hydrophobic; • The uptake of the [^{68}Ga]Ga-NODAGA-Bn-NH(CH$_2$)$_9$Tac and [^{68}Ga]Ga-THP-NH(CH$_2$)$_9$Tac radioconjugates in the brain was low and insufficient from the point of view of their application for the early diagnosis of AD; • Relatively high uptake of [^{68}Ga]Ga-THP-NH(CH$_2$)$_9$Tac in the lungs.	[35]
[^{99m}Tc]Tc-Hynic-(CH$_2$)$_2$2,3-dihydro-1H-cyclopenta[b] quinolone	Synthesis, physicochemical properties and docking studies of Hynic-(CH$_2$)$_n$-2,3-dihydro-1H-cyclopenta[b]quinolone derivatives, synthesis, physicochemical properties and biodistribution studies in rats of [^{99m}Tc]Tc-Hynic-(CH$_2$)$_2$-2,3-dihydro-1H-cyclopenta[b]quinolone radioconjugate • Derivative containing two methylene groups in the hydrocarbon chain was characterized by the highest activity towards BChE; • The highest uptake of radioconjugate was observed in the liver, followed by the kidneys, lungs and gastrointestinal tract; • Low uptake of the radioconjugate in the brain.	[36]

2.3. Molecular Modeling Studies of Tacrine Derivatives—Cholinesterase Interaction

The study of the crystal structure of *Torpedo californica* acetylcholinesterase (*Tc*AChE) made possible, for the first time at atomic resolution, the visualization of the acetylcholine (ACh) binding pocket [37–39]. The binding pocket is a narrow and deep gorge approximately 5 Å wide and 20 Å long and lined (in approximately 40–60%) with rings of 14 conserved aromatic residues: Y70, W84, F120, Y121, Y130, W233, W279, F288, F290, F330, F331, Y334, W432 and Y442 [28]. It penetrates into the enzyme more than halfway and expands near the bottom to form a cavity called the active binding site, containing the catalytic triad S200, E327 and H440 (Figure 7 in ref. [37]). Five amino acids, Tyr70, Asp72, Tyr121, Trp279 and Tyr334, located at the entrance to the gorge, constitute the peripheral anionic site (PAS) binding for AChE [38]. A large number of aromatic residues in the gorge walls and bases in the gorge bottom result in many different hydrophobic and "anionic" interaction sites in the binding pocket, located separately or overlapping with the active sites of the ACh-binding enzyme (the ACh binding site in AChE contains from six to nine negative charges). The aromatic nature of the gorge influences the high degree of binding of a given substrate and thus the high catalytic activity of the enzyme. *Tc*AChE has a very large dipole moment which is greatly influenced by the presence of five acidic amino acids located around the entrance to the gorge. The axis of the AChE dipole moment is oriented along the axis of the gorge's active sites. Along the gorge, along the entire length of the active sites, there is a potential gradient that can effectively pull the substrate appearing at the gorge mouth (Y121, F330) down the gorge.

Computational studies of the interaction of some tacrine derivatives as well as potential radiotracers based on them (Table 3) with acetyl- and/or butyrylcholinesterase were discussed in the works of Gniazdowska et al. [33,34], Koźmiński et al. [35] and Szymański et al. [36]. The conjugates and radioconjugates selected for the molecular docking studies are listed in Table 3.

Table 3. List and structure of conjugates and radioconjugates selected for research using computer calculations.

Radio(conjugates)	Molecular Docking Studies to	
	AChE [Ref.]	**BChE [Ref.]**
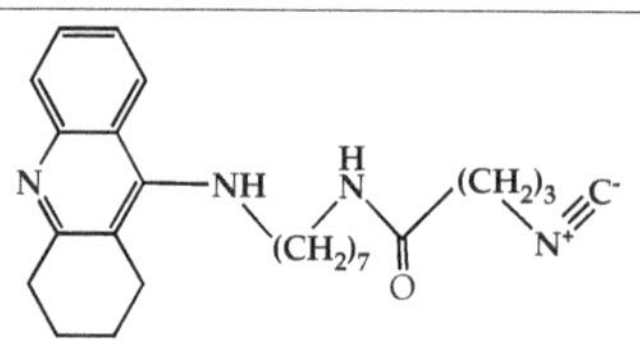	Figure 5 in [33]	-----
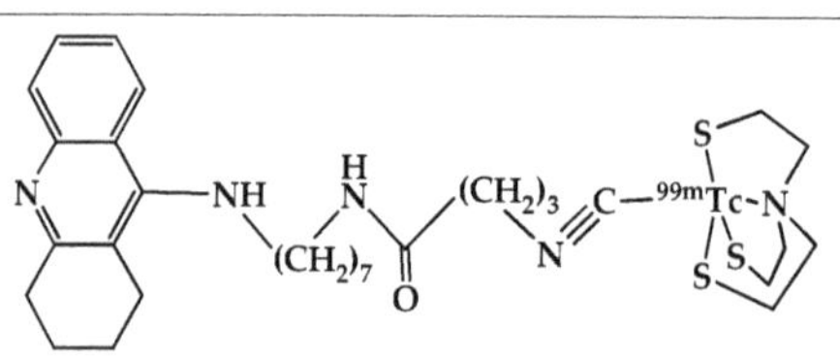	Figure 5 in [33]	-----
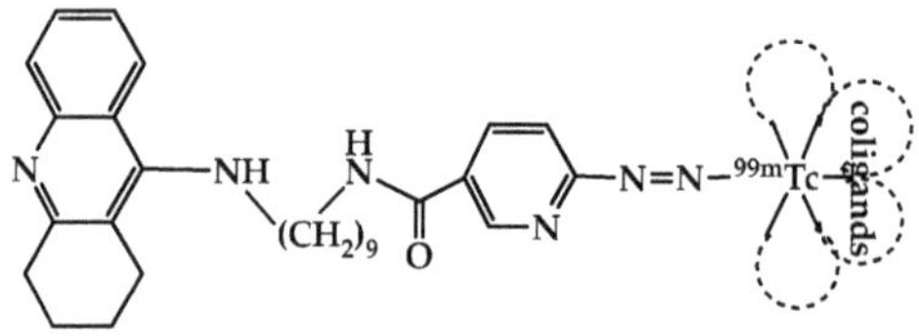	Figure 7 in [34]	-----

Table 3. *Cont.*

Radio(conjugates)	Molecular Docking Studies to	
	AChE [Ref.]	BChE [Ref.]
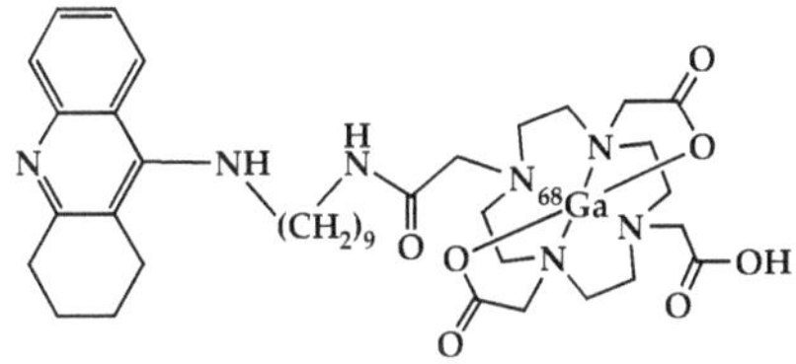	-----	Figure 8 in [34]
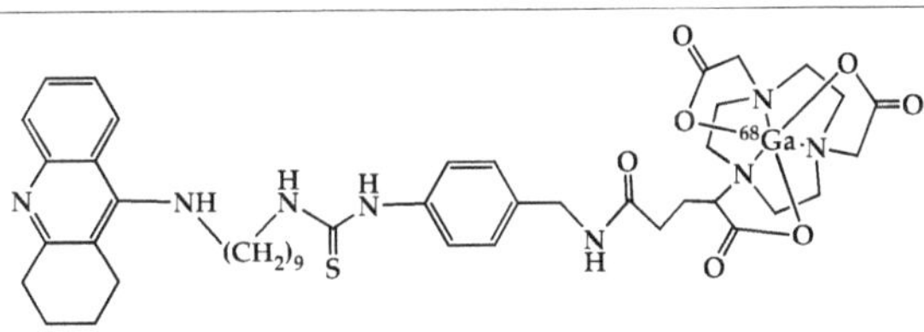	Figure 8 in [35]	-----
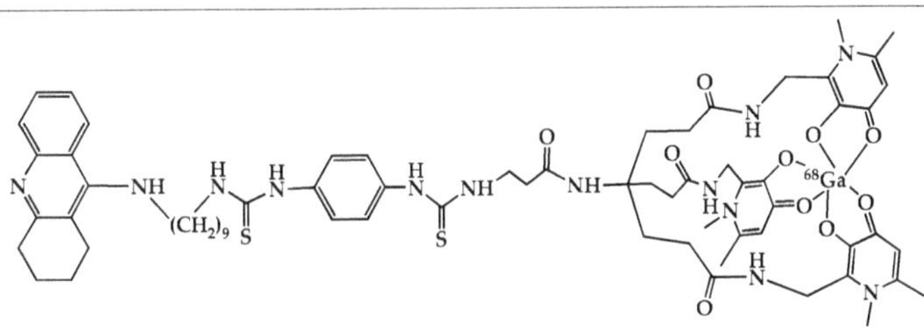	Figure 8 in [35]	------
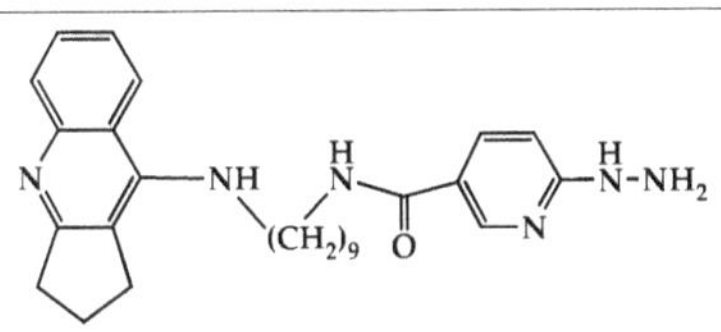	Figure 1 in [36]	-----
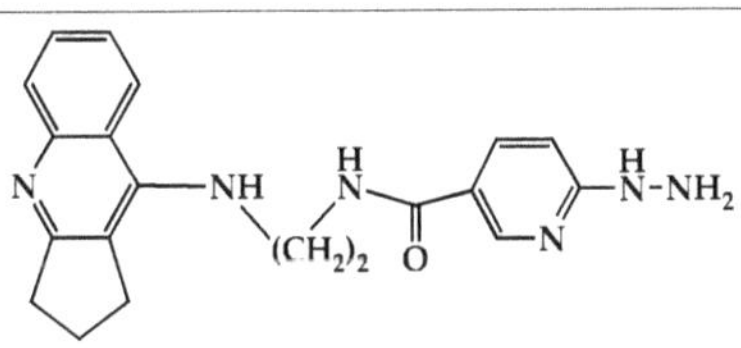	-----	Figure 2 in [36]

Molecular modeling studies made it possible to determine the structure of the inhibitor–cholinesterase system and to determine and locate individual interactions between the components of this system responsible for the action of the inhibitor on the enzyme. Detailed information about the nature and location of individual interactions of the inhibitor with the amino acids forming the enzyme's binding pocket is provided in the cited publications.

In general, all tested inhibitors had similar components, namely tacrine, a longer or shorter linker in the form of a hydrocarbon chain and, optionally, a radionuclide complex. Therefore, their fit into the cholinesterase binding pocket and their interactions with AChE and BChE were similar. A fragment of tacrine (as well as a tacrine analog with a cyclopentane ring) was located at the bottom of the gorge, and the main interaction here was the interaction with the catalytic triad. The hydrocarbon chain was located along the gorge and formed hydrophobic interactions with the aromatic rings of the amino acids present in the gorge wall. The radionuclide complex is usually located outside or at the entry to the gorge of the binding pocket and interacted through peripheral anionic sites.

3. Discussion

The works discussed in this article focused on the search for a diagnostic radiopharmaceutical for the earliest possible diagnosis of Alzheimer's disease. The biologically active molecule in these radiopharmaceuticals was the medicinal preparation tacrine (or its derivatives) that was approved by the FDA in 1993 for the treatment of Alzheimer's disease and was, in 2013, withdrawn due to harmful side effects. However, since radiopharmaceuticals are administered to the patient in microgram quantities and, in the case of diagnostic radiopharmaceuticals, their use is not frequent, the problem of the harmfulness of tacrine is not significant. However, due to the high ability of tacrine to cross the blood–brain barrier, its use as a vector in radiopharmaceuticals is justified.

As can be observed from the presented works, radiotracers based on the tacrine molecule, in which the radionuclide (^{3}H, ^{11}C, ^{14}C) is an isotope of one of the elements included in the tacrine molecule [9–12], easily cross the blood–brain barrier and accumulate in significant amounts in the brain. However, such radiopreparations are not easily available due to the too complicated procedures for synthesizing these radiopreparations, which are difficult or even impossible to perform in clinical conditions (in hospital laboratories).

The procedures for obtaining radioconjugates in the form of a tacrine solution, tacrine microemulsion and tacrine mucoadhesive microemulsion labeled (directly) with ^{99m}Tc and identifying them using the radio-TLC method [24,25] are relatively easy and can be performed in hospital laboratories. These radiopreparations accumulated relatively well in the brain, but regarding these compounds, there is no knowledge about their composition and structure, which, according to the authors of this review article, is a significant disadvantage. For these radiopreparations, intranasal administration has proven to be more effective than intravenous administration.

Radiopreparations based on a relatively minimally changed tacrine molecule also accumulated satisfactorily in the brain [26–30]. Modification of tacrine by adding a methyl group, fluorine or iodine did not reduce the ability of tacrine to cross the blood–brain barrier.

The most convenient procedure for the synthesis of diagnostic radiopharmaceuticals in clinical conditions is to perform a labeling reaction with generator radionuclides (^{99m}Tc, ^{68}Ga) of the active substance (tacrine or its derivative, often previously coupled with an appropriate chelator) included in the so-called kit [33–36] (kits are ready-made sets containing, in lyophilized form, the appropriate reagents in the appropriate quantities needed for the synthesis of a given radiopharmaceutical). However, the use of chelators significantly changes both the size of the final radiopreparation and its lipophilicity. Changes in these parameters do not have a significant impact on the inhibitory properties towards acetyl- and butyrylcholinesterase, but their effect in vivo is a significant reduction in the ability of the radiopharmaceutical to cross the blood–brain barrier and accumulate in the brain in the amounts necessary for imaging.

To sum up, it can be said that, despite many works, it has not been possible to find a radiopreparation based on tacrine or its derivatives that meets the requirements for radiopharmaceuticals. Perhaps it would be advisable to search for another biological molecule involved in the course of Alzheimer's disease from its earliest stage.

Author Contributions: Conceptualization, E.G. and P.K.; writing—original draft preparation, E.G. and P.K.; writing—review and editing, E.G. and P.K.; visualization, E.G. and P.K.; supervision, E.G.; project administration, E.G. and P.K. All authors have read and agreed to the published version of the manuscript.

Funding: This research received no external funding. This work has been supported by the statutory activity of the Institute of Nuclear Chemistry and Technology, Warsaw, Poland.

Institutional Review Board Statement: Not applicable.

Informed Consent Statement: Not applicable.

Data Availability Statement: No new data were created or analyzed in this study. Data sharing is not applicable to this article.

Conflicts of Interest: The authors declare no conflicts of interest.

Abbreviations

AChE	acetylcholinesterase
AD	Alzheimer's disease
APP	amyloid precursor protein
BChE, BuChE	butyrylcholinesterase
BFCA	bifunctional coupling agent
CN-BFCA	succinimidyl isocyanobutyric ester
CNS	central nervous system
DOTA	1,4,7,10-tetraazacyclododecane-1,4,7,10-tetraacetic acid
DOTAGA-Bn-NCS	2,2′,2″-(10-(1-carboxy-4-((4-isothiocyanatobenzyl)amino)-4-oxobutyl)-1,4,7,10-tetraazacyclododecane-1,4,7-triyl)triacetic acid
DTPA-CHX-Bn-NCS	[(R)-2-Amino-3-(4-isothiocyanatophenyl)propyl]-trans-(S,S)-cyclohexane-1,2-diamine-pentaacetic acid
FDA	US Food and Drug Administration
HYNIC	6-hydrazinonicotinamide
IN	intranasal
IR	infrared spectrum
IV	intravenous
MRI	magnetic resonance imaging
MS	mass spectrum
MTHA, N-methyl-THA	1,2,3,4-tetrahydro-9-methyl-amino acridine
NMR	nuclear magnetic resonance
NODAGA-Bn-NCS	2,2′-(7-(1-carboxy-4-((4-isothiocyanatobenzyl)amino)-4-oxobutyl)-1,4,7-triazonane-1,4-diyl)diacetic acid
NODAGA-NHS	2,2′-(7-(1-carboxy-4-((2,5-dioxopyrrolidin-1-yl)oxy)-4-oxobutyl)-1,4,7-triazonane-1,4-diyl)diacetic acid
NS_3	tris(2-mercaptoethyl)-amine
PET	positron emission tomography
RHPLC	radio-high performance liquid chromatography
RTLC	radio-thin layer chromatography
SPECT	single photon emission computer tomography
THA, Tac	1,2,3,4-tetrahydro-9-amino acridine
THP-Bn-NCS	N1,N7-bis((3-hydroxy-1,6-dimethyl-4-oxo-1,4-dihydropyridin-2-yl)methyl)-4-(3-(((3-hydroxy-1,6-dimethyl-4-oxo-1,4-dihydropyridin-2-yl)methyl)amino)-3-oxopropyl)-4-(3-(3-(4-isothiocyanatophenyl)thioureido)propanamido)heptanediamide
TLC	thin-layer chromatography
TME	tacrine microemulsion
TMME	tacrine mucoadhesive microemulsion
TS	tacrine solution

References

1. Mikiciuk-Olasik, E.; Szymański, P.; Żurek, E. Diagnostics and therapy of Alzheimer's disease. *Indian J. Exp. Biol.* **2007**, *45*, 315–325. [PubMed]
2. Declercq, L.D.; Vandenberghe, R.; Van Laere, K.; Verbruggen, A.; Bormans, G. Drug Development in Alzheimer's Disease: The Contribution of PET and SPECT. *Front. Pharmacol.* **2016**, *7*, 88. [CrossRef] [PubMed]
3. Knopman, D.S.; Amieva, H.; Petersen, R.C.; Chételat, G.; Holtzman, D.M.; Hyman, B.T.; Nixon, R.A.; Jones, D.T. Alzheimer disease. *Nat. Rev. Dis. Primers* **2021**, *7*, 33. [CrossRef] [PubMed]
4. Long, J.M.; Holtzman, D.M. Alzheimer Disease: An Update on Pathobiology and Treatment Strategies. *Cell* **2019**, *179*, 312–339. [CrossRef] [PubMed]

5. Sienski, G.; Narayan, P.; Bonner, J.M.; Kory, N.; Boland, S.; Arczewska, A.A.; Ralvenius, W.T.; Akay, L.; Lockshin, E.; He, L.; et a APOE4 disrupts intracellular lipid homeostasis in human iPSC-derived glia. *Sci. Transl. Med.* **2021**, *13*, eaaz4564. [CrossRef [PubMed]
6. Tackenberg, C.; Kulic, L.; Nitsch, R.M. Familial Alzheimer's disease mutations at position 22 of the amyloid β-peptide sequence differentially affect synaptic loss, tau phosphorylation and neuronal cell death in an ex vivo system. *PLoS ONE* **2020**, *15*, e0239584 [CrossRef] [PubMed]
7. Chen, Z.R.; Huang, J.B.; Yang, S.L.; Hong, F.F. Role of Cholinergic Signaling in Alzheimer's Disease. *Molecules* **2022**, *27*, 1816 [CrossRef] [PubMed]
8. Francis, P.; Palmer, A.; Snape, M. The cholinergic hypothesis of Alzheimer's disease: A review of progress. *J. Neurol. Neurosurg Psychiatry* **1999**, *66*, 137–147. [CrossRef] [PubMed]
9. Camps, P.; Munoz-Torrero, D. Cholinergic drugs in pharmacotherapy of Alzheimer's disease. *Mini Rev. Med. Chem.* **2002**, *2*, 11–25 [CrossRef]
10. Ferreira-Vieira, T.H.; Guimaraes, I.M.; Silva, F.R.; Ribeiro, F.M. Alzheimer's disease: Targeting the Cholinergic System. *Curr Neuropharmacol.* **2016**, *14*, 101–115. [CrossRef]
11. Colović, M.B.; Krstić, D.Z.; Lazarević-Pašti, T.D.; Bondžić, A.M.; Vasić, V.M. Acetylcholinesterase inhibitors: Pharmacology and toxicology. *Curr. Neuropharmacol.* **2013**, *11*, 315–335. [CrossRef] [PubMed]
12. Walczak-Nowicka, Ł.J.; Herbert, M. Acetylcholinesterase Inhibitors in the Treatment of Neurodegenerative Diseases and the Role of Acetylcholinesterase in their Pathogenesis. *Int. J. Mol. Sci.* **2021**, *22*, 9290. [CrossRef]
13. Ruangritchankul, S.; Chantharit, P.; Srisuma, S.; Gray, L.C. Adverse Drug Reactions of Acetylcholinesterase Inhibitors in Older People Living with Dementia: A Comprehensive Literature Review. *Ther. Clin. Risk Manag.* **2021**, *17*, 927–949. [CrossRef] [PubMed]
14. Jann, M.W.; Shirley, K.L.; Small, G.W. Clinical pharmacokinetics and pharmacodynamics of cholinesterase inhibitors. *Clin Pharmacokinet.* **2002**, *41*, 719–739. [CrossRef]
15. Crismon, M.L. Tacrine: First drug approved for Alzheimer's disease. *Ann. Pharmacother.* **1994**, *28*, 744–751. [CrossRef] [PubMed]
16. Riekkinen, P., Jr.; Kuikka, J.; Soininen, H.; Helkala, E.L.; Hallikainen, M.; Riekkinen, P. Tetrahydroaminoacridine modulates technetium-99m labelled ethylene dicysteinate retention in Alzheimer's disease measured with single photon emission computed tomography imaging. *Neurosci. Lett.* **1995**, *195*, 53–56. [CrossRef]
17. Stanciu, G.D.; Luca, A.; Rusu, R.N.; Bild, V.; Beschea Chiriac, S.I.; Solcan, C.; Bild, W.; Ababei, D.C. Alzheimer's Disease Pharmacotherapy in Relation to Cholinergic System Involvement. *Biomolecules* **2020**, *10*, 40. [CrossRef]
18. Holden, M.; Kelly, C. Use of cholinesterase inhibitors in dementia. *Adv. Psychiatr. Treat.* **2002**, *8*, 89–96. [CrossRef]
19. Mimica, N.; Presečki, P. Side Effects of Approved Antidementives. *Psychiatr. Danub.* **2009**, *21*, 108–113.
20. McNally, W.; Roth, M.; Young, R.; Bockbrader, H.; Chang, T. Quantitative whole-body autoradiographic determination of tacrine tissue distribution in rats following intravenous or oral dose. *Pharm. Res.* **1989**, *6*, 924–930. [CrossRef]
21. McNally, W.; Pool, W.F.; Sinz, M.W.; Dehart, P.; Ortwine, D.F.; Huang, C.C.; Chang, T.; Woolf, T.F. Distribution of tacrine and metabolites in rat brain and plasma after single- and multiple-dose regimens. Evidence for accumulation of tacrine in brain tissue *Drug Metab. Dispos.* **1996**, *24*, 628–633. [PubMed]
22. Mena, E.E.; Desai, M.C. High-affinity [^{3}H]THA (tetrahydroaminoacridine) binding sites in rat brain. *Pharm. Res.* **1991**, *8*, 200–203 [CrossRef]
23. Egan, J.A.; Nugent, R.P.; Filer, C.N. Tritium labelling and characterization of the cognition enhancing drug tacrine using several precursors. *Appl. Radiat. Isot.* **2002**, *57*, 837–840. [CrossRef] [PubMed]
24. Jogani, V.V.; Shah, P.J.; Mishra, P.; Mishra, A.K.; Misra, A.R. Nose-to-brain delivery of tacrine. *J. Pharm. Pharmacol.* **2007**, *59*, 1199–1205. [CrossRef]
25. Jogani, V.V.; Shah, P.J.; Mishra, P.; Mishra, A.K.; Misra, A.R. Intranasal mucoadhesive microemulsion of tacrine to improve brain targeting. *Alzheimer Dis. Assoc. Disord.* **2008**, *22*, 116–124. [CrossRef] [PubMed]
26. Bonnot, S.; Prenant, C.; Crouzel, C. Synthesis of 9-[^{11}C]methylamino-1,2,3,4-tetrahydroacridine, a potent acetylcholine esterase inhibitor. *Appl. Radiat. Isot.* **1991**, *42*, 690–691. [CrossRef]
27. Tavitian, B.; Pappata, S.; Bonnot-Lours, S.; Prenant, C.; Jobert, A.; Crouzel, C.; Di Giamberardino, L. Positron emission tomography study of [^{11}C]methyl-tetrahydroaminoacridine (methyl-tacrine) in baboon brain. *Eur. J. Pharmacol.* **1993**, *236*, 229–238. [CrossRef] [PubMed]
28. Traykov, L.; Tavitian, B.; Jobert, A.; Boller, F.; Forette, F.; Crouzel, C.; Di Giamberardino, L.; Pappata, S. In vivo PET study of cerebral [^{11}C] methyl- tetrahydroaminoacridine distribution and kinetics in healthy human subjects. *Eur. J. Neurol.* **1999**, *6*, 273–278. [CrossRef]
29. Nishioka, K.; Kamada, T.; Kanamaru, H. ^{14}C-labeling of a tetrahydroacridine, a novel CNS-selective cholinesterase inhibitor. *J. Label. Compd. Radiopharm.* **1992**, *31*, 553–560. [CrossRef]
30. Akula, M.R.; Kabalka, G.W. Synthesis of 7-[^{123}I]iodotacrine: A potential SPECT agent to map acetylcholinesterase. *J. Label. Compd. Radiopharm.* **1999**, *42*, 959–964. [CrossRef]
31. Szymański, P.; Zurek, E.; Mikiciuk-Olasik, E. New tacrine-hydrazinonicotinamide hybrids as acetylcholinesterase inhibitors of potential interest for the early diagnostics of Alzheimer's disease. *Pharmazie* **2006**, *61*, 269–273. [CrossRef]

32. Szymański, P.; Markowicz, M.; Mikiciuk-Olasik, E. Synthesis and biological activity of derivatives of tetrahydroacridine as acetylcholinesterase inhibitors. *Bioorg. Chem.* **2011**, *39*, 138–142. [CrossRef] [PubMed]
33. Gniazdowska, E.; Koźmiński, P.; Wasek, M.; Bajda, M.; Sikora, J.; Mikiciuk-Olasik, E.; Szymański, P. Synthesis, physicochemical and biological studies of technetium-99m labeled tacrine derivative as a diagnostic tool for evaluation of cholinesterase level. *Bioorg. Med. Chem.* **2017**, *25*, 912–920. [CrossRef] [PubMed]
34. Gniazdowska, E.; Koźmiński, P.; Halik, P.; Bajda, M.; Czarnecka, K.; Mikiciuk-Olasik, E.; Masłowska, K.; Rogulski, Z.; Cheda, Ł.; Kilian, K.; et al. Synthesis, physicochemical and biological evaluation of tacrine derivative labeled with technetium-99m and gallium-68 as a prospective diagnostic tool for early diagnosis of Alzheimer's disease. *Bioorg. Chem.* **2019**, *91*, 103136. [CrossRef] [PubMed]
35. Koźmiński, P.; Niedziałek, D.; Wieczorek, G.; Halik, P.K.; Czarnecka, K.; Rogut, A.; Cheda, Ł.; Rogulski, Z.; Szymański, P.; Gniazdowska, E. New Imaging Modality of COVID-19 Pneumonia Developed on the Basis of Alzheimer's Disease Research. *Int. J. Mol. Sci.* **2022**, *23*, 8405. [CrossRef] [PubMed]
36. Szymański, P.; Lázničková, A.; Lázniček, M.; Bajda, M.; Malawska, B.; Markowicz, M.; Mikiciuk-Olasik, E. 2,3-dihydro-1H-cyclopenta[b]quinoline derivatives as acetylcholinesterase inhibitors-synthesis, radiolabeling and biodistribution. *Int. J. Mol. Sci.* **2012**, *13*, 10067–10090. [CrossRef] [PubMed]
37. Dvir, H.; Silman, I.; Harel, M.; Rosenberry, T.L.; Sussman, J.L. Acetylcholinesterase: From 3D structure to function. *Chem. Biol. Interact.* **2010**, *187*, 10–22. [CrossRef] [PubMed]
38. Bajda, M.; Więckowska, A.; Hebda, M.; Guzior, N.; Sotriffer, C.A.; Malawska, B. Structure-Based Search for New Inhibitors of Cholinesterases. *Int. J. Mol. Sci.* **2013**, *14*, 5608–5632. [CrossRef]
39. Xu, Y.; Cheng, S.; Sussman, J.L.; Silman, I.; Jiang, H. Computational Studies on Acetylcholinesterases. *Molecules* **2017**, *22*, 1324. [CrossRef]

applied sciences

MDPI

Review

Sensorial Feedback Contribution to the Sense of Embodiment in Brain–Machine Interfaces: A Systematic Review

Diogo João Tomás [1,2], Miguel Pais-Vieira [3] and Carla Pais-Vieira [1,*]

1 Faculty of Health Sciences and Nursing (FCSE), Center for Interdisciplinary Research in Health (CIIS), Universidade Católica Portuguesa, 1649-023 Lisboa, Portugal; dtomas@uatla.pt
2 Escola Superior de Saúde Atlântica, 2730-036 Barcarena, Portugal
3 Department of Medical Sciences, Institute of Biomedicine (iBiMED), Universidade de Aveiro, 3810-193 Aveiro, Portugal; miguelpaisvieira@ua.pt
* Correspondence: cvieira@ucp.pt

Abstract: The sense of embodiment (SoE) is an essential element of human perception that allows individuals to control and perceive the movements of their body parts. Brain–machine interface (BMI) technology can induce SoE in real time, and adding sensory feedback through various modalities has been shown to improve BMI control and elicit SoEe. In this study, we conducted a systematic review to study BMI performance in studies that integrated SoE variables and analyzed the contribution of single or multimodal sensory stimulation. Out of 493 results, only 20 studies analyzed the SoE of humans using BMIs. Analysis of these articles revealed that 40% of the studies relating BMIs with sensory stimulation and SoE primarily focused on manipulating visual stimuli, particularly in terms of coherence (i.e., synchronous vs. asynchronous stimuli) and realism (i.e., humanoid or robotic appearance). However, no study has analyzed the independent contributions of different sensory modalities to SoE and BMI performance. These results suggest that providing a detailed description of the outcomes resulting from independent and combined effects of different sensory modalities on the experience of SoE during BMI control may be relevant for the design of neurorehabilitation programs.

Keywords: brain–machine interface; brain–computer interface; embodiment; sensorial feedback

Citation: Tomás, D.J.; Pais-Vieira, M.; Pais-Vieira, C. Sensorial Feedback Contribution to the Sense of Embodiment in Brain–Machine Interfaces: A Systematic Review. *Appl. Sci.* **2023**, *13*, 13011. https://doi.org/10.3390/app132413011

Academic Editor: Alexander N. Pisarchik

Received: 15 September 2023
Revised: 3 December 2023
Accepted: 4 December 2023
Published: 6 December 2023

1. Introduction

In the field of cognitive sciences, the ability that enables a person to feel their own body parts, initiate and control their own actions, and perceive mental states as their own is known as the sense of embodiment (SoE) [1–5]. SoE has been identified as a necessary component for achieving health outcomes and behaviors [6,7] and may be compromised under clinical conditions [8–11]. SoE is not limited to our own physical body but can also be induced through the perception and illusory control of a virtual or robotic body or body parts [12–18]. While there are various definitions of SoE concerning external bodies, in this study, we adopt the definition provided by Kilteni and colleagues [3]. According to these authors, SoE is a sense that arises when the properties of an external body are processed as if they were the properties of one's own biological body [3]. Furthermore, it has been demonstrated that some aspects of SoE can be achieved simply by being in control of objects that bear no human resemblance [17,19,20]. Measuring SoE has posed a challenge for empirical research, and several attempts have been made to find standardized psychometric measures [5,21,22]. In addition to subjective measures, there have been endeavors to measure embodiment through electrophysiological recordings [23,24], skin conductance responses (SCRs) [25–27], or body temperature measurements [28,29].

SoE can be broken down into three underlying components: the sense of self-location (SoL), the sense of ownership (SoO), and the sense of agency (SoA) [3]. While some studies do not differentiate between these components or employ different terminology, an analysis of the subjective questions used allows associating each of them with one of these three categories. According to Kilteni and colleagues (et al., 2012), the subjective experience of recognizing oneself as the agent of certain behaviors is described as SoA, the feeling that a body (or its parts) belongs to the person is referred to as SoO, and the experience of being situated in the space where one's body is located is denoted as SoL.

SoO can be considered in a broader sense, where it includes the feeling of mineness not only of the body (body ownership) and its parts (limb-ownership) but also of feelings and thoughts (see Braun et al., 2018, for a review) [30]. Meanwhile, the same authors highlight that SoA allows not only the distinguishing between self- and other-generated actions but also the intention to generate motor activity (i.e., motor imagery). Although there have been multiple studies on the neurophysiological basis of embodiment, its evaluation is usually complemented by the use of questionnaires [31–35].

There are several studies demonstrating that the illusory experience of the body increases with the use of a brain–machine interface (BMI) or brain–computer interface (BCI) [14,15,36–38]. However, to be able to understand how BMIs influence embodiment, it is necessary to define and classify BMIs. In general terms, BMI technology enables the use of brain activity (or a proxy) decoded in real time to control an external device [39,40]. The terms BMI and BCI are generally considered synonymous terms [41]. Here, we will adopt the BMI to refer to both BMI and BCI.

There are different types of BMIs and some of them can be categorized as active BMIs or as reactive BMIs [39]. The active BMI is a system that uses neural activity resulting from voluntary activity, as occurs during motor imagery (i.e., thinking about walking). Motor imagery-based BMIs (MI-BMIs) are the most used type of BMI. On the other hand, a reactive BMI is a system that uses brain signals resulting from a reaction to an external stimulus. A very common example of this is the steady-state visually evoked potentials (SSVEPs), where changes in brain activity are evoked in the visual cortex through a visual stimulus flickering at a specific frequency [39]. Other devices that interact with neural activity and embodiment to some degree, or are close to BMIs but do not constitute actual BMIs, will not be considered here. This is due to their potential to result in varying levels of user engagement and embodiment.

Previous studies have suggested that SoE can be increased if multimodal sensory stimulation (visual, tactile, auditory, etc.) is used [15,38,42–44]. In the same way, this increase in sensory stimulation plays an important role in improving performance during MI-BMI training sessions [17,35–47]. SoE increased via sensory feedback while using a BMI has been recognized as beneficial for more efficient MI-BMI training [13,17,45]. However, experimental studies that manipulate the increasing sensory feedback modalities' effect on SoE during BMI performance are scarce and necessary to assess the effectiveness and efficiency of using these technologies. Specifically, it is unclear whether increasing the number of sensory modalities during BMI increases SoE, as well as what is the contribution of each sensory modality. Studies describing the independent and joint contributions of the different sensory modalities to the SoE during BMI control may be relevant for neurorehabilitation protocols.

In this review, we will examine studies that have employed measures to assess SoE during use of BMIs that included single or multisensory feedback. More specifically, the purpose of the present study is to analyze the contribution of multisensory feedback to embodiment and encoding in studies using BMIs.

2. Methods

The systematic review was carried out according to the Reporting Items for Systematic Review and Meta-Analysis (PRISMA) 2020 checklist [48]. The protocol was registered in the International Prospective Register of Systematic Reviews (PROSPERO) in September 2022 (CRD42022348645).

2.1. Search Strategy and Selection

The search was carried out on PubMed, Web of Science, SCOPUS, and Cochrane databases. The search strategy was formulated based on combinations of three concepts: embodiment AND brain–machine interface AND sensorial feedback. Due to the diversity of nomenclature, the embodiment concept was also searched in the form of volition, ownership, agency, body experience, and presence. The brain–machine interface concept was also searched as brain–computer interface. And finally, the sensorial feedback concept was also searched in the form of sensory stimulation, multisensory, visual, tactile, haptic, vibrotactile, auditory, sound, temperature, virtual reality, and virtual immersion. Search results from each database were merged and sorted for the removal of duplicates. Afterwards, titles and abstracts were screened according to the inclusion/exclusion criteria. The screening process was performed by the authors. The full text of the selected papers was obtained for closer inspection. Any disagreement concerning whether to include a specific study was discussed among all the authors.

The studies selected in the review were based on the following inclusion criteria: (i) studies with the full text published in English; (ii) studies were original research; (iii) studies were only carried out with experiments related to humans; (iv) studies integrated SoE variables; (v) studies integrated BMIs (note that studies not using an actual BMI but only giving participants the impression of using it were also excluded from the search); and (vi) studies integrated at least one sensory feedback modality. Additionally, studies were excluded if they (i) were reviews, (ii) were conference papers, (iii) were not peer-reviewed material, or (iv) did not have accessible full text (also, see Figure 1).

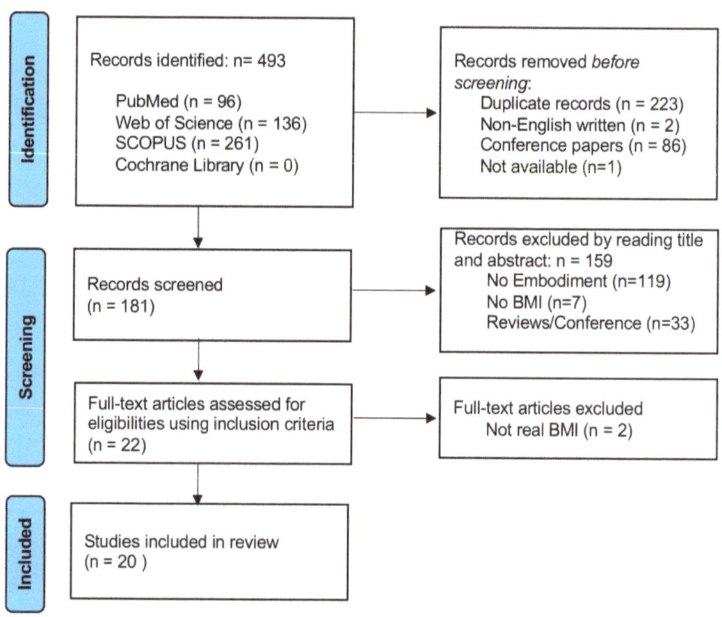

Figure 1. Flowchart of the selection process.

The searching strategy returned 493 articles, 223 of which were duplicates. A total of 2 were in foreign languages, 86 were conference papers, and 1 was not full-text available. From the remaining 181 articles searched, 119 did not have any type of sense of embodiment measures, 33 were reviews or conference papers, and 7 did not use brain–machine interface devices. After full-text reading, 2 papers were excluded for using just a simulation of a BMI. This resulted in a final list of 20 papers in our review. Details of the process are described in the flowchart (Figure 1).

2.2. Data Extraction and Analysis

For research articles that were included in the review, we extracted the following data systematically from each study: (1) sources of studies; (2) types of BMI; (3) SoE-reported measures; (4) modalities of sensorial feedback; and (5) BMI performance. The number of sensorial modalities used for feedback was counted. Also, the numbers of classes used for the task(s) employed in each study were also counted.

2.3. Assessment of Quality

Assessment of quality was performed as in previous studies [49,50]. Briefly, three different researchers independently read and scored each study (scores between 1 and 3) across the five dimensions: (1) research design; (2) methods and analysis; (3) generalizability; (4) relevance of focus; and (5) reliability of findings.

For dimension 1, research design, an evaluation of the experimental groups and variable manipulations (related to the topic of the present review) was conducted. Studies lacking a control group or that were unbalanced were scored as 2, while studies with control group or that were counterbalanced were scored as 3. For dimension 2, methods and analysis, the presence of proper statistical analysis was scored as 3. For dimension 3, generalizability, not only the size but also the existence of an equal number of male and female subjects, as well as age distribution, were considered. Studies with small sample sizes, heterogenous clinical presentations, and age or gender inequality received low scores (1 for single subject, 2 for <15 or only one gender, and 3 for larger and/or more representative samples). In dimension 4, the quality of the study, regarding the present review, was evaluated. An assessment was made regarding the extent to which each study, or its components, addressed the main questions, such as the effects of multisensory feedback on SoE (or SoA, or SoL, or SoO), as well as in BMI performance. Studies that focused solely on one patient were scored as 2. For dimension 5, reliability of findings, the extent to which the study findings can be trusted in answering the study question was considered. A score of 3 was assigned if the experiments conducted, the results obtained, their analysis, and their limitations effectively contributed to the conclusions drawn.

After scoring each paper, the mean score was calculated as 13.75 ± 1.08 (mean $\pm$ SD) with scores ranging between 11 and 14.67 points. For manuscripts with dimensions equally scored by all three researchers, the mean of the three values is presented (without the standard deviation, which was SD = 0.0). For dimension 3, generalizability, the researchers disagreed on the scores, so both the mean and standard deviation were presented.

3. Results

Of the twenty studies included in this review, two were carried out with only one participant (see Table 1).

The remaining studies had an average $N = 21.8 \pm 10.6$ subjects (mean and standard deviation), with a min of $N = 7$ and a max of $N = 40$ subjects (see Table 2).

The publication date ranged from 2009 to 2022 (see Tables 1–3). The score obtained for each paper regarding the 5 dimensions can be found in Table 3.

Table 1. Feedback modalities and SoE measures.

Author(s)	Type of BMI	Feedback Modality	Sense of Embodiment (SoE) Measures			Other Measures
			Sense of Self-Location	Sense of Ownership	Sense of Agency	
Perez-Marcos et al., 2009 [16]	EEG-based via MI	Visual	no	yes	yes	Proprioceptive drift EMG deltoid muscle
Legény et al., 2011 [51]	EEG-based via SSVEPs	Visual	yes	no	yes	--
Alimardani et al., 2013 [12]	EEG-based via MI	Visual (immersive)	no	yes	no	Skin conductance responses
Alimardani et al., 2014 [52]	EEG-based via MI	Visual (immersive)	no	yes	no	Skin conductance responses
Evans et al., 2015 [19]	EEG-based via MI	Visual	no	no	yes	--
Alimardani et al., 2016 [53]	EEG-based via MI	Visual (immersive)	no	yes	yes	--
Alimardani et al., 2016 [54]	EEG-based via MI	Visual (immersive)	no	yes	yes	Skin conductance responses
Vourvopoulos & Bermúdez i Badia, 2016 [55]	EEG-based via MI	Visual (immersive) + auditory	yes	no	no	--
Tidoni et al., 2017 [56]	EEG-based via P300	Visual (immersive) + haptic (vibratory)	yes	yes	yes	--
Tidoni et al., 2017 [18]	EEG-based via SSVEPs	Visual (immersive) + auditory	yes	yes	yes	--
Škola & Liarokapis, 2018 [17]	EEG-based via MI	Visual (immersive)	no	yes	yes	--
Penaloza et al., 2018 [45]	EEG-based via MI	Visual	no	yes	no	--
Škola et al., 2019 [57]	EEG-based via MI	Visual (immersive) + haptic (vibratory)	no	yes	yes	--
Juliano et al., 2020 [13]	EEG-based via MI	Visual/visual (immersive)	yes	yes	yes	--
Choi et al., 2020 [58]	EEG-based via MI	Visual (immersive)	yes	yes	no	--
Nierula et al., 2021 [14]	EEG-based via MI and SSVEPs	Visual (immersive) + auditory	no	yes	yes	--
Caspar et al., 2021 [36]	EEG-based via MI	Visual + auditory	yes	yes	yes	--
Ziadeh et al., 2021 [20]	EEG-based via MI	Visual (immersive) + auditory	no	yes	yes	Subjective proprioception
Serino et al., 2022 [37]	Intracortical	Visual + haptic (electrostimulation)	no	no	yes	--
Pais-Vieira et al., 2022 [15]	EEG-based via MI	Visual (immersive) + auditory + haptic (vibratory + thermal)	yes	yes	yes	--

Table 2. Summary of studies included.

Author(s)	Aims/Objectives of Study	Sample	Methods	Main Results
Perez-Marcos et al., 2009 [16]	To explore whether the control of a virtual arm through a non-invasive BCI can induce the illusion of ownership, proprioceptive displacement, and agency towards that arm, in the absence of tactile sensory stimulation.	$N = 16$ (healthy participants) Age: 26.1 ± 9.4 (Mean ± SD)	Two groups with different visual feedback conditions: Group 1: virtual hand moves congruently with the motor imagery attempt. Group 2: virtual hand moves randomly and independently of the participant's performance.	Sense of ownership and EMG deltoid activity higher in group 1. Sense of agency with high levels but not different between groups. Proprioceptive drift not significant in either of the two group's conditions.

Table 2. *Cont.*

Author(s)	Aims/Objectives of Study	Sample	Methods	Main Results
Legény et al., 2011 [51]	To study the usability of SSVEP-based BMIs in virtual environment navigation.	$N = 17$ (healthy participants) Age: 25.5 ± 4.3 (Mean ± SD)	Four experimental conditions: Condition 1: "arrow" visual trigger without real-time visual feedback of user's brain activity. Condition 2: "arrow" visual trigger with real-time visual feedback of user's brain activity. Condition 3: "butterfly" visual trigger without real-time visual feedback of user's brain activity. Condition 4: "butterfly" visual trigger with real-time visual feedback of user's brain activity.	Senses of self-location and agency significantly or near significantly higher in condition 4 than the other conditions.
Alimardani et al., 2013 [12]	To explore if sense of agency and body ownership illusions can be induced for a pair of BMI-operated human-like robotic hands without proprioceptive updates of real motions from operators' sensations.	$N = 40$ (healthy participants) Age: 21.13 ± 1.92 (Mean ± SD)	Two experimental conditions: Still condition: The robot's hands did not move at all throughout the whole session, although a subject performed motor imagery according to cues. Match condition: The robot's hands moved when the subject performed the MI. At the end of each test session, for both conditions, a syringe was injected into the robot's hand.	Sense of ownership higher in "match condition" compared with "still condition". Higher skin conductance response in "match condition" compared with "still condition" during the syringe injection.
Alimardani et al., 2014 [52]	To investigate the inducement of body ownership illusion for a pair of BMI-operated human-like robotic hands under different presentations of feedback.	$N = 40$ (healthy participants) Age: 21.13 ± 1.92 (Mean ± SD)	Two experimental conditions: Still condition: The robot's hands did not move at all throughout the whole session, although a subject performed motor imagery according to cues. Match conditions: The robot's hands moved only in those trials that the classification result was correct and in accordance with cue. Raw condition: The robot's hands moved according to the classification results in all trials. In case of wrong result that was not in accordance with cue, the robot's opposite hand moved. At the end of each test session, for both conditions, a syringe was injected into the robot's hand.	Sense of ownership higher in "match condition" than in "still condition" and "raw conditions". Sense of ownership higher in "raw condition" than in the "still condition" while operating the robot hands, but no significative differences between these two conditions when the robot's hand was injected. Higher skin conductance response in "match condition" compared with the other conditions but just statistically significant when compared with "still condition". Positive correlation between sense of ownership while operating the robot hands and the BMI's performance.

Table 2. *Cont.*

Author(s)	Aims/Objectives of Study	Sample	Methods	Main Results
Evans et al., 2015 [19]	To explore the sense of agency for BMI-mediated actions.	Study 1: $N = 8$ (healthy participants) Age: 26.5 ± 3.5 (Mean ± SD) Study 2: $N = 7$ (healthy participants) Age: 26.0 ± 2.3 (Mean ± SD)	Study 1: Control the right/left displacement of a virtual bar by imagining clasping the right/left hand under six different visual feedback delay conditions: 0 ms, 750 ms, 1500 ms, 2250 ms, 3000 ms, or 3750 ms. This feedback was also presented as congruent (displacement direction according to the intention) or incongruent (displacement direction opposite to the intention). Study 2: Control the right/left displacement of a virtual bar by imagining clasping the right/left hand under six different visual feedback delay conditions: 0 ms, 250 ms, 500 ms, 750 ms, 1000 ms, or 3750 ms.	Study 1: Sense of agency higher for congruent than incongruent feedback. For congruent feedback, sense of agency is higher when the delay is lower. For incongruent feedback, sense of agency is low and it is not dependent on the delay conditions. BMI performance cannot be explained using the level of sense of agency. Study 2: Sense of agency not significatively different between the conditions under 1000 ms. Lower sense of agency for the delay condition of 3750 ms compared with the others. No significative differences in BMI performances between conditions.
Alimardani et al., 2016 [53]	To assess the impact of embodiment on motor imagery learning during BMI control.	$N = 38$ (healthy participants) Age: 23.8 ± 8.2 (Mean ± SD)	Control of BMI-operated robotic arms in two followed sessions: Initial session with a positive visual feedback bias, and a subsequent session with feedback associated with the real performance. Geminoid group: Participants initially operated Geminoid's hands (human-like) and in a subsequent session proceeded to operation of the Arm Robot (robotic tweezers). Arm Robot group: Participants BMI-operated only the robotic tweezers in both sessions.	Sense of agency was greater in the positive visual feedback bias session compared to the real performance visual feedback for both groups. Sense of ownership, for Geminoid group, was significantly higher in the session with the human-like hands compared with the one with robotic tweezers. In the Arm Robot group, there were no significant differences between sessions.
Alimardani et al., 2016 [54]	To investigate the interference of proprioceptive afferences in the body ownership illusion when mismatching with visual feedback.	$N = 30$ (healthy participants) Age: 21.51 ± 1.73 (Mean ± SD)	To operate human-like Geminoid robot hands, participants performed two sessions in a random order: MoCap session: Subjects grasped their own right or left hand to control the robot's corresponding hand. BMI session: Subjects performed a right or left motor imagery task and controlled robot's hands without actual motions. In both sessions, the visual feedback had a certain amount of delay. At the end of each session, a syringe was injected into the robot's hand.	Higher senses of agency and body ownership in the BMI session. Skin conductance responses revealed that the operators' reactions to a painful stimulus (injection) were significantly stronger in the BMI sessions.

Table 2. *Cont.*

Author(s)	Aims/Objectives of Study	Sample	Methods	Main Results
Vourvopoulos & Bermúdez i Badia, 2016 [55]	To explore the role of motor priming in virtual reality in BMI-operated virtual arms.	$N = 9$ (healthy participants) Age: 27.0 ± 2.0 (Mean ± SD)	To perform MI of circular movements of arms for a garage door opening under three BCI conditions in a randomized order: VR condition: performing MI, receiving visual and auditory feedback trough a virtual environment. VR + MP condition: using real arm movements while performing MI, receiving visual and auditory feedback trough a virtual environment. Control condition: performing MI, receiving a visual standard feedback through arrows and bar.	VR + MP and VR conditions share high scores of sense of self-location. For BMI performances, no significative differences between VR + MP and VR with control condition. No significant correlation between BMI performance and sense of self-location.
Tidoni et al., 2017 [56]	To explore the role of proprioceptive feedback in healthy people and those living with SCI during a BCI-based social interaction task.	Study 1: $N = 8$ (healthy participants) Age: 27.0 ± 3.5 (Mean ± SD) Study 2: $N = 10 + 8$ (healthy participants) Age: 29.33 ± 2.87 (Mean ± SD) (SCI participants) Age: 28.0 ± 5.19 (Mean ± SD)	Study 1: Participants immersed into a virtual environment in two experimental conditions: MovI+: Vibration applied in right bicep's brachial tendon (inducing proprioceptive stimulation with illusion of downward extension of the elbow). MovI-: Vibration applied over the bone close to bicep's brachial tendon (proprioceptive stimulation without illusion). Study 2: Participants with vibration applied in right bicep's brachial tendon in two conditions: Virtual: controlling an avatar in a virtual environment. Robot: controlling a robot with the itself perspective.	Study 1: No significative differences in BMI's performances and SoE measures between conditions. High levels of sense of self-location and sense of agency. Study 2: Healthy participants: No significative differences in BMI's performances and SoE measures between conditions. High levels of sense of agency. SCI participants: SoE experience did not differ relative to healthy participants but had found a more variable performance in the control of the virtual avatar and the robotic surrogate.
Tidoni et al., 2017 [18]	To explore the use auditory combined with visual feedback in virtual navigation to the subjective experience in terms of BMI usability and feelings of ownership over the controlled surrogate.	$N = 14 + 3$ (healthy participants) Age: 25.8 ± 6.0 (Mean ± SD) (SCI participants) Age: 27.0 ± 4.0 (Mean ± SD)	Participants control a humanoid robot walking and grasping bottles. Visual feedback was given from its own perspective combined with four auditory stimulus conditions: Foot Sync: steps sound with the visual feedback. Foot Async: steps sound asynchronous with the visual feedback. Beep Sync: beep sound synchronous with the visual feedback. Beep Async: beep sound as asynchronous with the visual feedback.	Healthy participants: High levels of sense of agency and low levels of sense of embodiment and sense of self-location. Higher accuracy in the grasping bottles phase with footstep sound condition relative to a beep sound condition. No differences were found between synchronous and asynchronous. SCI participants: Reduced control of the robot when asynchronous auditory feedback was matched with the robot's movements.

Table 2. *Cont.*

Author(s)	Aims/Objectives of Study	Sample	Methods	Main Results
Škola & Liarokapis, 2018 [17]	To explore the use of a more realistic feedback during the MI-BMI training process in comparation to the traditional neurofeedback mediated via a simple symbolic representation.	N = 30 (healthy participants)	Experimental group: Training phase was placed into a virtual reality environment observed from a first-person view of a human-like avatar, and their rehearsal of MI actions was reflected by the corresponding movements performed by the avatar. Control group: Training phase instructions were delivered using the standard protocol with arrows, and feedback was displayed as extending blue bar, continuously changing according to the classifier decision.	Sense of agency was slightly higher in the experimental group than for the controls. Sense of ownership was higher for the controls, but with a very small difference. In both groups, there was a similar number of participants with scores that could be considered as "embodied participants group". This suggests that virtual reality experience during training did not affect ratings in the evaluation phase of the experiment. Similar tendency is present for the agency statements.
Penaloza et al., 2018 [45]	To investigate an alternative BMI training protocol that uses a human-like android robot (Geminoid HI-2) to provide realistic feedback.	N = 27 (healthy participants) Age: 21.5 ± 1.69 (Mean ± SD)	Two groups control an android robot (Geminoid HI-2) in a grasping hand task trough two different protocols: Classical Training Protocol (CTP): Calibration–Training–Evaluation. Android Feedback Training Protocol (AFTP): Pretraining–Training–Calibration–Evaluation (pretraining consists of rehearsing the kinesthetics of hand movements and memorizing the physical sensation).	Sense of ownership was significantly higher in the group with AFTP than the one with CTP. Strong correlation between AFTP group performance and sense of ownership. Moderate correlation between CTP group performance and sense of ownership.
Škola et al., 2019 [57]	To investigate the use of gamification in MI-BMI training.	N = 19 (healthy participants) Age: 26.0 ± 2.78 (Mean ± SD)	The gamified VR scene was set inside a cockpit of a spaceship containing a simplistic control panel. The objective was to trigger weapons aiming for the destruction of asteroids using MI of the left or right hand, depending on its source position. Feedback was provided using three modalities: (1) movements of the avatar, (2) vibrations, and (3) providing information about trial accuracy (score).	Positive, moderately strong rating of sense of ownership and sense of agency.
Juliano et al., 2020 [13]	To explore the role of embodiment on neurofeedback performance using HMD-VR versus a computer screen.	N = 12 (healthy participants) Age: 24.4 ± 2.7 (Mean ± SD)	Participants under three blocks of conditions: Block 1: controlling the virtual arm with brain activity on the computer (screen); Block 2: controlling the virtual arm with brain activity in a head-mounted display virtual reality (HMD-VR) system; Block 3: controlling the virtual arm with actual arm movements in a head-mounted display. (IMU): control.	Higher levels of embodiment in the HMD-VR condition. For the HMD-VR condition, a significant relationship between embodiment and neurofeedback performance was reported.

Table 2. *Cont.*

Author(s)	Aims/Objectives of Study	Sample	Methods	Main Results
Choi et al., 2020 [58]	To explore a novel control scheme in which virtually embodiable feedback is provided during control to enhance performance.	N = 14 (healthy participants)	Training phase: During the MI period, the virtual hands executed the movement corresponding to the task. Control conditions (EFCS and SCS): The device moved in a virtual route track based on the real-time EEG signals. Repeated left-hand grasping and right-hand grasping MIs were mapped to left rotation and right rotation of the device, respectively. Feedback was given in two different conditions: EFCS: virtual hands are shown and execute the movement that is classified; SCS: does not show the virtual hands.	Participants expressed great levels of sense of ownership and sense of self-location. They were able to perform MI better during the EFCS than during the SCS with statistical significance. Participants found the virtual hands to be helpful for performing MI during the training phase and during the EFCS. Significant positive linear relationships between classification accuracy and sense of ownership and sense of self-location were shown for EFCS. No statistically significant relationships were found for SCS.
Nierula et al., 2021 [14]	To investigate agency and responsibility by studying the control of movements of an embodied avatar, via BMI technology, in immersive virtual reality	N = 29 (healthy participants) Age: 21.5 ± 2.6 (Mean ± SD)	Participants went through three conditions: Observe: passive observation of the virtual arm performing the task; MI-BMI: control of the movement through motor imagery; SSVEP-BMI: control of the movement through steady-state visually evoked potentials.	Sense of agency was higher in MI-BMI than SSVEP-BMI. Sense of agency was higher in SSVEP-BMI than in "Observe". Sense of ownership was higher in MI-BMI than SSVEP-BMI and "Observe". Sense of ownership was not statistically different between SSVEP-BMI and "Observe". BMI performance in MI-BMI was slightly higher than SSVEP-BMI.
Caspar et al., 2021 [36]	To investigate whether using brain–machine interfaces influences the human sense of agency.	Study 1: N = 27 (healthy participants) Age: 23.78 ± 2.68 (Mean ± SD) Study 2: N = 30 (healthy participants) Age: 23.77 ± 2.76 (Mean ± SD)	Study 1: Participants had to press a keyboard button to produce a sound. They were then asked to estimate and report, in ms, the duration of the delay between their keypress and the resulting tone. This was proceeded using a real hand or controlling a robotic hand through BMI. Study 2: Same protocol as study 1, but participants were BMI-trained for two consecutive days, at the same hour of the day.	Study 1: The interval estimates in the real hand condition and the robotic hand condition were not significantly different. For the robotic hand, results indicated a higher score for sense of agency. Study 2: BMI performance higher on day 2 than on day 1. Sense of ownership, sense of self-location, and sense of agency did not differ between the two days.

Table 2. *Cont.*

Author(s)	Aims/Objectives of Study	Sample	Methods	Main Results
Ziadeh et al., 2021 [20]	To investigate whether higher levels of ownership from a humanoid hand in VR can enhance the perceived agency users feel over hand's movements during an online MI-BMI task.	$N = 22$ (healthy participants) Age: 24.0 (Mean)	Performing a virtual task (popping balloons). Group 1: First block popped with the virtual hands and the second block popped with virtual flying blocks. Group 2: First block popped with virtual flying blocks and the second popped with the virtual hands.	Similar BMI performances between virtual hands and blocks. Virtual hand induced higher sense of ownership and proprioception levels than blocks. Sense of ownership and performance significantly predicted sense of agency. Proprioception correlated with performance in the virtual hand but not the block's condition.
Serino et al., 2022 [37]	To explore sense of agency for intracortical brain–machine interfaces.	$N = 1$ (SCI participant) Age: 24	Experiment 1: Visual (V) was used to provide visual feedback, consisting of a life-sized virtual arm on a monitor superimposed over the participant's right arm, matching the location and dimensions of the participants real arm, which was occluded from view. Experiment 2: NMES (S) was used to provide somatosensory feedback; the patients upper limb muscles were electrically stimulated so they could feel, but not see, the selected movement. Experiment 3: Combined V and S to provide visual–somatosensory feedback. In half of the trials, sensory feedback was congruent with the cued action, while in the other half, it was incongruent.	Experiments 1 and 2: Congruent visual and congruent somatosensory feedback resulted in more frequent agency responses versus incongruent conditions. Confidence was modulated via somatosensory congruency (higher for somatosensory congruent than incongruent). The effect of visual feedback congruency on confidence ratings was not significant. Experiment 3: Somatosensory congruency was more effective in driving the sense of agency and the associated confidence. Ratings were higher when both feedback signals were congruent as compared to both being incongruent. When visual feedback was not congruent but somatosensory was congruent, higher levels of sense of agency and confidence were present when compared to the condition with congruent visual feedback but incongruent somatosensory feedback.
Pais-Vieira et al., 2022 [15]	To explore embodiment comfort levels during motor imagery training combined with immersive virtual reality in a spinal cord injury patient.	$N = 1$ (SCI participant) Age: 52	Walking and stopping with an avatar in a virtual environment. Protocol with three phases: (a) Habituation; (b) EEG baseline and neural data acquisition for classifier training; (c) Testing real-time decoding of neural activity without control of avatar.	High levels of senses of ownership, agency, and self-location. The participant could generate higher levels of neural commands associated with "Walk" and "Stop". Subjective reports describe this experience as being positive. In three sessions involving water scenarios, participant reported his legs feeling cold. Not exclusive of thermal feedback.

Table 3. Quality assessment.

References	Perez-Marcos et al., 2009 [16]	Legény et al., 2011 [51]	Alimardani et al., 2013 [12]	Alimardani et al., 2014 [52]	Evans et al., 2015 [19]	Alimardani et al., 2016 [53]	Alimardani et al., 2016 [54]	Vourvopoulos & Bermúdez i Badia, 2016 [55]	Tidoni et al., 2017 [56]	Tidoni et al., 2017 [18]	Škola & Liarokapis, 2018 [17]	Penaloza et al., 2018 [45]	Škola et al., 2019 [57]	Juliano et al., 2020 [13]	Choi et al., 2020 [58]	Nierula et al., 2021 [14]	Caspar et al., 2021 [36]	Ziadeh et al., 2021 [20]	Serino et al., 2022 [37]	Pais-Vieira et al., 2022 [15]
	Quality Assessment: (Classify from1 to 3; 1 = Low; 2 = Medium, and 3 = High)																			
1. How appropriate is the research design for addressing the question, or sub-questions, of this review (higher weighting for inclusion of a control group)?	3	3	3	3	3	3	3	3	3	3	3	3	2	3	3	3	3	3	2	2
2. How appropriate are the methods and analysis?	3	3	3	3	3	3	3	3	3	3	3	3	3	3	3	3	3	3	3	3
3. How generalizable are the findings of this study to the larger population with respect to the size and representativeness of the sample?	1.33 ± 0.58	1.33 ± 0.58	2.67 ± 0.58	2.67 ± 0.58	1.33 ± 0.58	2.67 ± 0.58	2.67 ± 0.58	1.33 ± 0.58	1.67 ± 0.58	2 ± 0.00	2.67 ± 0.58	2.67 ± 0.58	2 ± 0.00	1.67 ± 0.58	1.67 ± 0.58	2.67 ± 0.58	2.67 ± 0.58	2.33 ± 1.15	1 ± 0.00	1 ± 0.00
4. How relevant is the particular focus of the study (including conceptual focus, context, sample, and measures) for addressing the question or sub-questions of this review?	3	3	3	3	3	3	3	3	3	3	3	3	3	3	3	3	3	3	2	2
5. To what extent can the study findings be trusted in answering the study question(s)?	3	3	3	3	3	3	3	3	3	3	3	3	3	3	3	3	3	3	3	3
Total score (5–15)	13.33	13.33	14.67	14.67	13.33	14.67	14.67	13.33	13.67	14	14.67	14.67	13	13.67	13.67	14.67	14.67	14.33	11	11

We will start by describing the type of BMI technology used, then the type of questionnaire used to study SoE, as well as aspects related to the different nomenclature, followed by additional techniques used to objectively evaluate SoE (e.g., SCR). Then, a detailed analysis of sensory feedback modalities, SoE, and BMI performance will be made.

Regarding the type of BMI technology used, 15 studies were MI-based, where brain activity was captured using non-invasive electroencephalography (EEG). One study, a spinal cord-injured (SCI) patient [37], also used an active MI-BMI, but the activity was captured intracortically. Four studies used a reactive BMI, where one of them used P300 technology and the other three were based on SSVEPs (see Table 1). One of the studies that used SSVEP-BMI also used MI-BMI in the experimental design [14]. Here, the authors intended to explore the performance of the BMI and the effect on SoO and SoA when a BMI used neural activity from the sensorimotor areas (MI-BMI) or activity from the visual areas (SSVEP-BMI). Although the performance was slightly higher for the MI-BMI group, this difference was not statistically significant. Significative higher ratings of SoO and SoA were found in the MI-BMI group.

In all studies, subjective questionnaires (7-point or 11-point Likert rating) were used to measure embodiment variables (also see Table 4 for details). One study adopted a classification from 0 to 100 points [18]. In only two studies [19,37], the answers were given by choosing "yes" or "no". However, in one of these [37], the authors added a second request to rank from 0 to 100 on the degree of certainty in the given answer. The number of questions used in the questionnaires was also very variable. Of the analyzed studies, the number of questions to assess SoE variables ranged from 1 to 26.

Differences in the nomenclature of the various SoE components were present in the studies analyzed. As we have followed the classification proposed by Kilteni et al. (2012), it was necessary to specifically analyze all the questions posed in each of the studies to be able to fit them into the scope of SoO, SoL, or SoA (see Table 1). It was found that there is great heterogeneity in the terminology used. For example, in some studies, the authors used the term SoE when referring to "It was as if the virtual body was my body" and the term sense of control when referring to "I felt in control of avatar's actions" [18,46]. However, according to Kilteni et al., 2012, this should be included in the SoO and SoA categories, respectively. Among the 20 studies included in this review, 16 had SoO questions, 14 had SoA questions, and 8 studies had SoL questions in their questionnaires. Only five studies had questions covering the three components of the SoE. Other combinations of SoE components can be analyzed in Figure 2.

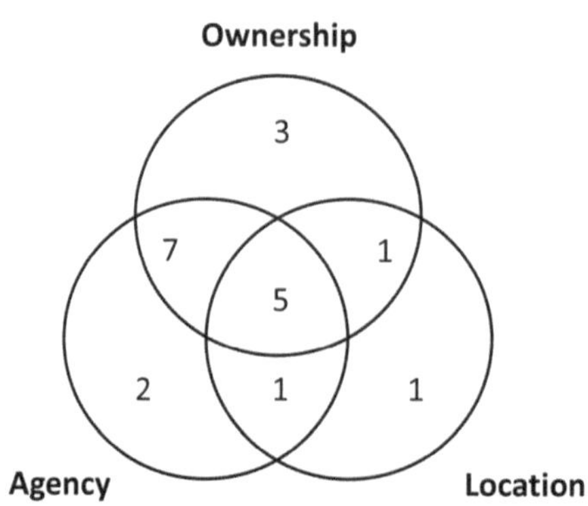

Figure 2. Number of studies per SoE dimension assessed.

Table 4. Embodiment and BMI evaluation.

Study	Neural Signal	Feedback	N Modalities	Compares Sensorial Modalities	BMI Classes	SoE Evaluation	BMI% (Mean/Median)	Embodiment and BMI Conclusion
Perez-Marcos et al., 2009 [16]	EEG-based via MI	Visual	1	No	2	Botvinick and Cohen, 1998 [42]	N/A	MI Visual fb SoO ↑, BMI% N/A
Legény et al., 2011 [51]	EEG-based via SSVEPs	Visual	1	No	3 (L, R, Forward)	Slater et al., 1998 [59]	N/A	SoO may improve ERD SoL↑, SoA↑, BMI% ↓
Alimardani et al., 2013 [12]	EEG-based via MI	Visual (immersive)	1	No	2	Two Qs Q1: "feel (...) your own hand received the injection? Q2: Feel as if they were your own hands?	N/A	Body ownership illusions can be induced without the correlation of multiple sensory modalities SoO ↑, BMI% N/A
Alimardani et al., 2014 [52]	EEG-based via MI	Visual (immersive)	1	Yes	2	Two Qs Q1: Feel (...) your own hand received the injection? Q2: Feel as if they were your own hands?	FakeP = 60.78 Raw = 49.22 Match = 54.37 FakeN = 50.47	MI improved with + bias feedback, SoO SoO ↑, BMI% =
Evans et al., 2015 [19]	EEG-based via MI	Visual	0 vs. 1	No	2	SoA " I was controlling the cursor". "Yes", "No"	Cong = 76→Incong~79; Visual = 76.7→None = 53.4	MI congruent Visual fb SoA ↑, BMI% =
Alimardani et al., 2016 [53]	EEG-based via MI	Visual (immersive)	1	Yes	2	(pre-) Botivnik and Cohen., 1998 [42]; (post-) Qs: Q2: (...) where your hands? Q3: (...) operation (...) was easier?	Geminoid: 1.31→1.08 Robot: 1.48→0.68	BMI's potential in inducing stronger agency-driven illusions SoO ↑, SoA ↑, BMI% ↑
Alimardani et al., 2016 [54]	EEG-based via MI	Visual (immersive)	1	No	2	Two Qs: (Q1) Could you operate the robot's hands according to your intentions? (Q2) Did you feel as if the robot's hands were your own hands? 7-point	SoA: Hum: S3 = 4.58→S4 = 3.05 ($p < 0.001$) Rob: S3 = 4.47→S4 = 3.21 ($p < 0.001$) SoO: Hum: S3 = 4.36→S4 = 2.53 ($p < 0.001$) Rob: S3 = 4.0→S4 = 3.53 ($p = 0.18$)	Improved BMI learning with visual humanoid SoO ↑, SoA ↑, BMI% ↑
Vourvopoulos & Bermúdez i Badia, 2016 [55]	EEG-based via MI	Visual (immersive) + auditory	2	Yes	2	Witmer and Singer 1998 [30]; Roberts et al., 2008 [60]	VRMP = 51.29 VR = 53.61 Control = 50.1	VR and MP can enhance the activation of brain patterns present during overt motor execution SoL N/A, BMI% =
Tidoni et al., 2017 [56] *	EEG based via P300	Visual (immersive) + haptic (vibratory)	2	Yes	9	Friedman et al., 2014; [34] Sanchez-Vives et al., 2010 [33]	BMI = 86.06 VR = 83.33 BMI = 95.00 VR Robot = 93.75	Proprioceptive feedback did not contribute to alter performance measures and body ownership sensations SoO =, SoA =, BMI =
Tidoni et al., 2017 [18] *	EEG-based via SSVEPs	Visual (immersive) + auditory	2	Yes	6 (5 + 0)	Wolpaw et al., 1998 [32]; Sanchez-Vives et al., 2010 [33]	(Dist. to bottle) Foot = 1.481 Beep = 1.975	Paired visual auditory (foot) improved BMI performance SoA ↑, SoO =, SoL =, BMI ↑

Table 4. *Cont.*

Study	Neural Signal	Feedback	N Modalities	Compares Sensorial Modalities	BMI Classes	SoE Evaluation	BMI% (Mean/Median)	Embodiment and BMI Conclusion
Škola & Liarokapis, 2018 [17]	EEG-based via MI	Visual (immersive)	1	No	2	12 Qs: SoO, SoA, other	VR = 58.3 MI-BMI = 52.9	SoA was higher, performance was higher in VR SoA ↑, SoO = N/A, BMI% ↑
Penaloza et al., 2018 [45]	EEG-based via MI	Visual	1	No	2	SoO: Did you feel that robot's hands were your own hands?	Android = 61.38 Classical = 52.38	MI robotic hand Visual fb SoO ↑, BMI% ↑
Škola et al., 2019 [57]	EEG-based via MI	Visual (immersive) + haptic (vibratory)	2	No	2	Botvinick and Cohen, 1998 [42]; Longo et al., 2008 [5]	MI = 75.84	SoO correlated with EEG modulation SoA N/A, SoO N/A, BMI% N/A
Juliano et al., 2020 [13]	EEG-based via MI	Visual/Visual (immersive)	1	Yes	2	Witmer and Singer, 1998 [42]	Screen = 80.95 VR = 83.33	VR fb improved embodiment but not BMI perf. SoE ↑, BMI% =
Choi et al., 2020 [58]	EEG-based via MI	Visual (immersive)	1	No	3	10 Qs: SoO, SoL	Embodied = 53.27 Standard = 39.99	Embodiable feedback generates SoO and SoL and improves BMI performance BMI% ↑(L/R)
Nierula et al., 2021 [14]	EEG-based via MI and SSVEPs	Visual (immersive) + auditory	2	Yes	2	7 Qs: my body, agency, responsibility	SSVEP = 90.9 MI = 87.4 ($p = 0.052$)	MI SoA ↑, SoO ↑ SSVEPs SoA ↑, SoO ↓, BMI% ↑
Caspar et al., 2021 [36]	EEG-based via MI	Visual + auditory	2	No	2	Kalckert and Ehrsson, 2012; Longo et al., 2008 [5]	Day 1 = 59.47 Day 2 = 61.72	Sensorimotor information may not be the most important cue for generating a sense of agency SoA =, SoL =, BMI N/A
Ziadeh et al., 2021 [20]	EEG-based via MI	Visual (immersive) + auditory	2	No	2	Skola, 2019 [57]	Hand = 53 Blocks = 54	Avatar increased SoO and SoA SoA ↑, SoO ↑, BMI% =
Serino et al., 2022 [37] *	Intracortical	Visual + haptic (electrostimulation)	2	Yes	4	Q1: sense of agency Q2: confidence	Visual Cong V = 93.8, incong = 5.2 Somat Cong. = 97.5, incong = 8.8	Vi inc. + Somatos. Cong., (somat. prevails) Somat. Cong. ↑ SoA, BMI ↑ (soma+ vis− versus soma− vis+)
Pais-Vieira et al., 2022 [15] *	EEG-based via MI	Visual (immersive) + auditory + haptic (vibratory + thermal)	4	Yes	2	Peck and Gonzalez-Franco, 2021	Sleeve = 82.50 No Sleeve = 73.50 ($p = 0.2857$)	Multimodal stimulation not detrimental for performance or embodiment SoA =, SoL =, SoO =, BMI% =

* Includes patients. N/A indicates Non Applicable or unable to access data. Up and Down arrows indicate an increase or decrease, respectively.

It is also common in studies related to SoE to include measures of disownership, defined as the experience that a body part does not belong to the subject [61]. In most cases, these measures were focused on assessing the participant's awareness of their actual body parts. An example of this is the study by Škola and Liarokapis (2018) when the participants were asked about the sense of proprioception related to awareness of the position of their own hands [17]. In two other studies, the authors were interested in assessing the influence of the apparatus illusion on the real body parts. In the work of Perez-Marcos et al. (2009), a proprioceptive drift was objectively measured by asking the participants to indicate, without looking, the location where they perceived the real hand to be placed after a virtual arm-moving task in the BMI [16]. Meanwhile, in the study of Ziadeh and colleagues [20], participants were questioned about the movement of the virtual hands over their own hands.

Apart from these subjective measures for SoE, three additional studies from the same group also assessed the SCR following a threatening stimulus to a non-real body part (injection in the robot hand) [12,52,54]. This procedure intended to assess more objectively how embodied the participant with the external body parts was during the BMI task. For this, the authors measured the physiologic changes (trough the assessment of SCR) in the real body during the BMI-performing task. Across the three studies, the SCR was higher in the experimental conditions with higher scores of reported SoO. In addition, this was also verified for SoA when it was included as a variable [54]. In another study, electromyography (EMG) was used to estimate the amount of muscle activity required to perform a virtual task if it occurred in the real world [16]. The group with higher EMG activity also reported higher levels of SoO.

Providing sensory feedback while performing BMI tasks is one of the presumed strategies used to improve BMI performance itself [62,63]. It is hypothesized that this agreement between stimuli and actions is relevant to the SoE experience. However, few studies explore the contribution of different sensory modalities to SoE. All studies included in this review used visual stimuli as their main sensory modality (see Figure 3). Although the auditory modality was present in six studies, and some form of haptic stimulus was present in four studies, none of them specifically quantified the influence of these stimuli on the SoE variables. Immersive visual feedback was present in 14 studies whereas the others used non-immersive visual feedback. Normally the non-immersive visual feedback was received through a computer screen. Only in one study, the visual feedback was non-virtual with participants observing a robotic hand moving in front of them [36]. Within the visual feedback modality, the immersive type has shown higher levels of SoE compared to the non-immersive type [13]. The congruence or incongruence of the visual stimulus with the MI action performed seems to have a strong impact on SoE. Congruent visual stimuli seem to increase SoO [12,16,52]. However, differences between studies were found for SoA. In one study, no significant differences were present in SoA levels between incongruent and congruent visual stimulus conditions [16]. In contrast, another study reported a negative impact of visual incongruence on SoA [19]. These differences may find some explanation regarding the type of visual stimulus that was provided. Although both used a non-immersive form of visual feedback, in the study by Perez-Marcos et al. (2009), the feedback was based on the movement of a virtual hand, while in Evans et al. (2015), the feedback congruence was associated with the displacement of a virtual bar on a screen. In this second study, it was also verified that larger delays negatively impact the SoA. However, for delays under 1000 ms, no significant differences were found [19]. This information becomes relevant to understand the acceptable limit of delay in this type of BMI technology, since any feedback addition has some amount of expected delay due to the normal computational processing. In the only study where the congruency of an auditory stimulus with a visual stimulus was explored, no significative impact on the SoE was found [18].

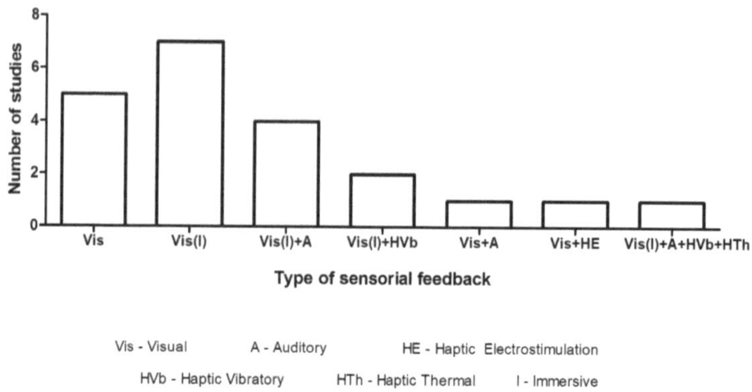

Figure 3. Number of studies per type of stimulation.

The realism of sensorial feedback seems to enhance more realistic responses of the participants that are exposed to it [15,44,64]. In a case study, the exposure of an SCI patient to a very realistic immersive virtual environment combined with auditory and thermal stimuli consistent with the scenario and with tactile vibratory stimuli coherent with the action (walking) has been shown to provide high levels of SoE [15]. Many attempts to provide realism to a BMI experience have been performed by several research groups. For example, it has been proposed that the control of robotic hands with a more human-like shape through MI-BMI may induce higher SoO than controlling robotic hands with the shape of mechanical tweezers [53]. Nevertheless, both conditions seem to induce high scores of SoA that do not differ significantly from each other. Also, in another study in which the MI-BMI task consisted in popping virtual balloons with virtual hands or with virtual blocks, both conditions showed high scores of SoA but did not differ among themselves [20]. In this same study, it was also found that the illusory induction of a sense of movement in one's own hands was greater in the virtual hands' condition than in the virtual blocks' condition.

The BMI tasks present in the studies analyzed typically involve training and evaluation phases. Interestingly, these studies tended to adopt realistic conditions in their evaluation phases, but the training phase for the acquisition of sensorimotor activity related to MI typically followed a standard protocol using simplistic arrows and a bar graph as visual feedback [65]. Exceptions to this are the studies by Škola and Liarokapis (2018) and Pais Vieira and colleagues (et al., 2022). Škola and Liarokapis (2018) also decided to explore if the introduction of a realistic scenario already in the training phase leads to a higher rate of SoE. However, they concluded that the more realistic experience during the training phase does not seem to significantly affect the SoO and SoA scores during the experience evaluation phase [17]. Meanwhile, Pais-Vieira and colleagues (et al., 2022) used a highly realistic scenario in the training and evaluation phases to ensure that spinal cord injury patients maintained high levels of engagement throughout the multiple sessions that constituted the experimental protocol [15].

Only four of the twenty studies included some type of haptic stimulus [15,37,56,57]. Three applied the stimulus in the form of vibration while one applied it in the form of electrostimulation. The vibratory feedback was applied in two studies, not intended to replicate the real tactile sensation expected in an action performed, but its application intended to be consistent with the timing of some event in the virtual scene. In the study of Pais-Vieira et al. (2022), the BMI task allowed an avatar of the subject taking steps. The vibration stimulus matched the moment that the sole of the avatar's foot touched the ground when walking and was delivered to the participant's forearm [15]. Meanwhile, in the study of Škola and colleagues (2019), the vibratory stimulus was applied to the participant's hand and was consistent with triggering a weapon from a spaceship. Both studies reported high levels of SoO and SoA despite the very different conditions tested [57]

Differences were found for the SoL where the apparatus of Pais-Vieira et al. (2022) seems to have also induced high levels of SoL. It should be noted, however, that multimodal feedback was used (visual, auditory, and tactile) and the contribution of each modality was not individually assessed.

The application of a vibratory stimulus as feedback to mimic the sensation induced by the action if it was truly performed was explored in the study of Tidoni, Gergondet, et al. (2017) [46]. A vibratory stimulus was applied in the right bicep's brachial tendon of the subjects, inducing the proprioceptive illusion of downward extension of the elbow while performing a similar movement with a virtual arm via MI-BMI. This condition was compared with a vibratory stimulus applied over the bone (not inducing a proprioceptive illusion). Despite the specificity of the stimulus application, no significant differences were found in SoE between the different conditions.

A very interesting case study using an SCI patient combined somatosensory feedback in the form of muscular electrostimulation with visual feedback through the visualization of a virtual hand [37]. Here, the authors explored the different combinations between the congruence and incongruence of the different stimuli with MI action during BMI. The authors reported that somatosensory congruency was more effective in driving SoA. Ratings were higher when both feedback signals were congruent as compared to both being incongruent. When visual feedback was incongruent but somatosensory feedback was congruent, higher levels of SoA were reported as compared to the condition where visual feedback was congruent but somatosensory was not.

The present review also aimed to explore the relationship between SoE variables and BMI performance. Of the 20 studies, 6 of them attempted to establish a relationship between reported SoE values and task performance in BMI. A positive relationship was found between SoO and BMI performance [13,45,52,58]. Also, the SoA showed a significant correlation with BMI performance [13,19]. Regarding SoL, two studies found a positive relationship with BMI performance [13,58] while another one found no significant correlation [55]. The studies analyzed here focused mainly on the effects of congruence, synchrony, and likeness of stimuli and scenarios in SoE.

A comparison between the number of sensory modalities, embodiment, and BMI performance does not support the existence of a clear relation between levels of embodiment, or its components, and BMI performance (refer to Table 4). Out of 17 studies where a comparison between BMI performance and SoE (or one of its components) values was possible, 11/17 = 64.7% involved a single modality. Moreover, 11/11 = 100% studies used had visual or visual immersive as the feedback sensorial modality. All of these eleven studies were associated with an increase in SoE or one of its components, but from the nine studies that reported both the values of embodiment and of BMI performance [13,17,19,45,51–54,58], five (5/9 = 55.56%) were associated with an increase in performance [17,45,53,54,58] and three (3/9 = 33.33%) [13,19,52] had no effect on performance. Only in one study (1/9 = 11.11%) [51], a detrimental effect in BMI performance occurred. Therefore, studies using a single feedback sensorial modality were all based on visual feedback, all reported increases in embodiment or one of its components, and approximately half reported improvements in BMI performance.

A total of eight studies included two sensorial modalities (8/20 = 40.00%) [14,18,20,36,37,55–57]. From these, 5/20 = 25.00% included visual (or visual immersive) and auditory feedback [14,18,20,36,55] and 3/20 = 15.00% included haptic feedback [37,56,57]. In two studies where visual feedback was paired with auditory feedback, an improvement in BMI performance and in SoE (or one of its components) occurred [14,18]. In the remaining three studies, either no difference in BMI [20,55] or in SoE [36] was reported. Meanwhile, in the three studies where haptic feedback was paired with visual feedback, no improvement in SoE nor in BMI performance was reported in one case [56], and no values were reported in another [57]. The third study with two haptic modalities [37] tested somatosensory and visual feedback and revealed that the congruent somatosensory feedback prevailed over incongruent visual feedback, namely increasing SoE and BMI performance. This study was performed in a single patient using intracortical recordings. Only one study included four different types of sensory feedback (visual, auditory, haptic vibratory, and haptic thermal) [15], but no difference in embodiment or in BMI performance was reported. It is noteworthy that this study was performed on a patient and included only 10 sessions.

Lastly, of the 20 studies analyzed here, gender bias was present in 12 of them. In three studies, the number of males more than doubled the number of females [19,20,51], and one studied only females ($N = 29$) [14]. This male gender bias most likely reflects the increased number of male patients previously reported [66] and highlights the need for more studies in the female population [67].

4. Discussion

This review has analyzed studies using BMIs to determine if an increase in the number of sensory modalities is associated with increased SoE and improved BMI performance. Most studies employed motor imagery-based BMIs through EEG recordings and included only one type of feedback, either visual or visual immersive. This type of feedback was consistently associated with increased SoE, but only 55.56% of the cases were associated with an improvement in BMI performance. Studies that combined two different types of sensory feedback either used visual (or visual immersive) and auditory feedback or alternatively, visual (or visual immersive) and haptic feedback in the forms of vibration or electrical stimulation. While most studies utilized one or two types of sensorial feedback in an EEG motor imagery-based BMI, a small number explored different approaches such as SSVEPs, intracortical recordings, or incorporated more than two types of sensory feedback. Additionally, a relatively limited number of studies were conducted in patients. Lastly, a noticeable bias towards male participants was observed. The studies analyzed in this review do not support the notion that an increased number of sensory modalities enhances SoE and BMI performance. However, they also highlight the fact that, to date, no study has systematically explored the influence of different sensorial modalities in SoE and BMI performance.

SoE, particularly the sense of ownership (SoO) over external objects, has been investigated using the rubber hand paradigm [42]. In this paradigm, users have one of their hands hidden but exposed to tactile stimuli while simultaneously observing a substitute rubber hand. During the experiment, the rubber hand receives the same stimulus at the same time as the participant's hidden hand, leading the participants to attribute the proprioceptive sensation to the observed stimulus rather than the one delivered to their own skin. Some researchers have proposed that the SoO experienced towards the rubber hand contributes to the SoA. In other words, participants who feel a strong SoO over the rubber hand also report a high perceived SoA, believing they could control the movements of the rubber hand if they desired [42,68]. Several studies included in this review reported a suggested connection between SoO and SoA [17,20,53,54]. However, previous research has shown that a strong SoO over a rubber hand can occur without feeling agency over its movements [69]. Also, visual representations resembling a human body or body part have been found to enhance the SoE compared to more abstract representations with subjects reporting feeling less embodied by a virtual block [20] or a robotic tweezer [53] than by a human-like hand.

Several research groups have made efforts to develop BMI tasks with more realistic actors such as robots or avatars. Many studies have focused on exploring the effects of congruent and incongruent sensory feedback on actions and their influence on SoE. This has been examined in relation to visual [12,16,19,37,52], auditory [18], or haptic [37,56] congruency/incongruency of feedback. These studies support the idea that the sense of agency is a fundamental component of embodiment processes and is influenced by sensorimotor congruence in the executed action, with sensory input playing a crucial role [19]. It appears that, as long as some congruent sensorial feedback is provided, the SoO [68] and SoA [70] can be induced in participants. In other words, these participants believe they are controlling a task through a BMI when, in fact, they are not. Therefore, it is possible to induce some SoO in an additional bodily part, such as a third hand, without losing the SoO in their real hands [68]. These studies suggest that the intention to control a BMI may recruit both SoO and SoA.

In general, congruent visual stimuli have been found to elicit higher levels of SoO [12,16,52] and SoA [19,37]. However, it is still inconclusive whether congruence of vibratory feedback [18] or auditory feedback stimuli [56] leads to a higher level of SoE. Only one study included in this review, which focused on an SCI patient, found that the congruence of the haptic feedback might have a greater influence on the induction of SoA than the congruence of the visual feedback [37]. However, it should be noted that the haptic stimulus used in that study involved muscular electrostimulation, which induces movement through muscular contraction and can be considered haptic–kinesthetic feedback. This type of feedback cannot be directly compared to haptic–tactile feedback, such as that resulting from vibration, as they have very different characteristics.

Interestingly, despite the importance of congruence in the feedback stimuli, the presence of visual feedback, even if incongruent, appears to have a more positive effect on SoE than its absence [52].

In addition to the feedback related to the action or the virtual/robotic actor, studies have also focused on the realism of the virtual scenario where that action took place. For example, the study by Pais-Vieira and colleagues [15] incorporated auditory and thermal stimuli that were coherent with a highly realistic virtual reality environment. Similarly, in the study by Legény and colleagues (et al., 2011), the visual elements necessary for the operation of an SSVEP-BMI were contextualized [51].

The results of the present review support the notion that visual sensorial feedback is beneficial for the SoE and that multisensory feedback combining visual and auditory or visual and haptic feedback tends to be beneficial for SoE, though not necessarily for BMI performance. As elegantly demonstrated in the single-SCI-patient study by Serino and colleagues [37], it is likely that the interplay of different sensorial modalities may be critical at specific points in time. Lastly, while the present review does not support the hypothesis that multisensory feedback necessarily improves SoE and BMI performance, the study of Pais-Vieira and colleagues [15], performed in a single SCI patient, suggests that including visual (immersive), auditory, vibratory, and thermal feedback is not detrimental to embodiment and BMI performance. However, any extrapolation of findings from these latter studies needs to be approached with caution due to the small number of patients and the fact that SCI patients already have an altered SoE.

After analyzing the studies associated with SoE and multisensory feedback during BMI control, it is proposed here that a detailed examination of the effects of each type, as well as the combination, of sensory modalities is crucial for our understanding of the neural basis of SoE (and SoO, SoA, and SoL) and how it relates to BMI performance. To achieve this, it is critical to systematically evaluate the effects of removing and adding each sensory modality, or combinations of modalities, in various types of tasks (SSVEPs, MI, P300) and with different types of neural signals (EEG, functional magnetic resonance (fMRI), intracortical recordings, etc.). We suggest that a series of experiments using within-subject designs could help control for individual differences in physiological parameters.

Additionally, this review highlights that only a limited number of studies have been conducted in SCI patients [15,18,37,56]. These studies, although conducted in a small number of patients (N = 1–8), allowed for the examination of SoE in pathological conditions and provided significant insights that could not otherwise be studied. Therefore, it is relevant for future studies to specifically address the role of multisensory feedback in SoE during BMI control in SCI and other patients. It is noteworthy that a large fraction of the BMIs analyzed here required users to engage in active motor imagery, with instructions to avoid making actual movements. However, in the context of rehabilitation, motor imagery BMIs are typically employed to promote or facilitate specific motor activities [15,37]. Consequently, users are instructed to attempt a set of pre-defined movements. This difference in goals should be considered in future studies examining the role of multisensory feedback in embodiment and BMI performance.

BMIs based on neural activity recorded invasively or non-invasively will inevitably r sult in significantly different decoding and experimental setup details, which may influenc SoE. The present review included only one study with intracortical neural recordings [37 revealing that the dynamics between sensory and motor cortices during BMI control ar crucial for the SoA, especially if visual feedback is incongruent. This study underscore the importance of recognizing that BMIs based in EEG recordings, while highly practica and reproducible, lack the ability to extract neural information (i.e., single- or multi-uni activity) with high spatial resolution.

Lastly, the concepts of SoE, SoO, SoL, and SoA can vary between authors leadin to different questionnaires [31–35]. Therefore, the present review must be cautiousl considered since the terms used by each author may present some degree of variation.

5. Conclusions

The number of BMI studies has significantly increased in the last two decades, bu the incorporation of SoE measurements in experimental designs remains relatively scarc The individual studies analyzed here suggest that greater realism, such as more immersiv scenarios, greater human similarities of the virtual/robotic avatar, and greater coherence o the feedback all contribute to higher levels of SoE and enhance the embodiment experienc Despite these individual results, the larger group of studies analyzed here does not suppor the notion that an increased number of sensorial modalities will lead to increased SoE an improved BMI performance. It should be noted, however, that no study has systematicall explored the influence of the different sensorial modalities in SoE and BMI performanc Therefore, we propose that it is necessary to perform experimental studies that separatel test the cumulative and isolated contributions of multimodal feedback in inducing SoE.

Author Contributions: Conceptualization, D.J.T., M.P.-V. and C.P.-V.; methodology, D.J.T. and C.P.-V validation, D.J.T., M.P.-V. and C.P.-V.; formal analysis, D.J.T. and C.P.-V.; writing—original draf preparation, D.J.T., M.P.-V. and C.P.-V.; writing—review and editing, D.J.T., M.P.-V. and C.P.-V visualization, D.J.T., M.P.-V. and C.P.-V.; supervision, M.P.-V. and C.P.-V.; project administration, D.J.T M.P.-V. and C.P.-V.; funding acquisition, M.P.-V. and C.P.-V. All authors have read and agreed to th published version of the manuscript.

Funding: Project CIIS I&D—UIDB/04279/2020—Base; Project iBiMED—UIDB/04501/2020—Base

Data Availability Statement: Not applicable.

Conflicts of Interest: The authors declare no conflict of interest.

References

1. Blanke, O.; Metzinger, T. Full-Body Illusions and Minimal Phenomenal Selfhood. *Trends Cogn. Sci.* **2009**, *13*, 7–13. [CrossRef [PubMed]
2. De Vignemont, F. A Self for the Body. *Metaphilosophy* **2011**, *42*, 230–247. [CrossRef]
3. Kilteni, K.; Groten, R.; Slater, M. The Sense of Embodiment in Virtual Reality. *Presence* **2012**, *21*, 373–387. [CrossRef]
4. Lee, K.M. Presence, Explicated. *Commun. Theory* **2004**, *14*, 27–50. [CrossRef]
5. Longo, M.R.; Schüür, F.; Kammers, M.P.M.; Tsakiris, M.; Haggard, P. What Is Embodiment? A Psychometric Approach. *Cognitio* **2008**, *107*, 978–998. [CrossRef] [PubMed]
6. Kim, S.Y.; Prestopnik, N.; Biocca, F.A. Body in the Interactive Game: How Interface Embodiment Affects Physical Activity an Health Behavior Change. *Comput. Hum. Behav.* **2014**, *36*, 376–384. [CrossRef]
7. Thorpe, G.; Arthur, A.; McArthur, M. Adjusting to Bodily Change Following Stoma Formation: A Phenomenological Stud *Disabil. Rehabil.* **2016**, *38*, 1791–1802. [CrossRef]
8. Fuentes, C.T.; Pazzaglia, M.; Longo, M.R.; Scivoletto, G.; Haggard, P. Body Image Distortions Following Spinal Cord Injury. *J Neurol. Neurosurg. Psychiatry* **2013**, *84*, 201–207. [CrossRef]
9. Lewis, J.S.; Kersten, P.; McCabe, C.S.; McPherson, K.M.; Blake, D.R. Body Perception Disturbance: A Contribution to Pain i Complex Regional Pain Syndrome (CRPS). *Pain* **2007**, *133*, 111–119. [CrossRef]
10. Lotze, M.; Moseley, G.L. Role of Distorted Body Image in Pain. *Curr. Rheumatol. Rep.* **2007**, *9*, 488–496. [CrossRef]
11. Pleger, B.; Ragert, P.; Schwenkreis, P.; Förster, A.F.; Wilimzig, C.; Dinse, H.; Nicolas, V.; Maier, C.; Tegenthoff, M. Pattern of Cortical Reorganization Parallel Impaired Tactile Discrimination and Pain Intensity in Complex Regional Pain Syndrome *Neuroimage* **2006**, *32*, 503–510. [CrossRef] [PubMed]

12. Alimardani, M.; Nishio, S.; Ishiguro, H. Humanlike Robot Hands Controlled by Brain Activity Arouse Illusion of Ownership in Operators. *Sci. Rep.* **2013**, *3*, 2396. [CrossRef] [PubMed]
13. Juliano, J.M.; Spicer, R.P.; Vourvopoulos, A.; Lefebvre, S.; Jann, K.; Ard, T.; Santarnecchi, E.; Krum, D.M.; Liew, S.L. Embodiment Is Related to Better Performance on a Brain–Computer Interface in Immersive Virtual Reality: A Pilot Study. *Sensors* **2020**, *20*, 1204. [CrossRef]
14. Nierula, B.; Spanlang, B.; Martini, M.; Borrell, M.; Nikulin, V.V.; Sanchez-Vives, M.V.; Taylor, J.; Farina, D. Agency and Responsibility over Virtual Movements Controlled through Different Paradigms of Brain−computer Interface. *J. Physiol.* **2021**, *599*, 2419–2434. [CrossRef] [PubMed]
15. Pais-Vieira, C.; Gaspar, P.; Matos, D.; Alves, L.P.; da Cruz, B.M.; Azevedo, M.J.; Gago, M.; Poleri, T.; Perrotta, A.; Pais-Vieira, M. Embodiment Comfort Levels During Motor Imagery Training Combined with Immersive Virtual Reality in a Spinal Cord Injury Patient. *Front. Hum. Neurosci.* **2022**, *16*, 909112. [CrossRef]
16. Perez-Marcos, D.; Slater, M.; Sanchez-Vives, M.V. Inducing a Virtual Hand Ownership Illusion through a Brain-Computer Interface. *Neuroreport* **2009**, *20*, 589–594. [CrossRef]
17. Škola, F.; Liarokapis, F. Embodied VR Environment Facilitates Motor Imagery Brain–Computer Interface Training. *Comput. Graph. Pergamon* **2018**, *75*, 59–71. [CrossRef]
18. Tidoni, E.; Gergondet, P.; Fusco, G.; Kheddar, A.; Aglioti, S.M. The Role of Audio-Visual Feedback in a Thought-Based Control of a Humanoid Robot: A BCI Study in Healthy and Spinal Cord Injured People. *IEEE Trans. Neural Syst. Rehabil. Eng.* **2017**, *25*, 772–781. [CrossRef]
19. Evans, N.; Gale, S.; Schurger, A.; Blanke, O. Visual Feedback Dominates the Sense of Agency for Brain-Machine Actions. *PLoS ONE* **2015**, *10*, e0130019. [CrossRef]
20. Ziadeh, H.; Gulyas, D.; Nielsen, L.D.; Lehmann, S.; Nielsen, T.B.; Kjeldsen, T.K.K.; Hougaard, B.I.; Jochumsen, M.; Knoche, H. "Mine Works Better": Examining the Influence of Embodiment in Virtual Reality on the Sense of Agency During a Binary Motor Imagery Task With a Brain-Computer Interface. *Front. Psychol.* **2021**, *12*, 806424. [CrossRef]
21. Gonzalez-Franco, M.; Peck, T.C. Avatar Embodiment. Towards a Standardized Questionnaire. *Front. Robot. AI* **2018**, *5*, 74. [CrossRef]
22. Schwind, V.; Knierim, P.; Haas, N.; Henze, N. Using Presence Questionnaires in Virtual Reality. In Proceedings of the Conference on Human Factors in Computing Systems—Proceedings, Association for Computing Machinery, Glasgow, Scotland, 4–9 May 2019. [CrossRef]
23. Franco, M.G. Neurophysiological Signatures of the Body Representation in the Brain Using Immersive Virtual Reality. 2014. Available online: http://hdl.handle.net/10803/359383 (accessed on 19 February 2023).
24. Alchalabi, B.; Faubert, J.; Labbe, D.R. EEG Can Be Used to Measure Embodiment When Controlling a Walking Self-Avatar. In Proceedings of the 2019 IEEE Conference on Virtual Reality and 3D User Interfaces (VR), Osaka, Japan, 23–27 March 2019. [CrossRef]
25. Armel, K.C.; Ramachandran, V.S. Projecting Sensations to External Objects: Evidence from Skin Conductance Response. *Proc. R. Soc. B Biol. Sci.* **2003**, *270*, 1499–1506. [CrossRef]
26. Ehrsson, H.H.; Rosén, B.; Stockselius, A.; Ragnö, C.; Köhler, P.; Lundborg, G. Upper Limb Amputees Can Be Induced to Experience a Rubber Hand as Their Own. *Brain* **2008**, *131*, 3443–3452. [CrossRef]
27. Tsuji, T.; Yamakawa, H.; Yamashita, A.; Takakusaki, K.; Maeda, T.; Kato, M.; Oka, H.; Asama, H. Analysis of Electromyography and Skin Conductance Response During Rubber Hand Illusion. In Proceedings of the 2013 IEEE Workshop on Advanced Robotics and its Social Impacts, Tokyo, Japan, 7–9 November 2013; pp. 88–93.
28. Kammers, M.P.M.; Rose, K.; Haggard, P. Feeling Numb: Temperature, but Not Thermal Pain, Modulates Feeling of Body Ownership. *Neuropsychologia* **2011**, *49*, 1316–1321. [CrossRef] [PubMed]
29. Llobera, J.; Sanchez-Vives, M.V.; Slater, M. The Relationship between Virtual Body Ownership and Temperature Sensitivity. *J. R. Soc. Interface* **2013**, *10*, 1–11. [CrossRef] [PubMed]
30. Braun, N.; Debener, S.; Spychala, N.; Bongartz, E.; Sörös, P.; Müller, H.H.; Philipsen, A. The senses of agency and ownership: A review. *Front. Psychol.* **2018**, *9*, 535. [CrossRef]
31. Witmer, B.G.; Singer, M.J. Measuring presence in virtual environments: A presence questionnaire. *Presence* **1998**, *7*, 225–240. [CrossRef]
32. Wolpaw, J.R.; Ramoser, H.; McFarland, D.J.; Pfurtscheller, G. EEG-based communication: Improved accuracy by response verification. *IEEE Trans. Rehabil. Eng.* **1998**, *6*, 326–333. [CrossRef] [PubMed]
33. Sanchez-Vives, M.V.; Spanlang, B.; Frisoli, A.; Bergamasco, M.; Slater, M. Virtual hand illusion induced by visuomotor correlations. *PLoS ONE* **2010**, *5*, e10381. [CrossRef]
34. Friedman, D.; Pizarro, R.; Or-Berkers, K.; Neyret, S.; Pan, X.; Slater, M. A method for generating an illusion of backwards time travel using immersive virtual reality—An exploratory study. *Front. Psychol.* **2014**, *5*, 1–15. [CrossRef]
35. Peck, T.C.; Gonzalez-Franco, M. Avatar embodiment. A standardized questionnaire. *Front. Virtual Real.* **2021**, *1*, 575943. [CrossRef]
36. Caspar, E.A.; de Beir, A.; Lauwers, G.; Cleeremans, A.; Vanderborght, B. How Using Brain-Machine Interfaces Influences the Human Sense of Agency. *PLoS ONE* **2021**, *16*, e0245191. [CrossRef] [PubMed]
37. Serino, A.; Bockbrader, M.; Bertoni, T.; Colachis Iv, S.; Solcà, M.; Dunlap, C.; Eipel, K.; Ganzer, P.; Annetta, N.; Sharma, G. Sense of Agency for Intracortical Brain-Machine Interfaces. *Nat. Hum. Behav.* **2022**, *6*, 565–578. [CrossRef]

38. Pais-Vieira, C.; Gaspar, P.; Matos, D.; Gago, M.; Azevedo, M.J.; Poleri, T.; Perrotta, A.; Paisvieira, M. Multimodal Visual, Auditory Thermal, and Tactile Feedback During Brain-Machine Interface Use by a Spinal Cord Injury Patient. In Proceedings of the Human Interaction & Emerging Technologies (IHIET-AI 2022): Artificial Intelligence & Future Applications, Nice, France, 21–23 April 2022 [CrossRef]
39. Mudgal, S.K.; Sharma, S.K.; Chaturvedi, J.; Sharma, A. Brain Computer Interface Advancement in Neurosciences: Applications and Issues. *Interdiscip. Neurosurg.* **2020**, *20*, 100694. [CrossRef]
40. Lebedev, M.A.; Nicolelis, M.A.L. Brain-Machine Interfaces: From Basic Science to Neuroprostheses and Neurorehabilitation *Physiol. Rev.* **2017**, *97*, 767–837. [CrossRef] [PubMed]
41. Wolpaw, J.R.; Millán, J.d.R.; Ramsey, N.F. Brain-Computer Interfaces: Definitions and Principles. *Handb. Clin. Neurol.* **2020** *168*, 15–23. [CrossRef] [PubMed]
42. Botvinick, M.; Cohen, J. Rubber Hands 'Feel' Touch That Eyes See. *Nature* **1998**, *391*, 756. [CrossRef] [PubMed]
43. Pazzaglia, M.; Galli, G.; Lewis, J.W.; Scivoletto, G.; Giannini, A.M.; Molinari, M. Embodying Functionally Relevant Action Sounds in Patients with Spinal Cord Injury. *Sci. Rep.* **2018**, *8*, 15641. [CrossRef]
44. Slater, M.; Khanna, P.; Mortensen, J.; Yu, I. Visual Realism Enhances Realistic Response in an Immersive Virtual Environment *IEEE Comput. Graph. Appl.* **2009**, *29*, 76–84. [CrossRef]
45. Penaloza, C.I.; Alimardani, M.; Nishio, S. Android Feedback-Based Training Modulates Sensorimotor Rhythms during Motor Imagery. *IEEE Trans. Neural Syst. Rehabil. Eng.* **2018**, *26*, 666–674. [CrossRef]
46. Tidoni, E.; Gergondet, P.; Kheddar, A.; Aglioti, S.M. Audio-Visual Feedback Improves the BCI Performance in the Navigational Control of a Humanoid Robot. *Front. Neurorobot.* **2014**, *8*, 20. [CrossRef] [PubMed]
47. Zhang, B.; Zhou, Z.; Jiang, J. A 36-Class Bimodal Erp Brain-Computer Interface Using Location-Congruent Auditory-Tactile Stimuli. *Brain Sci.* **2020**, *10*, 524. [CrossRef] [PubMed]
48. Page, M.J.; McKenzie, J.E.; Bossuyt, P.M.; Boutron, I.; Hoffmann, T.C.; Mulrow, C.D.; Shamseer, L.; Tetzlaff, J.M.; Akl, E.A. Brennan, S.E.; et al. The PRISMA 2020 Statement: An Updated Guideline for Reporting Systematic Reviews. *BMJ* **2021**, *372*, n71 [CrossRef]
49. Connolly, T.M.; Boyle, E.A.; MacArthur, E.; Hainey, T.; Boyle, J.M. A systematic literature review of empirical evidence on computer games and serious games. *Comput. Educ.* **2012**, *59*, 661–686. [CrossRef]
50. Vieira, C.; Pais-Vieira, C.; Novais, J.; Perrotta, A. Serious Game Design and Clinical Improvement in Physical Rehabilitation: Systematic Review. *JMIR Serious Games* **2021**, *9*, e20066. [CrossRef]
51. Legény, J.; Abad, R.V.; Lévuyer, A. Navigating in Virtual Worlds Using a Self-Paced SSVEP-Based Brain-Computer Interface with Integrated Stimulation and Real-Time Feedback. *Presence* **2011**, *20*, 529–544. [CrossRef]
52. Alimardani, M.; Nishio, S.; Ishiguro, H. Effect of Biased Feedback on Motor Imagery Learning in BCI-Teleoperation System. *Front. Syst. Neurosci.* **2014**, *8*, 52. [CrossRef]
53. Alimardani, M.; Nishio, S.; Ishiguro, H. The Importance of Visual Feedback Design in BCIs; from Embodiment to Motor Imagery Learning. *PLoS ONE* **2016**, *11*, e0161945. [CrossRef]
54. Alimardani, M.; Nishio, S.; Ishiguro, H. Removal of Proprioception by BCI Raises a Stronger Body Ownership Illusion in Control of a Humanlike Robot. *Sci. Rep.* **2016**, *6*, 33514. [CrossRef]
55. Vourvopoulos, A.; Bermúdez i Badia, S. Motor Priming in Virtual Reality Can Augment Motor-Imagery Training Efficacy in Restorative Brain-Computer Interaction: A within-Subject Analysis. *J. Neuroeng. Rehabil.* **2016**, *13*, 69. [CrossRef]
56. Tidoni, E.; Abu-Alqumsan, M.; Leonardis, D.; Kapeller, C.; Fusco, G.; Guger, C.; Hintermuller, C.; Peer, A.; Frisoli, A.; Tecchia, F.; et al. Local and Remote Cooperation with Virtual and Robotic Agents: A P300 BCI Study in Healthy and People Living with Spinal Cord Injury. *IEEE Trans. Neural Syst. Rehabil. Eng.* **2017**, *25*, 1622–1632. [CrossRef]
57. Škola, F.; Tinková, S.; Liarokapis, F. Progressive Training for Motor Imagery Brain-Computer Interfaces Using Gamification and Virtual Reality Embodiment. *Front. Hum. Neurosci.* **2019**, *13*, 329. [CrossRef]
58. Choi, J.W.; Huh, S.; Jo, S. Improving Performance in Motor Imagery BCI-Based Control Applications via Virtually Embodied Feedback. *Comput. Biol. Med.* **2020**, *127*, 104079. [CrossRef] [PubMed]
59. Slater, M.; Steed, A.; McCarthy, J.; Maringelli, F. The influence of body movement on subjective presence in virtual environments. *Hum. Factors Ergon. Soc.* **1998**, *40*, 469–477. [CrossRef] [PubMed]
60. Roberts, R.; Callow, N.; Hardy, L.; Markland, D.; Bringer, J. Movement imagery ability: Development and assessment of a revised version of the vividness of movement imagery questionnaire. *J. Sport Exerc. Psychol.* **2008**, *30*, 200–221. [CrossRef] [PubMed]
61. De Vignemont, F. Embodiment, Ownership and Disownership. *Conscious Cogn.* **2010**, *20*, 82–93. [CrossRef] [PubMed]
62. Shen, X.; Zhang, X.; Huang, Y.; Chen, S.; Yu, Z.; Wang, Y. Intermediate Sensory Feedback Assisted Multi-Step Neural Decoding for Reinforcement Learning Based Brain-Machine Interfaces. *IEEE Trans. Neural Syst. Rehabil. Eng.* **2022**, *30*, 2834–2844. [CrossRef] [PubMed]
63. Suminski, A.J.; Tkach, D.C.; Fagg, A.H.; Hatsopoulos, N.G. Incorporating Feedback from Multiple Sensory Modalities Enhances Brain-Machine Interface Control. *J. Neurosci.* **2010**, *30*, 16777–16787. [CrossRef]
64. Yu, I.; Mortensen, J.; Khanna, P.; Slater, M. Visual Realism Enhances Realistic Response in an Immersive Virtual Environment—Part 2. *IEEE Comput. Graph. Appl.* **2012**, *32*, 36–45. [CrossRef]

5. Renard, Y.; Lotte, F.; Gibert, G.; Congedo, M.; Maby, E.; Delannoy, V.; Bertrand, O.; Lécuyer, A. OpenViBE: An Open-Source Software Platform to Design, Test, and Use Brain-Computer Interfaces in Real and Virtual Environments. *Presence Teleoperators Virtual Environ.* **2010**, *19*, 35–53. [CrossRef]
6. McColl, M.A.; Charlifue, S.; Glass, C.; Lawson, N.; Savic, G. Aging, Gender, and Spinal Cord Injury. *Arch. Phys. Med. Rehabil.* **2004**, *85*, 363–367. [CrossRef]
7. Raguindin, P.F.; Muka, T.; Glisic, M. Sex and Gender Gap in Spinal Cord Injury Research: Focus on Cardiometabolic Diseases. A Mini Review. *Maturitas* **2021**, *147*, 14–18. [CrossRef] [PubMed]
8. Bashford, L.; Mehring, C. Ownership and Agency of an Independent Supernumerary Hand Induced by an Imitation Brain-Computer Interface. *PLoS ONE* **2016**, *11*, e0156591. [CrossRef] [PubMed]
9. Braun, N.; Emkes, R.; Thorne, J.D.; Debener, S. Embodied Neurofeedback with an Anthropomorphic Robotic Hand. *Sci. Rep.* **2016**, *6*, 37696. [CrossRef] [PubMed]
0. Lynn, M.T.; Berger, C.C.; Riddle, T.A.; Morsella, E. Mind Control? Creating Illusory Intentions through a Phony Brain-Computer Interface. *Conscious Cogn.* **2010**, *19*, 1007–1012. [CrossRef] [PubMed]

MDPI AG
Grosspeteranlage 5
4052 Basel
Switzerland
Tel.: +41 61 683 77 34

Applied Sciences Editorial Office
E-mail: applsci@mdpi.com
www.mdpi.com/journal/applsci

www.ingramcontent.com/pod-product-compliance
Lightning Source LLC
Chambersburg PA
CBHW041419210926
51850CB00013B/3316